Forestry Applications of Airborne I

Managing Forest Ecosystems

Volume 27

Series Editors:

Klaus von Gadow

Georg-August-University,
Göttingen, Germany

Timo Pukkala

University of Joensuu,
Joensuu, Finland
and

Margarida Tomé

Instituto Superior de Agronomía,
Lisbon, Portugal

Aims & Scope:

Well-managed forests and woodlands are a renewable resource, producing essential raw material with minimum waste and energy use. Rich in habitat and species diversity, forests may contribute to increased ecosystem stability. They can absorb the effects of unwanted deposition and other disturbances and protect neighbouring ecosystems by maintaining stable nutrient and energy cycles and by preventing soil degradation and erosion. They provide much-needed recreation and their continued existence contributes to stabilizing rural communities.

Forests are managed for timber production and species, habitat and process conservation. A subtle shift from *multiple-use management to ecosystems management* is being observed and the new ecological perspective of *multi-functional forest management* is based on the principles of ecosystem diversity, stability and elasticity, and the dynamic equilibrium of primary and secondary production.

Making full use of new technology is one of the challenges facing forest management today. Resource information must be obtained with a limited budget. This requires better timing of resource assessment activities and improved use of multiple data sources. Sound ecosystems management, like any other management activity, relies on effective forecasting and operational control.

The aim of the book series *Managing Forest Ecosystems* is to present state-of-the-art research results relating to the practice of forest management. Contributions are solicited from prominent authors. Each reference book, monograph or proceedings volume will be focused to deal with a specific context. Typical issues of the series are: resource assessment techniques, evaluating sustainability for even-aged and uneven-aged forests, multi-objective management, predicting forest development, optimizing forest management, biodiversity management and monitoring, risk assessment and economic analysis.

For further volumes:
http://www.springer.com/series/6247

Matti Maltamo • Erik Næsset • Jari Vauhkonen
Editors

Forestry Applications
of Airborne Laser Scanning

Concepts and Case Studies

 Springer

Editors
Matti Maltamo
School of Forest Sciences
University of Eastern Finland
Joensuu, Finland

Erik Næsset
Department of Ecology and Natural
  Resource Management
Norwegian University of Life Sciences
Ås, Norway

Jari Vauhkonen
Department of Forest Sciences
University of Helsinki
Helsinki, Finland

ISSN 1568-1319   ISSN 2352-3956 (electronic)
ISBN 978-94-017-8662-1   ISBN 978-94-017-8663-8 (eBook)
DOI 10.1007/978-94-017-8663-8
Springer Dordrecht Heidelberg New York London

Library of Congress Control Number: 2014935989

Printed on acid-free paper

Springer is part of Springer Science+Business Media (www.springer.com)

Preface

Use of airborne laser scanning to provide data for research and operational applications in management of forest ecosystems has experienced a tremendous growth since the mid-1990s and the amount of scientific publications resulting from this activity has increased rapidly. Yet there is no textbook available to bring together the results across this multitude of disciplines and synthesize on the state of the art. The aim of this book is to fill this gap by providing a unique collection of in-depth reviews and overviews of the research and application of airborne laser scanning in a broad range of forest-related disciplines. However, this book is more than just a collection of individual contributions – it consists of a well-composed blend of chapters dealing with fundamental methodological issues and contributions reviewing and illustrating the use of airborne laser scanning within various domains of application. There are numerous cross-references between the various chapters of the book which may be useful for readers who wish to get a more in-depth understanding of a particular issue.

We hope researchers, students, and practitioners will find this book useful. We also hope that colleagues will find the book of value as part of the curriculum in forestry schools and those schools offering courses in forest remote sensing and forest ecosystem assessments in a broader sense.

This book is the result of a collective effort by many good colleagues and friends. They are all listed by name as authors of the various chapters. In addition to the authors of the chapters, many researchers around the world have helped us by reviewing chapters and suggesting improvements. We would like to acknowledge these external reviewers for their efforts to improve this book: Gregory P. Asner, Mathias Disney, James W. Flewelling, Jari Kouki, Peter Krzystek, Mikko Kurttila, Tomas Lämås, Eva Lindberg, Steen Magnussen, Håkan Olsson, Pekka Savolainen, Svein Solberg, Göran Ståhl, and Valerie Thomas. The errors remaining are nevertheless attributable entirely to the authors.

Joensuu, Finland   Matti Maltamo
Ås, Norway   Erik Næsset
Helsinki, Finland   Jari Vauhkonen

Contents

Chapter 1
Introduction to Forestry Applications of Airborne Laser Scanning

Jari Vauhkonen, Matti Maltamo, Ronald E. McRoberts, and Erik Næsset

Abstract Airborne laser scanning (ALS) has emerged as one of the most promising remote sensing technologies to provide data for research and operational applications in a wide range of disciplines related to management of forest ecosystems. This chapter starts with a brief historical overview of the early forest-related research on airborne Light Detection and Ranging which was first mentioned in the literature in the mid-1960s. The early applications of ALS in the mid-1990s are also reviewed. The two fundamental approaches to use of ALS in forestry applications are presented – the area-based approach and the single-tree approach. Many of the remaining chapters rest upon this basic description of these two approaches. Finally, a brief introduction to the broad range of forestry applications of ALS is given and references are provided to individual chapters that treat the different topics in more depth. Most chapters include detailed reviews of previous research and the state-of-the-art in the various topic areas. Thus, this book provides a unique collection of in-depth reviews and overviews of the research and application of ALS in a broad range of forest-related disciplines.

J. Vauhkonen (✉)
Department of Forest Sciences, University of Helsinki, Helsinki, Finland
e-mail: jari.vauhkonen@helsinki.fi

M. Maltamo
School of Forest Sciences, University of Eastern Finland, Joensuu, Finland
e-mail: Matti.maltamo@uef.fi

R.E. McRoberts
Northern Research Station, U. S. Forest Service, Saint Paul, MN, USA

E. Næsset
Department of Ecology and Natural Resource Management, Norwegian University
of Life Sciences, Ås, Akershus, Norway
e-mail: Erik.naesset@umb.no

M. Maltamo et al. (eds.), *Forestry Applications of Airborne Laser Scanning: Concepts and Case Studies*, Managing Forest Ecosystems 27, DOI 10.1007/978-94-017-8663-8__1,
© Springer Science+Business Media Dordrecht 2014

1.1 Introduction

Satisfying national and international environmental goals and agreements requires heavy reliance on environmental mapping and monitoring. However, collection of information on the location and extent for many of the relevant characteristics of forest ecosystems by means of pure ground-based field inventories is prohibitively expensive. During the last decade, operational collection of such information has been revolutionized by the development of *Light Detection and Ranging* (LiDAR) – a technology producing distance measurements based on the return time of emitted light. LiDAR systems mounted on aircraft can be a cost-efficient means to obtain data on forest structure for vast geographical areas with high spatial resolution and high positional accuracy. The term *airborne laser* is frequently used to distinguish between systems that acquire LiDAR data from aircraft and systems using space-borne or terrestrial platforms. Most commercial airborne lasers, particularly those used for operational purposes, are equipped with a scanning device that distributes the emitted light across a wide corridor along the aircraft's flight path. Acquiring LiDAR data with these systems is known as *airborne laser scanning* (ALS).

A particular strength of ALS for forestry applications is its ability to accurately characterize the three-dimensional (3D) structure of the forest canopy. Such information is potentially more useful for forest inventories than the information from other remote sensing techniques (cf. Lefsky et al. 2001; Coops et al. 2004; Maltamo et al. 2006a). Height and density metrics derived from ALS data can be used to estimate the horizontal and vertical distribution of biological material for various forest growing stock and commercial timber surveys as well as ecological applications.

Research on the use of LiDAR data for forestry applications began in the mid-1970s with experiments using simple profiling instruments. Since approximately 1990, the emphasis has been development of tools for practical forest inventories that exploit the capability of ALS instruments to provide full areal coverage. ALS is currently incorporated as an essential component of operational forest inventories in multiple countries (Næsset 2007; Maltamo et al. 2011b; Woods et al. 2011). In addition, many countries and states are currently acquiring ALS data to construct detailed ground elevation models which then serve as a new data source for vegetation mapping and forest inventories (Nord-Larsen and Riis-Nielsen 2010). Nowadays, ALS data may be obtained for applications in a variety of fields from large numbers of providers and surveying companies, each with multiple systems.

Thus, ALS is recognized as a well-established and maturing discipline of both scientific and practical importance. Multiple scientific review articles (e.g. Lim et al. 2003; Næsset et al. 2004; Maltamo et al. 2007; Hyyppä et al. 2008; Koch 2010; McRoberts et al. 2010) and text book chapters (Koch et al. 2008; Packalén et al. 2008; Hyyppä et al. 2009) on ALS have been published in recent years. However, no full text book with a detailed overview on forestry applications is available. The aim of the present book is to fill this gap.

The overall purpose of this first chapter is to introduce and provide background for the subsequent chapters. Section 1.2 documents the early development of LiDAR for forestry applications with emphasis on ALS as that technology became available in the 1990s. Section 1.3 continues with a technological overview and description of basic techniques for using ALS data. The various applications described in Sect. 1.4 mainly refer to subsequent chapters. Finally, a brief overview of the structure of the book is presented.

1.2 Early Research on Application of Profiling LiDAR and ALS in Forestry

This section aims at documenting a few milestones important for the current development of ALS for forestry applications. The details are intentionally left to earlier reviews such as those by Lim et al. (2003) and Næsset et al. (2004) that reflect North American and Nordic perspectives, respectively. Nelson (2013) provides a comprehensive and historically interesting review on the early technical development from the mid-1970s to circa 1990.

The first studies on modeling forest attributes using ALS data continued previous research using data from airborne profiling LiDARs which, in turn, continued even earlier research using aerial photography. Nelson (2013) notes that the first mention of possible forestry applications was by Rempel and Parker (1964) who used an airborne profiling LiDAR system to obtain ground and tree heights. The first experiments to actually measure trees were reported by Solodukhin et al. (1976), who compared laser profiles of felled trees with tape measurements, later mounting the same instrument on an aircraft to acquire airborne profiles (Solodukhin et al. 1979). Similar studies using laser profilers were later conducted in North America (Nelson et al. 1984; Aldred and Bonner 1985; Maclean and Krabill 1986), and some studies developed methods for estimating forest volume and biomass from the airborne canopy profiles (Maclean and Martin 1984; Maclean and Krabill 1986).

Arp et al. (1982) also used a profiling system, along with aerial photography, to map forest height in Venezuela. The earliest uses of LiDAR data to support forest inventories were extensions of similar uses of photogrammetric data. Maclean and Martin (1984) reported $0.75 \leq R^2 \leq 0.87$ for a regression model of the relationship between timber volume and cross-sectional areas of canopy profiles obtained from aerial photography. The early LiDAR studies mimicked this approach using data from profiling LiDAR systems acquired for a narrow transect along the aircraft flight line. Aldred and Bonner (1985) used LiDAR profile data to estimate stand heights within 4.1 m of actual stand heights and to assign stands to crown density classes with 62 % accuracy. Maclean and Krabill (1986) obtained $R^2 \approx 0.9$ for a regression model of the relationship between gross merchantable volume and cross-sectional areas derived from laser profiles. Nelson et al. (1984) used data from a profiling system to characterize vertical forest canopy profiles. Means of LiDAR-based and photography-based estimates of tree heights along flight lines were comparable and linear models of the relationship between photo-interpreted canopy closure and LiDAR metrics produced $R^2 \approx 0.8$. Schreier et al. (1985) demonstrated the utility of LiDAR profiling data for distinguishing among coniferous trees, deciduous trees, and low-growing vegetation. Nelson et al. (1988) constructed linear models of relationships between height, mean volume per unit area, and mean biomass per unit area as response variables and height profile metrics derived from LiDAR profiling data as predictor variables and obtained $0.50 \leq R^2 \leq 0.60$. For forest inventory purposes, the important lesson learned from these early LiDAR profiling studies was that regression models using LiDAR metrics could be used to predict forest attributes of interest. The successors to LiDAR profiling systems were ALS systems with both small footprint diameters in the range 0.1–2.0 m and large footprint diameters in the range 5.0–10.0 m. Data from ALS systems are better suited than profiling data when full coverage is necessary.

An important step towards the commercial development of LiDAR measurements was the integration of Global Positioning System (GPS) and Inertial Navigation Systems (INS) in the 1990s which facilitated accurate positioning of the scanner and, consequently, the recorded data. Using some of the first commercial or pre-commercial systems equipped with GPS and INS (Næsset et al. 2004), multiple researchers reported relevant relationships between field-measured and ALS-based height metrics, typically quantified in terms of mean height (Nilsson 1996; Næsset 1997a; Magnussen and Boudewyn 1998; Magnussen et al. 1999). Further, simple stand level mean or total values of forest biophysical variables such as basal area (Means et al. 2000) and stand volume (Næsset 1997b; Means et al. 2000) were estimated from small-footprint ALS data with similar or better accuracies than those produced by pure sample-based ground methods.

The first studies focused on relationships between plot-level forest attributes based on field measurements and ALS-based height distributions. These approaches are characterized as *area-based* and are distinguished from *individual tree* approaches proposed a few years later by Brandtberg (1999) and Hyyppä and Inkinen (1999) that focused on detecting individual trees from surface models constructed using ALS data. Both area-based and individual tree approaches have been tested for ALS-based forest inventories, and variants and hybrid methods have also been developed (Sect. 1.3.2).

1.3 Technical Basics

1.3.1 General Principles of ALS

Overviews of theoretical and physical aspects of ALS that were published in the late 1990s (Baltsavias 1999; Wehr and Lohr 1999) still constitute useful sources of information. The physical principles of ALS are also described in Chap. 2 of this book. The basis of all ALS systems is emission of a short-duration pulse of laser light and measurement of the elapsed time between emission and detection of the reflected light back at the sensor. Using the speed of light, the elapsed time is converted to the distance from the point of emission to the underlying object from which the light was reflected. Because the position and the orientation of the sensor are recorded using a GPS and INS, ALS data constitute a set of 3D coordinates that represent the scanned surface from which the pulses were reflected.

ALS instruments used in forestry applications are typically small-footprint, discrete-return systems that record from one to a few echoes for each emitted pulse. The diameter of the emitted laser beam on the ground, characterized as the beam's footprint, depends on the beam divergence and the flying altitude. Small footprint proprietary systems typically have footprint diameters of 1 m or less and are often preferred for forestry applications because they facilitate accurate linkages between the scanning data and individual trees, plots, or forest stands. In contrast, large footprint systems typically have footprint diameters as great as 10s of meters. The spacing of pulses on the ground depends on parameters such as flying altitude and speed, pulse repetition frequency, viewing angle, and scanning pattern.

The emitted laser pulses generate different types of data as they are reflected from the ground and recorded at the sensor, depending on the interaction with the forest canopy (cf. Chap. 2) and whether the sensor has full-waveform capabilities or is a discrete return system. In simplified terms, full-waveform recording systems (Chap. 3) digitize the entire sequence of returned energy as a function of height, whereas most systems described in the following chapters are discrete return systems that record from one to a few (typically four to five) discrete echoes reflected from between the bare earth and the top of the canopy. The recorded data of discrete return systems depend on factors such as the algorithm used in real time to trigger and record echoes from the backscatter signal received at the sensor. Depending on the sensor, additional information for properties such as the intensities of echoes may be recorded.

1.3.2 Basic Inventory Techniques

1.3.2.1 Area-Based Approaches

A fundamental concept underlying the area-based approach is that the 3D point cloud consisting of the heights corresponding to all the echoes for a given surface area, for example a sample plot, contains information that can be used to characterize the ground surface and the vertical distribution of the biological material in the vegetation layers above that area. For most applications, the vertical ALS height distribution for a particular area ignores the horizontal position of individual echoes within the area in question but maintains heights for all echoes. The entire ALS height distribution, or the portion of it representing the vegetation, can then be used to calculate discrete metrics that are related to properties of the distribution of biological material. Metrics include parameters of the ALS height distribution such as the mean or percentiles and parameters related to the canopy density such as penetration rates. The ALS height distribution can also be approximated by continuous distribution functions.

Magnussen and Boudewyn (1998) showed that the proportion of small footprint ALS echo heights at or above a given reference height corresponded well with the proportion of leaf area above this height. Although leaf area is generally not an inventory variable of interest, a seminal contribution of this study was demonstration of the utility of ALS height percentiles as predictor variables. This innovative approach has subsequently been found useful for prediction of a wide range of forest attributes. Several extensions of this concept are relevant. First, ALS pulse densities for spatial units used for model calibration should be sufficiently large to produce reliable estimates of the parameters of the echo height distributions used as model predictor variables. Second, pulse densities should be sufficiently large to estimate these parameters for both spatial units used to calibrate models and spatial units used for areal estimation. The latter feature permits calibration of models using ground and ALS data for circular plots and estimation using square grid cells of the same area.

As noted by Magnussen and Boudewyn (1998), the number of pulses per spatial unit must produce reliable estimates of ALS metrics used to predict forest attributes. In this regard, minimum pulse densities have rarely been less than 0.1 pulses/m^2 (Næsset 1997b; Holmgren 2004), and minimum plot areas have rarely been less than 200 m^2 (Næsset 2002; Andersen and Breidenbach 2007; Gobakken and Næsset 2008; Breidenbach et al. 2008; Maltamo et al. 2011a, b). Although multiple reports of the effects of varying pulse densities on ALS-based volume and biomass estimates have been published, the results are not directly comparable because common plot areas were not used; similarly, reports of the effects of varying plot sizes are not directly comparable because common pulse densities were not used. However, when the results of these comparative studies are expressed in terms of pulses per plot, Thomas et al. (2006), Maltamo et al. (2006b), Breidenbach et al. (2008), and Gobakken and Næsset (2008) all reported that reducing pulse densities to 100–225 pulses per plot had no adverse effects on the quality of fit of volume or biomass models when using echo height deciles as predictor variables. Holmgren (2004), Magnusson et al. (2007), and Strunk et al. (2012) all reported that reducing the number of pulses per plot to 50 or fewer had no serious adverse effects on model fits or estimates. Thomas et al. (2006) compared not only qualities of fit but also the distributions of echo heights for pulse densities of 200 and 2,000 pulses per plot, and reported that the distributions were essentially the same.

Næsset (1997a) compared means of stand-level heights for 36 stands using tree measurements for approximately 15 plots per stand to three ALS-based means: arithmetic mean ALS height, mean weighted ALS height, and the within-stand mean over 15 × 15 m cells of the greatest ALS height per cell. The latter means deviated the least from the means based on tree measurements. Næsset (1997b) estimated stand volume from multiple linear and nonlinear models using three small footprint

ALS metrics as independent variables: mean stand height calculated as the mean of maximum echo heights for small grid cells, mean height of all ALS echoes, and canopy cover density calculated as the proportion of canopy echoes. R^2 values ranged from 0.45 to 0.89, depending on the site. A seminal contribution of these two studies was the idea of tessellating a spatial study area into grid cells, predicting the response variable for each grid cell, and calculating the mean over grid cells as an estimate for the entire study area (see further details in Chap. 11).

The insight of Magnussen and Boudewyn (1998) regarding the necessity of a sufficient number of pulses to characterize the vertical distributions of ALS echo heights, in combination with the grid cell approach proposed by Næsset (1997a, b), led Næsset and Bjerknes (2001) to select grid cells with the same areas as plots used to calibrate prediction models. Næsset (2004) provided a partial rationale for selecting spatial units of equal areas and a caution against using ALS metrics such as maximum height that depend on plot size. However, Zhao et al. (2009) proposed models using ALS metrics based on echo height distributions that can be calibrated using spatial units of one size and applied to spatial units of different sizes. Magnussen et al. (2013) proposed the use of cumulants (moments) of distributions of ALS canopy heights to predict plot-level distribution of forest attributes such as tree diameter, basal area, and volume. Despite these advances, the issue of whether models constructed using plots with particular areas and pulse densities can be applied without adverse effects to grid cells with different areas and pulse densities, particularly in a readily implementable and efficient manner, remains somewhat uncertain.

Numerous investigations of procedures for characterizing relationships between forest attributes and ALS metrics have been reported of which only those making unique contributions are cited herein. In a university-industry partnership, Means et al. (2000) used linear models and commercially available small footprint ALS data to predict height, basal area, and volume. Packalén and Maltamo (2006), Hudak et al. (2008), Latifi et al. (2010), and Andersen et al. (2011) investigated non-parametric approaches including nearest neighbors imputation. Maltamo et al. (2004), Breidenbach et al. (2008), and Vauhkonen et al. (2011) investigated linear mixed effects models with random stand-level intercepts, and Breidenbach and Kublin (2009) and Junttila et al. (2008) investigated Bayesian methods.

Lefsky et al. (2002), Næsset et al. (2005), Jensen et al. (2006), Næsset (2007), Breidenbach et al. (2008), and Asner et al. (2012) investigated using common models for relationships between forest attributes and LiDAR metrics for differing forest conditions. For response variables such as volume, biomass, and carbon, the prospects for common models were positive under the condition that the sensor systems and acquisition parameters were stable across study areas (Næsset 2009). However, for a comprehensive dataset consisting of 1,395 plot-level observations representing 10 different boreal forest areas, Næsset and Gobakken (2008) reported that the quality of fit of models to aboveground biomass data was significantly improved by including variables related to proportions spruce and proportion broadleaf trees (see also Chap. 11).

1.3.2.2 Individual Tree Approaches

For tree-level inventories, ALS data are typically used to detect individual treetops and to predict attributes of interest using sets of allometric models. Algorithms and techniques similar to those developed for aerial images can be used with ALS data (Hyyppä et al. 2008). Tree-level inventories consist of a sequence of steps that includes tree detection, feature extraction, and estimation of tree attributes. For these applications, the ALS data contributions include estimation of heights of the vegetation surface; estimation of physical properties such as height differences between potential tree tops during tree detection; estimation of individual tree heights and crown sizes used as input to the allometric models (e.g. Hyyppä and Inkinen 1999; Persson et al. 2002); and avoidance of problems related to variations in geometry and radiometry of spectral images (e.g. Mäkinen et al. 2006).

Detection of individual trees usually relies on a raster-based canopy height model (CHM) interpolated from the ALS height data (e.g. Hyyppä and Inkinen 1999; Persson et al. 2002; Popescu et al. 2003), although point-based techniques could be used, particularly for the segmentation (see Chap. 5). An important consideration is that all trees cannot usually be detected. The degree to which individual trees are successfully detected is affected by the detection algorithm and its parameterization (Kaartinen et al. 2012). In addition, detection success is strongly affected by forest conditions, particularly stand density and spatial pattern, which suggest problems related to both interlaced tree crowns and trees below the dominant canopy (Vauhkonen et al. 2012a). Methods for resolving these problems include selection of an appropriate level of filtering a priori (Heinzel et al. 2011; Ene et al. 2012), use of full-waveform data (Reitberger et al. 2009) and improved 3D algorithms (Lähivaara et al. 2014; Tang et al. 2013).

The estimation task requires that relevant information for estimating tree attributes of interest must be extracted from the ALS data. Tree-level estimation combines direct measurements, species-specific properties that can be predicted from the ALS data, and tree allometry in the form of dimensional relationships between plant parts. In addition to estimates of locations, the tree detection and delineation component typically produces heights, crown dimensions, and height and intensity distributions for individual tree segments (e.g. Holmgren and Persson 2004). Estimates for the latter properties are particularly useful for ALS-based species recognition (Chap. 7).

Use of allometric models to estimate attributes such as the diameter at-breast-height (DBH) poses problems, because relationships between measurable tree dimensions and DBH are far from deterministic but rather are affected by factors such as tree density and silvicultural history (Korpela 2004; Maltamo et al. 2007). Geometrically weighted regression methods have been investigated as a means of overcoming at least some of these problems (Salas et al. 2010). Propagation of errors in the ALS-based DBH estimates to stem taper model predictions can be avoided by estimating stem volumes directly from the ALS data using regression models (e.g. Takahashi et al. 2005) or nearest-neighbor techniques (e.g. Maltamo et al. 2009), which can use a large number of predictor variables (Vauhkonen et al. 2010).

Multiple tree-level studies have focused on individual tree detection, feature extraction, and estimation steps in this inventory procedure. Remarkably, however, few studies have reported plot- or stand-level estimates obtained from the full detection and estimation procedure (Korpela et al. 2007; Peuhkurinen et al. 2007, 2011; Vauhkonen et al. 2014). The accuracy of aggregated estimates depends on multiple factors including both tree-level detection and tree-level estimation errors. Undetected trees and errors in the allometric model predictions degrade the accuracies of plot- and stand-level estimates and seriously constrain single-tree applications in semi-natural forests (Korpela et al. 2007). Approaches that attempt to circumvent the effects of these error sources such as semi-individual tree detection are being developed (Breidenbach et al. 2010, Chap. 6). The individual tree approach may be most relevant for applications such as wood procurement planning for which unbiased estimators are not necessarily required (e.g. Peuhkurinen et al. 2007; Vauhkonen et al. 2014).

Finally, the complementary properties of the individual tree and area-based approaches should be noted. In addition to actual tree detection, data on individual trees can be generated by predicting diameter distributions with the area-based approach (Chap. 9), while combining these approaches has been found especially beneficial for reducing tree detection errors (Maltamo et al. 2004; Breidenbach et al. 2010; Lindberg et al. 2010). Breidenbach et al. (2012) and Vastaranta et al. (2012) proposed generating field reference data for the area-based approach by means of modified tree detection approaches. Vauhkonen et al. (2014) used an area-based mean diameter prediction for targeting detailed tree-level analyses for stands that are sufficiently mature for cutting. In addition to the earlier mentioned height distribution, other feature types can also be calculated at area and tree levels (e.g. Hyyppä et al. 2012; Vauhkonen et al. 2012b; Ørka et al. 2013), and for some applications, the addressed domain may even be difficult to identify (e.g. Vauhkonen et al. 2011; Mehtätalo and Nyblom 2012, Chap. 10).

1.4   Applications

The first studies of forestry applications of ALS focused on traditional forest inventory variables such as height, basal area and volume (e.g. Næsset 1997a, b; Hyyppä and Inkinen 1999; Means et al. 2000). The context for these studies was usually stand level management inventories using the area-based approach. Rather soon thereafter operational use of ALS for this purpose was initiated in Norway (Chap. 11) and subsequently in other countries. These first inventories were complemented by economic analyses of the costs and benefits of the use of ALS as an auxiliary source of information (Chap. 16). In Finland, because estimates of species-specific stand attributes are required, ALS data were combined with aerial image data (Chap. 12). Issues related to the co-registration of the different sources of data are discussed in Chap. 4. Nowadays, area-based forest inventories are increasingly conducted for different forest conditions such as plantations

(Chap. 13). Since circa 2008, use of ALS as a sampling mechanism has rapidly become more common (Chap. 14), especially for larger regional studies for which full ALS coverage obviously is cost prohibitive. Multiple ALS-assisted sampling approaches can still produce estimates with high precision. Thus, the currently most relevant forest inventory applications are for National Forest Inventories and tropical REDD + inventories. Additional ALS-assisted inventory applications include pre-harvest (e.g. Vauhkonen et al. 2014) and conservation inventories (Chaps. 17, 18, and 19).

The ability of ALS data to characterize the 3D structure of forest canopies facilitates numerous ecological applications. Some ecological parameters such as canopy cover (Chap. 20) and canopy gaps (Chap. 21) can be almost directly estimated from ALS data. While a highly accurate estimate for canopy cover can be calculated as a proportion of vegetation hits above a specified height limit, canopy gaps can be detected from a CHM. Measures associated with these variables are among the most obvious attributes that can be simply estimated from ALS data. Estimation of some other traditional forest inventory variables, such as biomass and volume, using ALS data can be based on allometric relationships, but may require considerably more complex, multi-step procedures (see Chap. 8).

Vertical forest structure, another key ecological parameter, can also be estimated from ALS data. Such information is useful and interesting if for no other reason than to distinguish canopy layers. However, this information can also be used as input to fire risk (Chap. 22) and habitat models (Chap. 17). To date, most ALS-assisted habitat models are for bird species (e.g. Hill et al. 2004), although recent studies have also focused on mammals (Melin et al. 2013). ALS has become an important information source for species level biodiversity studies (Chap. 18) that focus on characteristics such as richness, diversity, composition (e.g. Müller and Brandl 2009), and naturalness (e.g. Bater et al. 2009). Finally ALS data have been used to model deadwood attributes (Chap. 19) and to guide field inventories of deadwood (e.g. Pesonen et al. 2010).

1.5 Structure of the Book

ALS has gained widespread acceptance as an important and useful source of auxiliary information for describing the 3D structure of forests. Researchers and practitioners alike have found these data useful for forest inventory, ecological and environmental research. The overall intent of the book is to provide a com-prehensive, state-of-the-art review for these areas of application. The book is organized into three parts following this introduction. Because inventory, ecological and environmental disciplines share the same basic approach to extraction of information from the primary data, Part I (Chaps. 2, 3, 4, 5, 6, 7, 8, 9, and 10) focuses on methodological issues. However, most of the book is devoted to applications which are further organized into two parts. Part II (Chaps. 11, 12, 13, 14, 15, and 16) focuses on forest inventory applications at multiple geographical scales, from local, operational, wall-to-wall management inventories to regional, strategic, sampling-based inventories, and Part III (Chaps. 17, 18, 19, 20, 21, and 22) focuses on ecological and environmental applications. The application chapters of Parts II and III refer back to the general treatment of methodological issues in Part I whenever appropriate; thus, cross-references among chapters are frequent.

Because most chapters are written as state-of-the-art reviews, the book can be viewed as a comprehensive and unique review of a wide range of forestry applications of ALS. Many of the chapters focus on boreal forests simply because methods were initially developed for boreal conditions. The reviews provide a sufficiently comprehensive overview of recent research and applications that researchers, students and practitioners should all find the book to be a useful reference text.

References

Aldred AH, Bonner GM (1985) Application of airborne lasers to forest surveys. Information report PI-X-51, Technical Information and Distribution Center, Petawawa National Forestry Institute, Chalk River, Ontario, Canada, 62 pp

Andersen H-E, Breidenbach J (2007) Statistical properties of mean stand biomass estimators in a LIDAR-bases double sampling forest survey design. In: Proceedings of the ISPRS workshop laser scanning 2007 and SilviLaser 2007, Espoo, Finland, 12–14 Sept 2007, IAPRS, vol XXXVI, Part 3/W52, 2007, pp 8–13

Andersen H-E, Strunk J, Temesgen H, Atwood D, Winterberger K (2011) Using multi-level remote sensing and ground data to estimate forest biomass resources in remote regions: a case study in the boreal forests of interior Alaska. Can J Remote Sens 37(6):596–611

Arp H, Griesbach JC, Burns JP (1982) Mapping in tropical forests: a new approach using laser APR. Photogramm Eng Remote Sens 48:91–100

Asner GP, Mascaro J, Muller-Landau HC, Vieilledent G, Vaudry R, Rasamoelina M, Hall JS, van Breugel M (2012) A universal airborne LiDAR approach for tropical forest carbon mapping. Oecologia 168:1147–1160

Baltsavias EP (1999) Airborne laser scanning: basic relations and formulas. ISPRS J Photogramm Remote Sens 54:199–214

Bater CW, Coops NC, Gergel SE, LeMay V, Collins D (2009) Estimation of standing dead tree class distributions in northwest central forests using lidar remote sensing. Can J For Res 39:1080–1091

Brandtberg T (1999) Automatic individual tree-based analysis of high spatial resolution remotely sensed data. Doctoral thesis. Swedish University of Agricultural Sciences, Centre for Image Analysis, Uppsala, Sweden. Acta Uni Agric Suec Silv 118, 155 pp

Breidenbach J, Kublin E (2009) Estimating timber volume using airborne laser scanning data based on Bayesian methods. In: Proceedings of the IUFRO Division 4 conference extending forest inventories over space and time, Quebec City, Canada, 19–22 May 2009

Breidenbach J, Kublin E, McGaughey R, Andersen H-E, Reutebuch S (2008) Mixed-effects models for estimating stand volume by means of small footprint airborne laser scanner data. Photogramm J Finland 21:4–15

Breidenbach J, Næsset E, Lien V, Gobakken T, Solberg S (2010) Prediction of species specific forest inventory attributes using a nonparametric semi-individual tree crown approach based on fused airborne laser scanning and multispectral data. Remote Sens Environ 114:911–924

Breidenbach J, Næsset E, Gobakken T (2012) Improving k-nearest neighbor predictions in forest inventories by combining high and low density airborne laser scanning data. Remote Sens Environ 117:358–365

Coops NC, Wulder MA, Culvenor DS, St-Onge B (2004) Comparison of forest attributes extracted from fine spatial resolution multispectral and lidar data. Can J Remote Sens 30:855–866

Ene L, Næsset E, Gobakken T (2012) Single tree detection in heterogeneous boreal forests using airborne laser scanning and area based stem number estimates. Int J Remote Sens 33: 5171–5193

Gobakken T, Næsset E (2008) Assessing effects of laser point density, ground sampling intensity, and field sample plot size on biophysical stand properties derived from airborne laser scanner data. Can J For Res 38:1095–1109

Heinzel J, Weinacker H, Koch B (2011) Prior-knowledge-based single-tree extraction. Int J Remote Sens 32:4999–5020

Hill RA, Hinsley SA, Gaveau DLA, Bellamy PE (2004) Predicting habitat quality for Great Tits (*Parus major*) with airborne laser scanning data. Int J Remote Sens 25:4851–4855

Holmgren J (2004) Prediction of tree height, basal area and stem volume using airborne laser scanning. Scand J For Res 19:543–553

Holmgren J, Persson Å (2004) Identifying species of individual trees using airborne laser scanner. Remote Sens Environ 90:415–423

Hudak AT, Crookston NL, Evans JS, Hall DE, Falkowski MJ (2008) Nearest neighbor imputation of species-level, plot-scale forest structure attributes from LiDAR data. Remote Sens Environ 112:2232–2245. Corrigendum: Remote Sens Environ 113:289–290

Hyyppä J, Inkinen M (1999) Detecting and estimating attributes for single trees using laser scanner. Photogramm J Finland 16:27–42

Hyyppä J, Hyyppä H, Leckie D, Gougeon F, Yu X, Maltamo M (2008) Review of methods of small-footprint airborne laser scanning for extracting forest inventory data in boreal forests. Int J Remote Sens 29:1339–1366

Hyyppä J, Hyyppä H, Yu X, Kaartinen H, Kukko A, Holopainen M (2009) Forest inventory using small-footprint airborne LiDAR. In: Shan J, Toth CK (eds) Topographic laser ranging and scanning: principles and processing. CRC Press/Taylor & Francis Group, Boca Raton, pp 335–370

Hyyppä J, Yu X, Hyyppä H, Vastaranta M, Holopainen M, Kukko A, Kaartinen H, Jaakkola A, Vaaja M, Koskinen J, Alho P (2012) Advances in forest inventory using airborne laser scanning. Remote Sens 4:1190–1207

Jensen JLR, Humes KS, Conner T, Williams CJ, DeGroot J (2006) Estimation of biophysical characteristics for highly variable mixed-conifer stands using small-footprint lidar. Can J For Res 36:1129–1138

Junttila V, Maltamo M, Kauranne T (2008) Sparse Bayesian estimation of forest stand characteristics from ALS. For Sci 54:543–552

Kaartinen H, Hyyppä J, Yu X, Vastaranta M, Hyyppä H, Kukko A, Holopainen M, Heipke C, Hirschmugl M, Morsdorf F, Næsset E, Pitkänen J, Popescu S, Solberg S, Wolf BM, Wu J-C (2012) An international comparison of individual tree detection and extraction using airborne laser scanning. Remote Sens 4:950–974

Koch B (2010) Status and future of laser scanning, synthetic aperture radar and hyperspectral remote sensing data for forest biomass assessment. ISPRS J Photogramm Remote Sens 65:581–590

Koch B, Dees M, van Brusselen J, Eriksson L, Fransson J, Gallaun H, Leblon B, McRoberts RE, Nilsson M, Schardt M, Seitz R, Waser L (2008) Forestry applications. In: Li Z, Chen J, Baltsavias E (eds) Advances in photogrammetry, remote sensing and spatial information sciences: 2008 ISPRS congress book. Taylor & Francis Group, Boca Raton, pp 439–465

Korpela I (2004) Individual tree measurements by means of digital aerial photogrammetry. Doctoral thesis, University of Helsinki, Department of Forest Resource Management, Helsinki, Finland. Silva Fenn Monograph 3, 93 p

Korpela I, Dahlin B, Schäfer H, Bruun E, Haapaniemi F, Honkasalo J, Ilvesniemi S, Kuutti V, Linkosalmi M, Mustonen J, Salo M, Suomi O, Virtanen H (2007) Single-tree forest inventory using lidar and aerial images for 3D treetop positioning, species recognition, height and crown width estimation. In: Proceedings of the ISPRS workshop laser scanning 2007 and SilviLaser 2007, Espoo, Finland, 12–14 Sept 2007, IAPRS, vol XXXVI, Part 3/W52, 2007, pp 227–233

Lähivaara T, Seppänen A, Kaipio JP, Vauhkonen J, Korhonen L, Tokola T, Maltamo M (2014) Bayesian approach to tree detection based on airborne laser scanning data. IEEE Trans Geosci Remote Sens. doi:10.1109/TGRS.2013.2264548

Latifi H, Nothdurft A, Koch B (2010) Non-parametric prediction and mapping of standing timber volume and biomass in a temperate forest: applications of multiple optical/LiDAR-derived predictors. Forestry 83:395–407

Lefsky MA, Cohen WB, Spies TA (2001) An evaluation of alternate remote sensing products for forest inventory, monitoring, and mapping of Douglas-fir forests in western Oregon. Can J For Res 31:78–87

Lefsky MA, Cohen WB, Harding DJ, Parker GG, Acker SA, Gower ST (2002) Lidar remote sensing of above-ground biomass in three biomes. Glob Ecol Biogeogr 11:393–399

Lim K, Treitz P, Wulder M, St-Onge B, Flood M (2003) LiDAR remote sensing of forest structure. Prog Phys Geogr 27:88–106

Lindberg E, Holmgren J, Olofsson K, Wallerman J, Olsson H (2010) Estimation of tree lists from airborne laser scanning by combining single-tree and area-based methods. Int J Remote Sens 31:1175–1192

Maclean GA, Krabill WB (1986) Gross-merchantable timber volume estimation using an airborne lidar system. Can J Remote Sens 12:7–18

Maclean GA, Martin GL (1984) Merchantable timer volume estimation using cross-sectional photogrammetric and densitometric methods. Can J For Res 14:803–810

Magnussen S, Boudewyn P (1998) Derivations of stand heights from airborne laser scanner data with canopy-based quantile estimators. Can J For Res 28:1016–1031

Magnussen S, Eggermont P, LaRiccia VN (1999) Recovering tree heights from airborne laser scanner data. For Sci 45:407–422

Magnussen S, Næsset E, Gobakken T (2013) Prediction of tree-size distributions and inventory variables from cumulants of canopy height distributions. Forestry 86:583–595. doi:10.1093/forestry/cpt022

Magnusson M, Fransson JES, Holmgren J (2007) Effects on estimation accuracy of forest variables using different pulse density of laser data. For Sci 53:619–626

Mäkinen A, Korpela I, Tokola T, Kangas A (2006) Effects of imaging conditions on crown diameter measurements from high-resolution aerial images. Can J For Res 36:1206–1217

Maltamo M, Eerikäinen K, Pitkänen J, Hyyppä J, Vehmas M (2004) Estimation of timber volume and stem density based on scanning laser altimetry and expected tree size distribution functions. Remote Sens Environ 90:319–330

Maltamo M, Malinen J, Packalén P, Suvanto A, Kangas J (2006a) Nonparametric estimation of stem volume using airborne laser scanning, aerial photography, and stand-register data. Can J For Res 36:426–436

Maltamo M, Eerikäinen K, Packalén P, Hyyppä J (2006b) Estimation of stem volume using laser scanning based canopy height metrics. Forestry 79:217–229

Maltamo M, Packalén P, Peuhkurinen J, Suvanto A, Pesonen A, Hyyppä J (2007) Experiences and possibilities of ALS based forest inventory in Finland. In: Proceedings of the ISPRS workshop laser scanning 2007 and SilviLaser 2007, Espoo, Finland, 12–14 Sept 2007, IAPRS, vol XXXVI, Part 3/W52, 2007, pp 270–279

Maltamo M, Peuhkurinen J, Malinen J, Vauhkonen J, Packalén P, Tokola T (2009) Predicting tree attributes and quality characteristics of Scots pine using airborne laser scanning data. Silva Fenn 43:507–521

Maltamo M, Bollandsås OM, Næsset E, Gobakken T, Packalén P (2011a) Different plot selection strategies for field training data in ALS-assisted forest inventory. Forestry 84:23–31

Maltamo M, Packalén P, Kallio E, Kangas J, Uuttera J, Heikkilä J (2011b) Airborne laser scanning based stand level management inventory in Finland. In: Proceedings of the SilviLaser 2011 – 11th international conference of LiDAR applications for assessing forest ecosystems, Hobart, Australia, 16–20 Oct 2011

McRoberts RE, Cohen WB, Næsset E, Stehman SV, Tomppo EO (2010) Using remotely sensed data to construct and assess forest attribute maps and related spatial products. Scand J For Res 25:340–367

Means JE, Acker SA, Fitt BJ, Renslow M, Emerson L, Hendrix CJ (2000) Predicting forest stand characteristics with airborne scanning lidar. Photogramm Eng Remote Sens 66:1367–1371

Mehtätalo L, Nyblom J (2012) A model-based approach for ALS inventory: application for square grid spatial pattern. For Sci 58:106–118

Melin M, Packalén P, Matala J, Mehtätalo L, Pusenius J (2013) Assessing and modeling moose (Alces alces) habitats with airborne laser scanning data. Int J Appl Earth Obs Geoinfo 23:389–396

Müller J, Brandl R (2009) Assessing biodiversity by remote sensing and ground survey in mountainous terrain: the potential of LiDAR to predict forest beetle assemblages. J Appl Ecol 46:897–905

Næsset E (1997a) Determination of mean tree height of forest stands using airborne laser scanner data. ISPRS J Photogramm Remote Sens 52:49–56

Næsset E (1997b) Estimating timber volume of forest stands using airborne laser scanner data. Remote Sens Environ 51:246–253

Næsset E (2002) Predicting forest stand characteristics with airborne scanning laser using a practical two-stage procedure and field data. Remote Sens Environ 80:88–99

Næsset E (2004) Practical large-scale forest stand inventory using a small airborne scanning laser. Scand J For Res 19:164–179

Næsset E (2007) Airborne laser scanning as a method in operational forest inventory: status of accuracy assessments accomplished in Scandinavia. Scand J For Res 22:433–442

Næsset E (2009) Effects of different sensors, flying altitudes, and pulse repetition frequencies on forest canopy metrics and biophysical stand properties derived from small-footprint airborne laser data. Remote Sens Environ 113:148–159

Næsset E, Bjerknes K-O (2001) Estimating tree heights and number of stems in young forest stands using airborne laser scanner data. Remote Sens Environ 78:328–340

Næsset E, Gobakken T (2008) Estimation of above- and below-ground biomass across regions of the boreal forest zone using airborne laser. Remote Sens Environ 112:3079–3090

Næsset E, Gobakken T, Holmgren J, Hyyppä H, Hyyppä J, Maltamo M, Nilsson M, Olsson H, Persson Å, Söderman U (2004) Laser scanning of forest resources: the Nordic experience. Scand J For Res 19:482–499

Næsset E, Bollandsås OM, Gobakken T (2005) Comparing regression methods in estimation of biophysical properties of forest stands from two different inventories using laser scanner data. Remote Sens Environ 94:541–553

Nilsson M (1996) Estimation of tree heights and stand volume using an airborne lidar system. Remote Sens Environ 56:1–7

Nelson R (2013) How did we get here? An early history of forestry lidar. Can J Remote Sens 39:S6–S17. doi:10.5589/m13-011

Nelson R, Krabill W, Maclean G (1984) Determining forest canopy characteristics using airborne laser data. Remote Sens Environ 15:201–212

Nelson R, Krabill W, Tonelli J (1988) Estimating forest biomass and volume using airborne laser data. Remote Sens Environ 24:247–267

Nord-Larsen T, Riis-Nielsen T (2010) Developing an airborne laser scanning dominant height model from a countrywide scanning survey and national forest inventory data. Scand J For Res 25:262–272

Ørka HO, Dalponte M, Gobakken T, Næsset E, Ene LT (2013) Characterizing forest species composition using multiple remote sensing data sources and inventory approaches. Scand J For Res 28:677–688

Packalén P, Maltamo M (2006) Predicting the plot volume by tree species using airborne laser scanning and aerial photographs. For Sci 52:611–622

Packalén P, Maltamo M, Tokola T (2008) Detailed assessment using remote sensing techniques. In: Gadow K, Pukkala T (eds) Designing green landscapes, vol 15, Managing forest ecosystems. Springer, Dordrecht, pp 51–77

Persson Å, Holmgren J, Söderman U (2002) Detecting and measuring individual trees using an airborne laser scanner. Photogramm Eng Remote Sens 68:925–932

Pesonen A, Maltamo M, Kangas A (2010) The comparison of airborne laser scanning-based probability layers as auxiliary information for assessing coarse woody debris. Int J Remote Sens 31:1245–1259

Peuhkurinen J, Maltamo M, Malinen J, Pitkänen J, Packalén P (2007) Preharvest measurement of marked stands using airborne laser scanning. For Sci 53:653–661

Peuhkurinen J, Mehtätalo L, Maltamo M (2011) Comparing individual tree detection and the area-based statistical approach for the retrieval of forest stand characteristics using airborne laser scanning in Scots pine stands. Can J For Res 41:583–598

Popescu SC, Wynne RH, Nelson RF (2003) Measuring individual tree crown diameter with lidar and assessing its influence on estimating forest volume and biomass. Can J Remote Sens 29:564–577

Reitberger J, Schnörr C, Krzystek P, Stilla U (2009) 3D segmentation of single trees exploiting full waveform LIDAR data. ISPRS J Photogramm Remote Sens 64:561–574

Rempel RC, Parker AK (1964) An information note on an airborne laser terrain profiler for micro-relief studies. In: Proceedings of the 3rd symposium on remote sensing of environment, University of Michigan Institute of, Science and Technology, pp 321–337

Salas C, Ene L, Gregoire TG, Næsset E, Gobakken T (2010) Modelling tree diameter from airborne laser scanning derived variables: a comparison of spatial statistical models. Remote Sens Environ 114:1277–1285

Schreier H, Lougheed J, Tucker C, Leckie D (1985) Automated measurements of terrain reflection and height variations using airborne infrared laser system. Int J Remote Sens 6:101–113

Solodukhin VI, Zhukov AY, Mazhugin IN, Narkevich VI (1976) Metody Izuchenija Vertikal'nyh Sechenij Drevostoev (Method of study of vertical sections of forest stands). Leningrad Scientific Research Institute of Forestry, Leningrad, 55 pp (in Russian)

Solodukhin VI, Mazhugin IN, Zhukov AY, Narkevich VI, Popov YV, Kulyasov AG, Marasin LE, Sokolov SA (1979) Lazernaya aeros'emka profilei lesa (Laser aerial profiling of forest). Lesnoe Khozyaistvo 10:43–45 (in Russian)

Strunk J, Temesgen H, Andersen H-E, Flewelling JP, Madsen L (2012) Effects of pulse density and sample size on a model-assisted approach to estimate forest inventory variables. Can J For Res 38:644–654

Takahashi T, Yamamoto K, Senda Y, Tsuzuku M (2005) Predicting individual stem volumes of sugi (Cryptomeria japonica D. Don) plantations in mountainous areas using small-footprint airborne LiDAR. J For Res 10:305–312

Tang S, Dong P, Buckles BP (2013) Three-dimensional surface reconstruction of tree canopy from lidar point clouds using a region-based level set method. Int J Remote Sens 34:1373–1385

Thomas V, Treitz P, McGaughey JH, Morisson I (2006) Mapping stand-level forest biophysical variables for a mixedwood boreal forest using lidar: an examination of scanning density. Can J For Res 36:34–47

Vastaranta M, Kankare V, Holopainen M, Yu X, Hyyppä J, Hyyppä H (2012) Combination of individual tree detection and area-based approach in imputation of forest variables using airborne laser data. ISPRS J Photogramm Remote Sens 67:73–79

Vauhkonen J, Korpela I, Maltamo M, Tokola T (2010) Imputation of single-tree attributes using airborne laser scanning-based height, intensity, and alpha shape metrics. Remote Sens Environ 114:1263–1276

Vauhkonen J, Mehtätalo L, Packalén P (2011) Combining tree height samples produced by airborne laser scanning and stand management records to estimate plot volume in Eucalyptus plantations. Can J For Res 41:1649–1658

Vauhkonen J, Ene L, Gupta S, Heinzel J, Holmgren J, Pitkänen J, Solberg S, Wang Y, Weinacker H, Hauglin KM, Lien V, Packalén P, Gobakken T, Koch B, Næsset E, Tokola T, Maltamo M (2012a) Comparative testing of single-tree detection algorithms under different types of forest. Forestry 85:27–40

Vauhkonen J, Seppänen A, Packalén P, Tokola T (2012b) Improving species-specific plot volume estimates based on airborne laser scanning and image data using alpha shape metrics and balanced field data. Remote Sens Environ 124:534–541

Vauhkonen J, Packalen P, Malinen J, Pitkänen J, Maltamo M (2014) Airborne laser scanning based decision support for wood procurement planning. Scand J For Res. doi:10.1080/02827581.2013.813063

Wehr A, Lohr U (1999) Airborne laser scanning – an introduction and overview. ISPRS J Photogramm Remote Sens 54:68–82

Woods M, Pitt D, Penner M, Lim K, Nesbitt D, Etheridge D, Treitz P (2011) Operational implementation of a LiDAR inventory in Boreal Ontario. For Chron 87:512–528

Zhao K, Popescu S, Nelson R (2009) Lidar remote sensing of forest biomass: a scale-invariant estimation approach using airborne lasers. Remote Sens Environ 113:182–196

Part I
Methodological Issues

Chapter 2
Laser Pulse Interaction with Forest Canopy: Geometric and Radiometric Issues

Andreas Roncat, Felix Morsdorf, Christian Briese, Wolfgang Wagner, and Norbert Pfeifer

Abstract This chapter focuses upon retrieving forest biophysical parameters by extracting three-dimensional point cloud information from small-footprint full-waveform airborne laser scanner data. This full waveform gives the end user the possibility to gain control over range determination and the subsequent derivation of the point clouds. Furthermore, the attribution of physical parameters to the single points using these waveforms becomes additionally possible. The underlying physical principles form the begin of this chapter, followed by forward modeling of waveforms over simulated forested areas, the treatment of real waveforms and an example for validating the results of full-waveform analysis.

2.1 Introduction

Airborne Laser Scanning (ALS) has played a role of increasing importance in the management of forest ecosystems, as already pointed out in Chap. 1 of this volume. The principal outcome of ALS campaign are area-wide three-dimensional (3D) point clouds, derived by laser range measurements and knowledge about position and attitude of the instrument.

A. Roncat (✉) • W. Wagner • N. Pfeifer
Research Groups Photogrammetry and Remote Sensing, Department of Geodesy
and Geoinformation, Vienna University of Technology, Vienna, Austria
e-mail: Andreas.Roncat@geo.tuwien.ac.at

F. Morsdorf
Remote Sensing Laboratories, Department of Geography, University of Zurich – Irchel, Zurich,
Switzerland

C. Briese
Ludwig Boltzmann Institute for Archaeological Prospection and Virtual Archaeology,
Vienna, Austria

Research Groups Photogrammetry and Remote Sensing, Department of Geodesy
and Geoinformation, Vienna University of Technology, Vienna, Austria

M. Maltamo et al. (eds.), *Forestry Applications of Airborne Laser Scanning: Concepts
and Case Studies*, Managing Forest Ecosystems 27, DOI 10.1007/978-94-017-8663-8_2,
© Springer Science+Business Media Dordrecht 2014

Usually, the determination of the range is a "black box" to the end user. However, full-waveform (FWF) ALS instruments have become increasingly available throughout the past decade. These instruments sample the temporal profile (waveforms) of both the emitted laser pulse and of its echoes in a high frequency. As a consequence, they allow the end user to get control over range determination. Thus, studies of FWF data over forest canopies can be a means to gain fundamental knowledge and understanding of the interaction between laser pulses and vegetation. Furthermore, sophisticated analysis of radiometry for these waveforms enables the assignment of biophysical parameters to the target points, such as the backscatter cross-section and the target reflectance.

In this chapter, we give a detailed insight in the concept and the workflow of processing such FWF data in a general perspective as well as in the context of vegetation and forest studies. We start with presenting the physical principles in Sect. 2.2. Based on these principles and a three-dimensional model of the scene, FWF ALS return signals can be simulated. Section 2.3 presents approaches for forward modeling and provides several examples in the context of forest canopy analysis. However, in most cases the spatial distribution of the scatterers and their reflectance properties are not known in advance, so that they have to be reconstructed (see Sect. 2.4). In order to validate the results of this reconstruction from a geometric and radiometric point of view, terrestrial laser scanning (TLS) has proven to be a precise and efficient tool. An example for such a validation is given in Sect. 2.5. The last section concludes with a discussion and an outlook on directions for future work.

2.2   Physical Principles of Small-Footprint Full-Waveform ALS

Besides delivering three-dimensional point clouds in high resolution as basis for further analyses, ALS can also provide physical parameters of the scanned surfaces if temporal profiles of the transmitted laser pulse and of its echoes are recorded. This technique is known as full-waveform ALS (Wagner et al. 2004).

The relation of the transmitted laser power, P_t, to the echo power of its reflections, P_e, is given by the radar equation (Jelalian 1992):

$$P_e(t) = \frac{D_r^2}{4\pi R^4 \beta_t^2} P_t \left(t - \frac{2R}{v_g} \right) \sigma \, \eta_{\text{SYS}} \, \eta_{\text{ATM}} \tag{2.1}$$

with β_t denoting the beam width of the transmitted laser beam, R the range from the sensor to the reflecting surface, t the travel time, v_g the group velocity of the laser (approximately the speed of light in vacuum), σ the effective backscatter cross-section (in m^2), D_r the aperture diameter, η_{ATM} the atmospheric transmission

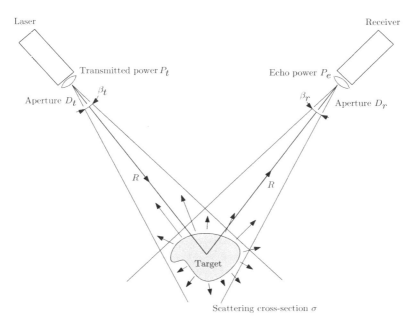

Fig. 2.1 Laser pulse interaction with an extended target (Wagner et al. 2006). In ALS, the laser and the receiver are very close together so that σ is called *backscatter* cross-section in the subsequent text

factor, and η_{SYS} the system transmission factor. The backscatter cross-section is a product of the target area ($A[m^2]$), the target reflectivity ($\rho[]$), and the factor $4\pi/\Omega$ describing the scattering angle of the target ($\Omega[sr]$) in relation to an isotropic scatterer:

$$\sigma = \frac{4\pi}{\Omega}\rho A \tag{2.2}$$

Figure 2.1 sketches the transmission of a laser pulse, the scattering process at a target and the recording of the echoes.

As an example of a small target in a forest canopy, we show the backscatter cross-section of a maple leaf in relation to its size and orientation to the laser ray (Fig. 2.2).

In the case of extended targets, the echo signal is a superposition of single echoes along the laser ray at different ranges R_i. Their respective time delay is $2R_i/v_g$ so that we can use time and range interchangeably in this context. If scatterers of equal reflectance are closer to each other than half the laser pulse length, then their echoes do not form separate maxima. While the use of advanced signal processing strategies enables for the separation of close targets below this threshold (Parrish et al. 2011; Jutzi and Stilla 2006), there is a certain minimal range difference ΔR

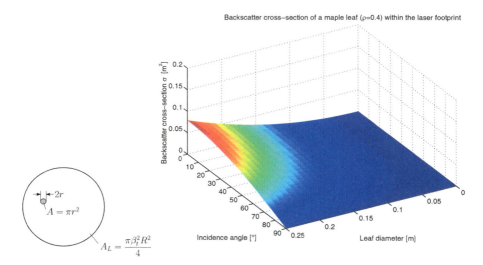

Fig. 2.2 Backscatter cross-section σ for a maple leaf (reflectance $\rho = 0.4$) within the area of the laser footprint A_L, in relation to its size and its orientation to the laser ray, expressed by the leaf radius r and the incidence angle ϑ, resp. The range was chosen to $R = 500\,\mathrm{m}$ and the beam divergence to $\beta_t = 0.5\,\mathrm{mrad}$

where targets cannot be separated any more and form a cluster. For such a cluster at a range $R_i \pm \Delta R$ we get (Wagner et al. 2006):

$$P_{e,i}(t) = \frac{D_r^2}{4\pi\beta_t^2}\eta_{\mathrm{SYS}}\,\eta_{\mathrm{ATM}} \int_{R_i-\Delta R}^{R_i+\Delta R} \frac{1}{R^4} P_t\left(t - \frac{2R}{v_g}\right)\sigma_i'(R)\mathrm{d}R \qquad (2.3)$$

with $\sigma_i'(R) = \mathrm{d}\sigma/\mathrm{d}R$ being the derivative of the backscatter cross-section w.r.t. the range. Since $\sigma_i'(R)$ is zero outside the interval $[R_i - \Delta R, R_i + \Delta R]$, the term

$$\int_{R_i-\Delta R}^{R_i+\Delta R} P_t\left(t - \frac{2R}{v_g}\right)\sigma_i'(R)\,\mathrm{d}R = \int_{-\infty}^{\infty} P_t\left(t - \frac{2R}{v_g}\right)\sigma_i'(R)\,\mathrm{d}R = P_t(t) \otimes \sigma_i'(t)$$

is the convolution ($\otimes$) of the transmitted laser power with the differential backscatter cross-section.

Assuming that $\Delta R \ll R_i$, Eq. (2.3) can be approximated by

$$P_{e,i}(t) \simeq \frac{D_r^2}{4\pi\beta_t^2 R_i^4}\eta_{\mathrm{SYS}}\,\eta_{\mathrm{ATM}} \int_{R_i-\Delta R}^{R_i+\Delta R} P_t\left(t - \frac{2R}{v_g}\right)\sigma_i'(R)\mathrm{d}R$$

$$= \frac{D_r^2}{4\pi\beta_t^2 R^4}\eta_{\mathrm{SYS}}\,\eta_{\mathrm{ATM}}\,P_t(t) \otimes \sigma_i'(t).$$

We cannot record $P_t(t)$ and $P_e(t)$ directly; instead, their convolution with the scanner's system response function $\Gamma(t)$ is recorded. This yields

$$P_{e,i}(t) \otimes \Gamma(t) = \frac{D_r^2}{4\pi\beta_t^2 R_i^4} \eta_{SYS} \, \eta_{ATM} \, P_t(t) \otimes \sigma_i'(t) \otimes \Gamma(t)$$

$$= \frac{D_r^2}{4\pi\beta_t^2 R_i^4} \eta_{SYS} \, \eta_{ATM} \, P_t(t) \otimes \Gamma(t) \otimes \sigma_i'(t)$$

because convolution is commutative. Thus, we can define $P_t(t) \otimes \Gamma(t)$ as *system waveform* $S(t)$, the quantity actually recorded when a copy of the transmitted laser pulse is stored. Summing up all n echoes of such a pulse, we get the *recorded* echo power $P_r(t)$ as

$$P_r(t) = \sum_{i=1}^{n} P_{e,i}(t) \otimes \Gamma(t) = \frac{D_r^2}{4\pi\beta_t^2} \eta_{SYS} \, \eta_{ATM} \sum_{i=1}^{n} \frac{1}{R_i^4} S(t) \otimes \sigma_i'(t). \qquad (2.4)$$

The primary output of ALS campaigns are three-dimensional area-wide point clouds. Full-waveform ALS allows for a precise determination for the range R_i of a target and subsequently for its three-dimensional position (See details on range determination in full-waveform data in Sect. 2.4). For the derivation of physical target properties, the first goal is to determine the backscatter cross-section σ_i. For this purpose, we re-group Eq. (2.4) and retrieve:

$$\sigma_i'(t) = \frac{4\pi\beta_t^2 R_i^4}{D_r^2 \eta_{SYS} \, \eta_{ATM}} P_r(t) \otimes^{-1} S(t).$$

The operator "$\otimes^{-1}$" denotes deconvolution, which is an ill-posed task and needs additional constraints for a stable solution (Tikhonov and Arsenin 1977). There are several deconvolution approaches in ALS research which we will treat in detail in Sect. 2.4.

After range determination and deconvolution, the term $(4\pi\beta_t^2)/(D_r^2 \eta_{SYS} \, \eta_{ATM})$ remains as unknown quantity whose elements cannot be easily separated and determined independently. As last step for the calculation of the backscatter cross-section, we need to solve the integral

$$\sigma_i = \int_{-\infty}^{\infty} \sigma_i'(t) dt.$$

The solution for σ_i and further radiometric quantities, known as *radiometric calibration*, is described in Sect. 2.4.

2.3   Forward Modeling of ALS Return Waveforms over Forested Areas

In the previous section, we have presented the calculation of the backscatter cross-section as result of a waveform deconvolution and integration. To foster a meaningful interpretation of such waveforms over forest canopies, we now synthesize them using radiative transfer models in virtual forest scenes.

The earliest modeling studies of vegetation ALS returns were carried out for large footprint systems (Sun and Ranson 2000; Ni-Meister et al. 2001; Koetz et al. 2006), such as LVIS (Land, Vegetation, and Ice Sensor) and the previously planned VCL (Vegetation Canopy Lidar) mission, which unfortunately never made it into space. Given the large footprint of these systems (10–25 m in diameter), the virtual scene was constructed using geometric primitives, such as cones or ellipsoids, to represent tree crowns. These geometric primitives were then filled with virtual plant material (a combination of the optical properties of leafs and branches) using a turbid media approach. However, it is very difficult to obtain the proper values for the parametrization of such turbid media from field data. In addition, the geometric representation using crown archetypes might be too coarse for real-world laser scanner data simulation (Calders et al. 2013). Consequently, for small footprint data, the model tree representations need to explicitly resolve the tree structure at the sub-footprint level, e.g. in the form of position and orientation of single leaves.

Driven by other application domains, such as ecological simulations and computer graphics, fractal models (or L-systems) of tree geometry have been developed in recent years, which explicitly describe the tree geometry from stem to the leaf or needle level. These three-dimensional models should provide a better geometric representation to individually simulate the effects of acquisition properties such as incidence angle, point density, terrain slope, laser footprint size, laser wavelength and canopy reflectance on the accuracy of biophysical vegetation data products.

To date, the most difficult task in simulating vegetation returns is to obtain a representative backscatter cross-section of the vegetation canopy. For this, one would need to know locations, sizes, orientations and reflectances of the scattering elements to compute ρ as in Eq. (2.2). If we consider small-footprint laser scanning, with footprint diameters normally in the order of some decimetres, the objects of relevance would be single leaves and branches within the canopy. Unfortunately, it is difficult to obtain realistic characterizations of actual vegetation canopies that are faithful at the single leaf level, including the location, orientation and optical properties of that leaf element. As opposed to the indirect parametrization of turbid media approaches, such information could at least directly be derived from field measurements, such as destructive sampling or terrestrial laser scanning.

2.3.1   Radiative Transfer Approaches

During the last decade, the number of models and approaches that have the capability to model small-footprint ALS returns has increased substantially. In one

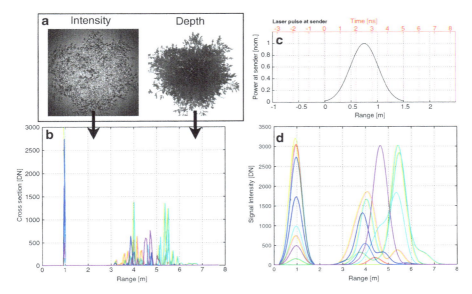

Fig. 2.3 Illustration of a waveform simulation based on a virtual tree model. The approach described here is based on an intensity and depth image derived from POVRAY (**a**). The differential backscatter cross-sections (**b**) derived from these images are convolved with a Gaussian-shaped laser pulse (**c**) to obtain the modeled laser scanner return waveforms (**d**)

of the first small-footprint simulation studies, Morsdorf et al. (2007) applied the open-source ray tracer POVRAY and detailed 3D tree models based on L-systems to simulate discrete-return ALS returns. The ray tracer is used to generate an intensity and a depth image, which are combined to form the differential backscatter cross-section $\sigma_i'(t)$. The differential cross-section is then convolved with a laser pulse of given shape and length ($P_t(t)$) to yield the simulated waveform (Fig. 2.3).

This approach has the drawback of only taking single scattering into account and Kotchenova et al. (2003) have shown that especially for the near-infrared wavelengths often used in LiDAR remote sensing, multiple scattering can have a significant effect on the lower parts of larger waveform returns. In addition, many of the well-established models previously applied in imaging spectroscopy and the modeling of passive optical imagery, have been adapted and extended to allow for the simulation of ALS waveforms. For instance, LIBRAT (Disney et al. 2006) is a Monte-Carlo-based radiative transfer model, that has been parametrized with forest scenes that are explicitly resolving the needle level of single shoots. These models have the advantage that they resolve multiple scattering within the vegetation canopy, although this is less of an issue for ALS remote sensing, since we measure in the so-called hot-spot configuration (sender and receiver are on the same optical path), where the contribution of multiple scattering is relatively low. It has been used by Disney et al. (2010) to simulate small-footprint ALS waveforms, see Sect. 2.3.3 for details.

Another model capable of simulating ALS returns is the well-established DART (Gastellu-Etchegorry et al. 1996) model, which was recently extended to have a forward simulation model for ALS returns (Rubio et al. 2009). FLIGHT (North 1996) is another radiative transfer model with a long tradition in the passive optical domain of remote sensing, which was enhanced with a laser mode and subsequently used by North et al. (2010) to simulate waveforms of the spaceborne GLAS (Geoscience Laser Altimeter System) instrument aboard ICESat.

2.3.2   Scene Construction

As described above, the theoretical concepts and tools to model ALS returns are well established nowadays. One critical obstacle for the wider use of these tools is the parametrization of the virtual, three-dimensional forest scene. Depending on the modeled sensor, the 3D scene needs to resolve different spatial scales. As stated above, for the early studies simulating large footprint lasers (Sun and Ranson 2000; Ni-Meister et al. 2001), the maximum level of detail in the canopy characterization was the tree/crown level. Thus, the models were easily parametrized by having a list of tree locations and dimensions to produce a representation to base the simulation upon. The distribution of the canopy elements within the crown was described using turbid media, without explicitly prescribing leaf or shoot locations and orientations.

A similar, but more detailed approach was used for the construction of the explicit 3D scenes in the RAMI IV experiment (Widlowski et al. 2008), using field measured tree locations and dimensions. Instead of using turbid media for the tree crowns, very detailed 3D tree models were used, which resolved the structure down to the shoot and needle level.

However, these detailed reconstructions were not modeled using real tree geometry, as this was very hard to obtain. Consequently, currently large effort is put on using TLS to provide detailed, realistic 3D models of vegetation elements, as e.g. outlined by Cote et al. (2009).

2.3.3   Applications

The motivation for simulation of ALS returns can be manifold, but generally there is an interest to study the variability of ALS vegetation returns and their derived products when changing some parameters of their acquisition. An example would be the gap fraction or fractional cover, which is an structural variable in many earth system models and can be provided by ALS at larger scales. Still, it is unclear how for instance the footprint size and wavelength would affect this variable. This is where modelling could help by establishing relationships between changes in sensor and survey configuration and the desired environmental variable.

So far, most applications of modeling of ALS vegetation returns have been focused on the simulation of full-waveform returns, as it is the default result of all models. Until now, only Morsdorf et al. (2007) and Disney et al. (2010) have simulated discrete return data. The largest problem for modeling discrete return data is to model the instrument's treatment of the incoming signal for return detection. To overcome this, Morsdorf et al. (2007) used actual ALS data and geometric-optical calibration targets to re-engineer the return detection method of the simulated ALS system. On the other hand, Disney et al. (2010) implemented a number of return detection methods to study the sensitivity of echo statistics (height) on the choice of echo detection. But even for full-waveform signals, the modeling often excludes most effects introduced by receiving the signal with a particular sensor.

The largest issue is the unavailability of the sensor response function of commercial ALS systems. The sensor response function introduces another convolution term to the radar equation; this has no significant influence as long as a linear transfer system can be assumed. In this case, we can apply Eq. (2.4). However, it is most likely that the instrument response function is specific for each particular ALS system and only a thorough laboratory calibration could provide this response function.

2.3.3.1   Sensor and Survey Configuration Effects

The first studies simulating small-footprint ALS were mostly looking at the effects sensor or survey settings would have on ALS waveforms. Morsdorf et al. (2007) looked into the effects of footprint size and laser wavelength on the distribution of discrete return data. While footprint size affected return distributions significantly, the impact of the laser wavelength (1,064 vs. 1,560 nm) on discrete return height statistics was much less evident. This was explained by the echo triggering methods applied, which will detect the rising edge of the signal, independent of the signal strength,[1] as long as it is above the noise level for a particular system. For the full-waveform data, however, the return energy was strongly related to canopy reflectance at the laser wavelength. A similar, but more extensive study was carried out by Disney et al. (2010) using a Monte Carlo ray tracing approach to study the effects of footprint size, canopy structure, scan angle, sampling density (point spacing) and signal triggering methods on average values of canopy height metrics obtained by small-footprint ALS. They found significant changes in the average canopy height for two different types of forest when changing the footprint diameter (between 0.1 and 1 m) and the scan angle, with a maximum incidence angle of 30°. For varying point density, triggering method and canopy structure, the changes in average canopy height were less significant or insignificant.

---

[1]Most commercial ALS systems employ adaptive thresholding to avoid "trigger walk", making discrete return data less susceptible to changes in object reflectance.

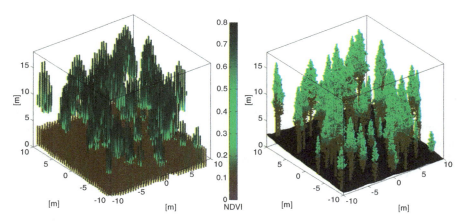

Fig. 2.4 Simulated NDVI profiles as potentially measured by an airborne multi-spectral ALS instrument (*left*). Three-dimensional representation of the virtual canopy (*right*)

2.3.3.2 Sensor and Product Prototyping

Another common application of radiative transfer models in remote sensing is the prototyping of new sensors or products. The large advantage of modeling environments is that almost everything can be easily measured in the virtual 3D scene, including properties that would be very hard to obtain in the field (e.g. true leaf area index). Thus, new sensors and new ways of product derivation can be fully tested for their feasibility in the virtual environment. As the costs for building airborne and spaceborne laser systems are quite high, a number of laser-based simulating studies have been devoted to this issue.

Morsdorf et al. (2009) illustrated the potential of a small-footprint, multi-spectral laser for the estimation of profiles of vegetation indices, e.g. such as NDVI (Normalized Differenced Vegetation Index). In a virtual forest stand comprised of detailed Scots pine trees derived from a tree growth model, they quantified how much a laser-based NDVI profile could be used to estimate the actual distribution of green and brown canopy elements. Figure 2.4 shows the resulting NDVI profiles along with the 3D structure of the simulated canopy. NDVI values are lower towards the lower parts of the canopy, which is caused by the higher amount of dead branches in those areas.

In another study, Hancock et al. (2012) tested the potential of dual-wavelength, large-footprint lasers to provide better estimates of the ground elevation opposed to using single-wavelength lasers. Single-wavelength, large-footprint lasers have problems in properly detecting the ground given sloped terrain and vegetation cover. According to the authors, dual-wavelength lasers can be used to get a more robust estimate of ground elevation, even on slopes. However, there needs to be a spectral gradient between the vegetation canopy and the ground (e.g. soil or litter).

2.4   Full-Waveform Data Processing and Model Inversion

In general, the position, sizes, orientations and reflectance characteristics of scattering objects are not given. Thus, these quantities have to be reconstructed, in our case from FWF ALS data. Raw FWF data are sampled values of the system waveform and of the recorded echo waveform, denoted by $S[t]$ and $P_r[t]$, respectively. The sampling interval is typically 1 ns whereas the bit depth is typically 8 or 12 bit (see Sect. 3.1). Figure 2.5 shows an example for $S[t]$ and $P_r[t]$, recorded with a Riegl LMS-Q560 instrument.

FWF data processing can be split into several tasks:

- Determination of the number of scatterers and
- Calculation of their distance from the scanner;
- Fitting the raw waveforms to continuous functions and
- Calculation of additional echo parameters;
- Deconvolution and
- Radiometric calibration

Throughout the last decade, a number of approaches for FWF data processing have been presented, e.g.:

- Range estimation using classical detectors (Wagner et al. 2004)
- Wiener-filter deconvolution (Jutzi and Stilla 2006)
- Gaussian Decomposition (Wagner et al. 2006)
- Correlation-based range estimation (Roncat et al. 2008)
- Generalized-Gaussian modeling (Mallet et al. 2009)
- Expectation/Maximization (EM) deconvolution (Parrish and Nowak 2009)
- B-spline deconvolution (Roncat et al. 2011a)

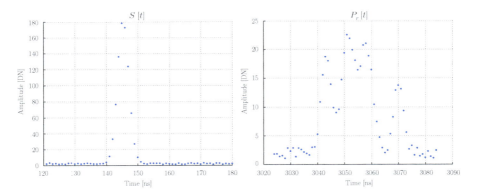

Fig. 2.5 Raw waveform measurements $S[t]$ and $P_r[t]$, recorded with a Riegl LMS-Q560 instrument in a sampling rate of 1 ns. The reconstructed continuous signal (Gaussian Decomposition) of this example can be seen in Fig. 2.6

Table 2.1 FWF data processing approaches and their outputs

Approach	# Echoes	Range	Ampl.	Cont.model	Add.echo params	Deconv.	Radiom.Cal.
Classical detectors	•	•	•	—	—	—	∼
Wiener filter	—	—	—	•	◇	•	◇
Gauss. decomp.	—	∘	•	•	∘	•	•
Correlation	•	•	—	—	◇	—	—
Gener. Gaussians	•	•	•	•	•	—	•
EM	•	•	—	—	—	•	◇
B-splines	◇	◇	—	•	◇	•	◇

Ampl. amplitude, *Cont.model* continuous model, *Add.echo params* additional echo parameters, *Deconv.* deconvolution, *Radiom.Cal.* radiometric calibration

bullet … primary results, ∘ …result based on initial estimates provided by another technique, ◇ … possible, but not originally intended, ∼ …approximately possible (e.g. with some further assumptions), — … not provided by the approach

These approaches aim at one or more of the above mentioned tasks. A summary is given in Table 2.1.

The Gaussian Decomposition approach allows for a very elegant solution of range determination and derivation of further echo attributes which additionally enable for the derivation of physical target parameters. This technique has therefore become a standard in FWF data processing and has been commonly used within vegetation and forestry studies (Chauve et al. 2007; Reitberger et al. 2008; Wagner et al. 2008; Höfle et al. 2012). We will thus have a detailed look at this technique in the subsequent paragraphs.

The goal is to fit Gaussian functions, i.e. scaled bell curves, to the sampled waveforms $S[t]$ and $P_r[t]$. While the temporal profile of the system waveform is very close to *one* Gaussian function, the recorded echo waveform may contain the added echoes of n scatterers, so that

$$S(t) = \hat{S} \exp\left(-\frac{(t - t_s)^2}{2s_s^2}\right) \text{ and } P_r(t) = \sum_{i=i}^{n} P_{r,i}(t) = \sum_{i=i}^{n} \hat{P}_{r,i} \exp\left(-\frac{(t - t_i)^2}{2s_{p,i}^2}\right),$$

(2.5)

respectively (Wagner et al. 2008). The fit is aimed at fulfilling $\sum(S[t] - S(t_i))^2 \rightarrow$ min. and $\sum(P_r[t] - P_r(t_j))^2 \rightarrow$ min. In case of Gaussian Decomposition this is a non-linear minimization problem which can e.g. be solved by the Levenberg-Marquardt approach (Wagner et al. 2006). Gaussian Decomposition is examplarily illustrated in Fig. 2.6 whereas Fig. 2.7 shows the distribution of the echo amplitudes $\hat{P}_i$ in two adjacent flight strips in a partly vegetated area around Schönbrunn palace in Vienna (Lehner and Briese 2010).

As input data additionally to the sampled waveforms, this adjustment approach needs the number of scatterers n and approximate values for the positions (t_s, t_i), the widths $(s_s, s_{p,i})$ and the amplitudes $(\hat{S}, \hat{P}_i)$ of the Gaussians (see Table 2.1).

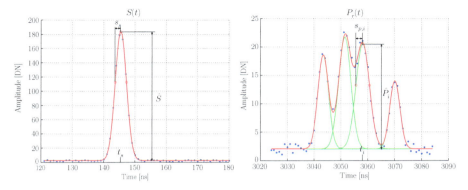

Fig. 2.6 Example of Gaussian Decomposition for the data shown in Fig. 2.5: *Left*: System waveform $S(t)$ (one Gaussian with parameters t_s, $\hat{S}$ and s_s), *right*: Recorded echo waveform $P_r(t)$ (*right*; sum of four Gaussians with parameters t_i, $\hat{P}_i$ and $s_{p,i}$)

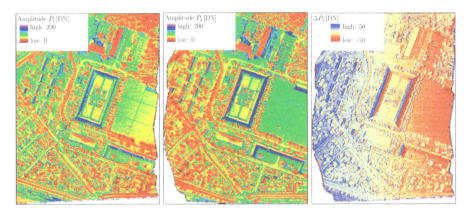

Fig. 2.7 Echo amplitudes $\hat{P}_i$ for two adjacent strips of the Schönbrunn campaign (Lehner and Briese 2010). *Left*: $\hat{P}_i$ in strip 5, *middle*: $\hat{P}_i$ in strip 6, *right*: difference in amplitude. The two strips are overlapping to around 50 %, therefore the difference is largest at the margins of the strips, mainly because of the range difference. Other big differences in amplitude can be found at tilted roof tops because of the different acquisition geometry

The number of scatterers can be retrieved e.g. by classical echo detectors, correlation techniques or statistical approaches (see references Wagner et al. 2004; Roncat et al. 2008; Mallet et al. 2009 and Table 2.1). They are capable to deliver approximate positions of the scatterers as well. The scattering process can only broaden the backscattered laser pulse[2] so that the width for a tentative echo i, $s_{p,i}$, has a lower bound given by the width of the system waveform, s_s. The latter can be retrieved

---

[2]Or leave its width unchanged in the case of direct reflection.

from the manufacturer's specifications. E.g. the Optech ALTM 3100 system has a pulse width of 8.0 ns, expressed as full width at half maximum (FWHM), when operated at a pulse repetition rate of 50 kHz (Chasmer et al. 2006; Mallet and Bretar 2009).[3] As mentioned earlier, this quantity is a limiting factor for the separation of two subsequent targets within the echo waveform: assuming equal signal strength, the minimal separation distance r_R is half the length of the transmitted pulse, i.e. $r_R = v_g \times \text{FWHM}/2$. In case of the above example of the Optech ALTM 3100 system, the minimal separation distance results to $r_R \simeq 8\,\text{ns} \times (30/2)\,\text{cm/ns} = 120\,\text{cm}$.

The relation of the standard deviation s of a Gaussian to its FWHM is

$$\text{FWHM} = 2\sqrt{2\ln 2}\, s \simeq 2.355\, s. \tag{2.6}$$

For the above mentioned system, this gives a value of $s_s \simeq 3.4\,\text{ns}$ for the system waveform.

Having solved the Gaussian decomposition for $S(t)$ and $P_r(t)$, we can re-write Eq. (2.3) as

$$\sigma_i = \frac{4\pi R_i^4 \beta_t^2}{D_r^2 \hat{S} s_s \eta_{\text{SYS}} \eta_{\text{ATM}}} \hat{P}_i s_{p,i}. \tag{2.7}$$

We retrieve the range R_i of the i-th echo from the difference in position between the echo and the system waveform, multiplied with the group velocity of the laser ray, v_g: $R_i = (t_i - t_s) v_g / 2$. This σ_i is a physical property independent of the transmitted laser pulse and has shown good performance in the classification of ground and vegetation echoes, especially when looking at the distribution of the single σ_i within the total backscatter cross-section σ (Wagner et al. 2008).

However, σ_i is still dependent on the size of the illuminated target area, $A_l = \pi R_i^2 \beta_t^2 / 4$. Normalizing w.r.t. A_l leads to the introduction of the *backscatter coefficient* $\gamma\,[\text{m}^2/\text{m}^2]$ (Wagner 2010):

$$\gamma_i = \frac{\sigma_i}{A_l} = \frac{4}{\pi R_i^2 \beta_t^2} \sigma_i = \frac{16 R_i^2}{D_r^2 \hat{S} s_s \eta_{\text{SYS}} \eta_{\text{ATM}}} \hat{P}_i s_{p,i}. \tag{2.8}$$

Figure 2.8 illustrates the relation of the parameters influencing σ and γ whereas Fig. 2.9 shows the values of the backscatter coefficient for the Schönbrunn example. It has to be considered that the illuminated target area can only be approximated in the case of an extended target (target surface bigger than the laser footprint).

---

[3]In the literature, varying pulse energies E_S and peak powers $\hat{S}$ are reported for this instrument w.r.t. the pulse repetition rate. E.g. in Chasmer et al. (2006), for pulse repetition rates of 33, 71 and 100 kHz, the respective FWHM resulted in 7.0, 10.8 and 14.9 ns, using Eq. (2.6) and $s = E_S/(\hat{S}\sqrt{2\pi})$. Næsset (2009) reports FWHM values of 10 ns at 50 kHz and 16 ns at 100 kHz pulse repetition rate.

Fig. 2.8 Geometric
parameters of σ and γ. While
σ changes proportional to R^2,
γ is independent of the range.
A is the target area actually
illuminated by the laser beam
(cf. Eq. (2.2)). For small
beam divergences β_t, the
laser footprint area A_l
perpendicular to the laser
beam can be approximated as
$A_l \simeq A \cos \vartheta$ with ϑ being
the angle between the local
surface normal $\mathbf{n}_\varepsilon$ and the
axis of the laser beam cone

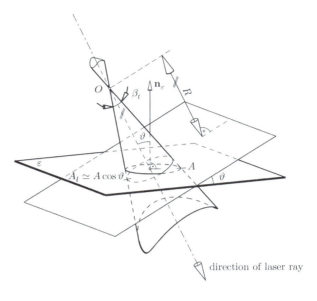

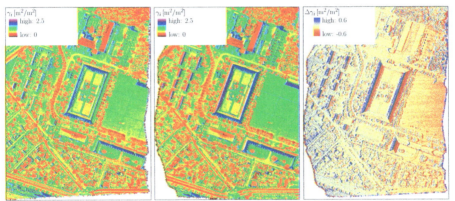

Fig. 2.9 Backscatter coefficient γ for two adjacent strips of the Schönbrunn example (cf. Fig. 2.7).
Left: γ in strip 5, *middle*: γ in strip 6, *right*: difference in γ (From Lehner and Briese (2010))

When the laser footprint only partly illuminates the target surface (e.g. leaves on a
tree or a building edge) the actual illuminated target area that corresponds to one
echo is unknown.

For practical computation of the radiometric calibration, the constant parameters
are separated from the others and summarized in the *calibration constant* C_{CAL}:

$$C_{\mathrm{CAL}} = \frac{16}{D_r^2 s_s \eta_{\mathrm{SYS}}}. \tag{2.9}$$

Empirical studies have shown that the amplitude of the system waveform $\hat{S}$ shows too high variation to be treated as constant (Bretar et al. 2009; Roncat et al. 2011b). It is therefore excluded from C_{CAL}. In practice, some of the instruments utilize gain control in order to adapt the output signal to the dynamic range of the sensor. This issue is discussed in Lehner et al. (2011). Further instrument details are typically not provided to the end user. The calibration constant is calculated using assumptions on the reflectivity of homogeneous regions (Wagner et al. 2006), artificial (Ahokas et al. 2006; Kaasalainen et al. 2007) or natural reference targets (Kaasalainen et al. 2009; Lehner and Briese 2010) of known reflectivity at the used laser wavelength. The atmospheric transmission factor may be determined using a standard model for radiative transfer in the atmosphere:

$$\eta_{ATM} = 10^{-\dfrac{2R_i a}{10,000}}$$

with a in [DB/km] as the atmospheric attenuation coefficient (in the range of 0.2 DB/km) (Höfle and Pfeifer 2007). The backscatter coefficient γ_i can be finally retrieved as

$$\gamma_i = \frac{\hat{P}_i}{\hat{S}} \frac{s_{p,i}}{\eta_{ATM}} R_i^2 C_{CAL}.$$

The use of γ instead of σ as feature for ALS point cloud classification has shown noticeable advantages, e.g. in the improved separation of grassland from road (Alexander et al. 2010). Höfle et al. (2012) pointed out that full-waveform parameters alone (normalized amplitude, σ and γ) may not be sufficient for proper point cloud classification; however, they are a good input for further classification tasks such as tree species classification (cf. Chap. 7).

For most surfaces, a diffuse reflectivity behaviour can be assumed due to the short wavelengths used in ALS. With γ_i determined as described before, we retrieve the diffuse Lambertian reflectance $\tilde{\rho}_i$ as (cf. Eqs. (2.2) and (2.8)) (Wagner 2010)

$$\tilde{\rho}_i = \frac{\gamma_i}{4\cos\vartheta_i} \tag{2.10}$$

with ϑ_i being the angle between the laser beam's direction and the local surface normal (cf. Fig. 2.8). The distribution of the diffuse reflectance in the Schönbrunn example is shown in Fig. 2.10. One can clearly see that the differences visible in σ and γ at tilted roof tops are reduced to a great extent. However, due to the less reliable estimation of the incidence angle in vegetation, the $\tilde{\rho}$ differences are much higher in the canopy. A second reason for that might be that Lambertian scattering cannot be assumed in vegetated areas.

Further practical results for the radiometric calibration of multi-spectral FWF ALS data have been published by Briese et al. (2012).

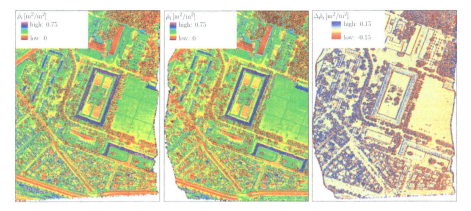

Fig. 2.10 Diffuse reflectance $\tilde{\rho}$ for two adjacent strips of the Schönbrunn example (cf. Fig. 2.7). *Left*: $\tilde{\rho}$ in strip 5, *middle*: $\tilde{\rho}$ in strip 6, *right*: difference in $\tilde{\rho}$ (From Lehner and Briese (2010))

2.5 Validation of Airborne FWF Data by Terrestrial Laser Scanning

Given a 3D point cloud over a forested area, stemming from an FWF ALS campaign and derived with the methods presented in the previous sections of this chapter, the question still remains how well the forest is represented by this point cloud; i.e. the canopy (cf. Chap. 6), the stems (cf. Chap. 8) and branches of the single trees, the understorey and the bare ground (see Sect. 3.3.1).

Several studies have already shown that TLS is a valuable tool for the retrieval of forest parameters on single trees and at the plot level (cf. Liang et al. 2012; Yu et al. 2013). In this section, we will present the validation of airborne full-waveform data by TLS following the example of Doneus et al. (2010). For that study, conducted originally for archaeological purposes, an FWF ALS dataset was simultaneously recorded with a TLS dataset in the surroundings of the former monastery "St. Anna in der Wüste" (Mannersdorf, Lower Austria). The study area consists of the monastery, meadows and forests with understorey. Figure 2.11 shows an overview of this site.

Data were acquired on November 10, 2009, with a Riegl LMS-Q680 airborne instrument and a Riegl VZ-400 terrestrial system. The registration of the ALS data was performed with a strip adjustment (Kager 2004) whereas the TLS data were co-registered with an automatic approach within the software package RiSCAN PRO (Riegl LMS 2013). These two datasets were aligned using tie patches on planar surfaces.

As a showcase example, Fig. 2.12 shows a section of the ALS and TLS point cloud around a single tree, together with corresponding waveforms. One can clearly see that the point clouds are well aligned and that the canopy is correctly represented by the FWF ALS points.

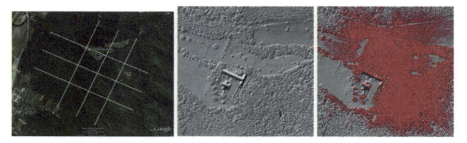

Fig. 2.11 Orthophoto of the study site "St. Anna in der Wüste" (*left*) with hillshade of the area around the former monastery, derived from FWF ALS data (*centre*) and overlaid TLS coverage in *red* (*right*) (From Doneus et al. (2010))

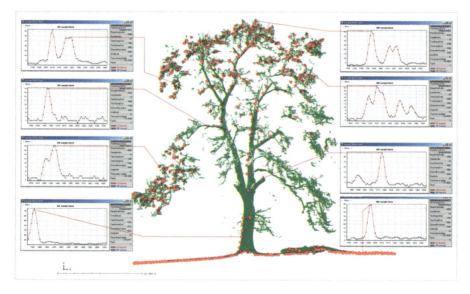

Fig. 2.12 Section of the TLS (*green*) and ALS (*red*) point cloud around a single tree, together with corresponding ALS waveforms. These waveforms represent amplitude over time which is the time lag since emission of the laser pulse. The displayed time range equals 60 ns at 1 GHz sampling rate (From Doneus et al. (2010))

Furthermore, the study concluded that the ALS points classified as terrain points lie on the terrain surface measured with TLS. In vegetation, there was typically one strong echo from the stem or a branch, with an echo width $s_{p,i}$ slightly broader than the ones from the terrain echoes. Moreover, in low vegetation, the echo width was in general higher than in regions with no understorey. This was due to the fact that low vegetation forms a group of small scatterers which are not separable any more in the echo waveform. Because of this, last echoes were typically some decimetres above the ground. Without using the FWF echo width, last echoes would have been

Fig. 2.13 TLS and FWF ALS points in low vegetation. Last ALS echoes are typically a few decimetres off-ground, only classifiable as non-terrain points using the echo width (From Doneus et al. (2010))

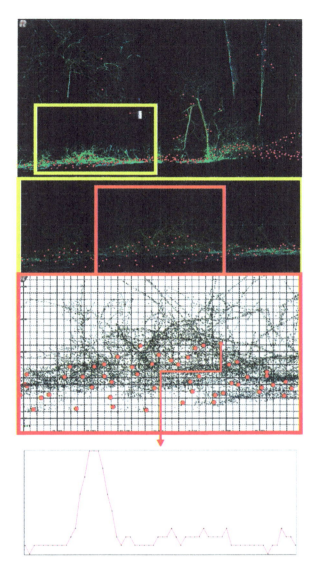

wrongly classified as ground echoes although they still belonged to vegetation. This fact gives further empirical evidence for the usability of the echo width as weighting parameter for DTM generation (Mücke et al. 2010). The situation in low vegetation is depicted in Fig. 2.13.

The example shows that TLS is capable of delivering high-resolution data on single-tree level and also a valuable tool for giving ground truth in the classification of FWF ALS points where an erroneous classification would lead to a wrong digital terrain model and consequently to a wrong estimation of the canopy height.

Especially in low vegetation in the understorey, the echo width was found to be a good indicator for reliable ground/off-ground classification.

2.6   Conclusion and Outlook

This chapter outlines the actual research in the field of geometric and radiometric information extraction from FWF ALS. While the geometric extraction (i.e. range per echo and subsequently the 3D echo position) can be solved by several different detection methods, the estimation of further physical parameters relies on additional assumptions (e.g. Gaussian reference pulse, extended single targets). The interaction process with complex (non-extended) targets in practical applications remains an ill-posed problem. Solution strategies for this issue will have to be developed in the future. In the meantime beside FWF ALS the first FWF TLS systems are also available. These systems will offer the possibility to advance the study of FWF data and have the advantage of an easier and repeatable setup of testing scenarios. Alongside the radiometric calibration and analysis of single-wavelength FWF ALS, systems of different laser wavelengths are already available (e.g. 532, 1064 and 1550 nm, see Briese et al. (2012)), but not yet as synchronously operated FWF systems. The exploration of multi-spectral ALS missions for vegetation mapping is one of the future research challenges in the field of FWF ALS.

The forward simulation of ALS waveforms will continue to significantly advance our understanding of the interaction of the laser pulse with vegetation canopies. More specifically, it will also help to assess systematic and non-systematic differences between ALS data obtained using different sensors and under different survey configurations. As many of the current ALS radiative-transfer models were originally made for passive optical imagery, a model based fusion of ALS and multi-/hyper-spectral data might become possible, for instance for a combined inversion of biophysical variables as presented by Koetz et al. (2007). Forward modeling and the possibility to measure everything in the virtual 3D scene will as well help in establishing robust and transferable, physically-based vegetation products without the need of laborious and error-prone calibration with field data. It is mandatory, however, that both system providers and surveying companies alike are more open with ancillary information about their systems and surveying campaigns and that this information is maintained throughout the ALS processing chain.

Acknowledgements Andreas Roncat has been supported by a Karl Neumaier PhD scholarship.

The Ludwig Boltzmann Institute for Archaeological Prospection and Virtual Archaeology is based on an international cooperation of the Ludwig Boltzmann Gesellschaft (Austria), the University of Vienna (Austria), the Vienna University of Technology (Austria), the Austrian Central Institute for Meteorology and Geodynamics, the office of the provincial government of Lower Austria, Airborne Technologies GmbH (Austria), RGZM (Roman-Germanic Central Museum Mainz, Germany), RA (Swedish National Heritage Board), VISTA (Visual and Spatial Technology Centre, University of Birmingham, UK) and NIKU (Norwegian Institute for Cultural Heritage Research).

References

Ahokas E, Kaasalainen S, Hyyppä J, Suomalainen J (2006) Calibration of the Optech ALTM 3100 laser scanner intensity data using brightness targets. Int Arch Photogramm Remote Sens Spat Inf Sci 36(1), 7 p

Alexander C, Tansey K, Kaduk J, Holland D, Tate NJ (2010) Backscatter coefficient as an attribute for the classification of full-waveform airborne laser scanning data in urban areas. ISPRS J Photogramm Remote Sens 65:423–432

Bretar F, Chauve A, Bailly JS, Mallet C, Jacome A (2009) Terrain surfaces and 3-d landcover classification from small footprint full-waveform lidar data: application to badlands. Hydrol Earth Syst Sci 13:1531–1545

Briese C, Pfennigbauer M, Lehner H, Ullrich A, Wagner W, Pfeifer N (2012) Radiometric calibration of multi-wavelength airborne laser scanning data. ISPRS Ann Photogramm Remote Sens Spat Inf Sci 1(7):335–340

Calders K, Lewis P, Disney M, Verbesselt J, Herold M (2013) Investigating assumptions of crown archetypes for modelling LiDAR returns. Remote Sens Environ 134:39–49

Chasmer L, Hopkinson C, Smith B, Treitz B (2006) Examining the influence of changing laser pulse repetition frequencies on conifer forest canopy returns. Photogramm Eng Remote Sens 72:1359–1367

Chauve A, Durrieu S, Bretar F, Pierrot-Deseilligny M, Puech W (2007) Processing full-waveform lidar data to extract forest parameters and digital terrain model: validation in an alpine coniferous forest. In: Proceedings of ForestSat conference'07, Montpeillier, France, p 5

Cote JF, Widlowski JL, Fournier RA, Verstraete MM (2009) The structural and radiative consistency of three-dimensional tree reconstructions from terrestrial lidar. Remote Sens Environ 113:1067–1081

Disney M, Lewis P, Saich P (2006) 3d modelling of forest canopy structure for remote sensing simulations in the optical and microwave domains. Remote Sens Environ 100:114–132

Disney M, Kalogirou V, Lewis P, Prieto-Blanco A, Hancock S, Pfeifer M (2010) Simulating the impact of discrete-return lidar system and survey characteristics over young conifer and broadleaf forests. Remote Sens Environ 114:1546–1560

Doneus M, Briese C, Studnicka N (2010) Analysis of full-waveform ALS data by simultaneously acquired TLS data: towards an advanced DTM generation in wooded areas. Int Arch Photogramm Remote Sens Spat Inf Sci 38(Part 7B):193–198

Gastellu-Etchegorry J, Demarez V, Pinel V, Zagolski F (1996) Modeling radiative transfer in heterogeneous 3-d vegetation canopies. Remote Sens Environ 58:131–156

Hancock S, Lewis P, Foster M, Disney M, Muller JP (2012) Measuring forests with dual wavelength lidar: a simulation study over topography. Agric For Meteorol 161:123–133

Höfle B, Pfeifer N (2007) Correction of laser scanning intensity data: data and model-driven approaches. ISPRS J Photogramm Remote Sens 62:415–433

Höfle B, Hollaus M, Hagenauer J (2012) Urban vegetation detection using radiometrically calibrated small-footprint full-waveform airborne lidar data. ISPRS J Photogramm Remote Sens 67:134–147

Jelalian AV (1992) Laser radar systems. Artech House, Boston

Jutzi B, Stilla U (2006) Range determination with waveform recording laser systems using a Wiener filter. ISPRS J Photogramm Remote Sens 61:95–107

Kaasalainen S, Hyyppä J, Litkey P, Hyyppä H, Ahokas E, Kukko A, Kaartinen H (2007) Radiometric calibration of ALS intensity. Int Arch Photogramm Remote Sens Spat Inf Sci 36(3/W52):201–205

Kaasalainen S, Hyyppä H, Kukko A, Litkey P, Ahokas E, Hyyppä J, Lehner H, Jaakkola A, Suomalainen J, Akujarvi A, Kaasalainen M, Pyysalo U (2009) Radiometric calibration of lidar intensity with commercially available reference targets. IEEE Trans Geosci Remote Sens 47:588–598

Kager H (2004) Discrepancies between overlapping laser scanning strips – simultaneous fitting of aerial laser scanner strips. Int Arch Photogramm Remote Sens Spat Inf Sci 35(B1):555–560

Koetz B, Morsdorf F, Sun G, Ranson KJ, Itten K, Allgöwer B (2006) Inversion of a lidar waveform model for forest biophysical parameter estimation. IEEE Geosci Remote Sens Lett 3:49–53

Koetz B, Sun G, Morsdorf F, Ranson K, Kneubühler M, Itten K, Allgöwer B (2007) Fusion of imaging spectrometer and lidar data over combined radiative transfer models for forest canopy characterization. Remote Sens Environ 106:449–459

Kotchenova SY, Shabanov NV, Knyazikhin Y, Davis AB, Dubayah R, Myneni RB (2003) Modeling lidar waveforms with time-dependent stochastic radiative transfer theory for remote estimations of forest structure. J Geophys Res 108:13

Lehner H, Briese C (2010) Radiometric calibration of full-waveform airborne laser scanning data based on natural surfaces. Int Arch Photogramm Remote Sens Spat Inf Sci 38(7B):360–365

Lehner H, Kager H, Roncat A, Zlinszky A (2011) Consideration of laser pulse fluctuations and automatic gain control in radiometric calibration of airborne laser scanning data. In: Proceedings of 6th ISPRS Student Consortium and WG VI/5 Summer School, Fayetteville State University, North Carolina, USA

Liang X, Litkey P, Hyyppä J, Kaartinen H, Vastaranta M, Holopainen M (2012) Automatic stem mapping using single-scan terrestrial laser scanning. IEEE Trans Geosci Remote Sens 50:661–670

Mallet C, Bretar F (2009) Full-waveform topographic lidar: state-of-the-art. ISPRS J Photogramm Remote Sens 64:1–16

Mallet C, Lafarge F, Bretar F, Roux M, Soergel U, Heipke C (2009) A stochastic approach for modelling airborne lidar waveforms. In: Bretar F, Pierrot-Deseilligny M, Vosselman G (eds) Int Arch Photogramm Remote Sens Spat Inf Sci 38(3/W8):201–206

Morsdorf F, Frey O, Koetz B, Meier E (2007) Ray tracing for modeling of small footprint airborne laser scanning returns. Int Arch Photogramm Remote Sens Spat Inf Sci 36(3/W52):294–299

Morsdorf F, Nichol C, Malthus T, Woodhouse IH (2009) Assessing forest structural and physiological information content of multi-spectral LiDAR waveforms by radiative transfer modelling. Remote Sens Environ 113:2152–2163

Mücke W, Briese C, Hollaus M (2010) Terrain echo probability assignment based on full-waveform airborne laser scanning observables. Int Arch Photogramm Remote Sens Spat Inf Sci 38(7A):157–162

Næsset E (2009) Effects of different sensors, flying altitudes, and pulse repetition frequencies on forest canopy metrics and biophysical stand properties derived from small-footprint airborne laser data. Remote Sens Environ 113:148–159

Ni-Meister W, Jupp DLB, Dubayah R (2001) Modeling lidar waveforms in heterogeneous and discrete canopies. IEEE Trans Geosci Remote Sens 39:1943–1958

North PRJ (1996) Three-dimensional forest light interaction model using a Monte Carlo method. IEEE Trans Geosci Remote Sens 34:946–956

North PRJ, Rosette JAB, Suarez JC, Los SO (2010) A Monte Carlo radiative transfer model of satellite waveform lidar. Int J Remote Sens 31(5):1343–1358

Parrish CE, Nowak RD (2009) Improved approach to lidar airport obstruction surveying using full-waveform data. J Surv Eng 135:72–82

Parrish CE, Jeong I, Nowak RD, Brent Smith R (2011) Empirical comparison of full-waveform lidar algorithms: range extraction and discrimination performance. Photogramm Eng Remote Sens 77:825–838

Reitberger J, Krzystek P, Stilla U (2008) Analysis of full waveform LIDAR data for the classification of deciduous and coniferous trees. Int J Remote Sens 29:1407–1431

Riegl LMS (2013) www.riegl.com. Homepage of the company RIEGL Laser Measurement Systems GmbH. Accessed Aug 2013

Roncat A, Wagner W, Melzer T, Ullrich A (2008) Echo detection and localization in full-waveform airborne laser scanner data using the averaged square difference function estimator. Photogramm J Finl 21:62–75

Roncat A, Bergauer G, Pfeifer N (2011a) B-Spline deconvolution for differential target cross-section determination in full-waveform laser scanner data. ISPRS J Photogramm Remote Sens 66:418–428

Roncat A, Lehner H, Briese C (2011b) Laser pulse variations and their influence on radiometric calibration of full-waveform laser scanner data. Int Arch Photogramm Remote Sens Spat Inf Sci 38(5/W12):137–142

Rubio J, Grau E, Sun G, Gastellu-Etchgorry J, Ranson K (2009) Lidar modeling with the 3D DART model. In: AGU Fall Meeting Abstracts, p A330

Sun G, Ranson K (2000) Modeling lidar returns from forest canopies. IEEE Trans Geosci Remote Sens 38:2617–2626

Tikhonov AN, Arsenin VY (1977) Solutions of ill-posed problems. V. H. Winston & Sons, Washington

Wagner W (2010) Radiometric calibration of small-footprint full-waveform airborne laser scanner measurements: basic physical concepts. ISPRS J Photogramm Remote Sens 65:505–513

Wagner W, Ullrich A, Melzer T, Briese C, Kraus K (2004) From single-pulse to full-waveform airborne laser scanners: potential and practical challenges. Int Arch Photogramm Remote Sens Spat Inf Sci 35(Part B3):201–206

Wagner W, Ullrich A, Ducic V, Melzer T, Studnicka N (2006) Gaussian decomposition and calibration of a novel small-footprint full-waveform digitising airborne laser scanner. ISPRS J Photogramm Remote Sens 60:100–112

Wagner W, Hollaus M, Briese C, Ducic V (2008) 3D vegetation mapping using small-footprint full-waveform airborne laser scanners. Int J Remote Sens 29:1433–1452

Widlowski JL, Robustelli M, Disney M, Gastellu-Etchegorry JP, Lavergne T, Lewis P, North PRJ, Pinty B, Thompson R, Verstraete M (2008) The RAMI On-line Model Checker (ROMC): a web-based benchmarking facility for canopy reflectance models. Remote Sens Environ 112:1144–1150

Yu X, Liang X, Hyyppä J, Kankare V, Vastaranta M, Holopainen M (2013) Stem biomass estimation based on stem reconstruction from terrestrial laser scanning point clouds. Remote Sens Lett 4:344–353

Chapter 3
Full-Waveform Airborne Laser Scanning Systems and Their Possibilities in Forest Applications

Markus Hollaus, Werner Mücke, Andreas Roncat, Norbert Pfeifer, and Christian Briese

Abstract Full-waveform (FWF) airborne laser scanning (ALS) systems became available for operational data acquisition around the year 2004. These systems typically digitize the analogue backscattered echo of the emitted laser pulse with a high frequency. FWF digitization has the advantage of not limiting the number of echoes that are recorded for each individual emitted laser pulse. Studies utilizing FWF data have shown that more echoes are provided from reflections in the vegetation in comparison to discrete echo systems. To obtain geophysical metrics based on ALS data that are independent of a mission's flying height, acquisition time or sensor characteristics, the FWF amplitude values can be calibrated, which is an important requirement before using them in further classification tasks. Beyond that, waveform digitization provides an additional observable which can be exploited in forestry, namely the width of the backscattered pulse (i.e. echo width). An early application of FWF ALS was to improve ground and shrub echo identification below the forest canopy for the improvement of terrain modelling, which can be achieved using the discriminative capability of the amplitude and echo width in classification algorithms. Further studies indicate that accuracies can be increased for classification (e.g. species) and biophysical parameter extraction (e.g. diameter at breast height) for single-tree- and area-based methods by exploiting the FWF observables amplitude and echo width.

M. Hollaus (✉) • W. Mücke • A. Roncat • N. Pfeifer
Research Groups Photogrammetry and Remote Sensing, Department of Geodesy and Geoinformation, Vienna University of Technology, Vienna, Austria
e-mail: Markus.Hollaus@geo.tuwien.ac.at

C. Briese
Ludwig Boltzmann Institute for Archaeological Prospection and Virtual Archaeology, Vienna, Austria

Research Groups Photogrammetry and Remote Sensing, Department of Geodesy and Geoinformation, Vienna University of Technology, Vienna, Austria

M. Maltamo et al. (eds.), *Forestry Applications of Airborne Laser Scanning: Concepts and Case Studies*, Managing Forest Ecosystems 27, DOI 10.1007/978-94-017-8663-8__3,
© Springer Science+Business Media Dordrecht 2014

3.1 Introduction

Full-waveform (FWF) airborne laser scanning (ALS) sensors and data have become increasingly available by the end of 2013. As there is no significant difference in the acquisition costs between discrete and FWF data, more and more contractors and public administrations decide in favour of the latter. Research on how the additional information contained in FWF ALS data can be exploited has also been going on for about a decade.

This chapter gives a review of the scientific outcomes concerning the exploitation of FWF-derived quantities for forestry applications. In Sect. 3.2, an overview of FWF recording systems is presented, followed by Sect. 3.3 exploring the potential of FWF data in comparison to discrete-return data. In Sect. 3.4, forestry-related applications are described, structured in (1) point cloud classification, (2) single-tree- and (3) area-based applications. The focus within (1) is on improved digital terrain model calculation in forested areas based on FWF features. In (2), the segmentation of individual tree crowns and tree species classification is discussed in detail. The area-based applications presented in section (3) cover growing stock estimation and structure type assessment for larger areas. The chapter ends with conclusions in Sect. 3.5.

3.2 FWF Recording Systems

The introduction of the first commercial small-footprint topographic ALS system capable of FWF recording dates back to the year 2004 with the Riegl LMS-Q560 (Riegl LMS 2013; Wagner et al. 2004), the TopEye Mark II and the Optech ALTM 3100 following shortly afterwards (Mallet and Bretar 2009). Before this date, some airborne bathymetric and experimental systems such as SHOALS and LVIS, respectively, had already been operable (Mallet and Bretar 2009).

In the beginning of the twenty-first century, the ALS community has seen great progress in hardware design, which lead to significantly increased pulse repetition rates, the resolution of range ambiguities[1] and, consequently, point clouds with strongly increased spatial density.

Beraldin et al. (2010) give a comprehensive set of instrument parameters relevant for ALS systems, some of which are of special interest when dealing with FWF recording systems:

Pulse length: the length of the emitted laser pulse is a limiting factor for the discrimination of two subsequent echoes (see Sect. 2.3).

---

[1]called *MPiA* (multiple pulses in air) by Leica Geosystems (2013), *CMP* (continuous multipulse) by Optech Inc. (2013) and *MTA* (multiple time-around) by Riegl LMS (2013).

Sampling interval: this quantity determines the highest frequency which can still be reconstructed from the recorded signal; according to the sampling theorem, this frequency amounts to the inverse of twice the sampling interval. Together with the pulse length, the sampling interval also influences the ranging accuracy.

Bit depth of amplitude recording: besides constraining the detection of weak echoes, it mainly affects the resolved radiometric details of radiometric calibration.

Laser wavelength: this applies mainly to radiometric calibration using natural targets or reflectivity assumptions in homogeneous areas (see Sect. 2.3).

Table 3.1 summarizes the system parameters of currently available FWF ALS systems.

3.3   Exploring the Potential of FWF Data

When compared to conventional (i.e. discrete echo recording) ALS systems, FWF laser scanners feature a number of enhancements (e.g. Adams et al. 2012; Liu et al. 2011), making them especially suitable for feature extraction or object classification purposes. Discrete ALS systems are capable of recording a fixed number of range measurements per emitted pulse (usually up to four), involving a certain threshold for return peak detection, which is inherent to the integrated (and most often proprietary) detection method (e.g. Thiel and Wehr 2004; Wagner et al. 2004). With the advent of FWF detection and storage in laser scanning systems, especially tuned echo extraction methods can be applied by the user in post-processing (Wagner et al. 2004). Detection thresholds can be adapted and the number of returns to be detected per each emitted pulse is (at least theoretically) no longer limited (cf. Sect. 3.2 and Mallet et al. 2010). As a consequence, FWF data provide a much higher number of returns per emitted pulse in contrast to conventional systems (e.g. Reitberger et al. 2008). Therefore denser point clouds are created, a fact which is of major importance for vegetation analysis, as the representation of the canopy structure is consequently more detailed.

To use the FWF observations to their full potential, one must understand how the backscattered waveforms are altered when the laser pulse interacts with different types of surfaces. The intensity of the backscattered pulse is dependent on a number of factors, some of which are very specific to the equipment in use (e.g. the reflectance of the surface at the laser scanner's wavelength) or unknown (e.g. the fraction of the laser pulse that interacted with the surface) (Ducic et al. 2006). Thus, radiometrically calibrating laser scanning intensity data is an important requirement before using it in subsequent classification tasks. It shall be noted that this refers to discrete (where some assumptions about the echo signal are necessary) as well as FWF data. Calibration approaches applied in the literature rely (1) on the fact that the target surface is flat and extended (e.g. Ahokas et al. 2006; Kaasalainen et al. 2007; Briese et al. 2008; Lehner and Briese 2010) or (2) on range normalization

Table 3.1 Currently available small-footprint topographic FWF ALS systems and their characteristic system parameters. Beam width refers to the angle where the laser power density has fallen to $1/e^2$ of its maximum

Manufacturer	Model	Scanning mech.	Max. PRR[a] (kHz)	Max. scan angle (°)	Mult. pulses	Max. range (m)
AHAB	Chiroptera	Palmer scan	400	±20	Yes	1,600
Leica	ALS60	Osc. mirror	120[b]	±37.5	Yes	6,300
Geosystems	ALS70 CM	Osc./rot. mirror	120[b]	±37.5	Yes	1,850
	ALS70 HP	Osc./rot. mirror	120[b]	±37.5	Yes	4,400
	ALS70 HA	Osc./rot. mirror	120[b]	±37.5	Yes	6,300
Optech	Aquarius[c]	Osc. mirror	70	±25	No	2,750
	Gemini	Osc. mirror	125[d]	±25	Yes	4,400
	Orion	Osc. mirror	125[d]	±30	Yes	4,600
	Pegasus	Osc. mirror	125[d]	±37.5	Yes	3,150
Riegl LMS	Q680i	Rot. pyramid	400	±30	Yes	3,000
	Q780	Rot. pyramid	400	±30	Yes	5,800

Manufacturer	Model	Pulse length	Sampl. interval (ns)	Beam width	Bit depth	Wavelength (nm)
AHAB	Chiroptera	4 ± 1 ns	0.55	0.55 mrad	N/A	1,064
Leica	ALS60	5 ns	1	0.22 mrad	8 bit	1,064
Geosystems	ALS70 CM	N/A	1	0.22 mrad	8 bit	1,064
	ALS70 HP	N/A	1	0.22 mrad	8 bit	1,064
	ALS70 HA	N/A	1	0.22 mrad	8 bit	1,064
Optech	Aquarius[c]	N/A	1	N/A	12 bit	1,064
	Gemini	N/A	1	0.35/1.13 mrad	12 bit	1,064
	Orion	N/A	1	0.35 mrad	12 bit	1,064/1,541
	Pegasus	N/A	1	0.35 mrad	12 bit	1,064
Riegl LMS	Q680i	4 ns	1	≤ 0.5 mrad	2 × 8 bit	1,550
	Q780	4 ns	1	≤ 0.25 mrad	2 × 8 bit	1,550

[a]Maximum pulse repetition rate
[b]The instrument's maximal PRR is higher but only 120 kHz can be recorded by the FWF digitizer (Leica Geosystems 2013)
[c]In topographic mode
[d]The instrument's maximal PRR is higher but only 125 kHz can be recorded by the FWF digitizer (Optech Inc. 2013)

(i.e. distance from scanner centre to target) of the recorded intensities (e.g. Höfle and Pfeifer 2007; Korpela et al. 2009; Luzum et al. 2004).

Vegetation usually represents very heterogeneous targets in relation to the laser's footprint, as the backscattering parts are distributed over the plant height and feature very diverse reflectance properties (e.g. leaves or needles, thin branches, thick branches, tree trunks). Consequently, the model assumptions for alternative (1) are only met in the case of very dense vegetation. Dense foliage may also represent a homogeneous and impenetrable target and will then be represented by a number of so-called single echoes (i.e. echoes without multiple reflections). Alternative (2) is especially effective in case of large height differences between targets (e.g. topographic variations in mountainous areas), cf. Eq. (2.4), but does not necessarily result in significant changes when applied to areas with low relief energy or differences in vegetation heights of a few tens of metres (e.g. tree heights of 30 m) when the data are acquired from 1,000 m flying height above ground or higher (cf. Morsdorf et al. 2010).

Given these challenging circumstances for radiometric calibration in vegetated areas, most studies make use of range calibration (e.g. Hopkinson and Chasmer 2009; Ørka et al. 2009; Holmgren and Persson 2004), therefore retaining the relationship between single echoes, but making their results hardly comparable with findings of other studies or measurements with other sensors.

Another way of ALS data calibration refers to the task of deriving the backscatter coefficient, which is calculated on the basis of the radar equation and includes area normalization by the laser footprint on the target (Lehner and Briese 2010) (cf. Sect. 2.3). It provides robust results, considering also differing flying altitude above ground, incidence angles or significantly different beam divergence of different sensors. Furthermore, assuming a Lambertian reflectance behaviour of the illuminated target allows for the estimation of the diffuse reflectance measure which enables the correction w.r.t different acquisition geometries (e.g. observation of a building root from two different ALS strips) (Lehner and Briese 2010; Briese et al. 2012).

Regardless of wavelength and surface reflectivity, returns from vegetated areas tend to have lower intensities than returns from artificial surfaces. This is mainly due to the fact that vegetation components commonly represent non-extended targets. Consequently, the available energy within the footprint is divided among a number of small scatterers within the footprint of the laser beam, leaving less energy for the single scattering element that reflects back to the sensor (Wagner et al. 2008). As a result, vegetated areas appear generally darker in intensity images, when compared to non-vegetated areas or artificial surfaces representing extended targets and producing single echoes (see Figs. 3.1c and 3.2).

The distribution of scattering elements inside the footprint also influences the width of the backscattered echo (i.e. the echo width). Height variations of small scatterers tend to widen the return pulse, leading to a systematic broadening of the echo width, which can be observed for vegetation targets (e.g. Ducic et al. 2006; Hollaus et al. 2009a; Lin and Mills 2009; Doneus et al. 2008; Höfle et al. 2008). The echo width is therefore a geometric measure that is sensitive to surface undulations smaller than the footprint size. Furthermore, similarities exist between geometric measures of echo distribution (e.g. vertical height variations of echoes or standard deviation of echoes with respect to an adjustment plane) and their respective echo widths in a defined neighbourhood (Hollaus et al. 2011) (see Fig. 3.3). In a broader sense, the echo width can therefore be referred to as a measure of surface roughness, which makes it a highly suitable discriminator for vegetated and non-vegetated areas (see Fig. 3.1d).

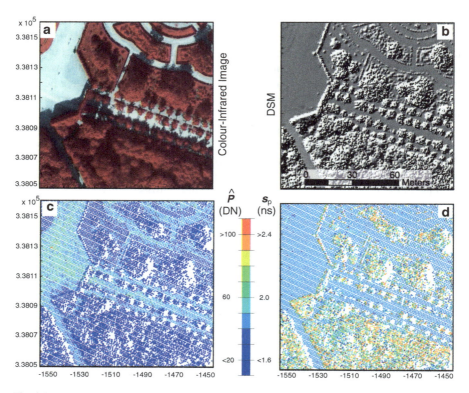

Fig. 3.1 (**a**) Colour-infrared aerial image, (**b**) digital surface model (DSM), visualizations of ALS point cloud colour-coded by (**c**) echo amplitude $\hat{P}$ (scaled in digital numbers DN; see Sect. 2.3) and (**d**) echo width s_P (standard deviation, scaled in nanoseconds; see Sect. 2.3). (Modified from Wagner et al. (2008))

Fig. 3.2 Profile view of ALS point clouds with coloured amplitude values derived from FWF data. The profile has a length of 100 m and a width of 5 m

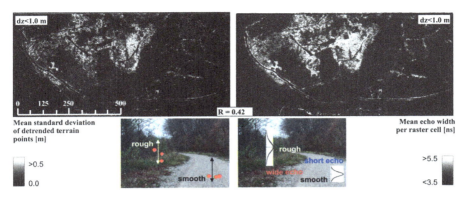

Fig. 3.3 Surface roughness raster layer (1 m horizontal resolution). *Left*: Mean standard deviation per raster cell of detrended terrain and near terrain echoes. *Right*: Mean echo width per raster cell derived from FWF ALS data (Modified from Hollaus et al. (2011))

3.4 Applications of FWF Data

3.4.1 *Improving Point Cloud Classification in Forested Areas Using FWF Airborne Laser Scanning*

In order to retrieve a reliable representation of the terrain surface in forested areas, a sophisticated point cloud classification into terrain and non-terrain points is essential (Kraus et al. 2004). Most forestry applications utilizing ALS data rely on such accurate classification and subsequently derived digital terrain models (DTMs) (Hollaus 2006; Hyyppä et al. 2008; Næsset 2007). Conventional methods for classifying the point cloud into terrain and non-terrain echoes, a process also called filtering, employ different geometric criteria for echo selection. These involve distances to prior computed surfaces of different scales or levels of detail (Axelsson 2000; Kraus and Pfeifer 1998; Pfeifer et al. 2001; Briese et al. 2002), relations of planimetric distance and height difference (Vosselman 2000) or orientation of normal vectors as homogeneity criterion in segmentation-based approaches (Tóvari and Pfeifer 2005).

However, in certain situations geometric criteria alone hardly suffice to distinguish ground echoes from echoes close to the terrain surface. This applies in particular to forested areas with dense understorey, as they pose two major problems. First, if the vegetation is very low, the range difference between two consecutive echoes may become too short for the sensor's detector to separate them (see Sect. 2.3). As a result only one target is identified and the measured distance (i.e. range) results from an overlap of the two actual reflections. The estimated three-dimensional (3D) point will then be located somewhere in between them.

Second, areas with dense lower vegetation often feature little to no penetration at all. This is especially crucial if the trend of the terrain surface changes significantly below the vegetation. In both cases, echoes from lower vegetation are very likely to be wrongly classified as ground points. A DTM based on such a point cloud runs right through the lowest vegetation levels and above the actual terrain (Pfeifer et al. 2004). These errors may be in the range of several decimetres and therefore critical for DTM-based applications where high accuracy is required (e.g. Doneus et al. 2008).

The usage of FWF ALS opens up new prospects for DTM generation to overcome the above mentioned difficulties. Apart from the higher point densities, FWF ALS point clouds include additional information on radiometric and geometric surface properties. While the echo amplitude contains information about the reflectivity of the surface hit by the laser beam (see Sect. 2.3), the echo width is influenced by the vertical distribution of small scattering elements within the footprint of the beam (Wagner et al. 2008). These relationships provide additional knowledge about surface properties and have been subject to analysis in several studies dealing with point cloud classification (Mallet and Bretar 2009). They can be integrated in the filtering process or used in a pre-processing step for ground point selection. The amplitude differences of grass or bare soil (usually high amplitudes), asphalt roads (usually low amplitudes) and rooftops (usually varying due to differing materials) were used by Alexander et al. (2010) for a decision-tree-based discrimination of surfaces.

According to the analysis of Wagner et al. (2008), echoes from the canopy, understorey or near ground vegetation have larger variations in vertical directions and consequently larger echo widths than the terrain. Based on this fact, Doneus and Briese (2006) used an empirically derived echo width threshold (see also Briese et al. 2007; Doneus et al. 2008), pre-selecting terrain echoes in the input point cloud for the hierarchic robust filtering (Pfeifer et al. 2001; Briese et al. 2002) and successfully removing echoes from piles of twigs and coarse woody debris (see Fig. 3.4). In Lin and Mills (2009), a point labeling process, determining terrain points by using a threshold for the echo width, was applied. This additional surface information was integrated in a DTM generation approach employing Axelsson's progressive densification method (Axelsson 2000). They found that without employing the echo width filter, the terrain height was significantly underestimated by the filtering strategy.

Rather than applying hard thresholds on the features of the ALS echoes which may be sensor specific, Mücke et al. (2010) proposed a point labeling strategy assigning weights to the single points indicating their likeliness to represent terrain (see Fig. 3.5). Based on the previous work by Wagner et al. (2008) and Mandlburger et al. (2007), the weights were derived based on the relationship of echo amplitude and width and are subsequently used in a modified version of hierarchic robust filtering. This approach does not eliminate points based on a priori determined thresholds, rather it flags the points while keeping them in the data set. Therefore the possibility of creating false negatives is reduced, meaning that reflections that might nevertheless stem from terrain are still included and not lost for terrain modeling.

Fig. 3.4 *Left*: aerial photograph of forested area with logging site. *Right*: zoomed view of photograph and overlaid 3D point cloud with applied echo width filtering. Echoes from piles of twigs have been removed (Modified from Doneus et al. (2008))

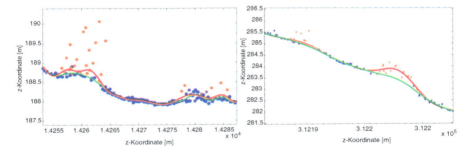

Fig. 3.5 Profile views of ALS point clouds with interpolated DTM with (*green*) and without (*red*) weights derived on the basis of echo amplitude and width. Colour-coding of the points refers to assigned weights (*blue* means higher weights, *orange* means lower weights). Both situations show that the terrain height is overestimated without the integration of FWF information (Modified from Mücke (2008))

3.4.2 Single-Tree-Based Applications

In ALS-based applications for forestry the incorporation of FWF data and attributes (i.e. amplitude, echo width) concentrates mainly on (1) delineation and three-dimensional modelling of single trees, (2) extraction of biophysical parameters (e.g. diameter at breast height, stem volume, biomass) and (3) species classification. Relevant literature has shown that already the higher amount of 3D points contained in FWF data is a huge benefit to forest applications. Through the more efficient detection of weak echoes (i.e. reflections of rather low backscattering energy), which are very likely to occur in the understorey, the characterization and depiction of the stratification of the foliage is much more detailed. Reitberger et al. (2009), who were interested in the detection of dominant and sub-dominant trees, found that the success rate increased by 20 % for trees in lower and intermediate layers when using FWF data due to the larger number of echoes. They used a 3D voxel-based segmentation technique referred to as normalized cut segmentation. Further incorporating the echo intensity and width, along with the 3D positions of the echoes in the clustering algorithm resulted in a total of 12 % more detected trees, improving the detection rate for sub-dominant trees by 16 %. Subsequent discrimination of the detected trees into needle and broadleaved species made the advantage of waveform data even more evident, when an overall accuracy of 95 % was achieved, in contrast to 80 % when using discrete data. Yao et al. (2012) used the same method as described by Reitberger et al. (2009), but extended it by automatic determination of diameter at breast height (DBH) and stem volume for the detected single trees. They reported a 50 % increase in accuracy using waveform metrics in their approach compared to using discrete ALS data (RMSE for stem volume and DBH of 16 and 9 % for FWF, in contrast to 38 and 21 % for discrete).

Heinzel and Koch (2011) identified echo intensity and width from FWF ALS data as highly significant parameters for species detection by means of exploratory data analysis. In Heinzel and Koch (2012) they made use of their earlier findings in a support vector classification of tree species, using rasterized FWF ALS data together with colour-infrared (CIR) images, true-colour (RGB) images and hyperspectral data. Out of a set of 464 features derived from the raw data to be used in the classification, their sensitivity analysis showed that the 14 most significant parameters came from FWF ALS and CIR data. They achieved an overall accuracy for species determination of 79.2 % using FWF data (including intensity and echo width information), compared to 47.3 % from 3D information alone, 46.8 % using only textural measures from RGB, 64.7 % employing hyperspectral metrics and 50.7 % exploiting solely CIR derived features. Using all input metrics together their evaluation showed an overall accuracy of 88 %. Also in Vaughn et al. (2012), the benefits of FWF data were shown for tree species classification. They could increase the classification accuracy for some tree species by 3.5–4 % by using FWF measures.

Hollaus et al. (2009a) explored the possibilities of FWF ALS data for discrimination of three dominant tree species (spruce, larch and beech) (see Fig. 3.6). Starting from an edge-based segmentation on the normalized digital surface model (nDSM), which was supposed to delineate single-tree crowns, they assigned geometric (e.g. echo height distributions, mean echo width and standard deviation of echo widths per segment) and radiometric (e.g. mean intensities and backscatter cross-sections) features to the segments for a decision tree classification. The echo width was identified as a valuable parameter for the distinction between spruce and larch, which were otherwise difficult to separate. This could be explained by biophysical properties of these two species and their differing representation in the incorporated leaf-off ALS data set. It was stated that a spruce features a rather dense and homogeneous "surface" of needles, yielding low standard deviations of echo widths, which describe the height variation of the scattering elements accumulated on a respective segment. A larch, on the other hand, loses large amounts of its needles during winter time and is therefore much more heterogeneous, consequently featuring higher standard deviations. The success rate for the three selected species

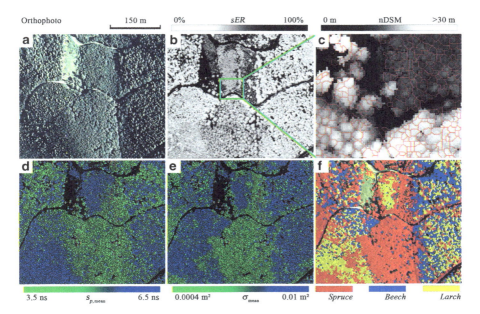

Fig. 3.6 (**a**) True-color orthophoto (© http://maps.live.de), (**b**) slope-adaptive echo ratio (sER), (**c**) nDSM overlaid with segmentation result and mean values per segment of the (**d**) echo width ($s_{p.\mathrm{mean}}$), (**e**) backscatter cross-section (σ_{mean}), and (**f**) final tree species classification result (From Hollaus et al. (2009a))

was 75 %, which increased to 83 % when only discrimination into coniferous and deciduous species was applied.

In contrast to the afore mentioned studies, Höfle et al. (2012) used calibrated FWF data (cf. Briese et al. 2008; Lehner and Briese 2010) for the detection of vegetation in urban areas. Also following an edge-based segmentation of the nDSM, they assigned attributes to the resulting polygon geometry, which included features like: mean and standard deviation of normalized point height, mean echo width, or mean backscatter cross-section and backscatter coefficient (cf. Chap. 2). These features were derived independently for either all echoes, only the first, the intermediate or the last echoes. Additionally, they computed a relation of the number of echoes to all echoes for the first, intermediate and last echoes as a measure for penetration. It was stated that this penetration measure, which was the least computationally expensive to derive, was the most significant classifier of vegetation segments (overall accuracy of 94 % using an artificial neural network for classification), a measure benefiting largely from the high point density provided by the FWF data set.

The authors were able to slightly increase the accuracy to 96 % by using a decision-tree classification based on combined feature sets of mean intensity, mean echo width per segment and metrics describing the height distribution of echoes inside a segment. It was also shown that vegetation areas in general correspond to high values of echo width and low values of signal amplitude. Furthermore, differences in the homogeneity of the echo width distributions for different echo types (i.e. first, intermediate, last) were observed. Single echoes from buildings, for example, were found to have more homogeneous echo widths than last echoes from trees, a knowledge which could also be used to further refine the classification results.

Single-tree-based methods basically describe the smallest entity of a forest, thus they stand and fall with the success of detecting the same, while stand-based methods are more robust due to the averaging over larger areas. FWF data have proven to be highly valuable in improving this process. However, the delineation of single trees is usually a complex and computationally extensive procedure and therefore not very suitable for large area applications. More robust methods, which are capable of handling large amounts of data, are used in area-based applications, which are introduced in the following section.

3.4.3 Area-Based Applications

In contrast to single-tree-based approaches, area-based ones use circular plots, grids or forest stands as reference unit (see Chap. 11). Commonly, ALS metrics describe the horizontal and vertical distribution of ALS echoes within a reference unit. They are used for a variety of statistical (e.g. Næsset 1997, 2002), (semi-) empirical (e.g. Hollaus et al. 2009b) or physical (e.g. Hyyppä et al. 2012) models for assessing forest parameters. An automated derivation of forest parameters using such regression models has already reached an operational status, but commonly uses only the 3D information (i.e. x, y, z coordinates of the echoes) from discrete, as well as FWF ALS data. Using the additional FWF information (e.g. echo width, backscatter cross-section) for area-based approaches is limited to few applications until now, either due to the limited availability of the data, or because of the lack of methods using this additional FWF information.

One of the first FWF ALS applications is described in Nilsson (1996) who used a helicopter-borne scanning laser for estimating stand volume for a coastal Scots pine stand. Similar to the publication by MacLean and Krabill (1986) that used the cross sectional areas derived from an Airborne Oceanographic Lidar profiler to estimate gross merchantable timber volume, Nilsson (1996) used the mean product of the waveform area and the laser height to estimate stand volume. The mean values were calculated for individual plots with a radius of 10 m. For the stand volume estimation a linear regression model was used and a R^2 of 0.78 was achieved.

In addition to the higher point density of FWF ALS data compared to discrete ones, the radiometric content (i.e. amplitude) has been shown as a beneficial property for tree species classification. For example Heinzel and Koch (2011) have shown that for a test site of $10\,km^2$ up to 6 tree species were classified with an overall accuracy of 57 % using 3 variables based on the intensity, the echo width and the total number of targets. The overall accuracy increased to 78 % if the classification was limited to the 4 main species, and to 91 % if the classification was limited to conifers and broadleaved trees, respectively. The incorporated ALS data were acquired with a Riegl LMS-Q560 scanner in the main growing season.

Wing et al. (2012) used intensities from ALS data acquired with a Leica ALS 50 Phase II laser system to predict understorey vegetation cover in an interior ponderosa pine forest. The explanatory ALS metrics were derived for $40.5\,m^2$ circular plots. The intensities were used to classify points associated with undesirable understorey components (e.g. non-vegetation). The intensity data for that study were acquired while using the variable gain setting and were therefore not calibrated. Even without calibration, the intensity information displayed great potential in distinguishing ALS points associated with the various understorey components. They found that echoes associated with live vegetation typically had intensity values within one standard deviation of the mean intensity value, and echoes associated with other understorey components were often outside this range (Wing et al. 2012). Making use of this knowledge, they successfully removed a large portion of non-vegetation understorey component points. Even though the employed Leica ALS 50 is not an FWF scanner, it can record intensity values for up to three echoes per shot. Therefore, the derived intensities are roughly comparable to the ones derived from an FWF scanner. Because of the recorded echo width information small-footprint FWF scanners provide more explanatory value for echo classification (cf. Sect. 3.4.1).

Several studies applying FWF ALS data for assessing forest parameters can be found in the literature (e.g. Kronseder et al. 2012; Rossmann et al. 2009), but in all of them only the 3D position information of the echoes is used to extract input parameters for regression models. In these studies the benefit of the FWF data is the higher echo density (Persson et al. 2005; Reitberger et al. 2008). Chauve et al. (2009) reported that tree crowns and undergrowth were more densely sampled due to the detection of weak and overlapping echoes within the backscattered waveforms, which allow a more detailed description of the structure of forest stands. For example Vetter et al. (2011) used detailed vertical structure information near the terrain surface derived from FWF ALS data for hydraulic surface roughness estimation within a forested area. Leiterer et al. (2012) developed a method for the physically-based extraction of canopy structure variables (e.g. the occurrence of various canopies in a vertical column and species composition) on grid level. They analysed different grid sizes to extract canopy structure types based on leaf-on and leaf-off data in a dense forest stand with high species diversity in Switzerland (see Fig. 3.7).

Lindberg et al. (2012) introduced a method to detect vertical vegetation structure to represent the shrub layer and one or several tree layers using discrete as well as FWF ALS data. The study was done in a hemi-boreal, spruce dominated forest in South-West Sweden. They used different methods to estimate the vegetation volume profile from the FWF data. The vegetation volume profile was defined as the volume of all tree crowns and shrubs in 0.1 m height intervals for 12 m radius plots. To assess the amount of vegetation at different heights above the ground, the ALS profiles

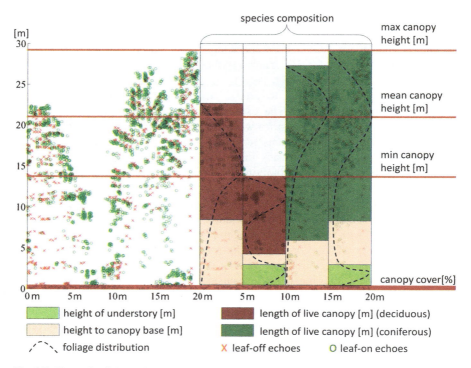

Fig. 3.7 Example of the grid-based vertical stratification of forest canopy structure (cell size 5 × 5 m²) (From Leiterer et al. (2012))

were rescaled using the estimated total vegetation volume. They concluded that the vegetation structure can be described more accurately from FWF data than from discrete ones.

Miura and Jones (2010) used different laser echo properties from FWF ALS data to characterize forest ecological structure for an Eucalyptus coastal forest and woodland study area in Tasmania, Australia. The applied regression analyses showed that the ALS derived variables were good predictors of field recorded variables describing forest structural types. They concluded that FWF ALS provide information on the complexity of habitat structure in an efficient and cost-effective manner.

Similar to this outcome, Neuenschwander et al. (2009) reported that FWF data provide detailed information about branches and leaves along the laser line-of-sight, which improved the classification accuracies of a landcover map by 15 % compared to high-resolution Quickbird multispectral imagery. They argued that spectral signatures are often similar for vegetation even though the vegetation types are structurally quite different. The analyses also showed that amplitudes derived from waveforms were selected most frequently for discriminating different tree types and densities.

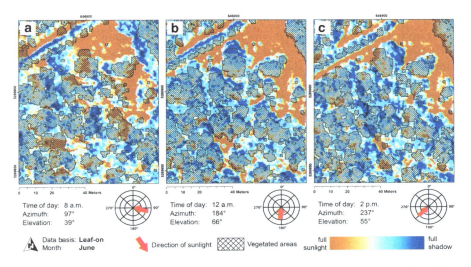

Fig. 3.8 Shadow raster maps based on leaf-on ALS data (selected vegetation echoes only) and the average sun positions in June (From Mücke and Hollaus (2011))

Mücke and Hollaus (2011) used high-density FWF ALS data (leaf-on and leaf-off) to model the light conditions in forests. The geometric information of the vegetation structure was derived directly from the 3D point cloud and was used to model the distribution of sunlight-absorbing or intercepting parts of the foliage which cast shadows on the surrounding understorey vegetation or ground. For modeling the light conditions, the photogrammetric monoplotting approach was adapted. They concluded that the resulting shadow raster maps provide valuable input for various biodiversity analyses, e.g. modeling the amount of available sunlight in vegetated areas (see Fig. 3.8).

3.5 Conclusion

FWF ALS offers a variety of additional features in comparison to discrete-return ALS while only demanding for moderate efforts in echo digitization. It can even be considered as the more natural approach since it records the signal traveling through the atmosphere. Discrete-return systems on the other hand only extract one parameter of this signal, namely its return time and therefore a distance, and in some cases an intensity value. With these digitized waveforms, practically all the information which the scanners themselves use for onboard echo extraction is available for further post-processing.

From a geometric point of view, more echoes are usually detectable in FWF data than in the discrete-return case, resulting in a higher point density. Furthermore, FWF allows for the calculation of radiometric echo parameters in the same spatial resolution. From these parameters, the echo width is of interest in vegetation and forest studies: (a) for an enhanced terrain/non-terrain classification of the points in order to reliably estimate the canopy height, and (b) for the classification of tree species. Furthermore, some studies have demonstrated the benefits of calibrated intensities for single-tree crown segmentation and tree species classification.

Even though the availability of FWF ALS data is increasing, the applications of e.g. echo width and backscatter cross-section in forestry are still limited. A severe limitation is often the availability of software tools allowing computations directly on the 3D point cloud considering also the point-based FWF attributes. Finally, the use of FWF information for assessing forest attributes is dependent on the capability of the ALS system to acquire the necessary information for calibrating the ALS data (e.g. amplitude and backscatter cross-section).

Acknowledgements Markus Hollaus has been supported by the project NEWFOR, financed by the European Territorial Cooperation "Alpine Space". Andreas Roncat has been supported by a Karl Neumaier PhD scholarship.

The Ludwig Boltzmann Institute for Archaeological Prospection and Virtual Archaeology is based on an international cooperation of the Ludwig Boltzmann Gesellschaft (Austria), the University of Vienna (Austria), the Vienna University of Technology (Austria), the Austrian Central Institute for Meteorology and Geodynamics, the office of the provincial government of Lower Austria, Airborne Technologies GmbH (Austria), RGZM (Roman-Germanic Central Museum Mainz, Germany), RA (Swedish National Heritage Board), VISTA (Visual and Spatial Technology Centre, University of Birmingham, UK) and NIKU (Norwegian Institute for Cultural Heritage Research).

References

Adams T, Beets P, Parrish C (2012) Extracting more data from LiDAR in forested areas by analyzing waveform shape. Remote Sens 4(3):682–702

Ahokas E, Kaasalainen S, Hyyppä J, Suomalainen J (2006) Calibration of the Optech ALTM 3100 laser scanner intensity data using brightness targets. Int Arch Photogramm Remote Sens Spat Inf Sci 36(1):T03–11

Alexander C, Tansey K, Kaduk J, Holland D, Tate NJ (2010) Backscatter coefficient as an attribute for the classification of full-waveform airborne laser scanning data in urban areas. ISPRS J Photogramm Remote Sens 65:423–432

Axelsson P (2000) DEM generation from laser scanner data using adaptive TIN models. Int Arch Photogramm Remote Sens 33(B4):110–117

Beraldin JA, Blais F, Lohr U (2010) Laser scanning technology. In: Vosselman G, Maas H-G (eds) Airborne and terrestrial laser scanning. Boca Raton, London, New York, CRC press, Taylor and Francis Group, chap 1, pp 1–42

Briese C, Pfeifer N, Dorninger P (2002) Applications of the robust interpolating for DTM determination. Int Arch Photogramm Remote Sens Spat Inf Sci 34(3A):55–61

Briese C, Doneus M, Pfeifer N, Melzer T (2007) Verbesserte DGM-Erstellung mittels full-waveform airborne laserscanning. In: Proceedings of the 3-Ländertagung DGPF, SGPBF, OVG, Basel (in German)

Briese C, Höfle B, Lehner H, Wagner W, Pfenningbauer M (2008) Calibration of full-waveform airborne laser scanning data for object classification. In: Proceedings of the SPIE: laser radar technology and applications XIII, Orlando

Briese C, Pfennigbauer M, Lehner H, Ullrich A, Wagner W, Pfeifer N (2012) Radiometric calibration of multi-wavelength airborne laser scanning data. ISPRS Ann Photogramm Remote Sensing and Spat Inf Sci 1(7):335–340

Chauve A, Vega C, Durrieu S, Bretar F, Allouis T, Pierrot-Deseilligny M, Puech W (2009) Advanced full-waveform lidar data echo detection: assessing quality of derived terrain and tree height models in an alpine coniferous forest. Int J Remote Sens 30:5211–5228

Doneus M, Briese C (2006) Digital terrain modelling for archaeological interpretation within forested areas using full-waveform laserscanning. In: Proceedings of the 7th international symposium on virtual reality, archaeology and cultural heritage VAST, Nicosia, Cyprus, pp 155–162

Doneus M, Briese C, Fera M, Janner M (2008) Archaeological prospection of forested areas using full-waveform airborne laser scanning. J Archaeol Sci 35:882–893

Ducic V, Hollaus M, Ullrich A, Wagner W, Melzer T (2006) 3D vegetation mapping and classification using full-waveform laser scanning. In: Proceedings of the international workshop 3D remote sensing in forestry, Vienna, pp 211–217

Heinzel J, Koch B (2011) Exploring full-waveform LiDAR parameters for tree species classification. Int J Appl Earth Obs Geoinf 13:152–160

Heinzel J, Koch B (2012) Investigating multiple data sources for tree species classification in temperate forest and use for single tree delineation. Int J Appl Earth Obs Geoinf 18:101–110

Höfle B, Pfeifer N (2007) Correction of laser scanning intensity data: data and model-driven approaches. ISPRS J Photogramm Remote Sens 62:415–433

Höfle B, Hollaus M, Lehner H, Pfeifer N, Wagner W (2008) Area-based parameterization of forest structure using full-waveform airborne laser scanning data. In: Proceedings of the SilviLaser 2008, Edinburgh, p 9

Höfle B, Hollaus M, Hagenauer J (2012) Urban vegetation detection using radiometrically calibrated small-footprint full-waveform airborne lidar data. ISPRS J Photogramm Remote Sens 67:134–147

Hollaus M (2006) Large scale applications of airborne laser scanning for a complex mountainous environment. PhD thesis, Vienna University of Technology

Hollaus M, Mücke W, Höfle B, Dorigo W, Pfeifer N, Wagner W, Bauerhansl C, Regner B (2009a) Tree species classification based on full-waveform airborne laser scanning data. In: Proceedings of the SilviLaser 2009, College Station, Texas, USA, pp 54–62

Hollaus M, Wagner W, Schadauer K, Maier B, Gabler K (2009b) Growing stock estimation for alpine forests in austria: a robust lidar-based approach. Can J For Res 39:1387–1400

Hollaus M, Aubrecht C, Höfle B, Steinnocher K, Wagner W (2011) Roughness mapping on various vertical scales based on full-waveform airborne laser scanning data. Remote Sens 3:503–523

Holmgren J, Persson Å (2004) Identifying species of individual trees using airborne laser scanner. Remote Sens Environ 90:415–423

Hopkinson C, Chasmer L (2009) Testing LiDAR models of fractional cover across multiple forest ecozones. Remote Sens Environ 113:275–288

Hyyppä J, Hyyppä H, Leckie D, Gougeon F, Yu X, Maltamo M (2008) Review of methods of small-footprint airborne laser scanning for extracting forest inventory data in boreal forests. Int J Remote Sens 29:1339–1366

Hyyppä J, Yu X, Hyyppä H, Vastaranta M, Holopainen M, Kukko A, Kaartinen H, Jaakkola A, Vaaja M, Koskinen J, Alho P (2012) Advances in forest inventory using airborne laser scanning. Remote Sens 4:1190–1207

Kaasalainen S, Hyyppä J, Litkey P, Hyyppä H, Ahokas E, Kukko A, Kaartinen H (2007) Radiometric calibration of ALS intensity. Int Arch Photogramm Remote Sens Spat Inf Sci 36(3/W52):201–205

Korpela I, Koskinen M, Vasander H, Holopainen M, Minkkinen K (2009) Airborne small-footprint discrete-return LiDAR data in the assessment of boreal mire surface patterns, vegetation, and habitats. For Ecol Manag 258:1549–1566

Kraus K, Pfeifer N (1998) Determination of terrain models in wooded areas with airborne laser scanner data. ISPRS J Photogramm Remote Sens 53:193–203

Kraus K, Briese C, Attwenger M, Pfeifer N (2004) Quality measures for digital terrain models. Int Arch Photogramm Remote Sens Spat Inf Sci 35(B2):113–118

Kronseder K, Ballhorn U, Böhm V, Siegert F (2012) Above ground biomass estimation across forest types at different degradation levels in Central Kalimantan using LiDAR data. Int J Appl Earth Obs Geoinf 18:37–48

Lehner H, Briese C (2010) Radiometric calibration of full-waveform airborne laser scanning data based on natural surfaces. Int Arch Photogramm Remote Sens Spat Inf Sci 38(7B):360–365

Leica Geosystems (2013) www.leica-geosystems.com. Homepage of the company Leica Geosystems. Accessed Aug 2013

Leiterer R, Morsdorf F, Schaepman M, Mücke W, Pfeifer N, Hollaus M (2012) Robust characterization of forest canopy structure types using full-waveform airborne laser scanning. In: Proceedings of the SilviLaser 2012, Vancouver, p 8

Lin Y, Mills J (2009) Integration of full-waveform information into the airborne laser scanning data filtering process. Int Arch Photogramm Remote Sens Spat Inf Sci 38(3/W8):224–229

Lindberg E, Olofsson K, Holmgren J, Olsson H (2012) Estimation of 3D vegetation structure from waveform and discrete return airborne laser scanning data. Remote Sens Environ 118:151–161

Liu Q, Li Z, Chen E, Pang Y, Li S, Tian X (2011) Feature analysis of lidar waveforms from forest canopies. Sci China Earth Sci 54:1206–1214

Luzum B, Starek J, Slatton K (2004) Normalizing ALSM intensities. Technical report Rep_2004-07-001, Geosensing Engineering and Mapping, Civil and Coastal Engineering Department, University of Florida, p 8

MacLean GA, Krabill WB (1986) Gross-merchantable timber volume estimation using an airborne LiDAR system. Can Journal Remote Sens 12:7–18

Mallet C, Bretar F (2009) Full-waveform topographic lidar: State-of-the-art. ISPRS J Photogramm Remote Sens 64:1–16

Mallet C, Lafarge F, Roux M, Soergel U, Bretar F, Heipke C (2010) A marked point process for modeling lidar waveforms. IEEE Trans Image Process 19:3204–3221

Mandlburger G, Briese C, Pfeifer N (2007) Progress in LiDAR sensor technology – chance and challenge for DTM generation and data administration. In: Proceedings of the 51st photogrammetric week, Stuttgart. Herbert Wichmann Verlag, pp 159–169

Miura N, Jones SD (2010) Characterizing forest ecological structure using pulse types and heights of airborne laser scanning. Remote Sens Environ 114:1069–1076

Morsdorf F, Mårell A, Koetz B, Cassagne N, Pimont F, Rigolot E, Allgöwer B (2010) Discrimination of vegetation strata in a multi-layered mediterranean forest ecosystem using height and intensity information derived from airborne laser scanning. Remote Sens Environ 114:1403–1415

Mücke W (2008) Analysis of full-waveform airborne laser scanning data for the improvement of DTM generation. Master's thesis, Institute of Photogrammetry and Remote Sensing, Vienna University of Technology

Mücke W, Hollaus M (2011) Modelling light conditions in forests using airborne laser scanning data. In: Proceedings of the SilviLaser 2011, Tasmania, p 8

Mücke W, Briese C, Hollaus M (2010) Terrain echo probability assignment based on full-waveform airborne laser scanning observables. Int Arch Photogramm Remote Sens Spat Inf Sci 38(7A):157–162

Næsset E (1997) Estimating timber volume of forest stand using airborne laser scanner data. Remote Sens Environ 61:246–253

Næsset E (2002) Predicting forest stand characteristics with airborne scanning laser using a practical two-stage procedure and field data. Remote Sens Environ 80:88–99

Næsset E (2007) Airborne laser scanning as a method in operational forest inventory: status of accuracy assessments accomplished in scandinavia. Scand J For Res 22:433–442

Neuenschwander AL, Magruder LA, Tyler M (2009) Landcover classification of small-footprint, fullwaveform lidar data. J Appl Remote Sens 3:033,544/1–033,544/13

Nilsson M (1996) Estimation of tree heights and stand volume using an airborne lidar system. Remote Sens Environ 56:1–7

Optech Inc. (2013) www.leica-geosystems.com. Homepage of the company Optech Inc. Accessed Aug 2013

Ørka HO, Næsset E, Bollandsås OM (2009) Classifying species of individual trees by intensity and structure features derived from airborne laser scanner data. Remote Sens Environ 113:1163–1174

Persson Å, Söderman U, Töpel J, Ahlberg S (2005) Visualization and analysis of full-waveform airborne laser scanner data. Int Arch Photogramm Remote Sens Spat Inf Sci 36(3/W19): 103–108

Pfeifer N, Stadler P, Briese C (2001) Derivation of digital terrain models in the SCOP++ environment. In: Torlegård K, Nelson J (eds) Proceedings of the OEEPE workshop on airborne laserscanning and interferometric SAR for detailed digital terrain models, Stockholm

Pfeifer N, Gorte B, Oude Elberink S (2004) Influences of vegetation on laser altimetry – analysis and correction approaches. Int Arch Photogramm Remote Sens Spat Inf Sci 36(8/W2):283–287

Reitberger J, Krzystek P, Stilla U (2008) Analysis of full waveform LIDAR data for the classification of deciduous and coniferous trees. Int J Remote Sens 29:1407–1431

Reitberger J, Schnörr C, Krzystek P, Stilla U (2009) 3D segmentation of single trees exploiting full waveform LIDAR data. ISPRS J Photogramm Remote Sens 64:561–574

Riegl LMS (2013) www.riegl.com. Homepage of the company RIEGL laser measurement systems GmbH. Accessed Aug 2013

Rossmann J, Schluse M, Buecken A, Krahwinkler P, Hoppen M (2009) Cost-efficient semi-automatic forest inventory integrating large scale remote sensing technologies with goal-oriented manual quality assurance processes. In: IUFRO division 4 – extending forest inventory and monitoring over space and time, Quebec City

Thiel KH, Wehr A (2004) Performance capabilities of laser scanners – an overview and measurement principle analysis. Int Arch Photogramm Remote Sens Spat Inf Sci 36(8/W2):14–18

Tóvari D, Pfeifer N (2005) Segmentation based robust interpolation – a new approach to laser data filtering. Int Arch Photogramm Remote Sens Spat Inf Sci 36(3/W19):79–84

Vaughn NR, Moskal LM, Turnblom EC (2012) Tree species detection accuracies using discrete point lidar and airborne waveform lidar. Remote Sens 4:377–403

Vetter M, Höfle B, Hollaus M, Gschöpf C, Mandlburger G, Pfeifer N, Wagner W (2011) Vertical vegetation structure analysis and hydraulic roughness determination using dense ALS point cloud data – a voxel based approach. Int Arch Photogramm Remote Sens Spat Inf Sci 38(5/W12):1–6

Vosselman G (2000) Slope based filtering of laser altimetry data. Int Arch Photogramm Remote Sens 33(B3):935–942

Wagner W, Ullrich A, Melzer T, Briese C, Kraus K (2004) From single-pulse to full-waveform airborne laser scanners: potential and practical challenges. Int Arch Photogramm Remote Sens Spat Inf Sci 35(Part B3):201–206

Wagner W, Hollaus M, Briese C, Ducic V (2008) 3D vegetation mapping using small-footprint full-waveform airborne laser scanners. Int J Remote Sens 29:1433–1452

Wing BM, Ritchie MW, Boston K, Cohen WB, Gitelman A, Olsen MJ (2012) Prediction of understory vegetation cover with airborne lidar in an interior ponderosa pine forest. Remote Sens Environ 124:730–741

Yao W, Krzystek P, Heurich M (2012) Tree species classification and estimation of stem volume and DBH based on single tree extraction by exploiting airborne full-waveform LiDAR data. Remote Sens Environ 123:368–380

Chapter 4
Integrating Airborne Laser Scanning with Data from Global Navigation Satellite Systems and Optical Sensors

Rubén Valbuena

Abstract Most forestry applications of airborne laser scanning (ALS) require simultaneous use of various data sources. This chapter covers a number of common issues that practitioners face when dealing with data fusion schemes. The first subsection points out the objectives that may be pursued when integrating different data sources, and the benefits that can be obtained from using diverse remote sensors onboard differing platforms. The next subsections are devoted to the two data sources that usually pose most problems in their spatial co-registration with ALS datasets: field inventory and aerial photographs. All data sources ultimately rely on global navigation satellite systems (GNSS) which are especially error-prone when operating under forest canopies. Positioning methods and spatial accuracy assessment applied to forest plot and individual tree surveying are presented, also including terrestrial laser scanning (TLS). Furthermore, procedures for digital elevation model (DEM) generation are reviewed in the context of their use in orthorectification, which is the most widespread method for fusion of ALS with optical sensors. Drawbacks of using orthophotos are identified, therefore suggesting alternatives: true-orthorectification, back-projecting ALS and image matching.

4.1 Introduction and Objectives

In this chapter, some fundamentals of remote sensing are reviewed with the purpose of assisting practitioners with the – sometimes difficult – task of integrating various data sources within the same project. Every positioning method is prone to errors, but some are more severe than others. It is also important to know whether a given error source leads to bias, or just isotropic uncertainty around a coordinate. Usually the ALS data serve as reference to which other data sources have to be

R. Valbuena (✉)
European Forest Institute HQ, Torikatu 34, 80100 Joensuu, Finland
e-mail: ruben.valbuena@efi.int

M. Maltamo et al. (eds.), *Forestry Applications of Airborne Laser Scanning: Concepts and Case Studies*, Managing Forest Ecosystems 27, DOI 10.1007/978-94-017-8663-8_4, © Springer Science+Business Media Dordrecht 2014

co-registered. The objective of this chapter is to guide on the choice of methods for georeferencing information that applies to each case. Section 4.2 reviews the different objectives that may be pursued in multi-source projects, specifying the benefits of including both airborne laser scanning (ALS) and optical sensors for forest assessment. Section 4.3 focuses on integrating the data acquired in the field, and therefore revises some principles of global navigation satellite systems (GNSS), detailing particularities of positioning sample plots under forest canopies. Methods for incorporating positions of individual trees are considered as well, including terrestrial laser scanning (TLS). Section 4.4 is devoted to fusion of ALS and optical sensors, reviewing some basics of photogrammetry. The procedures for digital elevation model (DEM) generation are discussed, pointing out the consequences of using them for optical image orthorectification. The last sub-section explains how back-projecting can be used for colouring ALS echo clouds if information on platform position at the time of exposure is available.

4.2 The Role of Additional Data Sources in ALS-Assisted Forest Inventory Projects

Most forestry applications of ALS depend on the combination of various data sources. In the simplest case, information surveyed from the field must be spatially registered with the ALS echo cloud. More complex projects may use ground-based TLS, or additional remote sensors onboard satellite or aircraft platforms: optical multispectral or hyperspectral imagery, synthetic aperture radar, etc (Fig. 4.1). In most cases, georeferencing these datasets is ultimately based on GNSS, such as the global positioning system (GPS) developed by the US Department of Defense, among others. The georeferencing accuracy of these datasets must be appropriate for the scale used to describe the forest variables (tree, plot or stand-level). This is a common issue for any ALS project, though forested environments add a number of particular challenges to data integration. Tree crowns obstruct the propagation of electromagnetic waves, thus critically affecting the accuracy and precision obtained by GNSS receivers situated under forest canopies. Moreover, the difficulty to obtain accurate DEMs in forested areas complicates the orthorectifying process of optical imagery. Errors in dataset co-registration introduce noise to the models, as predictors displace from their corresponding response. Their effect on estimated forest properties depends on the spatial heterogeneity of the variables considered, and the scale used. As all these complicating factors are relevant, they have to be taken into account at the planning stage, in order to ensure collection of data for forest inventory with an appropriate quality.

Incorporating field data is mandatory in most ALS forest assessment projects. ALS data are usually delivered with accurate coordinates, and no additional measures are therefore required prior to application relying solely on the ALS data, such as for example within hydrological or civil engineering applications, for instance to model flooding events. However, its use for forest inventory and

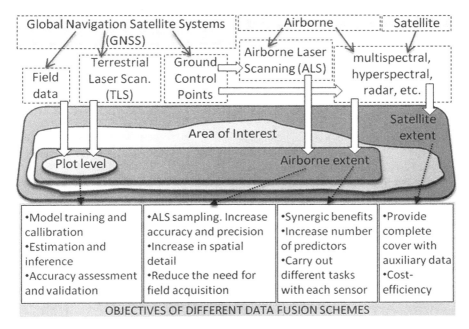

•Model training and calibration •Estimation and inference •Accuracy assessment and validation	•ALS sampling. Increase accuracy and precision •Increase in spatial detail •Reduce the need for field acquisition	•Synergic benefits •Increase number of predictors •Carry out different tasks with each sensor	•Provide complete cover with auxiliary data •Cost-efficiency

OBJECTIVES OF DIFFERENT DATA FUSION SCHEMES

Fig. 4.1 Levels of data combination involved in multisource ALS forest inventory projects. From *top* to *bottom*: platforms for data acquisition, sensors and types of information, spatial scale of each of them, and objectives achieved by data fusion

management, as well as most environmental and ecological applications, usually requires additional GNSS-surveyed datasets to be integrated in the same project. Field inventory data are certainly the most common data source to be included, as it may be mandatory for either training or validating the methods used. Due to differences among sensors and survey configurations, ALS predictive models generally need to be calibrated for each individual case (Næsset 2009). For this reason, surveying new GNSS-positioned field plots is required for most ALS-assisted forest inventory projects. GNSS errors in horizontal coordinates affect the positional co-registration between the ALS point cloud and the field data, lowering the accuracy of forest estimates (Gobakken and Næsset 2009; Mauro et al. 2011).

There may be a variety of reasons for including additional data sources in forestry applications of ALS. When diverse data sources are available, there are benefits gained from their synergies and complementary assets (Table 4.1). Approaches which use ALS estimations within forest stand boundaries obtained from aerial images are commonplace. Photointerpretation and manual delineation is the method most widely followed, though automated segmentation may be used as well (Leppänen et al. 2008). These boundaries can be used to divide the forest area into homogeneous strata, removing variability within individual strata for enhanced predictive modelling (e.g. Næsset 2004). Moreover, ALS echoes reach the ground and the understory, whereas optical imagery only collects information

Table 4.1 Summary of complementary properties of different sensors when fusing ALS with optical sensors onboard differing platforms: (a) airborne and (b) satellite

	Advantages	Disadvantages
(a) Airborne laser scanning vs. airborne optical sensors		
Airborne laser scanning	Direct data georeferring	Limited information on species diversity and forest health
	Provides information from under the dominant canopy	Requires prior stratification for better predictive modeling
Airborne optical sensors	Provides information on photosynthetic productivity	Need for perspective correction
		Unable to penetrate the canopy
(b) Airborne laser scanning vs. satellite sensors		
Airborne laser scanning	Accurate estimation of forest attributes	Tradeoffs between echo density and survey extent
	Detailed spatial information	High cost
Satellite optical sensors	Higher coverage at relatively low cost	Less precise estimates
		Sensor signal saturates at high biomass values
	Regular temporal recurrence and homogeneity in acquisition (global validity and baseline determination)	Lower spatial resolution

on the overstory (Baltsavias 1999). For this reason, tree height and crown shape properties can be obtained from ALS data, while species information may be better inferred from aerial photography (Persson et al. 2004; Leckie et al. 2005; Ørka et al. 2012). Multispectral and hyperspectral sensors can also be used to provide information about health conditions of the forests (Bright et al. 2012) and support water or chlorophyll content monitoring (Solberg et al. 2004). Adding predictors derived from different sensors may improve the explanatory power in the modelled relationships compared to using only one of them (Packalén and Maltamo 2006; Valbuena et al. 2013). However, ALS obtains positions with polar geometry while optical sensors are characterized by a perspective acquisition of incoming radiance, posing a challenge for data fusion.

The use of sensors onboard satellite platforms can add a number of advantages as well. Accurate field data can only be obtained for a limited number of sample plots, whereas remote sensors usually provide less detailed information which nevertheless covers larger areas at a lower cost. For this reason, large-scale forest assessment assisted by remote sensing must typically be based on multi-phase sampling schemes (e.g. McInerney et al. 2010; Andersen et al. 2011; Asner et al. 2013; Chap. 14). In remote sensing, sensor height determines a trade-off between total coverage and spatial and temporal resolution, so that advantages of airborne and satellite platforms complement each other (Table 4.1b). Satellite sensors acquire a large swath with high temporal recurrence, allowing sequential monitoring. Satellite multispectral sensors can provide an estimate of photosynthetic productivity that can be consistent globally, with the additional asset that carbon reference emission levels can be defined from imagery collections already acquired in the past (Cohen et al. 2010). However, they are seriously affected by the presence of cloud cover, and the estimations suffer from a saturation problem in highly stocked forest areas (Avitabile et al. 2012).

Three levels of data fusion may be involved in ALS projects (Fig. 4.1). Although the accuracy of elevation coordinates may be important for some applications, for most purposes the quality of the horizontal coordinates is the most relevant with regard to data fusion. First of all, the field data ought to be properly registered with respect to the ALS data by means of GNSS positioning. The performance of current GNSS receivers and operating methods has been evaluated in temperate and boreal forests (e.g. Næsset and Gjevestad 2008; Andersen et al. 2009; Valbuena et al. 2010), though the effect of complex tropical canopies on GNSS positioning is yet to be studied (d'Oliveira et al. 2012). The second level of data fusion is required when airborne optical sensors are to be included, either multispectral or hyperspectral, for example to determine the relative presence of different species (Packalén et al. 2009). When the aim is to assure that optical radiometric information corresponds with ALS returns reflected from a same feature on the ground, either of these alternatives may be followed: orthorectification, back-projection or image matching, in order of increasing accuracy of the horizontal coordinates. The last level of data combination is performed when satellite information is also included. To obtain accurate estimations in ALS projects, the errors of all data sources should be diminished. Should significant spatial mismatches be detected, methods for spatial adjustment among datasets may be considered (see methods for adjusting the field data in Sect. 4.3.3.2).

4.3 Incorporating Data Acquired in the Field into ALS-Based Forest Inventory Projects

Plot establishment, which is required for most forestry applications of ALS, relies on GNSS technologies for accurately determining the geographical position of the information acquired in the field. Forest measurements, acquired either at plot-level or for individual trees, is the most common type of field information to be integrated with ALS datasets, although field information may also originate from other remote sensing techniques, such as TLS or hemispherical photography (Lovell et al. 2003). GNSS receivers may also be positioned onboard harvesters or logging machines, obtaining the positions of selective cuttings (Holmgren et al. 2012) and skidroads (d'Oliveira et al. 2012). As a result, forest inventories based on ALS remote sensing are especially affected by issues concerning the accuracy of GNSS occupations obtained under dense canopies (Gobakken and Næsset 2009; Dorigo et al. 2010). The quality standards that must be attained in GNSS surveying will be determined by the scale and resolution of all the datasets integrated in an ALS-assisted forest inventory and management project. Plot size is an important factor, because small plots are more vulnerable to GNSS positional errors than large ones (Gobakken and Næsset 2009; Frazer et al. 2011; Mauro et al. 2011). In general, the level of accuracy which is acceptable depends on the actual application and characteristics of the data. For instance, when Tachiki et al. (2005) used GNSS for delineating stand boundaries, the same errors led to larger miscalculations in perimeter than area. Consequently, the effect of GNSS errors depends on the spatial variability of the forest property of interest, and the scale at which it is measured, estimated and validated, i.e., tree, plot or stand-level.

4.3.1 Basics in Global Navigation Satellite Systems (GNSS)

GNSS receivers observe and record electromagnetic signals from constellations of satellites: the American GPS, the Russian GLONASS, the European Galileo or the Chinese Compass/BeiDou (the latter ones will eventually be fully operational). The satellites broadcast their ephemerides, i.e. time and position in the sky, which are interpreted by the receiver to compute its distance to each of them. Horizontal coordinates can be computed with a minimum of three satellites by trilateration, whereas a fourth one is required for calculating the vertical coordinate. Observing a larger number of satellites increases the possibilities of obtaining a good geometrical position of satellites and therefore contributing to improved accuracy and precision (Habrich et al. 1999; Næsset et al. 2000). The wide range of receiver types available in the market may be roughly classified in order of increasing accuracy as recreational, mapping, survey and geodetic-grade. Most receivers currently observe epochs of the civilian code (coarse acquisition C/A), and the phase of the carrier frequency (so-called carrier phase). The distance to each satellite can be calculated from the code by pseudorange, i.e. the time difference between the receiver's clock and the moment when the satellite broadcasted the signal. Higher accuracy can be obtained when the receiver determines the carrier phase, however this requires that the initial phase ambiguity is solved (Næsset 1999).

Many error sources affect accurate range determination in GNSS surveys, and forest canopies add many obstacles such as complete blockage or attenuation of the signal and a strong multipath effect (Hasegawa and Yoshimura 2007). Multipath occurs when signals reflected from nearby objects lead to strong errors in distance measurement (Fig. 4.2). Many developments have provided solutions to all those issues. Dual-frequency receivers remove the ionospherical delay error, albeit with an elevated signal-to-noise ratio (Arslan and Demirel 2008). The choke-ring antenna (Jet Propulsion Laboratory, California) was designed to mitigate signals reflected from underneath, though signals reflected from above by tree crowns may still affect the accuracy of the position measurement (Valbuena et al. 2012).

Differential correction (DGNSS) is commonly applied for solving ephemeris and clock synchronism errors by observing the same signals simultaneously from a tandem of rover and base receivers, with the latter situated on a know position. Static observations can be acquired by a stand-alone receiver and corrected afterwards at post-processing stage. Alternatively, real-time kinematic (RTK) corrections can

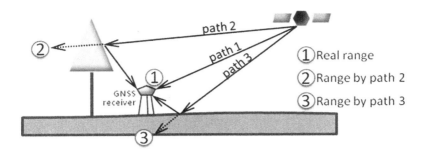

Fig. 4.2 The multipath effect. Signal reflections lead to uncertainty in range measurements

be applied while surveying with an established base receiver if continual radio contact with the rover is maintained. The baseline distance between rover and base affects the final accuracy, and therefore RTK provides best accuracy when operating under unobstructed conditions, i.e. out from the forest. However, establishing a dedicated base-station may be impractical for large-scale forest inventories. In such cases, post-processing differential corrections can be performed from independent local/regional/global networks of base stations (e.g. Andersen et al. 2009; Valbuena et al. 2010). Moreover, real-time corrections can also be obtained from satellite-based augmentation systems (SBAS) such as WAAS in north-America, EGNOS in Europe, SDCM in Russia, MSAS in Japan, or GAGAN in India. Other commercial SBAS services may also be considered, according to each study area and the accuracy requirements. SBAS are however affected by canopy blockage as well, as they also need to reach the GNSS receiver from the satellite. Roughly speaking, the accuracy obtained can improve from meters when measuring autonomously to just centimetres or even millimetres when applying these techniques for DGNSS (Næsset 2001; Næsset and Gjevestad 2008).

Three different types of DGNSS solutions may be obtained, depending on the quality and the time of observation. If continuous tracking is achieved, the differential correction procedure may be capable of solving the initial carrier phase ambiguity. This type of solution is called fixed-solution, and it achieves more accurate positioning than the other types of differential solutions: float and code-solutions. Repeated interruption of signal reception from satellites may prevent the receiver from solving phase ambiguities, since continuous tracking of the carrier phase is needed (Hasegawa and Yoshimura 2007). Thus, signal blockage caused by the forest canopy is a handicap for fixing DGNSS solutions. An approximation using both pseudorange and carrier phase, so-called float-solution, is then computed. When no approximation to phase ambiguities can be performed we obtain a code-solution computed from pseudorange only. The accuracy of the coordinates obtained by these different types of solutions computed by the differential correction may range from meters in code-solutions to centimetres in fixed-solutions (Næsset and Jonmeister 2002). As recording more epochs increases sample size, float-solutions are more reliable for long recording periods, reaching accuracies comparable to fixed-solutions (Næsset 2001). Whether a receiver situated under the canopy is capable of consistently obtaining fixed solutions is however a major issue in DGNSS surveying for ALS applications in forestry (Valbuena et al. 2010).

The roving receiver is usually set at a certain height above ground (usually around two meters), in order to widen the observed horizon for its antenna and also to avoid multipath. In rough terrain, signals from the ground can be further masked by applying a cut-off angle. Moreover, static GNSS observations result in multiple epochs which are recorded during the observing time at a rate specified by the operator. The final GNSS occupation is usually computed by the receiver's software by means of epoch averaging, also involving algorithms for differential correction and outlier filtering. The choice of an optimal antenna height, logging rate and cut-off angle depends on which conditions, in each case, present the highest limit to GNSS surveying's reliability. If field conditions impede observing the minimum number of satellites required, the logging rate can be set to record at lower frequencies, allowing a larger time window to observe satellites. On the other hand, when operating in floating mode, a higher frequency may be preferred to assure a large sample size of epochs which may increase the precision (Næsset 2001). Elevation angle masks may be chosen in the presence of rocky terrain, or they may be avoided when mountains blocking the horizon are a limiting condition (Valbuena et al. 2010). The antenna is commonly elevated higher above the ground as it may significantly increase the number of observed satellites at high latitudes (Arslan and Demirel 2008), or avoid the obstruction of vegetation, e.g. in the presence of dense understory. On the other hand, Valbuena et al. (2012) suggested that distancing the receiver from the canopy bulk may be advisable in a self-pruning forest or wet conditions, advising to lower the antenna heights in such cases.

4.3.1.1 Determining Accuracy and Precision

Absolute accuracy of a surveyed position's coordinates $p = (x, y, z)$ can only be known by comparing them against their reference ones $p_{ref} = (x_{ref}, y_{ref}, z_{ref})$, obtained with a more accurate methodology. In the case of GNSS, a closed-traverse total station survey is the common procedure for quality control. Bias is estimated by calculating absolute errors (e_p) at each coordinate direction:

$$e_p = \sqrt{\left(p - p_{ref}\right)^2}. \tag{4.1}$$

Nonetheless, determining the accuracy of a large number of GNSS occupations is usually impractical and expensive. Alternatively, a figure of precision (σ_p) may be provided by the receiver according to the GNSS epochs' dispersion around their average (each log $t = 1, \ldots, m$ recorded along the observation time is an epoch):

$$\sigma_p = \sqrt{\sum_{t=1}^{m} (p_t - \overline{p})^2 / (m - 1)}. \tag{4.2}$$

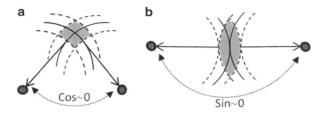

Fig. 4.3 Anisotropic accuracy in trilateration and positional dilution of precision. (**a**) In trilateration, the uncertainty in range measurements (*dashed lines*) leads to an overall precision in the final positioning (*shaded area*). (**b**) This area depends on the relative positions of satellites (*dotted arrows*), being larger for trilateration angles of lower sine

Another common way of describing GNSS precision is given by the dimensionless parameter positional dilution of precision (PDOP). PDOP is a measure of the geometry of available satellites' relative positions, and decreases when conditions are favourable:

$$\text{PDOP} = \sqrt{\sigma_x^2 + \sigma_y^2 + \sigma_z^2}. \tag{4.3}$$

where σ_x^2, σ_y^2 and σ_z^2 depend on the cofactor matrix resulted from trilateration, and therefore change according to the position of satellites. As a result, an error in ranging propagates to a larger uncertainty in final coordinates when the satellites are situated in a near-straight line with the GNSS receiver (Fig. 4.3).

Many studies have been carried out to evaluate the performance of GNSS receivers under forest canopies. Practitioners who need to evaluate a subcontracted GNSS survey may wish to hone their ability to distinguish when accuracy and precision are computed against an independent ground-truth (e.g. Liu and Brantigan 1995; Sigrist et al. 1999; Næsset and Jonmeister 2002; Andersen et al. 2009; Valbuena et al. 2012), or merely against the mean position of epochs recorded (e.g. Tachiki et al. 2005; Zengin and Yeşil 2006). GNSS positioning is usually prone to bias when operating under forest conditions, since multipath necessarily leads to overestimating the range measurement (Fig. 4.2). Therefore, an independent set of highly accurate reference points is needed to determine absolute accuracy as distances between GNSS-occupied coordinates and their reference positions (Eq. 4.1). In contrast, the precision computed by the receiver for that same GNSS occupation is the standard deviation of the epochs recorded around their mean (Eq. 4.2) (Hasegawa and Yoshimura 2007; Valbuena et al. 2010). To properly reflect on the meaning of a given precision measurement, it should be remembered that twice that distance from their average is needed for enclosing 95 % of the epochs observed. If many positions $i = 1, \ldots, n$ are occupied, the proper assessment of their overall performance should be done by means of their root mean squared error

(RMSE) (Sigrist et al. 1999), which takes into account their bias and variability in comprehensible units of measure. RMSE allows comparing groups surveyed under different conditions (e.g. Næsset 2001):

$$RMSE_p = \sqrt{\left(\sum_{i=1}^{n} e_{p_i}/n\right)^2 + \sum_{i=1}^{n}\left(e_{p_i} - \overline{e_{p_i}}\right)^2 / (n-1)}. \qquad (4.4)$$

Andersen et al. (2009) and Valbuena et al. (2010) showed that precision and PDOP values provided by the receivers may disagree with real accuracies, especially at short observing times. Therefore, obtaining the reference positions is compulsory for truly determining the accuracy of GNSS occupations. Control datasets are usually generated by means of ground topographic traverse surveying. In forested environments, traverses allow to link terrain positions under tree crowns with reference DGNSS occupations obtained in the absence of canopy's influence. A traverse consists of determining the positions of visually-connected points by the distances between them and their bearing angles, usually with a total station. Measuring backsight and foresight at each point determines the angle between consecutive segments, and their corresponding distances. Best conditions are obtained when a polygonal traverse can be closed and all surveyed points are determined by least-squares adjustment of these redundant measures (Wolf and Brinker 1994). This allows for internal assessment of error propagation, resulting in a traverse survey with an accuracy superior to that of under-canopy DGNSS surveying.

4.3.2 GNSS for Positioning Plot-Level Information

The effects of GNSS errors on ALS-assisted forest inventory estimates depend mainly on the response variable and ALS metrics selected for the models, that tie the ALS metrics to the biophysical property observed on a plot (Gobakken and Næsset 2009), the choice of GNSS receiver and plot size (Mauro et al. 2011), and the spatial heterogeneity of canopy structure at the study area (Frazer et al. 2011). A number of authors have tested the performance of several types of GNSS receivers and positioning procedures, and the effect of diverse canopy types (e.g. Liu and Brantigan 1995; Deckert and Bolstad 1996; Sigrist et al. 1999; Næsset 1999, 2001). Provided that there are sufficiently long observation periods, the accuracies obtained generally tend to converge among systems and conditions (Næsset et al. 2000; Valbuena et al. 2010). The most practical approach is therefore to record GNSS epochs at plot centre while forest mensuration is carried out, allowing enough time for the receiver to log several hundred epochs. Næsset and Gjevestad (2008) suggested that differential correction must be a mandatory practice for the observation periods which would be realistic in field data collection for ALS forest inventory. Obtaining accurate coordinates matters more if small plots are located within spatially heterogeneous forest areas (Frazer et al. 2011; Mauro et al. 2011). For this reason, Gobakken and Næsset (2009) found the ALS estimates to be more affected by GNSS positioning errors in mature forests with relatively fewer stems than young planted stands with more even tree spacing. It has been suggested that recording times may be optimized at each plot according to dominant height, basal area (Næsset and Jonmeister 2002), canopy cover, volume, stand density or leaf area index (Valbuena et al. 2012). Overall, an occupation time of 10 min should be sufficient in practice for positioning 9–15 m-radius circular plots with current survey-grade GNSS receivers. This works as a general rule of thumb in fairly uniform boreal and temperate forest conditions, and for the actual accuracy required for ALS area-based applications. In dense tropical forests accurate positioning may be more challenging, and the effect of GNSS errors may be diminished by using larger plot sizes (e.g. Asner et al. 2013). The use of hand-held receivers is discouraged for this duty.

In clear sky conditions, frequent logging rates decrease the uncertainty of the positions obtained, since a larger number of observations is obtained (Eq. 4.2). However, this practice may be counterproductive when the GNSS receiver is set under dense forest cover, as numerous obstacles increase the probability for the signal to be interrupted (Hasegawa and Yoshimura 2007). It is an advised practice to lower the logging rate, especially when detecting that the receiver hardly fixes any solution. This would allow a longer window of time for the receiver to fix the initial phase ambiguity. Næsset (2001) found that float dual-frequency solutions may achieve lower accuracies than fixed single-frequency ones. Even with a large number of epochs recorded, float solutions will be worse than fixed ones in the presence of bias. The continuous tracking of satellites is essential for fixing the phase ambiguity, and most receivers can deduce satellite trajectory when the observation is temporarily interrupted, being more likely to detect the signal back afterwards. The receiver should therefore be kept functioning during the whole campaign, even when no epochs are being recorded at a forest plot, unless battery availability is a limitation. The operator may well allow the receiver to find satellites at a gap in the canopy at the beginning of the survey, and also at several intervals during a working day if eventually detecting a low availability of satellites.

In mountainous areas, the obstruction of the horizon significantly limits the number of satellites observed (Deckert and Bolstad 1996). In DGNSS, there is not much practical use in choosing a base station situated at the other side of a ridge, if that prevents rover and base receivers from simultaneously observing the same satellites. The effect of the surrounding topography can be an important determinant in the choice of stations for differential correction. Only one base receiver situated in the main aspect direction of the foreslope may obtain higher accuracies than a network of stations observing satellites in the backslope, out of the rover's horizon (Valbuena et al. 2010). This criterion prevails over the baseline distance, unless exceeding 50–100 km (Andersen et al. 2009). Moreover, receivers incorporating a cut-off mask and a choke-ring antenna have been found to efficiently mitigate the effect of steep terrain. Hence, antenna heights no longer limit the accuracy obtained critically. However, other benefits can be obtained from raising the antenna up to 4 m from the ground, as increasing the observing horizon may be crucial in study areas in mountainous terrain or high latitudes (Arslan and Demirel 2008).

In such conditions, when the number of satellites observed is a limiting factor, the best setting is to raise the antenna height and apply no cut-off. Nevertheless, as indicated above, approximating the GNSS receiver to the crown bulk of the canopy may compromise the accuracy obtained, especially in wet conditions (Sigrist et al. 1999; Valbuena et al. 2012).

External validation of GNSS positioning by traverse surveying can be reserved for only a sample of all GNSS occupations, as long as the whole range of forest conditions occurring at the study area are present. On the other hand, evaluating the quality of forest plot establishment by means of descriptors obtained by GNSS receivers during the occupation may be sufficient for most area-based inventory applications of ALS. The best ones are, in order of reliability (c.f. Næsset and Jonmeister 2002): the type of solution obtained (fixed, float or code), the precision around the mean provided by the receiver, and the PDOP. However, some of these are receiver-dependent (Valbuena et al. 2010), and it is difficult to know exactly the built-in algorithms included in the instruments, for instance outlier filtering. These descriptors are therefore only to be compared when obtained by the same receiver.

4.3.3 Subplot-Level Positioning of Individual Tree Data

Incorporating tree-wise information into an ALS project is a demanding task, mostly in terms of field acquisition costs. It is therefore important to consider beforehand whether comparing against individual tree positions is really required, as the reliability of individual tree detection approaches for forest parameter estimation from ALS can in many cases be adequately evaluated at the plot or stand-level. The decision depends mainly on whether the targeted forest property requires spatially explicit information, for example in studies on tree competition or spatial pattern distributions. Although most forest management planning and forecasting systems have traditionally been based on stand-level inventory, there are increasingly more methods based on tree lists or species-wise diameter distributions.

A number of techniques can be used to obtain individual tree positions in the field (see e.g. Vauhkonen et al. 2012). Frequently, individual trees are positioned relatively to reference landmarks occupied by GNSS or within a traverse survey. Depending on the survey equipment and the amount of resources available for field data acquisition, the positioning of individual trees may be grounded on each of the three basic positioning methods: ranging, triangulation or trilateration (Fig. 4.4). In the most common case, (a) bearings and distances may be determined using a compass and a measuring tape, or a total station, which usually requires an additional operator holding the prism (see e.g. Valbuena et al. 2012). The reference landmark may also be the position where to set a TLS instrument, using its resulting point cloud as a source to extract the positions of individual trees (Henning and Radtke 2006). (b) Field photogrammetry applies triangulation of stereo images to determine individual tree positions (Forsman et al. 2012). Alternatively, (c) using direct ultrasonic transponder trilateration may allow carrying out tree position surveying by a single operator alone (Lämås 2010; Holmgren et al. 2012).

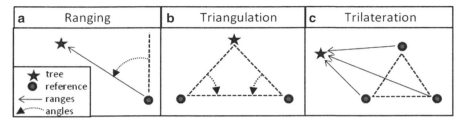

Fig. 4.4 Methods for determining spatial positions. The position of an individual tree is determined relative to a reference, using a different method depending on what can be measured in each case. (**a**) If a distance and a bearing angle are available, the position can be determined by simple trigonometry. (**b**) If angles can be known but not ranges, triangulation can be used from two reference observations. (**c**) If distances can be determined but not angles, three range observations are needed for trilateration. In any of the cases, the absolute (or relative) positions of reference landmarks must be known (*dashed lines*). Any positioning method falls within either of these categories: as ranging is used by ALS, TLS and total stations, triangulation by photogrammetry and theodolites, and trilateration by GNSS

4.3.3.1 Fusion of Terrestrial Laser Scanning with ALS

Combining ALS with TLS may be based on measures at plot-level (Lovell et al. 2003; Hilker et al. 2010). TLS surveys have the capacity for automatically determining accurate positions of individual stems, obtaining accuracies of few centimetres (Henning and Radtke 2006; Maas et al. 2008). For this reason, the combination of ALS and TLS can be carried out at the level of individual trees as well. In order to avoid trees shadowing each other, a whole plot can be covered from many reference landmarks (Hilker et al. 2012). Lindberg et al. (2012) extracted individual stem positions and attributes from TLS and used them for training the ALS data. They found that TLS can be a reliable alternative to traditional field collection in ALS-based forest assessment, and a step toward harmonizing procedures and reducing human subjectivity. The weight of TLS sensors must nevertheless be born in mind, as well as the fact that dense forest conditions may require some manual clearing in the areas surrounding the scanner, as excessive noise challenges most stem detection algorithms.

4.3.3.2 Methods for Adjusting Field Information to ALS Data

The accuracy of relative stem positions offered by TLS or total station surveying is superior to the accuracy offered by any alternative in GNSS positioning under forest canopies. Moreover, due to slanted trunks, tree positions may differ significantly depending on whether they are determined on the ground or by detecting tree tops from above via remote sensing. A systematic method should be used to determine which stem positions represent the same tree in both the ALS and the field data.

Korpela et al. (2007) proposed a practical two-phase method based on both remote sensing and field survey. The positions of tree tops in the dominant canopy can be determined from photogrammetry or ALS in the first phase. The second phase assumes those tree positions as reference and the rest of the trees are positioned relative to them in the field, assuring redundancy for trilateration and least-squares adjustment. The outcome is a reasonably good match between the two datasets achieved directly. As an alternative to adjusting the positions of one dataset against another, individual trees can be linked virtually, based on their relative positions and taking into account e.g. the error of each positioning method (Fritz et al. 2011), or their estimated maximum crown width (Persson et al. 2004).

Furthermore, it may also be the case that mismatches between two datasets are due to any of the other causes listed throughout this chapter: GNSS errors, orthorectification, coordinate transformations, etc. If individual tree positions are available, there are still good chances to manually adjust the data to match the positioning of the ALS point cloud or canopy height model. Field plot coordinates can be rotated and translated according to several criteria such as tree species and height (Dorigo et al. 2010). One solution may also be obtained by minimizing height-weighted residual distances or maximizing likelihood with crown width-based gaussian kernels (Olofsson et al. 2008).

4.4   Fusion of ALS and Optical Sensors

A number of applications which can be developed by using a combination of ALS with optical sensors are detailed in several chapters of this book. In general, using both ALS and optical sensors improves the results obtained with any of them alone (Holmgren et al. 2008; Packalén et al. 2009; McInerney et al. 2010; Valbuena et al. 2013). However, this kind of approach faces important challenges in relation to achieving a correct spatial adjustment of the information derived from diverse sensors (Honkavaara et al. 2006). The precision of the data recorded by remote sensors has to be accompanied by their accurate georeferencing, otherwise the potential synergies among sensors are lost. It is not uncommon to encounter problems of data mismatching (see Valbuena et al. 2011; Bright et al. 2012), and it is therefore advised to reflect on the possible sources of georeferencing errors that may affect the datasets involved in forest applications of ALS.

In case of using aerial or satellite imagery in an ALS-based forest inventory project, the user should understand the characteristics of the optical sensor and the post-processing tasks that have been carried out to obtain the final product. The objective must be to acknowledge the geometric and radiometric properties of the images, which will determine the forest characteristics that can (or cannot) be inferred from them. Aerial photographs are less affected by atmospheric effects than satellite imagery, but are more affected by anisotropy of forest canopy reflectance (Tuominen and Pekkarinen 2004). Questions on radiometric properties of optical sensors are well covered elsewhere, and hereby only the processes that affect the geometric properties of optical products will be described. Practitioners should check for two image properties: (1) the spatial resolution, which defines the minimum size of objects to be identifiable; and (2) the accuracy obtained in georeferencing the position of those objects. These are always important in any remote sensing application, but the high positional accuracy obtained by the ALS sensor makes them critical. For this reason, when an orthoimage is to be used along with ALS data, even if only for manual delineation of forest stands, the accuracy of its georeferencing should be well documented.

4.4.1 Perspective Acquisition in Optical Sensors

Optical sensors acquire information in a perspective projection and thus tree crowns are observed from above at nadir, but sideways elsewhere. Moreover, because optical sensors are passive instruments, the acquired data depend on the bidirectional reflectance of sunlight on objects, and thus a same tree shows differing spectral characteristics in different pictures, according to its relative position within each one (Tuominen and Pekkarinen 2004). The radiation acquired arrives at the sensor at different nadir angles (θ), i.e. the angle between the projective ray which carried the radiometric information and the nadir position. Due to perspective, equal segments within the photo represent different distances on the ground, depending on whether they are at nadir or close to the picture's border. For these reasons, in order to use optical imagery along with ALS data, the sensor perspective has to be corrected so that elements in the picture are located with a map-like orthogonal projection. Pixel size in an image is defined by the size of each individual charge-coupled device (CCD) sensor and the altitude of the platform. Hence, higher spatial resolutions are offered when a sensor is operated at lower altitude, but anisotropy and within-picture scale differences due to high variability in nadir angles becomes larger. Satellite imagery is therefore less affected by perspective and it is usually purchased as an already orthorectified product, whose positional errors have been assured by the vendor to be lower than pixel size. The end-user may consequently assume the position of all pixels as correctly georeferenced in satellite imagery, and they usually need to pay more attention to their radiometric characteristics (e.g. atmospheric effects). This chapter will therefore be more focused on the problems arising when aerial images are to be integrated in an ALS project for forest inventory, as the quality of their geometrical properties must be ensured.

The model describing the perspective acquisition of optical sensors is grounded on the condition of collinearity, for which an element situated on the terrain and its position in the picture are in line with the centre of projection (Wolf 1983). Accordingly, relations between segments defining distances to the centre of projection must be proportional at sensor and terrain scales, which are provided by the interior and exterior orientation (IO and EO), respectively. Two IO horizontal distances are obtained as the distances from photo-coordinates of a pixel (X_{ph}, Y_{ph}) to the picture's centre or principal point (X_{pp}, Y_{pp}), while the vertical distances to the centre of projection is the focal length[1] of the lens (f). These are properly obtained only after the camera is calibrated, so that the imagery is usually supplied as an already distortion-free product (otherwise, the distortion model and principal point displacement ought to be determined for the same focal length used in the survey). The EO provides these same relations at terrain scale, as distances from the GNSS-determined position of the onboard sensor at the time of exposure (X_0, Y_0, Z_0) to the position of the target element (X_t, Y_t, Z_t), e.g. a tree top or an element on the ground. Absolute coordinates for EO should be expressed in a common reference system, whereas the size of the CCD sensor is needed to calculate the IO photo coordinates. Regardless of the units and reference system used, the ratios between each horizontal distance and the vertical one must be equal at camera and terrain scales, respectively, at left (IO) and right (EO) sides of the collinearity equations (Wolf 1983; Schenk 1999):

$$\frac{X_{pp} - X_{ph}}{f} = \frac{m_{11}(X_t - X_0) + m_{12}(Y_t - Y_0) + m_{13}(Z_t - H_0)}{m_{31}(X_t - X_0) + m_{32}(Y_t - Y_0) + m_{33}(Z_t - H_0)}, \qquad (4.5)$$

$$\frac{Y_{pp} - Y_{ph}}{f} = \frac{m_{21}(X_t - X_0) + m_{22}(Y_t - Y_0) + m_{23}(Z_t - H_0)}{m_{31}(X_t - X_0) + m_{32}(Y_t - Y_0) + m_{33}(Z_t - H_0)}. \qquad (4.6)$$

As the position of the sensor does not coincide with the coordinate system used, its EO attitude angles must be used for rotating the segments by means of a rotation matrix $\mathbf{M}_{3\times3} = [m_{ij}]$. The roll, pitch and yaw of sensor's platform, which are provided by the inertial navigation system in aerial photography, are respectively the rotation angles around along-track, across-track and vertical axes (ω, φ, κ). If all these angles are determined to be clock-wise positive, the rotation matrix can be constructed as (Wolf 1983; Schenk 1999):

$\mathbf{M} =$

$$\begin{bmatrix} \cos\varphi \cdot \cos\kappa & \cos\omega \cdot \sin\kappa + \sin\omega \cdot \sin\varphi \cdot \cos\kappa & \sin\omega \cdot \sin\kappa - \cos\omega \cdot \sin\varphi \cdot \cos\kappa \\ -\cos\varphi \cdot \sin\kappa & \cos\omega \cdot \cos\kappa - \sin\omega \cdot \sin\varphi \cdot \sin\kappa & \sin\omega \cdot \cos\kappa + \cos\omega \cdot \sin\varphi \cdot \sin\kappa \\ \sin\varphi & -\sin\omega \cdot \cos\varphi & \cos\omega \cdot \cos\varphi \end{bmatrix}$$

$$(4.7)$$

The collinearity equations are the general model, with variations, for correcting the perspective acquisition of optical sensors. They can be used for multiple applications applying calculations at either direction, i.e. not only to calculate terrain locations of elements in the picture but also for rendering positions on the ground from the camera perspective, a practice called synthetic rendering (Forkuo and

---

[1]More precisely denominated camera constant as most cameras consist of a system of compound lenses, though the collinearity model simplifies it to a pin-point exposure of a single lens with equivalent focal distance.

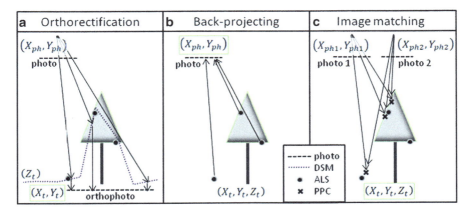

Fig. 4.5 Schematic representation of alternatives for positioning information from optical sensors: (**a**) orthorectification, (**b**) back-projecting ALS and (**c**) image matching. *Arrows* are projective rays defined by Eqs. 4.5 and 4.6. The key input parameters for the equations, and the unknowns (enclosed in a *grey rectangle*) that apply in each case are shown

King 2005). For this reason, Eqs. 4.5 and 4.6 form the basis for all the techniques which allow for the information of optical sensors to be integrated within an ALS project (Fig. 4.5), i.e., imagery orthorectification, ALS back-projection, or photogrammetric point clouds (PPCs) generated by image matching. They can also be used for other applications related to data integration, as for instance St-Onge (2008) and Armenakis et al. (2010) used synthetic rendering as a quality control test for evaluating and correcting systematic errors in the relative GNSS-based georeferencing of ALS datasets and imagery.

By using orthophotos, an ALS point cloud may be coloured by attaching to each echo the digital number (DN) of the image pixel located vertically below the ALS echo according to its horizontal coordinates. It is common practice to use DTM-rectified photos for this purpose, therefore carrying the ensuing positioning errors for those DNs (Valbuena et al. 2008). The product of back-projecting is also a coloured ALS point cloud, but in that case the positional accuracy of the DNs is at the scale of the pixel size (Valbuena et al. 2011). To date, the positional errors of DNs located in image matching PPCs has not been contrasted against the other methods, but the author advices applying this method for data fusion. Using several scenes, redundancy in Eqs. 4.5 and 4.6 can be solved by least-squares adjustment, leading to a sub-pixel accuracy in horizontal coordinates (X_t, Y_t) which is superior to any other method. On the other hand, elevation coordinates (Z_t) are still better determined by ALS direct range measurement, while its uncertainty in PPCs depends on the angle used for triangulation (Fig. 4.6).

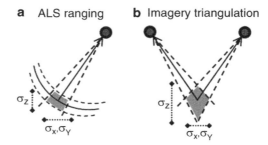

Fig. 4.6 Anisotropic accuracy in positioning. (**a**) The uncertainty in range and scan angle measurements leads to higher accuracy in vertical than horizontal coordinates in ALS (Baltsavias 1999). (**b**) In image matching most uncertainty is in depth, and therefore in PPCs the accuracy of elevations is lower than horizontal coordinates

4.4.2   ALS-Assisted Orthorectification

The process of orthorectification consists of computing the absolute coordinates (X_t, Y_t) of each pixel within an image by applying its IO and EO in Eqs. 4.5 and 4.6, and assuming its elevation (Z_t) according to a DEM. Thus, the quality of the final positioning of features depicted in the image largely depends on the type of DEM used – digital terrain model (DTM) or digital surface model (DSM) –, and the procedure used for generating it. While a DTM represents the ground surface, a DSM details other elements, such as tree crowns and canopy gaps. Traditionally, DTMs have been used for orthophoto production (Baltsavias and Käser 1998). They were produced by manual photogrammetric measurement, i.e. deducing Z_t by triangulating the positions of the same element in overlapping stereo images (following the same principle as in Fig. 4.4b). Alternatively, automated image matching techniques can also be used for mass point detection (e.g. Zhang and Gruen 2004), allowing to obtain DSMs whose quality differs among methods (St-Onge and Achaichia 2001; Waser et al. 2008). As optical cameras are unable to observe positions from underneath the vegetation, the capacity of ALS to obtain ground echoes offered great advancements for DTM generation in forested areas (Baltsavias 1999).

4.4.2.1   Precise DTM Generation

Three processing steps are involved in DEM generation from an ALS point cloud (c.f. Axelsson 1999). First, echoes are *classified*, according to their relative geometry or type, into ground, vegetation, buildings, or other features that reflected them. Second, the echoes from a certain class may be *filtered*, as for example only those classified as ground are used for DTM generation. The last step is the DEM-*modelling* itself, which may involve several tasks for interpolation, such as tessellation or rasterization. The optimal filter for ground classification varies among vegetation types and terrain roughness conditions (Sithole and Vosselman 2004), as well as characteristics of the ALS acquisition such as pulse density (Liu et al. 2007). The classification can be done according to whether they are first/last echoes, or simply as ground/vegetation based on the relative geometry (Axelsson 2000). The quality of this classification of ALS echoes into either ground or vegetation is the factor most affecting DTM and DSM generation (Hollaus et al. 2006). Most procedures have shown the capacity of ALS surveys to generate much more accurate DTMs than photogrammetry (Kraus and Pfeifer 1998). Reported RMSEs of elevation coordinates obtained from ALS-derived DTMs are generally within the range of 0.15–0.35 m in forested areas. Its accuracy concerns the reliability for forest inventory directly, as tree heights are commonly considered above the DTM. In practice, the DTM is used as reference, either to compute heights above ground of individual ALS echoes, or to obtain a raster-type canopy height model by subtraction of DSM minus DTM. It must also be taken into account that DTM quality is affected by the density of the forest canopy (Reutebuch et al. 2003), and therefore the accuracy is generally lower in deciduous than coniferous forests (Hodgson and Bresnahan 2004). The success of classification and interpolation is most challenged when vegetation grows on rough terrain (Axelsson 2000), and therefore elevation errors are also higher in steeper slopes (Hodgson and Bresnahan 2004). For these reasons, correct estimation of tree heights in the presence of dense understory and mountainous relief may still be subject to further methodological refinements (Gatziolis et al. 2010).

Due to the use of DTMs for orthophoto generation, problems have been found when integrating them with ALS data in forestry applications, as sometimes the position of tree crowns is different in each dataset (Valbuena et al. 2008; Bright et al. 2012). As the presence of trees is not modelled in the DTM, the Z_t position of tree tops is presumed lower than it actually is, and they appear leaning over canopy gaps in the final orthoproduct. The magnitude of this positional error mainly depends on the nadir angle θ, while tree height and terrain slope are also significant factors (Valbuena et al. 2011). Consequently, elements at the border of the picture are more affected than those close to the nadir point (X_{pp}, Y_{pp}), and the error is larger for tall trees growing in mountainous terrain.

4.4.2.2 True-Orthorectification

A so-called true-orthophoto can be generated by applying the height from a DSM as Z_t, rather than using the DTM as in traditional orthorectification, presumably obtaining a better correction of the objects' positions. Procedures for true-orthophoto generation aim at solving the described geometrical errors encountered in orthophoto production (Schickler and Thorpe 1998). An ideally perfect DSM which truly models the presence of trees could be used to solve the deficiencies found in orthophotos, therefore providing a smaller geolocation error between optical and ALS data. Valbuena et al. (2011) found errors in their true-orthophotos to be mainly related to the uncertainty of the DSM used. Hence, the final quality of a true-orthophoto is completely dependent on whether the technique used for modelling tree canopy truly represents the trees depicted in the picture.

Smoothing techniques may be required pre-processing steps when a DSM is intended to be used as a basis for orthorectification. Image quality can be degraded significantly if the DSM contains too much detail. For this reason, St-Onge (2008) applied a cavity filling procedure to a high-resolution DSM. The inclusion of breaklines, i.e. sudden changes in the trend between neighbouring pixels, which is common when modelling buildings in urban areas (Schickler and Thorpe 1998), is impractical for forested areas. Moreover, true-orthorectification usually includes a visibility analysis, e.g. a depth-buffer algorithm, in search for blind spots occurring behind elements significantly taller than wide. This prevents imputing the DNs of a given image to areas where no information was actually sensed from the camera's perspective at the moment of exposure. In theory, true-orthorectification procedures must suffice to achieve a correct reposition of tree crowns in their real coordinates, but practice is however different. Techniques which are well implemented in urban areas can be difficult to adapt to forest environments, and therefore studies involving forestry applications are scarce (Küchler et al. 2004; Waser et al. 2008; Valbuena et al. 2008).

Besides of its geometric quality, it is also worth checking the radiometric properties of a final orthorectified product, as they may also be affected during a number of procedures involved in orthophoto creation. The final product is usually supplied as a mosaic created from many individual exposures. Smoothing techniques for seamless mosaics are common at this stage, as well as methods for radiometric normalization (Baltsavias and Käser 1998). To ensure the final orthoproduct to be a single raster image with internally-consistent geometry, a final resampling stage is usually carried out. Resampling can also induce some additional geometric displacement, besides of modifying the original DNs in the final orthophoto. When a visibility analysis takes place, checking whether occluded areas have been filled with real DNs from other pictures, or synthetic DNs have been derived from neighbouring pictures (e.g. St-Onge 2008) during the mosaicking and resampling stages would be beneficial.

4.4.3   Back-Projecting ALS

When integrating the optical data with the ALS, as an alternative to using an orthorectified product, individual ALS echoes can be rendered from the perspective of the original unrectified scenes (Elmstrom et al. 1998). This technique is called back-projected ALS, as the original DN corresponding to each echo is retrieved back to its original position in the ALS cloud for its further use as a map-like orthogonally-projected product. This way, the colour information from the optical sensor is attached to each echo (Fig. 4.7), and therefore becomes available for its direct use in forestry applications. As the back-projected ALS is a point cloud in

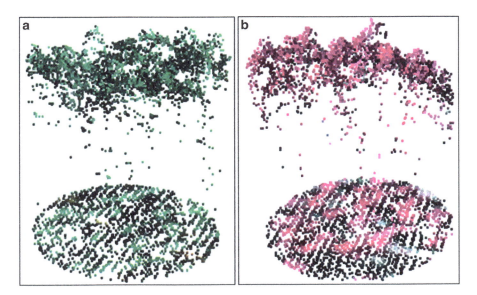

Fig. 4.7 View in 3D perspective of two forest plots' ALS back-projected onto pansharpened (**a**) true colour rgb-composite and (**b**) false colour infrared images. Their detail illustrates how the resulting accuracy may allow distinguishing areas of differing bidirectional reflection within individual tree crowns

vector format, mosaicking and resampling procedures are unnecessary, avoiding errors and artefacts. For this reason, back-projecting has been found superior to orthorectification for integrating the information from optical data and ALS datasets (Valbuena et al. 2011). Persson et al. (2004) and Holmgren et al. (2008) implemented this technique pursuing the fusion of ALS-delineated crowns and aerial photography for species classification. Packalén et al. (2009) successfully improved prediction models by including species mixture information deduced from the optical DNs of back-projected point clouds. Ørka et al. (2012) applied this technique for individual tree species classification using combinations of ALS metrics with intensity and multispectral predictors.

Although mismatches between optical and ALS information may still be found in back-projected products, they are usually small in magnitude. Valbuena et al. (2011) considered the errors induced by atmospheric refraction and Earth's curvature. Atmospheric refraction has a minimal effect in aerial surveying, although ALS systems include built-in corrections that could be implemented while back-projecting as well. The effect of the Earth's curvature implied that back-projecting ALS is best carried out prior to obtaining projected coordinates. Packalén et al. (2009) outlined a number of additional sources for mismatch in back-projected ALS. Tree crowns may be moved by the wind and therefore be detected at different positions by each sensor. Moreover, an ALS echo situated on a blind spot behind a tree, in an area occluded from the perspective at the camera's position, would erroneously by associated with a DN which does not correspond to the element which reflected it. For this reason, echoes situated under the canopy lack a counterpart in the optical imagery. This effect can be diminished by filtering the point cloud as a first step, so that only single and first echoes are selected for their back-projection. Also, the occurrence of blind spots can be reduced by back-projecting each ALS echo only to the picture from which the nadir angle θ is narrowest.

4.5 Concluding Remarks

Using other sources of information along with ALS can be highly beneficial for forest assessment. However, coordinate errors and ensuing mismatching among data sources are fairly common issues to deal with in forestry applications of ALS. For this reason, good planning in advance is recommended as most trouble may be avoided during data acquisition, or by requesting specific data from the provider. This information includes methods and networks employed for differential correction in GNSS positioning, or the EO obtained in photogrammetric surveys, for instance. The positional accuracy of field data should be in accordance with the objectives pursued in each case. Canopy conditions affect GNSS errors and the subsequent inaccuracy in ALS forest estimations. Thus, GNSS surveying should be tailored to the specific properties of the forest area under study. The positional errors found in orthophotos are difficult to track without knowing the platform's position during acquisition. If such information is available, it is recommended to consider alternatives for data fusion: back-projecting ALS or image matching. The latter provides higher accuracy but requires greater overlap between images.

Acknowledgments The author thanks José Antonio Manzanera and Susana Martín (Technical University of Madrid), Petteri Packalén (University of Eastern Finland) and the editors for their revision and useful comments, and Niina Valbuena (European Forest Institute) for language revision. Rubén Valbuena's work is funded by Metsähallitus (Finnish Forest Service) Grant awarded by the Foundation for European Forest Research (FEFR).

References

Andersen H, Clarkin T, Winterberger K, Strunk J (2009) An accuracy assessment of positions obtained using survey- and recreational-grade global positioning system receivers across a range of forest conditions within the Tanana valley of interior Alaska. West J Appl For 24:128–136

Andersen H, Strunk J, Temesgen H, Atwood D, Winterberger K (2011) Using multilevel remote sensing and ground data to estimate forest biomass resources in remote regions: a case study in the boreal forests of interior Alaska. Can J Remote Sens 37:596–611

Armenakis C, Gao Y, Sohn G (2010) Co-registration of lidar and photogrammetric data for updating building databases. ISPRS Arch 36, Part 4–8–2–W9:96–100

Arslan N, Demirel H (2008) The impact of temporal ionospheric gradients in Northern Europe on relative GPS positioning. J Atmos Sol-Terr Phys 70:1382–1400

Asner GP, Mascaro J, Anderson C, Knapp DE, Martin RE, Kennedy-Bowdoin T, van Breugel M, Davies S, Hall JS, Muller-Landau HC, Potvin C, Sousa W, Wright J, Bermingham E (2013) High-fidelity national carbon mapping for resource management and REDD+. Carbon Balance Manage 8(1):art.7

Avitabile V, Baccini A, Friedl MA, Schmullius C (2012) Capabilities and limitations of Landsat and land cover data for aboveground woody biomass estimation of Uganda. Remote Sens Environ 117:366–380

Axelsson P (1999) Processing of laser scanner data—algorithms and applications. ISPRS J Photogramm Remote Sens 54:138–147

Axelsson P (2000) DEM generation from laser scanner data using adaptive TIN models. Int Arch Photogramm Remote Sens 33(Part B4):110–117

Baltsavias EP (1999) A comparison between photogrammetry and laser scanning. ISPRS J Photogramm Remote Sens 54:83–94

Baltsavias EP, Käser C (1998) DTM and orthoimage generation—a thorough analysis and comparison of four digital photogrammetric systems. Int Arch Photogramm Remote Sens 32:42–51

Bright BC, Hicke JA, Hudak AT (2012) Estimating aboveground carbon stocks of a forest affected by mountain pine beetle in Idaho using lidar and multispectral imagery. Remote Sens Environ 124:270–281

Deckert C, Bolstad PV (1996) Forest canopy, terrain, and distance effects on global positioning system point accuracy. Photogramm Eng Remote Sens 62:317–321

Cohen WB, Yang Z, Kennedy R (2010) Detecting trends in forest disturbance and recovery using yearly Landsat time series: 2. TimeSync – Tools for calibration and validation. Remote Sens Environ 114:2911–2924

d'Oliveira MVN, Reutebuch SE, McGaughey RJ, Andersen H (2012) Estimating forest biomass and identifying low-intensity logging areas using airborne scanning lidar in Antimary State Forest, Acre State, Western Brazilian Amazon. Remote Sens Environ 124:479–491

Dorigo W, Hollaus M, Wagner W, Schadauer K (2010) An application-oriented automated approach for co-registration of forest inventory and airborne laser scanning data. Int J Remote Sens 31:1133–1153

Elmstrom MD, Smith PW, Abidi MA (1998) Stereo-based registration of Ladar and color imagery. P Soc Photo-Opt Ins, pp 343–354

Forkuo EK, King B (2005) Automatic fusion of photogrammetric imagery and laser scanner point clouds. Int Arch Photogramm Remote Sens 35:921–926

Forsman M, Börlin N, Holmgren J (2012) Estimation of tree stem attributes using terrestrial photogrammetry. Int Arch Photogramm Remote Sens Spat Inf Sci XXXIX-B5:261–265

Frazer GW, Magnussen S, Wulder MA, Niemann KO (2011) Simulated impact of sample plot size and co-registration error on the accuracy and uncertainty of LiDAR-derived estimates of forest stand biomass. Remote Sens Environ 115:636–649

Fritz A, Weinacker H, Koch B (2011) A method for linking TLS- and ALS-derived trees. SilviLaser 2011, University of Tasmania, Hobart, Australia

Gatziolis D, Fried JS, Monleon VS (2010) Challenges to estimating tree height via lidar in closed-canopy forests: a parable from Western Oregon. For Sci 56:139–155

Gobakken T, Næsset E (2009) Assessing effects of positioning errors and sample plot size on biophysical stand properties derived from airborne laser scanner data. Can J For Res 39:1036–1052

Habrich H, Gurtner W, Rothacher M (1999) Processing of GLONASS and combined GLONASS/GPS observations. Adv Space Res 23:655–658

Hasegawa H, Yoshimura T (2007) Estimation of GPS positional accuracy under different forest conditions using signal interruption probability. J For Res 12:1–7

Henning JG, Radtke PJ (2006) Detailed stem measurements of standing trees from ground-based scanning lidar. For Sci 52(1):67–80

Hilker T, van Leeuwen M, Coops N, Wulder M, Newnham G, Jupp D, Culvenor D (2010) Comparing canopy metrics derived from terrestrial and airborne laser scanning in a Douglas-fir dominated forest stand. Trees-Struct Func 24:819–832

Hilker T, Coops NC, Culvenor DS, Newnham G, Wulder MA, Bater CW, Siggins A (2012) A simple technique for co-registration of terrestrial LiDAR observations for forestry applications. Remote Sens Lett 3:239–247

Hodgson ME, Bresnahan P (2004) Accuracy of airborne lidar-derived elevation: empirical assessment and error budget. Photogramm Eng Remote Sens 70:331–339

Hollaus M, Wagner W, Eberhöfer C, Karel W (2006) Accuracy of large-scale canopy heights derived from LiDAR data under operational constraints in a complex alpine environment. ISPRS J Photogramm Remote Sens 60:323–338

Holmgren J, Persson A, Soderman U (2008) Species identification of individual trees by combining high resolution LIDAR data with multi-spectral images. Int J Remote Sens 29:1537–1552

Holmgren J, Barth A, Larsson H, Olsson H (2012) Prediction of stem attributes by combining airborne laser scanning and measurements from harvesters. Silva Fenn 46(2):227–239

Honkavaara E, Ahokas E, Hyyppä J, Jaakkola J, Kaartinen H, Kuittinen R, Markelin L, Nurminen K (2006) Geometric test field calibration of digital photogrammetric sensors. ISPRS J Photogramm Remote Sens 60:387–399

Korpela I, Tuomola T, Välimäki E (2007) Mapping forest plots: an efficient method combining photogrammetry and field triangulation. Silva Fenn 41:457–469

Kraus K, Pfeifer N (1998) Determination of terrain models in wooded areas with airborne laser scanner data. ISPRS J Photogramm Remote Sens 53:193–203

Küchler M, Ecker K, Feldmeyer-Christe E, Graf U, Küchler H, Waser LT (2004) Combining remotely sensed spectral data and digital surface models for fine-scale modelling of mire ecosystems. Comm Ecol 5:55–68

Lämås T (2010) The Haglöf PosTex ultrasound instrument for the positioning of objects on forest sample plots. SLU Arbetsrapport 296, Skoglig resurshushållning. Umeå. Sweden

Leckie DG, Gougeon FA, Tinis S, Nelson T, Burnett CN, Paradine D (2005) Automated tree recognition in old growth conifer stands with high resolution digital imagery. Remote Sens Environ 94:311–326

Leppänen VJ, Tokola T, Maltamo M, Mehtätalo L, Pusa T, Mustonen J (2008) Automatic delineation of forest stands from lidar data. In: Hay GJ, Blaschke T, Marceau D (eds) GEOBIA 2008 – Pixels, Objects, Intelligence. GEOgraphic Object Based Image Analysis for the 21st century, Calgary, Alberta, Canada

Lindberg E, Holmgren J, Olofsson K, Olsson H (2012) Estimation of stem attributes using a combination of terrestrial and airborne laser scanning. Eur J For Res 131:1–15

Liu CJ, Brantigan RD (1995) Using differential GPS for forest traverse surveys. Can J For Res 25:1795–1805

Liu X, Zhang Z, Peterson J, Chandra S (2007) LiDAR-derived high quality ground control information and DEM for image orthorectification. Geoinformatica 11:37–53

Lovell JL, Jupp DLB, Culvenor DS, Coops NC (2003) Using airborne and ground-based ranging lidar to measure canopy structure in Australian forests. Can J Remote Sens 29:607–622

Maas HG, Bienert A, Scheller S, Keane E (2008) Automatic forest inventory parameter determination from terrestrial laser scanner data. Int J Remote Sens 29:1579–1593

Mauro F, Valbuena R, Manzanera JA, García-Abril A (2011) Influence of Global Navigation Satellite System errors in positioning inventory plots for tree-height distribution studies. Can J For Res 41:11–23

McInerney DO, Suárez-Mínguez J, Valbuena R, Nieuwenhuis M (2010) Forest canopy height retrieval using LiDAR data, medium-resolution satellite imagery and kNN estimation in Aberfoyle, Scotland. Forestry 83:195–206

Næsset E (1999) Point accuracy of combined pseudorange and carrier phase differential GPS under forest canopy. Can J For Res 29:547–553

Næsset E (2001) Effects of differential single- and dual-frequency GPS and GLONASS observations on point accuracy under forest canopies. Photogramm Eng Remote Sens 67:1021–1026

Næsset E (2004) Practical large-scale forest stand inventory using a small-footprint airborne scanning laser. Scand J For Res 19:164–179

Næsset E (2009) Effects of different sensors, flying altitudes, and pulse repetition frequencies on forest canopy metrics and biophysical stand properties derived from small-footprint airborne laser data. Remote Sens Environ 113:148–159

Næsset E, Gjevestad JG (2008) Performance of GPS precise point positioning under conifer forest canopies. Photogramm Eng Remote Sens 74:661–668

Næsset E, Jonmeister T (2002) Assessing point accuracy of DGPS under forest canopy before data acquisition, in the field and after postprocessing. Scand J For Res 17:351–358

Næsset E, Bjerke T, Ovstedal O, Ryan LH (2000) Contributions of differential GPS and GLONASS observations to point accuracy under forest canopies. Photogramm Eng Remote Sens 66:403–407

Olofsson K, Lindberg E, Holmgren J (2008) A method for linking field-surveyed and aerial-detected single trees using cross correlation of position images and the optimization of weighted tree list graphs. SilviLaser 2008, Edinburgh, UK

Ørka HO, Gobakken T, Næsset E, Ene L, Lien V (2012) Simultaneously acquired airborne laser scanning and multispectral imagery for individual tree species identification. Can J Remote Sens 38:125–138

Packalén P, Maltamo M (2006) Predicting the plot volume by tree species using airborne laser scanning and aerial photographs. For Sci 52:611–622

Packalén P, Suvanto A, Maltamo M (2009) A two stage method to estimate species-specific growing stock. Photogramm Eng Remote Sens 75:1451–1460

Persson A, Holmgren J, Söderman U, Olsson H (2004) Tree species classification of individual trees in Sweden by combining high resolution laser data with high resolution near-infrared digital images. Int Arch Photogramm Remote Sens XXXVI(W2):204–207

Reutebuch SE, McGaughey RJ, Andersen HE, Carson WW (2003) Accuracy of a high-resolution LIDAR terrain model under a conifer forest canopy. Can J Remote Sens 29:527–535

Schenk T (1999) Digital photogrammetry. TerraScience, Laurelville

Schickler W, Thorpe A (1998) Operational procedure for automatic true orthophoto generation. Int Arch Photogramm Remote Sens 32:527–532

Sigrist P, Coppin P, Hermy M (1999) Impact of forest canopy on quality anti accuracy of GPS measurements. Int J Remote Sens 20:3595–3610

Sithole G, Vosselman G (2004) Experimental comparison of filter algorithms for bare-Earth extraction from airborne laser scanning point clouds. ISPRS J Photogramm Remote Sens 59:85–101

Solberg S, Næsset E, Lange H, Bollandsås OM (2004) Remote sensing of forest health. ISPRS Arch XXXVI, 8–W2:161

St-Onge BA (2008) Methods for improving the quality of a true orthomosaic of Vexcel UltraCam images created using a lidar digital surface model. SilviLaser 2008, Edinburgh, UK

St-Onge A. Achaichia N (2001) Measuring forest canopy height using a combination of LIDAR and aerial photography data. Int Arch Photogramm Remote Sens 34:131–137

Tachiki Y, Yoshimura T, Hasegawa H, Mita T, Sakai T, Nakamura F (2005) Effects of polyline simplification of dynamic GPS data under forest canopy on area and perimeter estimations. J For Res 10:419–427

Tuominen S, Pekkarinen A (2004) Local radiometric correction of digital aerial photographs for multi source forest inventory. Remote Sens Environ 89:72–82

Valbuena R, Fernández de Sevilla T, Mauro F, Pascual C, García-Abril A, Martín-Fernández S, Manzanera JA (2008) Lidar and true-orthorectification of infrared aerial imagery of high Pinus sylvestris forest in mountainous relief. SilviLaser 2008, Edinburgh, UK

Valbuena R, Mauro F, Rodríguez-Solano R, Manzanera JA (2010) Accuracy and precision of GPS receivers under forest canopies in a mountainous environment. Span J Agric Res 8:1047–1057

Valbuena R, Mauro F, Arjonilla FJ, Manzanera JA (2011) Comparing airborne laser scanning-imagery fusion methods based on geometric accuracy in forested areas. Remote Sens Environ 115:1942–1954

Valbuena R, Mauro F, Rodriguez-Solano R, Manzanera JA (2012) Partial least squares for discriminating variance components in global navigation satellite systems accuracy obtained under Scots pine canopies. For Sci 58:139–153

Valbuena R, De Blas A, Martín Fernández S, Maltamo M, Nabuurs GJ, Manzanera JA (2013) Within-species benefits of back-projecting laser scanner and multispectral sensors in monospecific Pinus sylvestris forests. Eur J Remote Sens 46:401–416

Vauhkonen J, Ene I, Gupta S, Heinzel J, Holmgren J, Pitkänen J, Solberg S, Wang Y, Weinacker H, Hauglin KM, Lien V, Packalén P, Gobakken T, Koch B, Næsset E, Tokola T, Maltamo M (2012) Comparative testing of single-tree detection algorithms under different types of forest. Forestry 85:27–40

Waser LT, Baltsavias EP, Ecker K, Eisenbeiss H, Feldmeyer-Christe E, Ginzler C, Küchler M, Zhang L (2008) Assessing changes of forest area and shrub encroachment in a mire ecosystem using digital surface models and CIR aerial images. Remote Sens Environ 112:1956–1968

Wolf PR (1983) Elements of photogrammetry. McGraw-Hill, New York

Wolf PR, Brinker RC (1994) Elementary surveying. Harper Collins, New York, pp 248

Zengin H, Yeşil A (2006) Comparing the performances of real-time kinematic GPS and a handheld GPS receiver under forest cover. Turk J Agric For 30:101–110

Zhang L, Gruen A (2004) Automatic DSM generation from linear array imagery data. Int Arch Photogramm Remote Sens 35:128–133

Chapter 5
Segmentation of Forest to Tree Objects

Barbara Koch, Teja Kattenborn, Christoph Straub, and Jari Vauhkonen

Abstract This chapter reviews the use of airborne LiDAR data for the segmentation of forest to tree objects. The benefit obtained by LiDAR data is typically related to the use of the third dimension, i.e. the height data. Forest and stand objects may be segmented based on physical criteria, for example height and density information, while a further delineation to different timber types would require leaf-off data or an additional data source such as spectral images. Most forest applications of the LiDAR data are based on using digital surface models, but especially tree-level segmentation may benefit from a combination of raster and point data, and can be performed solely on point data. Finally, there are several established techniques for tree shape reconstruction based on the segmented point data.

5.1 Introduction

Segmentation in general means identifying and grouping objects based on statistical similarities or similar features in data. This procedure has a statistical background, but not necessarily any relation to the thematic content. The use of segmentation has a long history in the processing of 2D remote sensing data (e.g. Blaschke 2010), and well-known pixel-, edge- and region-based methods described in general image

B. Koch (✉) • T. Kattenborn
Department of Remote Sensing and Landscape Information Systems,
University of Freiburg, Freiburg im Breisgau, Germany
e-mail: barbara.koch@felis.uni-freiburg.de

C. Straub
Bayerische Landesanstalt für Wald- und Forstwirtschaft, Hans von Carl Carlowitz Platz,
85354 Freising, Germany

J. Vauhkonen
Department of Forest Sciences, University of Helsinki, Helsinki, Finland

M. Maltamo et al. (eds.), *Forestry Applications of Airborne Laser Scanning: Concepts and Case Studies*, Managing Forest Ecosystems 27, DOI 10.1007/978-94-017-8663-8__5,
© Springer Science+Business Media Dordrecht 2014

analysis text books (e.g. Gonzalez and Woods 2008) are applied. Since the most common methods used with airborne laser scanning data (referred to as LiDAR data in the following text) are also fundamentally similar to those based on high-resolution aerial images (Hyyppä et al. 2008), the present chapter mainly omits detailed descriptions of the basic algorithms and focuses on the utility and special features produced by the LiDAR data towards the segmentation task.

The main advantage of the LiDAR data as an input for the segmentation and object building is the use of the third dimension in the resolution and accuracy which cannot be provided by other (3D) remote sensing methods. Most of the forestry applications are related to the segmentation of single crowns, while the use of LiDAR data for the segmentation of stands, or between forest and non-forest, has been in a more narrow focus in the last years. However, to automatically process large forest areas, a combination of data and methods for hierarchical segmentation would be required, i.e. the use of spectral or microwave data for forest–non-forest or stand type segmentation (according to e.g. crown closure or species), and LiDAR data for more detailed segmentation within the segmented strata, for example. A premise behind this approach is a possibility to provide detailed (up to tree-level) information for the estimations for larger areas. Based on this background, the chapter is divided into following main sections:

1. segmentation of forest objects,
2. segmentation of forest stand objects, and
3. segmentation of single tree objects.

The segmentation may be based directly on the point cloud, or more typically, on a raster image interpolated from the data. While the point data provide information on the position of each reflected and registered echo, the rasterized data include averaged or classified information (e.g. the highest or lowest reflection within a pixel). The use of the raster data thus results in a loss of information, but reduces the computational burden involved in processing point data. Most forest applications of the LiDAR data are based on *digital surface models* (DSMs, the top of the canopy), *digital terrain models* (DTMs, the bare ground), and *normalized DSMs* (nDSMs, obtained as the difference between the DSM and DTM, thus representing vegetation height), the latter being also referred to as *canopy* or *crown height models* (CHMs). A number of possible algorithms exist for surface modeling (e.g. Lloyd and Atkinson 2002). The produced (normalized) vegetation height values are denoted by dZ in the following text.

Approaches combining various data sources are not discussed in detail, even though multispectral, hyperspectral, and/or radar data could have a high importance in producing additional information for forest and stand segmentation. Instead, to follow the scope of the entire book, the focus is mainly on demonstrating the possibilities of using LiDAR data for the segmentation of the different objects.

5.2   Forest Segmentation

Mapping forest area boundary is required as such by many applications related to sustainable forest and landscape management, such as change statistics and the estimation of stem volume, biomass and carbon stock. The forest boundary are also needed prior to any more detailed (stand or tree level) segmentation. Up-to-date information on the forest boundary does often not exist or may need improvement especially in areas in which the forest cover is changing rapidly.

Not many papers focus on the delineation of the forest boundary based on LiDAR data. One of the few studies was carried out by Eysn et al. (2012) and is based on a single crown approach. The approach, entitled *tree triples method*, defines the crown cover based on the sum of the crown area of three neighboring trees combined with the area of their convex hull. This approach is especially useful for open areas in which it is more easy to delineate single trees but more difficult to define a forest border. While Eysn et al. (2012) used exclusively LiDAR data, for example, Wang et al. (2007, 2008a) presented an approach combining spectral information from aerial images and curvature features from airborne LiDAR data. For the segmentation they used a JSEG algorithm (Deng and Manjunath 2001), which divides the segmentation in a color quantization and a spatial segmentation parts. In addition, they compared the texture between crown height models, green vegetation index, and the JSEG-segmentation to refine the forest boundary.

Straub et al. (2008) developed a procedure for forest area segmentation, which can be applied in coniferous, deciduous and mixed forests of different complexity based solely on aerial LiDAR. The procedure is implemented in the TreesVis software (Weinacker et al. 2004b) and is presented below as an example algorithm for forest delineation. The approach is divided into two main parts:

1. Delineation of regions covered by vegetation.
2. Classification of the vegetation regions into different vegetation classes:

   (a) forest areas,
   (b) connected groups of trees outside forests, and
   (c) single trees outside forests.

The procedure makes use of the multiple returns of the point cloud. A DSM and a DTM are derived from the points using an "Active Surface Filtering Algorithm" (Weinacker et al. 2004b). Within a tolerance zone defined by the maximum distance beneath the DSM and the maximum distance above the DTM the echoes are classified into:

1. ground points (within the zone above the DTM),
2. surface points (within the zone below the DSM), and
3. intermediate points (points with height values between the tolerance zone above the DTM and the zone below the DSM).

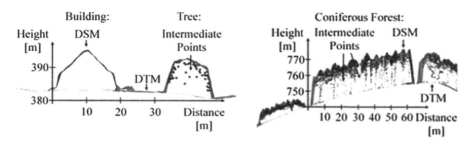

Fig. 5.1 Cross-sections of point clouds for building, ground and vegetation (Straub et al. 2008)

The "intermediate points" will be found in nearly all cases only within the vegetation and not within man-made objects (like buildings) or ground points (Fig. 5.1). The irregularly distributed points are arranged on a grid with a cell size of 1 m^2 (adaptable to different resolutions) within the extent of the full study area. Two types of local density images are created:

1. *AllPointsImage*, in which the cell values represent the number of all points/echoes of the dataset, and
2. *IntermediatePointsImage*, in which the cell values represent the number of "intermediate points/echoes".

Due to the fact that more points will be found in overlapping flight strips compared to single ones, the cell values of the *IntermediatePointsImage* are normalized with those of the *AllPointsImage* as $g' = g_1/g_2$, where g' is the output value of the "normalized image". A median filter with a 5×5 pixel window is used to smooth the normalized image. Finally a global threshold (defined by empirical tests) is applied to extract regions which are covered by vegetation. Connected vegetation pixels are grouped together into vegetation objects. Morphological closing with a circular structuring element (radius 3 m) is used for boundary smoothing and to fill small holes. In this way vegetation (trees and bushes) is delineated from ground, open grass areas and man-made objects.

To classify the vegetated area into forest, tree and bush areas, the vegetation features which can be measured or estimated from ALS data are applied. Such features are (1) height of the vegetation, (2) tree crown cover, (3) size of the vegetation region, and (4) width of the vegetation region. Forest vegetation is defined according to minimum values set for these attributes, and objects not fulfilling the defined minima are classified as non-forest vegetation.

The mentioned features are computed with the following approaches (the threshold values used by Straub et al. 2008 are given):

1. The vegetation height based on nDSM with a pixel wise computation. A threshold of 3 m is defined as a minimum height for forest. The forest pixels are intersected with the regions classified as vegetation before. The resulting regions represent potential vegetation pixels above 3 m.

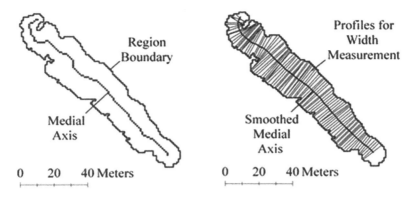

Fig. 5.2 Width measurement with profiles along the medial axis of a vegetation region

2. The local tree crown cover of a vegetation area characterized by a grid of 20×20 m (adaptable) with the vegetation pixels above 3 m. The result is a mixture of regular and individually shaped grid cells. Within each cell, the tree crown cover is computed by first extracting potential canopy gaps by thresholding all pixels below the 3 m of the nDSM. By subtracting the surface percentage of gaps from each individual cell the area covered by tree crowns is derived and expressed as a percentage of total ground area of each grid cell. Only cells with a tree crown cover of at least 50 % are accepted as forest vegetation.

3. The size of a vegetation region (a group of connected pixels) calculated from the number of pixels multiplied with the corresponding ground resolution of a pixel, with a minimum size of 1,000 m^2.

4. The width of the irregularly shaped vegetation regions measured with the help of profiles along the medial axis obtained by "skeletonization" (Soille 2003). The skeleton is computed by fitting circles into a vegetation region with the largest radius possible. A circle C has the largest radius in the input region R if there is no other circle in R that is a superset of C. The skeleton is derived from the center points of those maximal circles. There will be at least two points on the region boundary to which a center point will have the same shortest distance (Steger et al. 2008).

Due to the fact that vegetation objects can have high variations in their shape, post-filtering is necessary to remove irrelevant branches from the skeleton. Based on the skeleton the longest connected chain of pixels is computed which finally is a single one-pixel-wide center line without other branches.

Raster to vector conversion is used to convert the one pixel-wide center lines into vector lines which are smoothed using a method which projects the contour points onto a local regression line (a least-squares approximating line) fitted to a defined number of original contour points. For each contour point of the vectorized and smoothed center line a profile is generated to measure the local width of the vegetation object (Fig. 5.2).

Finally the median of all width measurements is computed as an average width for each object and used for classification. A minimum width of 30 m was defined in this example for forest vegetation. Objects classified as non-forest vegetation are further classified into single tree objects or a connected group of trees based on the size and shape of the area. Finally vegetation objects are slightly modified. Areas surrounded by forest vegetation without trees and smaller than 0.5 ha are extracted and merged with the forest mask.

The accuracy assessment based on a point raster method 10 by 10 m with 66 points proved for forest areas in this example an accuracy of 99 %. For group of trees and single trees the accuracy was lower and between 60 and 70 %. However this was more a problem of the not good enough co-registration to hit with the verification point raster the crown areas as reported by Straub et al. (2008).

5.3   Stand Segmentation

A 'stand' refers to the basic unit of forest management, for which reason the assessment and delineation of such units is the basis for all operational management measures. A forest stand is "*a geographically contiguous parcel of land whose site type and growing stock is homogeneous*" (Koivuniemi and Korhonen 2006). Stands are thus mainly defined based on growing stock features, but in practice also operational and organizational considerations have an influence on stand delineation; however, the segmentation based on the LiDAR data can be based solely on physical criteria such as height and density. The problems related to stand segmentations based on similar physical criteria will likely increase along with the implementation of management forms favoring uneven-aged and multi-species forest structures, but currently most commercially managed forest stands pose a similar tree age and dominant tree type.

Probably one of the first approaches using LiDAR data alone for the stand segmentation was carried out by Diedershagen et al. (2004), while later approaches have combined multispectral and LiDAR data (Pascual et al. 2008; Packalén et al. 2008). Leppänen et al. (2008) tested an automatic approach using colour infrared aerial photographs and LiDAR data. The segmentation was based on a region growing algorithm and height, density and hardwood pixel percentage, aiming at producing stands with an equal timber type. They compared the result with a manual aerial image interpretation, and reported that the segmented units derived from this process were unacceptably small (0.6–0.8 ha) for forest operations under the Nordic conditions. However, the density and height data were found reasonable for segmenting and clustering different timber classes to come up with operational units.

Tiede et al. (2004) used an object-based semi-automatic mapping approach to combine LiDAR and multi-spectral data. They used an initial image segmentation followed by object-relationship modeling of forest development stages. They achieved an overall accuracy of 63 %, identifying problems with younger development stages. A combined approach of aerial photography and LiDAR-based CHMs for an automatic segmentation of forest stands was also carried out by Mustonen et al. (2008), according to whom the combination of the CHM and multi-spectral image did not improve the result compared to the use of CHM alone. The segmentation was based on producing homogeneous diameter and height distributions. Wu et al. (2013) used a hybrid segmentation approach for the stand segmentation. They first extracted a three-band image containing height, density and species features. Based on the image, a mean shift algorithm was applied to generate raw forest stands, which were then refined by a spectral clustering algorithm. The results produced by the developed approach outperformed the reference methods for the segmentation.

An example approach that is described as a detailed example of LiDAR-based stand segmentation (Koch et al. 2009) is based on the use of DTMs, DSMs, and nDSMs. The extraction of forest stand boundaries is divided into several steps, which each can be used independently allowing to focus on those steps which are required or possible under given frame conditions. The presented semi-automatic, stepwise approach combining different segmentation methods in a series of modules starts with a semi-automatic segmentation of forest roads and is followed by a fully automated segmentation of forest areas into stands according to height, stand type and density.

In many cases in intensively managed forest areas, the stand boundaries are following forest roads. Therefore a method to extract forest roads semi-automatically from the DTM is first used. Due to the fact that roads normally have lower slope values compared to the surrounding terrain in shaped areas, this information can be used for delineating road segments automatically (Koch et al. 2009). The procedure is mainly applicable in mountainous areas and thus in terrain with >15 % inclination. In flat areas the forest road an automatic extraction from laser data is not possuble but has to be extracted form other data sources. In the second step the actual segmentation of the forest stands is based on stand type, crown cover and height class, determined as follows:

Stand type: The definition employs the different reflectance and penetration rates between deciduous and coniferous vegetation under leaf-off conditions. A comparison of two nDSMs calculated under leaf-off conditions (Fig. 5.3) shows less height differences for broadleaved than coniferous stands. However, an ideal leaf-off situation is not obtained, for example, in the case of evergreen broadleaved species. Therefore a combination of LiDAR and spectral data is recommended if the classification of broadleaved and coniferous stands is required (see also Chap. 7).

Crown cover is defined by the tree crowns covering the ground using the method described by Straub et al. (2008). The canopy density is estimated based on the nDSM which represents canopy heights for each x–y position. A threshold operation (selection of pixels with height values within a defined interval) is used to extract potential crown regions with the height values between 50 and 100 % of the top height per plot. The ratio of the size of extracted crown regions to the plot size is used as an estimate for the canopy density.

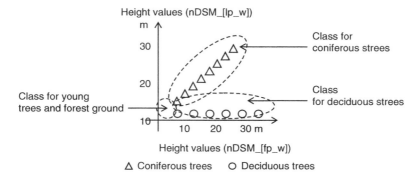

Fig. 5.3 The differences between deciduous and coniferous stands in a nDSM generated under a leaf-off period

Height estimation: The top height H_t, i.e. the mean height of the 100 *thickest* trees per ha, is often used to characterize a forest stand. As the height of trees is correlated with their stem diameters, the top height is determined as the average height of the 100 *tallest* trees per ha (Dees et al. 2006). The 90th percentile computed from the dZ values is used as an estimate for stand height following several corresponding studies (Næsset and Bjerknes 2001; Means et al. 2000; Rieger et al. 1999). In Straub et al. (2008) this variable yielded a strong correlation of 0.87 with the top height derived from the field data.

For a better characterization of the different stand features the forest area is divided into grid cells for which different cell sizes can be selected. A grid size of 20 m turned out to be most appropriate for applications in central European forests. The calculation of the above mentioned stand features besides the pre-segmentation by roads are carried out for each grid cell. This will provide a thematic map layers of forest type (species), crown closure and height classes. In a final step the different layers are combined in a hierarchical classification process. After the classification, neighboring grid cells belonging to the same classes are merged into larger segments. In the final step areas smaller than a defined minimum size are erased or merged with the neighboring segments. This stand segmentation is not purely based on physical consideration but tries to follow the criteria used by forest organizations for stand delineation and therefore differs from the approach suggested by Leppänen et al. (2008).

5.4 Segmentation of Single Trees

5.4.1 An Overview

The detection and delineation of the individual trees are often carried out as two (separate) steps. Only the former is in fact required to derive an estimated position and height for the detected trees, which can be readily used for modeling other tree attributes. However, subsequent crown delineation allows extracting information on the crown dimensions and a further modeling of the crown shape, for example.

With some degree of adaptations, the majority of the tree detection studies are addressed in the image domain (e.g. Hyyppä and Inkinen 1999; Hyyppä et al. 2001a; Persson et al. 2002; Popescu et al. 2002, 2003, and many others). The tree tops are typically detected with an approach based on searching local height maxima. The differences between the algorithms are typically related to adjusting the CHM smoothing in order to obtain a desired number of local maxima in varying canopy conditions. In the local maxima search, the applied window size has a major impact on the tree detection result. The window size and the degree of smoothing can be set according to prior information from preceding field measurements (Popescu et al. 2003), as a function of the height values of the CHM (Pitkänen et al. 2004) or based on other information such as estimated stand density (Ene et al. 2012).

The tree crown segments are typically formed around the local maxima using well-known watershed or region growing techniques (e.g. Vincent and Soille 1991). However, also techniques based on template matching (Holmgren and Wallerman 2006; Holmgren et al. 2012) and clustering of the point data (e.g. Morsdorf et al. 2004) are used. The section below presents the most typical techniques based on the use of raster (Sect. 5.4.2), vector (Sect. 5.4.3) and hybrid (Sect. 5.4.4) data, the latter including both raster and vector data sources and *a priori* information. An additional Sect. (5.4.5) is dedicated to tree shape reconstruction techniques following the actual segmentation.

5.4.2 *Raster-Based Methods*

5.4.2.1 Treetop Detection

To apply local maxima detection on a CHM, the latter has to be extracted from the laser point cloud, interpolated and smoothed. The smoothing typically results in a loss of detail, trimming upper branches, and filling holes in the CHM. The smoothing process is however required to identify a correct number of local maxima as starting point for the tree segmentation process. Together with the penetration of the laser pulses through the canopy layer, the CHM underestimates the true height of the canopy surface. To compensate for these effects, Solberg et al. (2006) implemented a residual height adjustment method, in which the median of the differences calculated between control-measured tree heights and the associated local maxima were used to obtain a residual percentile. The corresponding value of the residual percentile was added to the dZ-values of the entire CHM, resulting in a residual height adjusted CHM. However, the degree of the height underestimation may vary over the CHM, and for instance, the edges of the crowns may be more strongly affected and more residuals may occur above the corrected CHM. Thus, this procedure primarily aims at correcting the height values of the tree tops.

All methods based on CHM smoothing require a decision on the applied smoothing factor. A strong smoothing normally leads to an under-representation of local height maxima relative to the treetops, while a weak smoothing leads to an over-representation of the maxima. Even though smoothing factors can be optimized for given stand structures, this step implies frequent interactions by the user and restricts the automatic process. If the stands are uneven-aged the problem gets more complicated because no optimal solution can be provided for all trees within the stand.

A solution which may improve the results of the single tree segmentation for uneven-aged stands and large areas with different stand structures is a prior stratification into height classes (Koch et al. 2006). In this approach the smoothing of the height model is adapted to different height classes within the stand. Based on the smoothed CHM a pixel counts as a local maximum, if all of its neighbors (in a 4-connected neighborhood) have got a lower height-value or if all neighbors of some connected pixels with equal height (a "plateau") have got a lower height-value. Correspondingly, Pitkänen et al. (2004) increased the intensity of the smoothing window as a stepwise function of the height of the CHM. The range for Gaussian standard deviations (σ) defining the smoothing intensity was adjusted manually to obtain a feasible number of local maxima at the both ends of the tree height range, whereas the σ in between were linearly interpolated. After this height based filtering, the tree crowns can be separated by normal segmentation methods, e.g. watershed segmentation.

As all these approaches are based on analyzing local maxima, they lack the ability to detect trees which are not represented in the CHM. For example trees in the understory are overtopped by tall trees or branches of the dominant trees. The detection of those trees that are visible in the CHM also depends on the sensitivity of the applied algorithm. Various tree detection approaches were tested in international algorithm comparison studies by Kaartinen et al. (2012) and Vauhkonen et al. (2012). While the findings of Kaartinen et al. (2012) point the applied algorithm as a major factor towards the success rates in the tree detection, Vauhkonen et al. (2012), including a considerably wider range of forest types in the comparison, found that the tree detection result was more dependent on tree density and spatial distribution of trees (i.e. clustering) than on the algorithm itself.

5.4.2.2 Segmentation and Post-processing of the Result

One of the most popularly used raster-based approaches for the segmentation is the watershed or pouring algorithm implemented in many image processing software products. The pouring algorithm starts "flowing water" from a defined maximum height towards the lower heights and the area is split into regions according to the water flow. The watershed algorithm has a similar but inverse principle (Soille

1999): the regions are extended, as long as neighboring pixels with lower or same height value exist. Overlapping regions in the "height-valleys" are finally distributed evenly to all involved tree regions. The algorithm produces a 2D-approximation of the tree crown shape.

The segmentation of crowns with the pouring algorithm works reasonably well for highly uniform stands. However, the result may include segments not resembling tree crowns, i.e. regions too small to be trees, non-tree-like shapes, unusual spatial relationships, and combinations of tree groups or canopy gaps. For example, Solberg et al. (2006) restricted their region growing algorithm by including rules for polygon convexity, when considering those directions where the regions could grow. Alternatively, post-processing the segments by split and merge rules based on allowable segment dimensions could possibly improve the result. Another splitting or merging criterion could be a topological character like the minimum distance between the tree tops. To avoid problems in very young stands with small trees, the number of possible merges of adjacent region can be restricted.

Geometrical reasoning criteria can be used, for example identifying elliptical groups with a combination of a minimal area and the regions anisometry (the quotient of both radii of a fitted ellipse). For example, if the length of the region is at least 2.5 times its width with at least three times the respective minimal area for its height and tree species class, it can be most probably judged as a group of trees. Such congregations can be disjoined analogous to Straub et al. (2008), who used the approach of Heipke (2001), which has been developed for tree groups within settlements. For each tree group the biggest inner circle was consecutively detected and subtracted, until the area of the circle fell below the defined double minimum area for the given height and species class. Correspondingly, the circles could be expanded according to similar criteria.

Geometric tree crown models or templates (Holmgren and Wallerman 2006; Holmgren et al. 2012) have also directly been used for tree crown segmentation. In these studies, a correlation surface was created as the maximum pixelwise correlation between the CHM and geometric tree crown models, defined as generalized ellipsoids of revolution (Pollock 1996). Both the CHM and the correlation surface were used in the segmentation, and additional splitting and merging criteria were defined according to the geometric models.

The pouring and watershed algorithms involve inherent problems related to defining the final segments. The segment boundaries may not overlap, causing potential problems in dense stands, whereas in more open stands, the problem is to detect the minimum edge of the crowns. An algorithm to separate the actual crown edge from neighboring canopy gaps or adjacent understorey trees may be required. An algorithm based on searching vectors within the segmented regions to determine the crown edge (Hyyppä et al. 2001b; Friedländer 2002) reduces the area of a tree crown but does not enlarge it. Starting from the tree's top, a vector to each border point of the segmented region is calculated. Proceeding in one pixel wide steps on each vector, the slope of the tree crown at each of these points is measured as height difference between two points. If this height difference or slope exceeds a certain threshold the vector breaks and a new border point is generated. The crown edge is moved inside of the region. Occasionally occurring outliers are removed afterwards.

5.4.2.3 Object-Based Methods

The use of object-based algorithms was followed by Tiede and Hoffmann (2006), who segmented tree crowns using a two stage approach. In the first stage, a non-tree/tree classification is carried out in order to later focus on more complex object creating functions. The object creation process is limited only to those areas classified as canopy. The canopy is the next step broken down into pixel-sized objects for local maxima detection. A weak point for this method is the requirement to set an appropriate search radius for each stand. Within this radius the local maxima are considered as tree tops and seed points for a region growing algorithm. Different stopping criteria for the region growing are used to define the crown edge. One is the height difference and another is the maximum crown area. Normally the procedure is finished by a clean-up process to fill the holes within the crown region. This algorithm provided good results for non-complex stand structures. However, the definition of the radius requires user-interaction, which limits the use of this approach for large or diverse areas.

5.4.3   Point Cloud Based Methods

5.4.3.1   k-means Clustering Techniques

Among the vector-based methods, clustering is one of the most often used approaches for segmentation. Several clustering mechanisms exist, among which k-means is the most popular iterative partitioning approach. Several attempts to partition ALS data into clusters (Jain et al. 1999) and in particular single tree crowns have been reported recently (Morsdorf et al. 2004; Cici et al. 2008; Doo-Ahn et al. 2008; Reitberger et al. 2009). The k-means method requires seed points, which are typically derived as smoothed CHM-based local maxima (Morsdorf et al. 2004). It is noteworthy that the k-means method works well when a data set has "compact" or "isolated" clusters (Mao and Jain 1996). Therefore, more adaptive alternatives have been developed for different forest structures. Gupta et al. (2010) showed that it is advantageous to scale down the dZ values of both the seed points and normalized raw points. This helped in minimizing the squared error function, which was the ultimate objective of the k-means method. Gupta et al. (2010) obtained good results by not smoothing the CHM, but removing the superfluous local maxima using a search algorithm based on a distance threshold, which was adaptable to the stand characteristics.

Li et al. (2012) describe a different method to avoid the inherent errors and uncertainties due to the CHM interpolation. They used the highest points within a threshold distance as seed points and grew the cluster within a threshold moving downwards. However, they assumed an always existing spacing between the crown tops of the trees to find a correct seed point for starting the process. As the threshold, they suggested a distance adapted to the crown size. In addition, they used a convex hull-based crown shaping index to improve the detection. The method works well for forest stands with homogeneous crown sizes, but in case of natural or uneven-aged stands the performance of the algorithm is unknown.

5.4.3.2 Voxel Based Single Tree Segmentation

The inspiration for the 3D single tree modeling comes from the inspection of horizontal distribution of forest canopies. Wang et al. (2008b) projected a normalized point cloud to 2D canopy layers of different height levels to describe the distribution of the tree crown reflections along z-axis. For this approach a local voxel space was defined by projecting the normalized points to a 2D horizontal plane. This projection starts with the voxels of the top layer which comprises the highest point and moves downwards layer by layer. The resolutions for the layers can be adapted to the data quality and stand characteristics. Wang et al. (2008b) showed that for data with 5–12 points per m^2 from an even-aged or multi-storey old-grown forest, a resolution between 0.5 and 1 m provided the best results.

The clusters on the horizontal projection image at each layer represent the distribution of reflections from tree crowns in the corresponding height level. The basic idea is to trace the reflections from top to bottom. At each layer the tree contours are delineated based on hierarchical opening and closing processes using a set of predefined structuring elements. Considering the projection image, a higher gray value of a pixel in the 2-D layer is set to represent a higher amount of points in the corresponding voxel. Thus a higher significance is assigned to the pixel with a higher gray value by keeping a larger neighborhood around the pixel. The morphological process starts with the brightest pixels on the projection image as seed pixel. The lower gray value the pixels have, the smaller structuring element is used for closing and the bigger structuring element is used for opening. Finally, potential regions from different gray value levels at same neighborhoods are merged.

The uppermost part of the individual crowns is normally easy to delineate because of the concentration of LiDAR reflections at the tree tops. The delineation is more difficult when proceeding to lower layers due to a lower number of reflections and conjunct neighboring crowns. To solve this problem an improvement of the hierarchical morphological algorithm is necessary. To face the problem, crown contours from the higher height level are copied to the next layer and only expanded according to the cluster features on the projection image. The enlargement stops when the neighboring regions conjunct. The utilization of reference regions will not influence the emergence of new tree tops at sub height level due to the parallel performance of the process in and out-side the reference regions.

5.4.4   Combining Raster, Point and a Priori Information for Tree Object Building

5.4.4.1   Adapting Tree Detection and Segmentation Algorithms with a Priori Information

As summarized from the previous sections, a major missing source of information that could benefit both tree detection and segmentation is the expected crown size and stand density. The studies by Heinzel et al. (2011) and Ene et al. (2012) have developed the use of such information to find the appropriate raster or voxel resolution and a proper smoothing factor for the CHM.

Ene et al. (2012) developed an adaptive method for the CHM generation and single tree delineation. They adjusted the filter size and the CHM resolution according to prior information obtained in the form of area-based stem number estimates. Assuming that trees are located according to a homogenous Poisson process, one can calculate an expected tree-to-tree distance for optimizing the CHM resolution and filter size. For assessing the CHM resolution, Ene et al. (2012) proposed two different methods, in which a set of CHMs in varying resolutions was created as a starting point. To obtain a feasible CHM resolution, they considered two approaches based on a trialing with *a priori* information on the stem number.

The approach followed by Heinzel et al. (2011) integrated a granulometry method in the single tree segmentation and estimation of tree crown size. The method builds upon the principles of grey-scale granulometry (Dougherty 1992; Chen and Dougherty 1994), in which images are analyzed by a series of basic morphological operations (dilation, erosion, opening, closing) followed by differential calculus. Structuring elements (SEs), which detect or measure objects, can be considered as templates of the objects to be observed. A series of SEs with varying size are analyzed on the same image, and the size at which most of the texture, measured as grey values, disappears refers to the desired size of the observed objects. Earlier, granulometry has been tested in tree crown applications by Soille (1999, 2003), who used aerial photographs, restricted the attempts to a small subset of two homogenous stands, and did not provide a verification of the results.

Heinzel et al. (2011) used color-infrared (CIR) images with the nDSM as intensity images reflecting high pixel values on the illuminated top of the crown as well as nearly constant low grey values at the borders. Under these conditions the image grey values are applicable for texture analyses. An algorithm by Straub et al. (2008) was used to remove non-tree objects from the image data in advance and the CIR values of the ground regions visible between the crowns were set to zero according to a height threshold based on the nDSM. A moving window of varying size was then iteratively adapted to the size of the observed crowns, and in each iteration only one crown size class was stored to the raster image.

This approach achieved reliable estimations of the crown sizes from a full automated texture analysis. The investigations by Heinzel et al. (2011) also proved the granulometry based method to be operational. The mean error for the test sites amounted to approximately 4.7 pixels, corresponding to a 1.2 m divergence from the reference crown size measured in the field. When considering the total spectrum of 20 crown size classes ranging from 0 to 25 m, and the fact that the texture features from the vegetation are highly variable, the described error appears to be relatively small. The granulometric method is thus a promising approach to extract information for finding an appropriate smoothing for raster-based single tree segmentation.

5.4.4.2   Combined Image and Point Cloud Analyses

Some approaches combine raster and point data to improve the segmentation of single trees (Reitberger et al. 2009; Höfle and Hollaus 2010). Reitberger et al. (2009) used a normalized cut approach first presented by Shi and Malik (2000) for segmentation based on full waveform LIDAR point data. The approach is based on graph partitioning and criteria to measure the between- and within-group dissimilarity. As a first step followed by Reitberger et al. (2009), a crude segmentation was performed with the watershed algorithm (Vincent and Soille 1991). Importantly, this segmentation was run to a highly smoothed CHM to produce an under-segmented result, which leaves room for further refinement within the segmented regions. The reflections extracted from the full-waveform data were arranged in voxels, which were cut to regions according to the graph partitioning idea to maximize the within-segment similarity and correspondingly minimize that between the segments. The similarity between the voxels was measured by point distribution, echo width and intensity. The approach produced a considerably higher detection rate for the small trees compared to using watershed segmentation alone, yet with a cost of likely false detections.

Höfle and Hollaus (2010) used an edge based approach to the segmentation. After generating nDSMs with a pixel size of 0.5 m they searched for concave edges between objects. An edge cutting algorithm was combined with constraints on normalized heights and occurrences of multiple reflections within cells. A ratio of the amount of first and intermediate echoes to the amount of last and single echoes and a height threshold were used as criteria to delineate vegetation. Based on the edge detector applied in a predefined (e.g. 7 m) window, a final edge map was derived by cutting the area into segments. The segments were finally combined with the information based on the echo and height thresholds to identify vegetation segments. The segment geometry and topology of the vegetation objects were further classified, producing a correct segmentation of more than 95 % of the trees in urban areas. The method is also documented as robust, fast and transferable.

5.4.4.3   Integration of Echo Features for Segmentation

In earlier studies, the tree detection and segmentation is typically based on the first echoes and namely on their coordinate values. Hyyppä et al. (2012) and

Rutzinger et al. (2008) employed last pulses and other echo features in addition to the coordinates, respectively, for the segmentation. The former generated the surface models based on the last pulse data, suggesting better chances to separate neighboring trees. An improvement of 6 % compared to the use of first pulse data was reported.

Rutzinger et al. (2008) describe objects using the full information obtained from full-waveform LiDAR data flown under leaf-off conditions. This includes occurrence and distribution measures, and waveform describing features such as amplitude, echo width and the number of echoes. After echo labeling they used a fixed neighborhood (0.5 m) for calculating the features. For the calculation of the point features roughness criteria derived from the standard deviation of a plane fitting residuals in a fixed distance 3-D neighborhood was used. For the segmentation all points sorted descending according to their roughness were used as seed points for the segmentation. The used homogeneity criteria were the echo width and the tolerance setting of the user. If an echo width of the considered point was within a predefined tolerance, then the point was accepted as a part of the segment and as the next seed point. The segment cut was also limited by 3D maximum growing distance and segment maximum size.

Although some of the criteria required site-specific adapting, the echo width as a criterion for region growing to separate vegetation from non-vegetation is physically reasonable and seems to be robust. For the classification of the segments into vegetation and non-vegetation a classification tree built mostly from mean segment features was used. Classification accuracies >90 % were reported for these classes.

5.4.5   Tree Shape Reconstruction

5.4.5.1   Convex Hull

After single tree segmentation and object building, the modeling of the 3D shape of the tree is of interest for many applications such as the tree species recognition (cf. Chap. 7). There are different methods available of which some will be shortly presented to give an idea how the tree shape reconstruction can be approached. The basis for all tree shape reconstructions is a successful segmentation of the single trees; for this process, additional refinement in horizontal and vertical directions may be required.

The shape of the point cloud clusters representing tree crowns can be geometrically reconstructed by means of the convex hull (Morsdorf et al. 2004; Koch et al. 2009), which corresponds to an outer boundary of a triangulation of point data (Fig. 5.4). For example Preparata and Shamos (1985) and O'Rourke (1994) describe algorithms for the computation of the convex hull. The convex hull can be produced from point data in 2D and higher dimensions, and it is applicable if the surface is convex or completely visible from an interior point. The facets of the convex

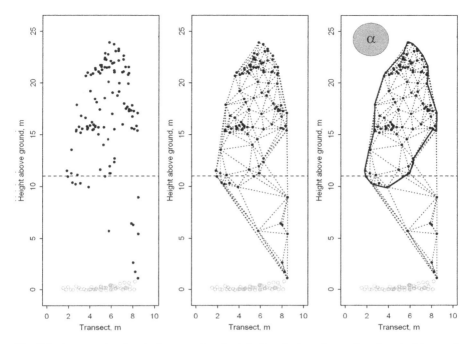

Fig. 5.4 An example of a single tree point cloud (*left*) and triangulations based on it, illustrated in 2-D for ease of visualization (Vauhkonen 2010b). The outer boundary of the triangulation in the *middle* figure corresponds to the convex hull of the point cloud, whereas the outer boundary illustrated by the *solid line* in the *right-hand* figure corresponds to a single connected component extracted from an alpha shape determined by the given alpha value (*filled circle*). A field-measured crown base height is illustrated using a *dashed, horizontal line* and ground hits using *grey circles*

hull correspond to the outer boundary of a triangulation of the input points. The major idea is to compute a convex envelope (polytope) based on the construction of "an unambiguous, efficient representation of the required convex shape". The complexity of the algorithms depends on the number of input points and the number of those points belonging to the convex hull. In LiDAR studies, Gupta et al. (2010) used the Quick hull (QHull) algorithm because this algorithm has an output sensitive performance (in terms of the number of extreme points), reduced space requirements and a floating-point error handling. The QHull accounts for round-off errors and it returns "thick" facets defined by two parallel hyperplanes. The outer planes contain all input points and the inner planes exclude all output vertices. The QHull algorithm removes facets that are not clearly convex by merging with neighbors.

5.4.5.2 Alpha Shapes

As an alternative for the convex hull approach, the number of facets belonging to the minimum convex polygon may be restricted to obtain a more detailed shape.

A particularly useful approach to perform this restriction is the concept of 3D alpha shapes (Edelsbrunner and Mücke 1994), in which a predefined parameter alpha is used as a size-criterion to determine the level of detail in the obtained triangulation (Fig. 5.4). The convex hull of the point data thus corresponds to an alpha shape computed using an infinitely large alpha value, whereas with very small values, the shape reverts to the input point set. The alpha value thus defines the level of detail in the obtained shape and, along with the applied point density, determines whether the resulting shape is formed from a solid structure or from cavities, holes or even separate components (see Edelsbrunner and Mücke 1994).

The benefit of this approach lies in various applications, which require a detailed characterization of the 3D crown shape and/or crown attributes following an initial segmentation. For example, Vauhkonen (2010a) showed that by iterating the alpha values, the horizontal segmentation of living crowns of Scots pine trees could be further vertically delimited, allowing the determination of tree crown attributes such as the crown base height. Tree crown objects formed by the alpha shape technique were used in predicting the species and stem attributes of Scandinavian trees (Vauhkonen et al. 2008, 2009), the crown volume derived in this way also being found the strongest predictor of tree stem attributes (Vauhkonen et al. 2010). The presented approach can be expected to be sensitive to the applied point density (Vauhkonen et al. 2008), for which reason the related applications are expected to benefit from a presence of high density point data. Examples beyond discrete-return ALS data are given by Reitberger et al. (2009), Rentsch et al. (2011), and Yao et al. (2012) using full-waveform ALS data, and by Rutzinger et al. (2010) using terrestrial (mobile) laser scanning data.

5.4.5.3 Superquadrics

The shape modeling of segmented single trees can also be performed by the geometry of extended superquadrics (Weinacker et al. 2004a). In mathematics the superquadrics belong to the family of geometric shapes resembling ellipsoids and quadrics, which resemble many shapes like cubes, cylinders, and spindles in varying levels of detail (Fig. 5.5). Superquadrics are highly flexible and therefore popular in geometric modeling especially in computer graphics (Barr 1981; Chevalier et al. 2003). This flexibility makes superquadrics also interesting for modeling crown shapes. Weinacker et al. (2004a) used the superquadrics for crown shape modeling improving flexibility by integrating deformations (Jaklic et al. 2000) like "tapering" (causes dilation or compression along one direction), "displacement" (planes parallel a distinct plane can be displaced, the z-coordinate was not but the total height was changed), "bending" (the length of the axis is not changed but the planes are rotated in different angles), "cavity" (convex objects can be changed in concave ones) and "torsion" (torsion around the z-axis). The function to be minimized is described in Weinacker et al. (2004a). For the fitting process based on

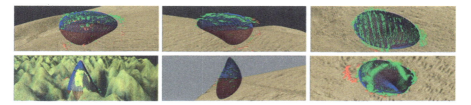

Fig. 5.5 Modeling a broad-leaved (*above*) and a conifer tree crown using superquadrics based on LiDAR point data. The *green* and *red* points show the points which have been used for the crown modeling

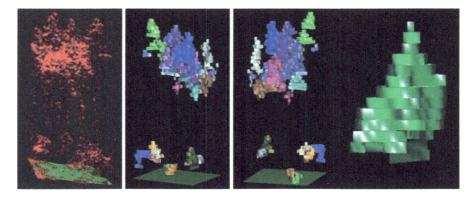

Fig. 5.6 Tree crown shape modeling using a prismatic 3D crown model

the segmented single tree point clouds all parameters can be introduced separately as stochastic. A more robust fitting and a priori weighting of each laser point is also possible. Therefore the algorithm is very flexible and a high number of possible parameter combinations can be used to achieve the best adaption.

5.4.5.4 Prismatic 3D Models

Finally, each detected tree crown may be described by an array of 2D tree crown regions in different layers at different height levels. Since the layers in voxel space have a certain thickness, 3D prisms can be constructed for the 2D crown regions in different layers with the thickness of layers as the height of the prisms. Groups of 3D prisms at different height levels are then derived for each individual crown, and a prismatic 3D crown model can be reconstructed by combining all the crown prisms (Fig. 5.6).

5.5   Concluding Remarks

To conclude, there are several currently established techniques that allow forest to tree level segmentation and object building based on airborne LiDAR data. Furthermore, the segmented objects can be further detailed or combined in hierarchical, multi-scale procedures, for which the example algorithms and techniques presented in this chapter provide a starting point.

Both area and single tree level approaches are used for forest area (Sect. 5.2) and individual stand (Sect. 5.3) delineations. Open forests and regeneration areas constitute challenges towards defining forest area, as a clear measure for the tree assembly constituting a forest area may not be identifiable based on the analyzed height, density and spatial pattern metrics. Stand level segmentation can be performed successfully, if a physical property like similar height and crown size distribution structure is considered as an adequate criterion for separating the individual stands. Compared to visual delineations, the results produced automatically are typically more fine-grained and may require an additional aggregation step to produce operational units. In uneven-aged forest areas and when non-physical criteria like administrative units need to be considered, the identification of the stand borders becomes more difficult and may require additional external information.

Individual tree crown delineations (Sect. 5.4) are most commonly based on raster-image analysis techniques such as local maxima detection and watershed segmentation. However, detecting the local height maxima from raster-based CHMs inherently misses trees below the dominant canopy, and techniques developed to adjust the degree of the CHM smoothing cannot fundamentally overcome this limitation. The detection of the small trees may be improved by means of a local refinement (Reitberger et al. 2009) or full analysis based on the 3D point clouds (Lähivaara et al. 2014; Tang et al. 2013). The latter techniques have been developed only recently and their operational applications may be currently restricted by the computational burden involved. Furthermore, due to the transmission losses occurring in the upper canopy (cf. Korpela et al. 2012) and the fact that forest structure also affects the detection of the dominant trees (Vauhkonen et al. 2012), even the best-case information obtainable regarding the lower-storey trees may only be indicative. The detection of "semi-individual" tree crowns (Chap. 6) is an example approach to compensate for the segmentation errors.

Despite the limitations described in the previous paragraph, the segmented dominant trees constitute useful information for applications such as wood procurement planning (see Vauhkonen et al. 2014). Tree crown reconstruction techniques presented in Sect. 5.4.5 enable further delineation of the tree crowns in the vertical direction, allowing further applications towards modeling tree growth and forest development, forest light interaction, and photo-realistic visualization, to name but a few.

References

Barr AH (1981) Superquadrics and angle preserving transformations. IEEE Comput Graph Appl 1:11–23

Blaschke T (2010) Object based image analysis for remote sensing. ISPRS J Photogramm Remote Sens 65:2–16

Chen Y, Dougherty ER (1994) Gray-scale morphological texture classification. Opt Eng 33:2713–2722

Chevalier L, Jaillet F, Baskurt A (2003) Segmentation and superquadric modeling of 3D objects. In: Proceedings of the European conference on computer vision, J WSCG03 11(1), 2003. ISSN 1213–6972

Cici A, Kevin T, Nicholas JT, Sarah S, Jörg K (2008) Extraction of vegetation for topographic mapping from full-waveform airborne laser scanning data. In: Proceedings of the SilviLaser 2008, Edinburgh, U.K., 17–19 September, pp 343–353

Dees M, Straub Ch, Koch B, Weinacker H (2006) Combining airborne laser scanning and GIS data to estimate timber volume of forest stands based on yield models. Project report, University of Freiburg, Freiburg, Germany

Deng Y, Manjunath BS (2001) Unsupervised segmentation of color-texture regions in images and video. IEEE Trans Pattern Anal Mach Intell 23:800–810

Diedershagen O, Koch B, Weinacker H (2004) Automatic segmentation and characterisation of forest stand parameters using Airborne LIDAR data multispectral and FOGIS data. Int Arch Photogramm Remote Sens Spat Inf Sci XXXVI:208–212

Doo-Ahn K, Woo-Kyun L, Hyun-Kook C (2008) Estimation of effective plant area index using LiDAR data in forest of South Korea. In: Proceedings of the SilviLaser 2008, Edinburgh, U.K., 17–19 September, pp. 237–246

Dougherty ER (1992) An introduction to morphological image processing. SPIE Optical Engineering Press, Center for Imaging Science, Rochester Institute of Technology, Rochester

Edelsbrunner H, Mücke EP (1994) Three-dimensional alpha shapes. ACM Trans Graph 13:43–72

Ene L, Næsset E, Gobakken T (2012) Single tree detection in heterogeneous boreal forests using airborne laser scanning and area based stem number estimates. Int J Remote Sens 33:5171–5193

Eysn L, Hollaus M, Schadauer K, Pfeifer N (2012) Forest delineation based on airborne LiDAR data. Remote Sens 4:762–783

Friedländer H (2002) Die Anwendung von flugzeuggetragenen Laserscannerdaten zur Ansprache dreidimensionaler Strukturelemente von Waldbeständen, Ph.D. dissertation, University of Freiburg, Freiburg, Germany, 64 p

Gonzalez R, Woods RE (2008) Digital image processing, 3rd edn. Prentice Hall, Englewood Cliffs, 954 p. ISBN 9780131687288

Gupta S, Weinacker H, Koch B (2010) Comparative analysis of clustering-based approaches for 3-D single tree detection using airborne fullwave Lidar data. Remote Sens 2010:968–989

Heinzel J, Weinacker H, Koch B (2011) Prior-knowledge-based single-tree extraction. Int J Remote Sens 32:4999–5020

Heipke C (2001) Digital photogrammetric workstations – a review of the state-of-the-art for topographic application. GIM Int 15(4):35–37

Höfle B, Hollaus M (2010) Urban vegetation detection using high density full-waveform airborne lidar data – combination of object-based image and point cloud analysis. Int Arch Photogramm Remote Sens Spat Inf Sci XXXVIII-7B:281–286

Holmgren J, Wallerman J (2006) Estimation of tree size distribution by combining vertical and horizontal distribution of LIDAR measurements with extraction of individual trees. In: Proceedings of the workshop on 3D remote sensing in forestry, 14–15 Feb 2006. University of Natural Resources and Applied Life Science, Vienna, pp. 168–173

Holmgren J, Barth A, Larsson H, Olsson H (2012) Prediction of stem attributes by combining airborne laser scanning and measurements from harvesters. Silva Fenn 46:227–239

Hyyppä J, Inkinen M (1999) Detecting and estimating attributes for single trees using laser scanner. Photogramm J Finland 16:27–42

Hyyppä J, Kelle O, Lehikoinen M, Inkinen M (2001a) A segmentation-based method to retrieve stem volume estimates from 3-D tree height models produced by laser scanners. IEEE Trans Geosci Remote Sens 39:969–975

Hyyppä J, Schardt M, Haggren H, Koch B (2001b) HIGH-SCAN: the first European-Wide attempt to derive single tree information from Laserscanner data. Photogramm J Finland 17:58–69

Hyyppä J, Hyyppä H, Leckie D, Gougeon F, Yu X, Maltamo M (2008) Review of methods of small-footprint airborne laser scanning for extracting forest inventory data in boreal forests. Int J Remote Sens 29:1339–1366

Hyyppä J, Yu X, Hyyppä H, Vastaranta M, Holopainen M, Kukko A, Kaartinen H, Jaakkola A, Vaaja M, Koskinen J, Alho P (2012) Advances in forest inventory using airborne laser scanning. Remote Sens 4:1190–1207

Jain AK, Murty MN, Flynn PJ (1999) Data clustering: a review. ACM Comput Surv 31(3):264–323

Jaklic A, Leonardis A, Solina F (2000) Segmentaion and recovery of superquadrics. Kluwer Academic Publisher, Dordrecht, 266 p. ISBN 0-7923-6601-8

Kaartinen H, Hyyppä J, Yu X, Vastaranta M, Hyyppä H, Kukko A, Holopainen M, Heipke C, Hirschmugl M, Morsdorf F, Næsset E, Pitkänen J, Popescu S, Solberg S, Wolf BM, Wu J-C (2012) An international comparison of individual tree detection and extraction using airborne laser scanning. Remote Sens 4:950–974

Koch B, Heyder U, Weinacker H (2006) Detection of individual tree crowns in airborne LIDAR data. Photogramm Eng Remote Sens 72:357–363

Koch B, Straub C, Dees M, Wang Y, Weinacker H (2009) Airborne laser data for stand delineation and information extraction. Int J Remote Sens 30:935–936

Koivuniemi J, Korhonen KT (2006) Inventory by compartments. In: Kangas A, Maltamo M (eds) Forest inventory – methodology and applications, vol 10, Managing forest ecosystems. Springer, Dordrecht, pp 271–278

Korpela I, Hovi A, Morsdorf F (2012) Understory trees in airborne LiDAR data – selective mapping due to transmission losses and echo-triggering mechanisms. Remote Sens Environ 119:92–104

Lähivaara T, Seppänen A, Kaipio JP, Vauhkonen J, Korhonen L, Tokola T, Maltamo M (2014) Bayesian approach to tree detection based on airborne laser scanning data. IEEE Trans Geosci Remote Sens. doi:10.1109/TGRS.2013.2264548

Leppänen VJ, Tokola T, Maltamo M, Mehtätalo L, Pusa T, Mustonen J (2008) Automatic delineation of forest stands from LIDAR data. In: Proceedings of the GEOBIA 2008 – Pixels, objects, intelligence GEOgraphic object based image analysis for the 21st Century, Calgary, Alberta, Canada, 5–8 Aug 2008, 6 p

Li W, Guo Q, Jakubowski M, Kelly M (2012) A new method for segmenting individual trees from the lidar point cloud. Photogramm Eng Remote Sens 78:75–84

Lloyd CD, Atkinson PM (2002) Deriving DSMs from LiDAR data with kriging. Int J Remote Sens 23:2519–2524

Mao J, Jain AK (1996) A self-organizing network for hyperellipsoidal clustering. IEEE Trans Neural Net 7:16–29

Means JE, Acker SA, Fitt BJ, Renslow M, Emerson L, Hendrix CJ (2000) Predicting forest stand characteristics with airborne laser scanning LIDAR. Photogramm Eng Remote Sens 66:1367–1371

Morsdorf F, Meier E, Kotz B, Itten KI, Dobbertin M, Allgöwer B (2004) Lidar-based geometric reconstruction of boreal type forest stands at single tree level for forest and wildland fire management. Remote Sens Environ 92:353–362

Mustonen J, Packalén P, Kangas A (2008) Automatic delineation of forest stands using a canopy height model and aerial photography. Scand J For Res 23:534–545

Næsset E. Bjerknes KO (2001) Estimating tree heights and number of stems in young forest stands using airborne laser scanner data. Remote Sens Environ 78:328–340

O'Rourke J (1994) Computational geometry in C. Cambridge University Press, Cambridge

Packalén P, Maltamo M, Tokola T (2008) Detailed assessment using remote sensing techniques. Designing green landscapes. Springer, Dordrecht, pp 53–77

Pascual C, Garda-Abril A, Garcia-Montero LG, Martin-Femandez S, Cohen WB (2008) Object-based semi-automatic approach for forest structure characterization using lidar data in heterogeneous *Pinus sylvestris* stands. For Ecol Manag 255:3677–3685

Persson Å, Holmgren J, Söderman U (2002) Detecting and measuring individual trees using an airborne laser scanner. Photogramm Eng Remote Sens 68:925–932

Pitkänen J, Maltamo M, Hyyppä J, Yu X (2004) Adaptive methods for individual tree detection on airborne laser based canopy height model. In: Proceedings of the ISPRS Working Group VIII/2: "Laser-scanners for forest and landscape assessment". University of Freiburg, Germany, pp 187–191

Pollock RJ (1996) The automatic recognition of individual trees in aerial images of forests based on a synthetic tree crown image model. PhD thesis, University of British Columbia, Canada

Popescu SC, Wynne RH, Nelson RH (2002) Estimating plot-level tree heights with lidar: local filtering with a canopy-height based variable window size. Comput Electr Agric 37:71–95

Popescu SC, Wynne RH, Nelson RF (2003) Measuring individual tree crown diameter with lidar and assessing its influence on estimating forest volume and biomass. Can J Remote Sens 29:564–577

Preparata FP, Shamos MI (1985) Computational geometry – an introduction. Springer, New York. ISBN 978-1-4612-7010-2

Reitberger J, Schnörr C, Krzystek P, Stilla U (2009) 3D segmentation of single trees exploiting full waveform LIDAR data. ISPRS J Photogramm Remote Sens 64:561–574

Rentsch M, Krismann A, Krzystek P (2011) Extraction of non-forest trees for biomass assessment based on airborne and terrestrial LiDAR data. Photogramm Image Anal 6952:121–132, Springer, Berlin/Heidelberg

Rieger W, Eckmüllner O, Müllner H, Reiter T (1999) Laser-scanning from the derivation of forest stand parameters. ISPRS workshop: mapping forest structure and topography by airborne and spaceborne lasers, La Jolla, CA

Rutzinger M, Höfle B, Pfeifer N (2008) Object detection in airborne laser scanning data – an integrative approach on object-based image and point cloud analysis. In: Blaschke T, Lang S, Hay G (eds) Object-based image analysis – spatial concepts for knowledge-driven remote sensing applications, Springer, Berlin, pp 645–662

Rutzinger M, Pratihast AK, Oude Elberink S, Vosselman G (2010) Detection and modeling of 3D trees from mobile laser scanning data. In: Proceedings of the ISPRS TCV mid-term symposium, Newcastle upon Tyne, 6 p

Shi J, Malik J (2000) Normalized cuts and image segmentation. IEEE Trans Pattern Anal Mach Intell 22:888–905

Soille P (1999) Morphological image analysis: principles and applications. Springer, New York

Soille P (2003) Morphological image analysis: principles and applications, 2nd edn. Springer, Berlin/New York, 316 pp

Solberg S, Næsset E, Bollandsås OM (2006) Single tree segmentation using airborne laser scanner data in a heterogeneous spruce forest. Photogramm Eng Remote Sens 72:1369–1378

Steger C, Ulrich M, Wiedemann C (2008) Machine vision algorithms and applications. Wiley, Weinheim, pp 137–138

Straub C, Weinacker H, Koch B (2008) A fully automated procedure for delination and classification of forest and non-forest vegetation based on full waveform laser scanner data. In: Proceedings of the XXI ISPRS congress Beijing, China, 3–8 July 2008. Int Arch Photogramm Remote Sens XXVII part B 8, WG VII711:1013–1020

Tang S, Dong P, Buckles BP (2013) Three-dimensional surface reconstruction of tree canopy from lidar point clouds using a region-based level set method. Int J Remote Sens 34:1373–1385

Tiede D, Hoffmann C (2006) Process oriented object-based algorithms for single tree detection using laser scanning. In: Proceedings of the 3-D remote sensing in forestry, University of Natural Resources and Applied Life Sciences (BOKU), Vienna, Austria, pp 151–156

Tiede D, Blaschke T, Heurich M (2004) Object-based semi-automatic mapping of forest stands with laser scanner and multispectral data. Int Arch Photogramm Remote Sens XXXVI(Part8/W2):328–333

Vauhkonen J (2010a) Estimating crown base height for Scots pine by means of the 3-D geometry of airborne laser scanning data. Int J Remote Sens 31:1213–1226

Vauhkonen J (2010b) Estimating single-tree attributes by airborne laser scanning: methods based on computational geometry of the 3-D point data. Diss For 104, 44 p

Vauhkonen J, Tokola T, Maltamo M, Packalén P (2008) Effects of pulse density on predicting characteristics of individual trees of Scandinavian commercial species using alpha shape metrics based on airborne laser scanning data. Can J Remote Sens 34:S441–S459

Vauhkonen J, Tokola T, Packalén P, Maltamo M (2009) Identification of Scandinavian commercial species of individual trees from airborne laser scanning data using alpha shape metrics. For Sci 55:37–47

Vauhkonen J, Korpela I, Maltamo M, Tokola T (2010) Imputation of single-tree attributes using airborne laser scanning-based height, intensity, and alpha shape metrics. Remote Sens Environ 114:1263–1276

Vauhkonen J, Ene L, Gupta S, Heinzel J, Holmgren J, Pitkänen J, Solberg S, Wang Y, Weinacker H, Hauglin KM, Lien V, Packalén P, Gobakken T, Koch B, Næsset E, Tokola T, Maltamo M (2012) Comparative testing of single-tree detection algorithms under different types of forest. Forestry 85:27–40

Vauhkonen J, Packalen P, Malinen J, Pitkänen J, Maltamo M (2014) Airborne laser scanning based decision support for wood procurement planning. Scand J For Res. doi:10.1080/02827581.2013.813063

Vincent L, Soille P (1991) Watersheds in digital spaces: an efficient algorithm based on immersion simulations. IEEE PAMI 13:583–598

Wang Z, Boesch R, Ginzler C (2007) Color and LiDAR data fusion. Int Arch Photogramm Remote Sens Spat Inf Sci [CD] 36, 1/W51:4

Wang Z, Boesch R, Ginzler C (2008a) Integration of high resolution aerial images and airborne LiDAR data for forest delineation. ISPRS XXI congress Beijing, China, 3–8 July 2008, Int Arch Photogramm Remote Sens XXVII part B 8, WG VII711:1203–1208

Wang Y, Weinacker H, Koch B (2008b) A Lidar point cloud based procedure for vertical canopy structure analysis and 3D single tree modelling in forest. Sensors 2008:3938–3950

Weinacker H, Koch B, Heyder U, Weinacker R (2004a) Development of filtering, segmentation and modelling modules for LIDAR and multispectral data as a fundamental of an automatic forest inventory system. Int Arch Photogramm Remote Sens Spat Inf Sci XXXVI:90–95

Weinacker H, Koch B, Weinacker R (2004b) TREESVIS – a software system for simultaneous 3D-real-time visualization of DTM, DSM, laser row data, multi-spectral data, simple tree and building models. Int Arch Photogramm Remote Sens Spat Inf Sci XXXVI:90–95

Wu Z, Heikkinen V, Hauta-Kasari M, Parkkinen J, Tokola T (2013) Forest stand delineation using a hybrid segmentation approach based on airborne laser scanning data. Lect Notes Comput Sci 7944:95–106

Yao W, Krzystek P, Heurich M (2012) Tree species classification and estimation of stem volume and DBH based on single tree extraction by exploiting airborne full-waveform LiDAR data. Remote Sens Environ 123:368–380

Chapter 6
The Semi-Individual Tree Crown Approach

Johannes Breidenbach and Rasmus Astrup

Abstract The individual tree crown (ITC) approach is a popular method for estimating forest parameters from airborne laser scanning data. One disadvantage of the approach is that errors in tree crown detection can result in estimates of forest parameters with considerable systematic errors. The semi-ITC approach is one method to reduce such systematic errors. In this chapter, we present different variations of the semi-ITC approach and review their application. Two variations of the semi-ITC approach are applied in a case study and compared with the ITC and the area-based approach. One of the semi-ITC approaches is based on the k nearest neighbors (kNN) method used to estimate forest parameters. In the case study, we analyze how different distance metrics and numbers of neighbors influence the accuracy and precision of forest parameter estimates at plot level and stand level.

6.1 Introduction

Forest inventories can be used to provide estimates of natural resources on national, regional, and local scales in order to meet reporting requirements and to support decision-making processes in policy and management. Traditionally, forest inventories are based on extensive field work. The combination of field data and airborne laser scanning (ALS) has proven valuable when providing estimates with high precision especially for management inventories that provide information on small domains such as stands.

The first studies attempting to use ALS data in a forest inventory context (Nilsson 1996; Næsset 1997; Means et al. 2000) applied the area-based approach (ABA). In the ABA, the response variable is an aggregated value over a sample plot such as

J. Breidenbach (✉) • R. Astrup
National Forest Inventory, Norwegian Forest and Landscape Institute, Ås, Norway
e-mail: Johannes.Breidenbach@skogoglandskap.no

M. Maltamo et al. (eds.), *Forestry Applications of Airborne Laser Scanning: Concepts and Case Studies*, Managing Forest Ecosystems 27, DOI 10.1007/978-94-017-8663-8_6,
© Springer Science+Business Media Dordrecht 2014

113

timber volume per ha or mean tree height. The predictor variables are characteristics of the ALS height distribution at the sample plot such as the mean height, height percentiles, or proportion of returns within a certain height layer (Næsset 2002). The predictor variables in the ABA are often denoted as height and density metrics (see Chap. 1 for further details).

Due to the visibility of single trees in high-resolution ALS data, Hyyppä and Hyyppä (1999), Hyyppä and Inkinen (1999), Borgefors et al. (1999), and Hyyppä et al. (2001b) started using automatically detected tree crowns to estimate forest properties. The aim of the individual tree crown approach (ITC), which is also known as individual tree detection (ITD) or single tree approach, is to derive tree attributes of interest from trees or tree crowns detected in ALS data (see Chap. 1). The ITC approach is conceptually similar to earlier approaches used in high-resolution photography (Gougeon 1995; Gougeon and Leckie 2003; Hyyppä et al. 2008) and basically consists of five steps:

1. Detection of tree crowns in canopy height models or point clouds covering the areas of interest (AOIs) for which estimates of forest parameters are required.
2. Linking detected crowns with trees observed on field plots with known tree locations. The primary assumption in the traditional ITC approach is that one field-measured tree can be linked to one crown detected in the ALS data.
3. Fitting statistical models that regress field-measured tree characteristics against metrics derived from the detected tree crowns. Such metrics usually include the area of the detected crown and the maximum ALS height within the detected crown.
4. Application of the fitted model to the detected tree crowns within AOIs such as stands in order to estimate tree characteristics of interest.
5. Typically, the estimates for crown segments within each stand are averaged or summed in order to estimate mean or total forest characteristics within the stand.

Popular methods for crown detection in canopy height models are watershed algorithms (Chap. 5). Since the canopy height model is segmented, the detected tree crowns are often denoted "segments".

Usually, only dominant trees are correctly identified in step 1 in the sense that exactly one field-measured tree is within one detected crown segment (Persson et al. 2002). Frequently, only "correctly identified" segments with one linked field tree are used to fit regression models in step 3. However, the number of field-measured trees within a segment influences the response variable such as timber volume or dbh (diameter at breast height) of the segment. In general, it can be said that omitting observations from a regression model based on the response variable, for example empty segments or segments with several field trees, will result in biased regression models.

Hyyppä and Inkinen (1999) and Hyyppä et al. (2001a) estimated the dbh for detected crowns in stands using the fitted model in step 3. Using the estimated dbh and maximum ALS height, timber volume was estimated for each segment using existing volume models. The authors reported considerable systematic errors for estimates on stand level using the ITC approach, which largely resulted from errors in the crown segmentation. Hyyppä and Inkinen (1999) suggested the use of a correction factor to compensate for the systematic error. The semi-ITC approach—described in detail in the section below—is another method to eliminate, or at least reduce, such systematic errors. It should be noted that the semi-ITC approach is not necessarily more precise than the ABA, but provides estimates with a higher spatial resolution.

In general, the data requirement of ITC approaches is greater than for the ABA because high-density ALS data (>1 return per m^2) are necessary and the coordinates of single trees need to be recorded in the field. However, the ITC approach may be more intuitive for forest practitioners because the single tree is the smallest unit of interest in operational management. Additionally, in mixed-species forests, ITC approaches may have an advantage over the ABA when it comes to tree species classification. Furthermore, the ITC approach provides tree coordinates that may be useful in certain applications.

In the next section (Sect. 6.2) we give a short overview of developments around the ITC approach. Thereafter (Sect. 6.3) we describe the semi-ITC approach. The results of a case study, in which different variations of the ITC and semi-ITC approach were compared to the ABA, are presented and discussed in Sects. 6.4 and 6.5. While the ITC approach may be useful in many different contexts such as the estimation of environmental variables or wood quality, in this chapter we focus on traditional forest inventories that provide information about timber volume or biomass. Forest inventories are usually based on field samples and statistical estimators to infer population-level characteristics. In this regard, ALS data analyzed with a (semi-) ITC approach may help to improve the precision of estimates. This chapter covers primarily the modeling part of the analyses. We refer to Chap. 14 and Flewelling (2008, 2009) where sampling-related issues are covered.

6.2   Development of the ITC Approach

A large number of studies have modified or advanced the ITC approach, either adopting one or several of the steps mentioned above, or estimating different response variables. Only a few examples are summarized in this section.

- Most work has focused on improving the automated detection of trees in ALS data (e.g., Kaartinen et al. 2012). Especially algorithms that adapt to certain forest structures such as crown sizes or the number of trees appear promising (e.g., Heinzel et al. 2011; Ene et al. 2012; Lindberg et al. 2013). Chapter 5 gives an overview of methods for single tree detection and segmentation from ALS surface models or point clouds that can be used to detect trees or their crowns in step 1.
- A somewhat extreme approach was to omit step 3 and instead use existing relationships between crown properties that were obtained from the detected tree

crowns to estimate tree parameters of interest in step 4 (Hyyppä et al. 2005; Vastaranta et al. 2011). Many growth models, for example, contain sub-models for the relationship of the crown diameter and the dbh which could be exploited. However, usually the measuring method of the data used to fit the growth model differs from the method used to measure the crown diameter from the segments. While this approach is tempting because no field data are needed for model fitting, extreme care during the validation of the results is necessary, as it is very likely to arrive at biased estimates. Transferred to the ABA, this approach would mean that a documented relationship between ALS metrics and a response variable is utilized without fitting a new model (Suvanto and Maltamo 2010).

• Instead of a parametric regression in step 3, nonparametric regression techniques can be used to describe the relationship between tree parameters of interest and predictor variables obtained from the ALS data within the segments (e.g., Maltamo et al. 2009; Vauhkonen et al. 2010; Breidenbach et al. 2010; Yu et al. 2011). This approach is especially useful if the response is multivariate (e.g., simultaneous prediction of tree height, crown height, and timber volume) or if it is desirable to use many predictor variables (Vauhkonen et al. 2010). In addition to metrics derived from ALS returns within the tree segment, Maltamo et al. (2009), Vauhkonen et al. (2010), and Peuhkurinen et al. (2011) also used metrics similar to those used in ABA, computed from the ALS returns in a small neighborhood about the ITC.

6.3   The Semi-ITC Approach

6.3.1   Background

The detection of trees and their crowns is most successful in open-spaced homogenous coniferous forests. Nonetheless, even in forests well-suited for automated tree detection algorithms, only a proportion of the trees are "correctly" identified in the sense that one field-measured tree can be linked to exactly one identified crown segment (e.g., Persson et al. 2002; Koch et al. 2006). In general, two types of errors (omission and commission) can occur:

1. Field-measured trees may be missed during automated segmentation because several trees can be clustered within one segmented tree crown. Especially small trees under or next to a dominant tree often will not be detected. Such commission errors are sometimes called "under-segmentation."
2. Commission errors usually result from "over-segmentation" when one tree crown is split into several segments. A less common reason for commission errors may be the segmentation of non-tree objects.

The frequency of the errors depends on the properties of the crown detection algorithm and the ALS data as well as the forest structure (Vauhkonen et al. 2012).

The reasons for systematic errors of estimates based on the conventional ITC approach are twofold. First, and most importantly, the above-described omission and commission errors usually will not compensate for each other. The number of stems is usually underestimated because suppressed trees are less likely to be detected (Maltamo et al. 2004). Second, the correctly identified trees are the largest and most dominant trees (Persson et al. 2002). The allometry of the detected trees thus differs from the missing trees, which results in an overestimation of the mean tree size for the detected trees due to a bias in the regression. In a study by Peuhkurinen et al. (2011), underestimation of stem number and the overestimation of mean stem size compensated for each other almost exactly in the estimates of timber volume and basal area on plot-level.

The systematic errors in ITC estimates motivated Maltamo et al. (2004) and Lindberg et al. (2010) to combine the ABA and ITC approach. In their studies, the diameter distribution predicted using the ABA was utilized to augment the diameter distribution resulting from the ITC approach, as suggested by Hyyppä and Inkinen (1999). Yu et al. (2010) and Vastaranta et al. (2011) included a plot-level correction for the systematic error of the ITC estimates. Vastaranta et al. (2011) found 20 sample plots to be sufficient to estimate the correction factor.

Other ways for reducing systematic errors of ITC approaches are semi-ITC approaches (Hyyppä et al. 2005; Flewelling 2008; Breidenbach et al. 2010). The prefix "semi" supposedly weakens the following word "individual" because it indicates that an automatically detected tree crown can be a cluster that may contain no, one, or several field-measured trees (Breidenbach et al. 2010). A better term for the semi-ITC approach could be TCA (tree crown approach), from which the word "individual" is omitted. Nonetheless, the term semi-ITC (or semi-ITD) has become somewhat established and is therefore also used in this chapter.

In contrast to the other above-described methods, semi-ITC approaches aim to reduce or prevent systematic errors on the segment level, not on the plot level. Another method that reduces systematic errors has been described by Mehtätalo (2006). His method compensates for effects of overlapping tree crowns. In his example, the method improved ITC stem count estimates based on simulated ALS data.

6.3.2   Variations of the Semi-ITC Approach

Semi-ITC approaches are special cases of the ITC approach in the sense that some of the 5 steps described in the Introduction section (Sect. 6.1) are modified.

Flewelling (2008, 2009) described what we refer to as the parametric semi-ITC approach. In the parametric semi-ITC approach, all trees within a segment are considered (step 2). In step 3, logistic regressions are used to estimate the probabilities that an ITC represents 0, 1, 2, ..., n trees. Conditional regressions are fit to estimate the diameters and heights of the various numbers of trees. In application (step 4), the logistic regressions and conditional regressions are combined to estimate the probabilistic outcome for each detected tree crown; expectations are calculated and summed to obtain estimated stand tables for each stand. Flewelling also used fused CIR (color infrared) data to separate further the ITC outcomes and resultant stand tables by species group. The regression equations were fit in such a way as to be unbiased for basal area and tree count by species group. Special steps were taken to retain the full variability of the diameter and height outcomes. Otherwise, the diameter regressions would have caused the inferred diameter distribution to be too narrow.

Compared to the above-described approach, a simplified variation of the parametric semi-ITC approach is the tree cluster approach proposed by Hyyppä et al. (2005, 2006). In the approach, all trees within a segment are considered (step 2). In cases where there are several trees within a segment, the variable of interest such as timber volume is aggregated (added) within a segment. The aggregated variable of interest is then regressed against predictor variables derived from the segments (step 3). In order to arrive at unbiased results, it is important to use also the empty segments with no linked trees in the regression. In this approach it is assumed that all trees within one segment belong to the same tree species. Since this assumption may often not hold in the case of natural forests, the tree cluster approach is suited for estimates independent of tree species.

Breidenbach et al. (2010) have described what we hereafter call the nonparametric semi-ITC approach, in which each segment is attributed with a tree list in step 2. The tree list includes tree properties, such as the timber volume of field-measured trees associated with a segment. The list can be empty for segments that do not contain a tree or it can consist of one or several trees. If there are several trees within a segment, they may be different species. In step 3, a nonparametric k nearest neighbor model (kNN) is "fitted" using the aggregated tree properties. The kNN model fulfills the same purpose as the linear and logistic models in the parametric semi-ITC approach. In step 4, the tree lists associated with the reference segments used to fit the kNN model are imputed to the target segments in stands. Since tree lists are imputed, any property measured for trees in the field could be estimated using one kNN model.

Since parametric and nonparametric methods have advantages and disadvantages, approaches combining both methods are feasible too. For example, a kNN method could be used determining how many trees a semi-ITC segment contains, and use regression models to estimate the sizes of the associated trees.

6.3.3   Studies Applying the Semi-ITC Approach

Hyyppä et al. (2005, 2006) used the tree cluster approach to estimate timber volume within segments. They found the tree cluster approach to be superior to the ITC approach on the segment level. The estimates were not aggregated on the plot level or stand level.

Breidenbach et al. (2010) compared the most similar neighbor (MSN) technique (Moeur and Stage 1995) and random forests (Breiman 2001) to determine the distance in the feature space[1] between a target and a reference segment. The predictor variables were obtained from ALS and multispectral images. The response was species-specific timber volume and the number of neighbors (k) was set to one. The semi-ITC approach did not result in systematic errors and outperformed the ABA slightly for overall timber volume but more clearly for species-specific timber volume. MSN was found to be a better suited distance metric than random forests.

The nonparametric semi-ITC approach was used by Packalén et al. (2011, 2013) to estimate the spatial distribution of trees. The semi-ITC approach was found to be superior to both the ITC approach and ABA but the accuracy was in general low. Also Vauhkonen et al. (2011), found the nonparametric semi-ITC approach superior to other approaches considered for timber volume estimation. However, the differences between the approaches were minor and the ITC approach was better suited for species prediction. The latter was also reported by Ørka et al. (2013).

Breidenbach et al. (2012) used the nonparametric semi-ITC approach to generate artificial sample plots in areas of high-density ALS data. The Euclidean distance metric was used instead of MSN. An aggregation of response variables from field measured trees within a segment to "fit" the kNN model was therefore not necessary. The artificial plots supplemented the field plots in the ABA based on MSN using low-density ALS data. The authors found that the use of the artificial plots generated using the semi-ITC approach helped to improve estimates of timber volume.

Holmgren et al. (2012) applied the nonparametric semi-ITC approach to estimate stem volume, mean tree height, mean diameter, and stem number that had been measured with harvesters. The nonparametric semi-ITC approach was also used to compare the performance of new and existing tree detection algorithms (Lindberg et al. 2013). Vauhkonen et al. (2013) compared ITC and semi-ITC (termed "tree-list imputation") for wood procurement planning. While the semi-ITC estimates of timber volume by species were usually less biased than the ITC estimates, the variances were not notably different in most cases.

Wallerman et al. (2012) used the nonparametric semi-ITC approach to estimate forest parameters from photogrammetric point clouds instead of ALS. Due to the availability of digital aerial images and improved photogrammetric software, it is foreseeable that this kind of application will increase considerably in the future.

The above-cited list of studies using the semi-ITC approach may not be complete. Nonetheless, the simplicity and wide acceptance of kNN methods seems to have favored the selection of the nonparametric semi-ITC approach over the parametric semi-ITC approach so far. This is somewhat surprising, since the tree cluster approach is also easily implemented.

---

[1]The feature space is spanned by the selected predictor variables. The distance is thus not geographic in nature but determined by the similarity of the predictor variables.

6.4   A Case Study

In the case study presented here, four approaches are compared to estimate tree
biomass on plot level and stand level: ITC, the nonparametric semi-ITC, the
(parametric) tree cluster approach, and the (parametric) area-based approach. The
study area is located in the municipality of Lardal in southern Norway. For
the nonparametric semi-ITC approach, different nearest neighbors distance metrics
(raw Euclidean, Mahalanobis, MSN, and random forest) as implemented in the R
package yaImpute (Crookston and Finley 2008) and different numbers of neighbors
($k = 1, \ldots, 10$) are tested. For an introduction to kNN methods, we refer to the
review paper by Eskelson et al. (2009).

6.4.1   Field Data

A total of 30 forested Norwegian National Forest Inventory (NFI) sample plots were
located in the study area. The NFI plots in this part of the country are distributed
along a 3×3 km grid. Each circular plot has a size of 250 m^2 and all trees with a dbh
>5 cm are measured for dbh, species and location. The location of the plot center
coordinate is measured with survey-grade GPS equipment resulting in accuracies
usually better than 0.5 m. The tree height is measured on a subsample of 10 trees per
plot. The heights of the remaining trees are estimated based on a tariff method using
the measured trees for calibration at the plot level. Single-tree biomass is estimated
using the models by Marklund (1988). The biomass models are species-specific and
require dbh and tree height as input parameters. The model for birch was used for
all deciduous trees. The NFI sample plots were measured between 2005 and 2009
and were used to fit or train the statistical models. The proportions of spruce, pine
and deciduous species on the NFI plots were 74 %, 6 %, and 20 %, respectively.
Table 6.1 gives an overview of further characteristics of the NFI data.

For validation, in 2012, between 5 and 7 sample plots were measured in 30
randomly selected compact stands of between 1 and 3 ha in size. Compact stands
were selected to minimize problems resulting from detecting stand borders during
field work. The compactness of the stands was determined based on their area to
perimeter ratio which had to be >0.2. Stand borders were determined from aerial
images in a recent forest management inventory. It was known from the inventory
that large-volume stands (≥ 150 m^3/ha) were twice as common as small-volume

Table 6.1   Characteristics of NFI tree-level and plot-level parameters

Parameter	Mean	Standard deviation	Maximum
Tree-level dbh (cm)	12	67	41
Tree-level height (m)	10	4	24
Tree-level biomass (kg)	86	129	920
Plot-level biomass (Mg/ha)	84	73	249

Table 6.2 Characteristics of tree-, plot- and stand-level parameters of validation data

Parameter	Mean	Standard deviation	Maximum
Tree-level dbh (cm)	16	83	51
Tree-level height (m)	13	5	31
Tree-level biomass (kg)	163	191	1,528
Plot-level biomass (Mg/ha)	111	83	338
Mean stand-level biomass (Mg/ha)	124	66	257

stands (<150 m^3/ha). In order to reflect the whole range of stands, the sampling probability of small-volume stands was increased by two. A total of 186 sample plots were available for validation. The measurements on the validation plots were done in accordance with the NFI protocol described above. However, tree positions were not recorded. The proportions of spruce, pine, and deciduous species on the validation plots were 73 %, 7 %, and 20 %, respectively. Table 6.2 gives an overview of further characteristics of the validation data.

6.4.2 ALS Data

Discrete return ALS data with up to three intermediate returns were acquired between 21 and 25 May 2009 using Optec Gemini sensors (ALTM05SEN180 and ALTM04SEN161) with a half scan-angle of 12° (Gjessing and Werner 2009). The average flying height and speed of the fixed-wing aircrafts was 690 m above ground and 80 m s^{-1}, respectively. The flight and sensor settings resulted in an average density of approximately 10 pulses per m^2 and a footprint size of 13 cm. Elevations were normalized to heights above ground by the data provider. ALS intensities were not used in the study.

6.4.3 Segmentation

Using the software FUSION (McGaughey 2010), a digital canopy height model (CHM) with a cell size of 35 cm was calculated from the highest ALS return within a cell. All return types were considered. A simple watershed algorithm was then applied to the inverted CHM where the canopy height was above 2 m. Segments smaller than 0.5 m^2 were deleted. For more details on the segmentation procedure, we refer to Breidenbach et al. (2012).

A total of 884 segments were located wholly within the 30 NFI sample plots. The segment size was in the range 0.6–18.7 m^2, with a median of 1.6 m^2. These reference segments were imputed in the nonparametric semi-ITC approach (Sect. 6.4.5.2). These segments were also used in the tree cluster approach (Sect. 6.4.5.3). A subset of these segments was used in the ITC approach as described in Sect. 6.4.5.1. Of the 884 segments, 64 % were empty in the sense that no field-measured tree was located within their boundary. Further, 25 %, 5 %, and 5 % of the segments had one, two, and three or more field-measured trees located within their boundary. A total of 498 trees were located within 315 of the segments and a maximum of 9 trees were located within one segment.

A total of 9,114 segments had their centroids within the validation plots and were used to estimate the biomass on these plots using the ITC and semi-ITC approaches. The segment size was in the range 0.6–41.1 m^2 with a median of 1.6 m^2.

6.4.4 Explanatory Variables

Height and density metrics are well-known explanatory variables in the ABA. Height metrics are statistics that describe the distribution of the heights of ALS returns within a sample plot or some other area. For example, height metrics can be the mean (H_{mean}), standard distribution (H_{SD}), or percentiles of the height distribution (H_{min}, H_{P10}, ..., H_{P90}, H_{max}). Density metrics are the proportion of returns within certain height bins or above or below a certain pre-determined height. In the case study, density metrics within 5 m vertical bins were calculated. For example, D_{5-10} denotes the proportion of returns within 5–10 m and $D_{>mean}$ denotes the proportion of returns above the mean. In this case study, the explanatory variables derived from ALS raw data are based on all return types (i.e., first, intermediate and last returns).

Similar to the ABA, metrics describing the height and density distribution of the returns within each segment can be calculated. It should be noted that the metrics of the segments are even more closely intercorrelated than those used in the ABA because the segments are smaller than sample plots and consequently the metrics are based on fewer returns. In addition to the explanatory variables known in the ABA, the segment's area (A_{seg}) or variables derived thereof such as the crown diameter of the segment are important predictor variables for ITC and semi-ITC approaches. Selection of explanatory variables was based on previous experience and iterative comparisons of different model alternatives.

6.4.5 Specifics of the ITC, Semi-ITC, and Area-Based Approaches

6.4.5.1 ITC Approach

Similarly to Hyyppä and Inkinen (1999), Persson et al. (2002), and Peuhkurinen et al. (2011), a linear model based on segments wholly within a NFI sample plot with one "correctly identified" field measured tree was fitted in step 3. However, instead of estimating the segment's dbh independent of species in step 3 and using the estimated dbh with existing biomass models, we used the tree biomass within a segment as the response variable in step 3. This procedure was followed because the study site consisted of heterogeneous forests with several tree species. Only the segment area and its interaction with the largest return height within the segment turned out to be significant explanatory variables. This resulted in the model

$$y = \beta_0 + \beta_1 A_{seg} + \beta_2 H_{max} + \beta_3 A_{seg} H_{max} + \varepsilon, \tag{6.1}$$

where ε is a random error. The estimated standard distribution of the model was $\widehat{\sigma} = 60.2$ kg or 64.7 % of the mean where $\widehat{\sigma} = \sqrt{\frac{1}{n-1} \sum_{i=1}^{n} (y_i - \widehat{y}_i)^2}$ and $\widehat{y} =$ the estimated biomass of the segment. A total of $n = 225$ segments were used to fit the model. The coefficient of determination was $R^2 = 0.75$, while the estimated coefficients were $\widehat{\beta}_0 = -19.43, \widehat{\beta}_1 = -18.17, \widehat{\beta}_2 = 8.39, \widehat{\beta}_3 = 2.18$. The hierarchical structure of the data—several segments were clustered within one plot—and the heteroskedasticity of the residuals were ignored.

6.4.5.2 Nonparametric Semi-ITC Approach

The explanatory variables A_{seg}, H_{max}, H_{SD}, and $D_{>mean}$ were used to "fit" kNN models using the 884 segments that were wholly within a NFI sample plot. As opposed to Breidenbach et al. (2010), no segments were excluded due to an unexpected relationship between measured height and maximum ALS height. In a pre-analysis it was found that excluding "outlying" segments surprisingly resulted in larger root mean squared errors (RMSEs) on plot level.

To fit kNN models based on the distance metrics MSN and random forests, the sum of the single tree biomass within the segment independent of species and the sum of the single tree biomass for spruce trees within the segment were used as response variables. The mean of the nearest neighbors was used for estimates based on more than one nearest neighbor (k > 1).

6.4.5.3 Tree Cluster Approach

The tree cluster approach (Hyyppä et al. 2005, 2006) was used to provide an example of the parametric semi-ITC approach. The linear regression model was based on the same 884 segments as were used in the nonparametric semi-ITC approach (all segments wholly within a NFI sample plot). If several trees were within a segment, the biomass of the trees within a segment was summed. In contrast to the approach used by Hyyppä et al. (2005, 2006), also empty segments were used in the regression. The biomass of empty segments was assumed to be zero. The structure of the model was the same as Eq. (6.1), with $\widehat{\sigma} = 58.6$ kg or 138 % of the mean biomass within a segment at the NFI sample plots including all empty segments and $R^2 = 0.70$. The estimated coefficients were $\widehat{\beta}_0 = 24.53, \widehat{\beta}_1 = -3.75, \widehat{\beta}_2 = -13.18, \widehat{\beta}_3 = 2.81$. Given these regression parameters, negative estimates can occur. Measures could be taken to prevent negative estimates but this was not considered necessary in our case.

6.4.5.4 Area-Based Approach

The following model was fit to estimate the biomass observed on the 30 NFI plots using the ABA

$$ y = \beta_0 + \beta_1 \, H_{mean} + \beta_2 \, H_{P25} + \beta_3 \, H_{P90} + \beta_4 \, D_{5-10} + \beta_5 \, H_{P90} D_{5-10} + \varepsilon, $$

where $\widehat{\sigma} = 592.81$ kg or 24.3 % of the mean biomass observed at the NFI sample plots and $R^2 = 0.89$. All estimates were in kg and then converted to Mg/ha. The estimated coefficients were $\widehat{\beta}_0 = 382.54, \widehat{\beta}_1 = 503.04, \widehat{\beta}_2 = 280.02, \widehat{\beta}_3 = -92.33, \widehat{\beta}_4 = -7158.20, \widehat{\beta}_5 = 1019.92$.

6.4.5.5 Plot-Level and Stand-Level Estimates

The ITC and semi-ITC models were used to estimate the biomass of all segments with a centroid within the validation plots in step 4. The validation plots were therefore the areas of interest, not complete stands. The estimates for segments associated with a sample plot were aggregated to the plot-level by summing them in order to obtain the biomass estimate of the sample plot.

The estimates on the stand level were the averaged estimates for the sample plots within a stand. Therefore, we assume that a stand only consists of the area covered by the sample plots within the stand. The sample plots thus allow a full census of the stand-level biomass. It should be noted that this approach is different to the operational application where stand-level estimates are obtained by segmenting the full stand and applying the regression models to all segments within a stand. The area of a validation stand is the sum of the areas of the validation plots within the stand. Our validation stands are therefore rather small compared to real stands.

6.4.6 Goodness-of-Fit Criteria

The alternative approaches are compared based on the root mean squared error

$$ RMSE = \sqrt{\frac{1}{n} \sum_i (y_i - \widehat{y}_i)^2} $$

and the systematic error (also denoted bias[2])

$$\frac{1}{n}\sum_i (y_i - \widehat{y}_i)$$

where y_t is the observed biomass, $\widehat{y}_i$ is the estimated biomass, n is the number of observations and $i = 1, \ldots, n$. Relative RMSE and systematic errors were obtained by dividing the RMSE and systematic errors by the mean observed biomass (Table 6.2) multiplied by 100.

6.4.7   Results and Discussion

6.4.7.1   Comparison of the ITC, Semi-ITC, and Area-Based Approaches

The ITC approach resulted in RMSEs of 111.8 % and 104.0 % on plot level and stand level, respectively. The large RMSE was mostly driven by the large systematic error of −90.4 % and −88.3 % on plot level and stand level, respectively. The absolute value of the systematic error increases with increasing biomass (Figs. 6.1a and 6.2a). The negative systematic error results from the fact that each segment is attributed with a biomass estimate for a complete tree with the crown properties of the segment although many segments only cover a crown part without stem. The number of segments that cover only parts of crowns increases with crown size. Therefore, the absolute value of the systematic error increases with increasing biomass at the sample plot.

Since the nonparametric semi-ITC approach imputes reference segments that can contain none, one, or several trees, it does not predict a biomass for every segment. As a consequence, the systematic error is reduced to 7.1 % and 7.2 % on plot level and stand level, respectively (Figs. 6.1b and 6.2b). In fact, in our case study, a positive systematic error was visible. This means that the observed biomass in the field tended to be slightly larger than the estimated value. The biomass was especially underestimated in stands within the medium range of observed biomass. The smaller systematic error compared to the ITC approach also results in smaller RMSEs of 29.4 % and 14.4 % on plot level and stand level, respectively.

RMSEs of 27.9 % and 16.8 % respectively on plot level and stand-level were obtained using the tree cluster approach, which is one example of a parametric semi-ITC approach (Figs. 6.1c and 6.2c). Systematic errors were practically zero. However, underestimates in stands in the medium range were cancelled out by

---

[2]If we assume that the biomass obtained from field measurements is the true biomass or at least very close to it, we could use the term bias. However, biomass was not measured. Instead, biomass models were used to estimate the tree biomass from dbh and height measurements. Furthermore, height models were used to estimate the height of some trees. Without any assumptions, the term systematic error is technically more correct in our case.

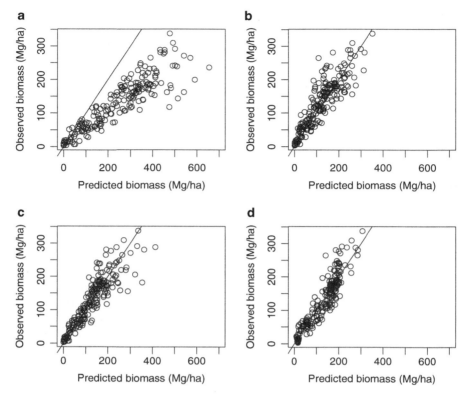

Fig. 6.1 Observed versus predicted tree biomass on plot level. (**a**) ITC. (**b**) Nonpar. semi-ITC. (**c**) Tree cluster approach. (**d**) ABA

overestimates in stands with large biomass ranges. Overestimation in stands with large biomass may have resulted from the fact that large segments were located on these plots which were not in the reference data.

Compared to the other methods, the ABA resulted in the smallest RMSEs of 22.4 % and 12.7 % on plot level and stand level, respectively (Figs. 6.1d and 6.2d). Systematic errors were practically irrelevant. This result shows that a small number of (NFI) sample plots can be sufficient to fit models for forest management inventories, a finding that is supported by several previous studies (e.g., Peuhkurinen et al. 2011; Næsset 2002).

Hardly any systematic errors were reported in the semi-ITC study by conducted Breidenbach et al. (2010). The reason for the slight remaining systematic error in the semi-ITC approaches thus requires further research. Given that the systematic error did not occur when the semi-ITC approaches were applied to the NFI plots,[3]

---

[3]Leave-one-plot-out cross validation was applied. Detailed results are not presented here.

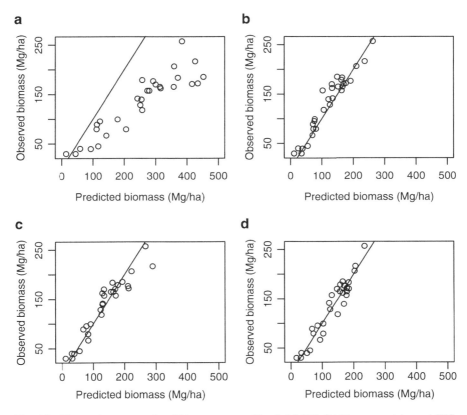

Fig. 6.2 Observed versus predicted biomass on stand level. (**a**) ITC. (**b**) Nonparametric semi-ITC. (**c**) Tree cluster approach. (**d**) ABA

one explanation for the systematic error could be that the growth in the 3–8 years time difference between the measurement of the NFI and validation plots affected the semi-ITC approaches more than the ABA. The fact that especially stands in the medium range, which have the strongest biomass growth, were underestimated supports this hypothesis. In the absence of the remaining systematic error, the semi-ITC approaches would be slightly more precise than the ABA.

Although many studies have further developed automatic tree detection and segmentation algorithms (see Chap. 5), only few have analyzed how the improved methods influence estimates of forest parameters on plot level or stand level and how the accuracies compare to the ABA. As RMSEs cannot be compared across study sites, since they are influenced by the forest structure, we next attempt a more qualitative comparison of our results with other studies.

As opposed to Hyyppä and Inkinen (1999), who found that the ITC approach underestimates timber volume, the findings by Hyyppä et al. (2001a) and our own results show that the ITC approach overestimates timber volume and tree biomass, respectively. Reasons for differences between the studies may be that different tree crown segmentation algorithms were used and that we did not apply existing models to estimate biomass from an estimated dbh but rather estimated biomass directly. It should also be noted that we did not smooth the surface model before segmentation as is often done in ITC studies. Hyyppä and Inkinen (1999) and Hyyppä et al. (2001a) assumed that the systematic error resulted from the fact that only dominant trees were detected in their study, which suggests that their algorithm resulted in less over-segmentation than ours. In our study, RMSEs on stand level are not comparable with those of Hyyppä and Inkinen (1999) and Hyyppä et al. (2001a) because in the latter studies the variance of the field data were subtracted from the RMSE.

As in our case study, Yu et al. (2010) and Vastaranta et al. (2011) reported that the ITC approach, if not corrected for the systematic error, resulted in considerably larger RMSEs than the ABA. After including a correction for the systematic error on the area-level, the ITC approach was slightly more precise than the ABA. In contrast to Yu et al. (2010) and Vastaranta et al. (2011), Peuhkurinen et al. (2011) found the ABA to be slightly more precise than the ITC approach for timber volume estimates on plot level, but also described situations in which the ITC approach may be better.[4] A small systematic error in the ITC volume estimates was not significantly different from zero, although the number of stems was significantly underestimated (Peuhkurinen et al. 2011). The reason for the small systematic error of the ITC estimate of timber volume by Peuhkurinen et al. (2011) is that the mean segment volume was overestimated and the number of stems was underestimated. In their case, these two systematic errors had opposite signs and almost canceled each other out entirely.

Whereas more over-segmentation than under-segmentation occurred in the study by Vauhkonen et al. (2013), the opposite is the case in our study. In both cases (more over-segmentation or more under-segmentation), the semi-ITC approach helps to reduce the systematic errors.

Bortolot (2006) described an approach where plot-level metrics are calculated from tree clusters detected on the plot. The approach proved to better than the ITC approach and may be an alternative to the semi-ITC approach in cases where field-measured tree positions are not available or when the spatial resolution of the ALS data is low.

6.4.7.2   Influence of Distance Metrics and Number of Neighbors in the Nonparametric Semi-ITC Approach

Since the nonparametric semi-ITC approach uses the kNN method, the number of neighbors used for the prediction and choice of the distance metric influences the accuracy and precision of the estimates. Of all the distance metrics compared

---

[4]Plots consisted of several sub plots such that their plot-level results were obtained in a similar way to the stand-level estimates in our study.

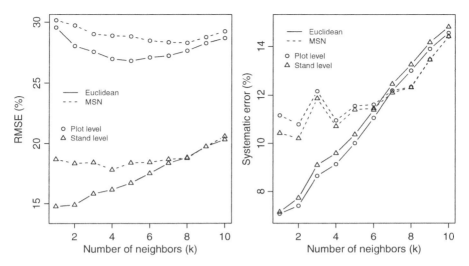

Fig. 6.3 Influence of the number of neighbors used for prediction (k) and the distance metric (Euclidean and most similar neighbors) on the RMSE (*left-hand graph*) and systematic error of nonparametric semi-ITC estimates (*right-hand graph*)

(Euclidean, Mahalanobis, MSN, and random forest), the Euclidean distance resulted in the smallest RMSEs on both plot and stand level. A comparison of RMSEs calculated with the Euclidean distance metric and the next-best metric, MSN, is shown in Fig. 6.3. Similar results have been reported by McRoberts (2012), who attributed the advantage of the Euclidean distance metric over others to the fact that it does not overfit the data.

An increase in the number of neighbors to 5 improved the RMSEs of the semi-ITC approach on plot level based on the Euclidean distance metric (Fig. 6.3). Similar observations were made by Vauhkonen et al. (2011, 2013). However, interestingly, the RMSE on stand level constantly increased with increasing k. An increased number of neighbors (k) means that the average of several segments is used for the prediction in semi-ITC. As stated by Moeur and Stage (1995), the use of one nearest neighbor ($k = 1$) usually maintains the original variance of the response variable (or the covariance between several response variables in a multivariate setting). This variability is lost with increasing k (McRoberts 2009). Our case study suggests that this loss of the original variability results in a lower level of precision when aggregating the results on stand level. The RMSEs on plot and stand level are thus competing criteria for selecting the optimal number of neighbors (McRoberts 2012). It is also interesting to note that the systematic error, as opposed to the random error, does not decrease by aggregating estimates on the stand level (Fig. 6.3). In our case, the most likely reason why the systematic error too increased with k is that the extrapolation bias was promoted with an increasing number of neighbors.

6.5 Conclusions

Semi-ITC approaches help to reduce the systematic error of ITC approaches caused by errors in crown segmentation. However, the assumption that each detected tree crown relates to exactly one field tree is abandoned in semi-ITC approaches. For the nonparametric semi-ITC approach, a simple distance metric (Euclidean) and the use of one nearest neighbor ($k = 1$) gave the best results in the case study, especially on aggregated levels such as forest stands. The tree cluster approach (a parametric semi-ITC approach) can be a good alternative if tree species are not of interest. Compared with the area-based approach (ABA), systematic errors are more likely to occur if ITC or semi-ITC approaches are applied. Furthermore, compared with the ABA, estimates of higher precision cannot always be realized using ITC or semi-ITC approaches. Whether an ITC, semi-ITC, or area-based approach is superior depends on many factors such as the segmentation algorithm, the statistical methods, the available ALS and field data, the variable of interest, and the forest structure. Regardless of which approach is used to estimate forest parameters, the estimates should always be carefully validated on all scales of interest.

Semi-ITC approaches have not yet been studied extensively. We see their potential especially with respect to response variables that strongly vary on small scales that are difficult to model using the ABA. Response variables that fulfill this criterion include biomass (see also Chap. 8), diameter distributions (Chap. 9) and tree species (Chap. 7), although this strongly depends on the forest structure. A rigid comparison with the ABA should always be included. We see this as a challenge for future studies.

Acknowledgements We thank Dr. Jim Flewelling, Seattle Biometrics, USA, for improving the description of the parametric semi-ITC approach and many other useful comments that considerably improved this chapter. Dr. Ronald E. McRoberts, Northern Research Station, St. Paul, USA, Dr. Christoph Straub, Bavarian State Institute of Forestry, Freising, Germany, and Mr. Johannes Rahlf, Norwegian Forest and Landscape Institute, Ås, Norway, are thanked for their valuable comments on an early version of the manuscript. In addition we thank Dr. Jari Vauhkonen and Dr. Barbara Koch for their review statements. We acknowledge the help of Mr. Wiley Bogren, Norwegian Forest and Landscape Institute, Ås, Norway, who assisted in improving the language of this chapter.

References

Borgefors G, Brandtberg T, Walter F (1999) Forest parameter extraction from airborne sensors. Int Arch Photogramm Remote Sens 32:151–158
Bortolot ZJ (2006) Using tree clusters to derive forest properties from small footprint LiDAR data. Photogramm Eng Remote Sens 72:1389–1397
Breidenbach J, Næsset E, Lien V, Gobakken T, Solberg S (2010) Prediction of species specific forest inventory attributes using a nonparametric semi-individual tree crown approach based on fused airborne laser scanning and multispectral data. Remote Sens Environ 114:911–924

Breidenbach J, Næsset E, Gobakken T (2012) Improving k-nearest neighbor predictions in forest inventories by combining high and low density airborne laser scanning data. Remote Sens Environ 117:358–365

Breiman L (2001) Random forests. Mach Learn 45:5–32

Crookston NL, Finley AO (2008) yaImpute: an R package for k-NN imputation. J Stat Softw 23:1–16

Ene L, Næsset E, Gobakken T (2012) Single tree detection in heterogeneous boreal forests using airborne laser scanning and area-based stem number estimates. Int J Remote Sens 33:5171–5193

Eskelson BNI, Temesgen H, Lemay V, Barrett TM, Crookston NL, Hudak AT (2009) The roles of nearest neighbor methods in imputing missing data in forest inventory and monitoring databases. Scand J For Res 24:235–246

Flewelling JW (2008) Probability models for individually segmented tree crown images in a sampling context. In: Proceedings of the SilviLaser 2008 conference, Edinburgh, UK

Flewelling JW (2009) Forest inventory predictions from individual tree crowns: regression modeling within a sample framework. In: McRoberts RE, Reams GA, Van Deusen PC, McWilliams WH (eds) Proceedings of the eighth annual forest inventory and analysis symposium; 2006 October 16–19; Monterey, CA. Gen. Tech. Report WO-79. Washington, DC: U.S. Department of Agriculture, Forest Service. 408 p

Gjessing I, Werner M (2009) LIDAR rapport, Vestfold 2009. Blom ASA, Lardal

Gougeon FA (1995) A crown-following approach to the automatic delineation of individual tree crowns in high spatial resolution aerial images. Can J Remote Sens 21:274–284

Gougeon FA, Leckie DG (2003) Forest information extraction from high spatial resolution images using an individual tree crown approach. PFC information report, Victoria, BC, Canada

Heinzel JN, Weinacker H, Koch B (2011) Prior-knowledge-based single-tree extraction. Int J Remote Sens 32:4999–5020

Holmgren J, Barth A, Larsson H, Olsson H (2012) Prediction of stem attributes by combining airborne laser scanning and measurements from harvesters. Silva Fenn 46:227–239

Hyyppä H, Hyyppä J (1999) Comparing the accuracy of laser scanner with other optical remote sensing data sources for stand attributes retrieval. Photogramm J Finl 16(2):5–15

Hyyppä J, Inkinen M (1999) Detecting and estimating attributes for single trees using laser scanner. Photogramm J Finl 16:27–42

Hyyppä J, Kelle O, Lehikoinen M, Inkinen M (2001a) A segmentation-based method to retrieve stem volume estimates from 3-D tree height models produced by laser scanners. IEEE Trans Geosci Remote Sens 39:969–975

Hyyppä J, Schardt M, Haggrén H, Koch B, Lohr U, Scherrer HU, Paananen R, Luukkonen H, Ziegler M, Hyyppä H, Pyysalo U, Friedländer H, Uuttera J, Wagner S, Inkinen M, Wimmer A, Kukko A, Ahokas A, Karjalainen M (2001b) HIGH-SCAN: the first European-wide attempt to derive single-tree information from laserscanner data. Photogramm J Finl 17:43–53

Hyyppä J, Mielonen T, Hyyppä H, Maltamo M, Yu X, Honkavaara E, Kaartinen H (2005) Using individual tree crown approach for forest volume extraction with aerial images and laser point clouds. In: Vosselman G, Brenner C (eds) ISPRS workshop laser scanning 2005, Enschede, The Netherlands. Int Arch Photogramm Remote Sens Spat Inf Sci, pp 12–14

Hyyppä J, Yu X, Hyyppä H, Maltamo M (2006) Methods of airborne laser scanning for forest information extraction. In: Koukal T, Schneider W (eds) Workshop on 3D remote sensing in forestry, Vienna, Austria

Hyyppä J, Hyyppä H, Leckie D, Gougeon F, Yu X, Maltamo M (2008) Review of methods of small-footprint airborne laser scanning for extracting forest inventory data in boreal forests. Int J Remote Sens 29:1339–1366

Kaartinen H, Hyyppä J, Yu X, Vastaranta M, Hyyppä H, Kukko A, Holopainen M, Heipke C, Hirschmugl M, Morsdorf F, Næsset E, Pitkänen J, Popescu S, Solberg S, Wolf B, Wu J-C (2012) An international comparison of individual tree detection and extraction using airborne laser scanning. Remote Sens 4:950–974

Koch B, Heyder U, Weinacker H (2006) Detection of individual tree crowns in airborne lidar data. Photogramm Eng Remote Sens 72:357–363

Lindberg E, Holmgren J, Olofsson K, Olsson H, Wallerman J (2010) Estimation of tree lists from airborne laser scanning by combining single-tree and area-based methods. Int J Remote Sens 31:1175–1192

Lindberg E, Holmgren J, Olofsson K, Wallerman J, Olsson H (2013) Estimation of tree lists from airborne laser scanning using tree model clustering and k-MSN imputation. Remote Sens 5:1932–1955

Maltamo M, Mustonen K, Hyyppä J, Pitkänen J, Yu X (2004) The accuracy of estimating individual tree variables with airborne laser scanning in a boreal nature reserve. Can J For Res 34:1791–1801

Maltamo M, Peuhkurinen J, Malinen J, Vauhkonen J, Packalén P, Tokola T (2009) Predicting tree attributes and quality characteristics of Scots pine using airborne laser scanning data. Silva Fenn 43:507–521

Marklund L (1988) Biomass functions for pine, spruce and birch in Sweden. Rapport-Sveriges Lantbruksuniversitet, Institutionen foer Skogstaxering, Sweden

McGaughey R (2010) Fusion. Manual version 2.90. USDA, Pacific North-West Research Station, Seattle

McRoberts RE (2009) Diagnostic tools for nearest neighbors techniques when used with satellite imagery. Remote Sens Environ 113:489–499

McRoberts RE (2012) Estimating forest attribute parameters for small areas using nearest neighbors techniques. For Ecol Manage 272:3–12

Means JE, Acker SA, Fitt BJ, Renslow M, Emerson L, Hendrix CJ (2000) Predicting forest stand characteristics with airborne scanning lidar. Photogramm Eng Remote Sens 66:1367–1371

Mehtätalo L (2006) Eliminating the effect of overlapping crowns from aerial inventory estimates. Can J For Res 36:1649–1660

Moeur M, Stage AR (1995) Most similar neighbor: an improved sampling inference procedure for natural resource planning. For Sci 41:337–359

Næsset E (1997) Determination of mean tree height of forest stands using airborne laser scanner data. ISPRS J Photogramm Remote Sens 52:49–56

Næsset E (2002) Predicting forest stand characteristics with airborne scanning laser using a practical two-stage procedure and field data. Remote Sens Environ 80:88–99

Nilsson M (1996) Estimation of tree heights and stand volume using an airborne lidar system. Remote Sens Environ 56:1–7

Ørka HO, Dalponte M, Gobakken T, Næsset E, Ene LT (2013) Characterizing forest species composition using multiple remote sensing data sources and inventory approaches. Scand J For Res 28:677–688

Packalén P, Vauhkonen J, Kallio E, Peuhkurinen J, Pitkänen J, Pippuri I, Maltamo M (2011) Comparison of the spatial pattern of trees obtained by ALS based forest inventory techniques. In: SilviLaser 2011, 11th international conference on LiDAR applications for assessing forest ecosystems, University of Tasmania, Hobart, Australia, 16–20 October 2011

Packalén P, Vauhkonen J, Kallio E, Peuhkurinen J, Pitkänen J, Pippuri I, Strunk J, Maltamo M (2013) Predicting the spatial pattern of trees by airborne laser scanning. Int J Remote Sens 34:5154–5165

Persson A, Holmgren J, Söderman U (2002) Detecting and measuring individual trees using an airborne laser scanner. Photogramm Eng Remote Sens 68:925–932

Peuhkurinen J, Mehtätalo L, Maltamo M (2011) Comparing individual tree detection and the area-based statistical approach for the retrieval of forest stand characteristics using airborne laser scanning in Scots pine stands. Can J For Res 41:583–598

Suvanto A, Maltamo M (2010) Using mixed estimation for combining airborne laser scanning data in two different forest areas. Silva Fenn 44:91–107

Vastaranta M, Holopainen M, Yu X, Haapanen R, Melkas T, Hyyppä J, Hyyppä H (2011) Individual tree detection and area-based approach in retrieval of forest inventory characteristics from low-pulse airborne laser scanning data. Photogramm J Finl 22:1–13

Vauhkonen J, Korpela I, Maltamo M, Tokola T (2010) Imputation of single-tree attributes using airborne laser scanning-based height, intensity, and alpha shape metrics. Remote Sens Environ 114:1263–1276

Vauhkonen J, Packalén P, Pitkänen J (2011) Airborne laser scanning-based stem volume imputation in a managed, boreal forest area: a comparison of estimation units. In: Proceedings of SilviLaser 2011, 11th international conference on LiDAR applications for assessing forest ecosystems, University of Tasmania, Hobart, Australia, 16–20 October 2011

Vauhkonen J, Ene L, Gupta S, Heinzel J, Holmgren J, Pitkänen J, Solberg S, Wang Y, Weinacker H, Hauglin KM (2012) Comparative testing of single-tree detection algorithms under different types of forest. Forestry 85:27–40

Vauhkonen J, Packalén P, Malinen J, Pitkänen J, Maltamo M (2013) Airborne laser scanning based decision support for wood procurement planning. Scand J For Res. doi:10.1080/02827581.2013.813063

Wallerman J, Bohlin J, Fransson JES (2012) Forest height estimation using semi-individual tree detection in multi-spectral 3D aerial DMC data. In: Geoscience and Remote Sensing Symposium (IGARSS), Munich, Germany, 22–27 July 2012. IEEE international, New York, USA, pp 6372–6375

Yu X, Hyyppä J, Holopainen M, Vastaranta M (2010) Comparison of area-based and individual tree-based methods for predicting plot-level forest attributes. Remote Sens 2:1481–1495

Yu X, Hyyppä J, Vastaranta M, Holopainen M, Viitala R (2011) Predicting individual tree attributes from airborne laser point clouds based on the random forests technique. ISPRS J Photogramm Remote Sens 66:28–37

Chapter 7
Tree Species Recognition Based on Airborne Laser Scanning and Complementary Data Sources

Jari Vauhkonen, Hans Ole Ørka, Johan Holmgren, Michele Dalponte, Johannes Heinzel, and Barbara Koch

Abstract Species-specific information is important for many tasks related to forest management. We review the use of airborne laser scanning (ALS) and complementary data for providing this information. The main ALS-based information is related to structural features, intensity of the echoes, and waveform parameters, whereas spectral information may be provided by fusing data from different sensors. Various types of classifiers are applied, the current emphasis being in non-linear or otherwise complex techniques. The results are successful with respect to the main species, whereas the overall accuracy depends on the desired level of detail

J. Vauhkonen (✉)
Department of Forest Sciences, University of Helsinki, Helsinki, Finland
e-mail: jari.vauhkonen@helsinki.fi

H.O. Ørka
Department of Ecology and Natural Resource Management, Norwegian University
of Life Sciences, Ås, Norway

J. Holmgren
Department of Forest Resource Management, Swedish University of Agricultural Sciences,
Umeå, Sweden

M. Dalponte
Department of Sustainable Agro-Ecosystems and Bioresources, Research and Innovation Centre,
Fondazione Edmund Mach, San Michele all'Adige, Italy

J. Heinzel
Department of Remote Sensing and Landscape Information Systems, University of Freiburg,
Freiburg, Germany

European Commission – Joint Research Centre, Institute for the Protection and Security
of the Citizen, Global Security and Crisis Management, Ispra, Italy

B. Koch
Department of Remote Sensing and Landscape Information Systems, University of Freiburg,
Freiburg, Germany

M. Maltamo et al. (eds.), *Forestry Applications of Airborne Laser Scanning: Concepts
and Case Studies*, Managing Forest Ecosystems 27, DOI 10.1007/978-94-017-8663-8__7,
© Springer Science+Business Media Dordrecht 2014

in the classification. We expect fusion approaches combining ALS and especially hyperspectral data to become more common and further improvements by the development of advanced sensor technology.

7.1 Introduction

Information on tree species is required by forest management systems that use species-specific growth and yield models or involve different treatment schedules depending on species. A species recognition step is also required if the prediction of the other tree attributes like stem diameter is based on species-specific equations; in that case the selection of the correct allometric model is crucial towards the obtained accuracies (Korpela and Tokola 2006). Distinguishing between the species has been one of the biggest challenges in the ALS-based inventories and ALS itself was not originally considered an adequate data source with this respect. However, the most recent research suggests that a high degree of species-specific information may be extracted also from ALS data.

One of the earliest attempts to predict tree species based on ALS data was reported by Törmä (2000), who concluded that *"only 3D-coordinates are not enough to estimate stand tree species proportions. The alternatives to get better results are either to classify single trees and compute tree species proportions from these or to use also intensity information by using aerial images or laser which can measure intensity"*. Following this study, the first tree-level analyses were carried out by Brandtberg et al. (2003) and Holmgren and Persson (2004), while the studies by Persson et al. (2004), Hill and Thomson (2005), Koukoulas and Blackburn (2005), and Packalén and Maltamo (2006) combined ALS with spectral image data to provide species information at various scales. Furthermore, Reitberger et al. (2006) proposed the use of full-waveform recording sensors to produce additional information to the discrete-return data. The number of papers published in this research area has rapidly increased since around 2008, but the main principles and data sources have remained more or less the same compared to these initial approaches. The most recent studies focus on advanced sensors and information extraction and estimation techniques to improve the predictions.

Most of the research in the field of the ALS-based species recognition has been conducted in Scandinavia, Central Europe and North America. Boreal forests located in the northern part of Europe are generally characterized by a smaller number of species compared to temperate forest types in Central Europe and Northern America. For certain applications in Scandinavian boreal forests it may be sufficient to distinguish between conifers and deciduous trees or at the utmost between spruce and pine (Holmgren and Persson 2004; Ørka et al. 2009a). This is reasonable since the latter two species are economically most important for Scandinavian forestry.

Besides the studies concerted on Scandinavian conditions there are also some investigations which have developed methods to distinguish a larger number of tree species. In this context and in a taxonomically correct way, the term "species" should be substituted by "genera" because the differentiation between individual taxonomic species is usually not feasible by means of remotely sensed data. Whether based on ALS or spectral data only very few approaches have been undertaken so far to separate taxonomic species within the same genera. Examples of those studies are Hughes et al. (1986) and Dalponte et al. (2008, 2012).

As these differences in the species composition and forest structure naturally affect the methodology used in the different study areas, we divide the rest of this chapter according to following: First, we present a framework and an overview of the common methodology used in the species classification task regardless of the species composition. Second, we describe the potential sources for the species-specific information based on ALS and additional data using the classification of individually delineated Scandinavian trees as an example case. This case may be considered as a simple problem (only three species classes), but it still represents the different aspects of the problem, since a successful classification of these vegetation types usually requires information from multiple data sources. We then generalize these information sources to area-based, species-specific estimation applications and shortly extend the discussion to the classification of several genera in Central Europe, North America and tropics. Finally we present conclusions and future outlooks regarding species-specific estimation based on ALS and complementary data.

7.2   An Overview of the Species-Specific Estimation Task

There are different strategies how to treat the species in the estimation tasks. The simplest and least detailed approach is to determine the stand-level main species based on some external information, such as visual interpretation of aerial photographs (e.g. Næsset 2002). In that case, the ALS-based estimation is focused on the total stand attributes, yet the result also includes the determination of the species. If the species are predicted, the prediction can be done separately from other tree/stand attributes, or simultaneously with the other attributes. Instead of producing a comprehensive species map, the interest may be in detecting occurrences of economically or ecologically important species (e.g. Säynäjoki et al. 2008).

All the previously mentioned problems can be basically solved both at the single tree and area (e.g. plot or stand) levels. It is notable that tree species itself is rarely the actual end-product expected from an ALS-based forest inventory. Instead, forest inventories typically aim at predicting *species-specific forest attributes* at a certain area, which constitute continuous properties and may require an aggregation of predictions from the initial computation units (tree segments or grid cells) to the desired geographical unit. Despite the previously mentioned differences, the species-specific estimation tasks often share similar properties. The schematic diagram in Fig. 7.1 attempts to cover those methodological similarities in species-specific estimations. We acknowledge that additional and other categories could be

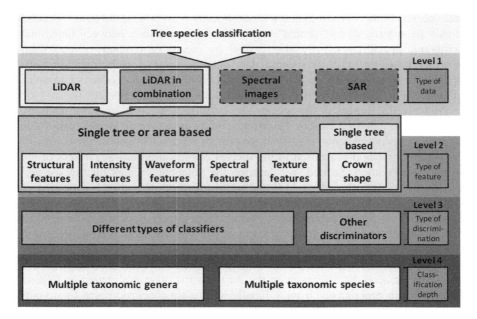

Fig. 7.1 Schematic diagram describing a general species classification task

formulated and alternative terms used. However, we try to generalize the problem as much as possible; for example, over different spatial scales (tree, plot, forest levels) and various classification vs. estimation based approaches.

The diagram contains four subsequent levels which refer to the major aspects of the classification approaches. The first one includes the type of data used. For the scope of this textbook, ALS-based methods play the major role and can be divided into a sole use of the ALS data and a combination with multi- or hyperspectral images. Considering species classification, the use of other airborne data types like SAR is rare and only few authors like Sato and Koike (2003) have worked on it. The first studies using terrestrial laser scanning (TLS) for the same purpose have also been published recently (Puttonen et al. 2010). However, due to the non-area wide nature of the TLS data, the analysis of these data results in different types of applications, in terms of scale, than the airborne data covered here.

The second level consists of predictive features derived from the ALS data and complementary data sources. A premise for the discrimination between species is that their differences can be adequately described in the feature space determined by the extracted features. The principal information sources include the (*i*) structural properties in the 3D point clouds and (*ii*) measures of the pulse energy reflected back, i.e. the intensity data. Using (*iii*) full-waveform recording sensors, the information about the geometry of the reflections produces additional information. Even though the intensity values are somewhat related to the reflectance properties of the target, the ALS data need to be coupled with spectral data to truly measure these properties. The information in such images includes (*iv*) spectral and (*v*) textural features. Few authors like Vauhkonen et al. (2008, 2009), Barilotti et al. (2009), and Heinzel and Koch (2012) have derived textural features also from ALS based surface models. The last group, (*vi*) crown shape features, is related to individually delineated trees, being therefore less applicable for area based approaches.

The use of the classification features derived at the tree level requires the prior extraction of the single trees either in 3D from the point cloud or in 2.5D from a digital surface model (cf. Chap. 5). Failures of the single tree extraction directly reduce the quality of the species differentiation. Until now, classifications based on crown and shape models have resulted in an adequate accuracy only when classifying a few species with strong differences in the crown shape and for this reason they have mainly been applied in Scandinavian boreal forests (e.g. Holmgren and Persson 2004). For the classification of several species with minor shape differences these approaches may not be applicable. However, the shape properties might contribute to a better detection of subtle differences between species of nearly similar appearance in tandem with other feature groups.

The third level of the diagram refers to the method of discrimination between tree species. These are methods which assign the observations to classes corresponding to tree species, being mainly different statistical classifiers or machine learning systems. After extracting the features, producing the species-specific predictions corresponds to a general classification or estimation problem described in the literature related to pattern recognition and machine learning (e.g. Fukunaga 1990; Bishop 2006; Theodoridis and Koutroumbas 2009). Therefore, this part is addressed here very briefly.

As outlined earlier, the species classification accuracies may benefit from combining features of different types (e.g. categorical, continuous) and data from multiple sources. One cannot usually assume a linearly separable classification problem or make similar assumptions regarding the distribution of the features, for which reason the use of different non-linear or non-parametric classification techniques like support vector machines (SVMs) or nearest neighbor (NN) algorithms has become especially popular. Furthermore, since the dimensionality of the feature space extracted from the airborne data may be very high, the generalization properties of a classifier may be further improved by employing feature reduction strategies. Different statistical approaches to produce species-specific information from ALS and complementary data sources are compared by Hudak et al. (2008), Korpela et al. (2010b), Niska et al. (2010), Vauhkonen et al. (2010) and Ørka et al. (2012), whereas Brandtberg et al. (2003), Holmgren and Persson (2004), Packalén and Maltamo (2007) and Ørka et al. (2009a, 2012) present different feature selection/reduction approaches applied for species-specific estimation.

The fourth and last level of the diagram divides the so-called classification depth into two subtypes. It is distinguished between classification approaches which aim at differentiating between genera and those aiming at even more detailed differentiation between species. As explained in the introduction, the terms genera and species are often used outside their taxonomic meaning in the remote sensing studies. The Scandinavian species classification approaches typically aim at separating the groups of Scots pine (*Pinus sylvestris*), Norway spruce (*Picea abies*) and a broad group of deciduous trees, mainly consisting of birches (*Betula* spp.). In Central Europe, an example genera-level classification could aim at separating spruce, pine, beech (*Fagus* spp.) and oak (*Quercus* spp.), whereas a complete classification of the last species, for example, would require separating red oak (*Quercus rubra*), sessile oak (*Quercus petraea*) and common oak (*Quercus rubur*).

7.3   Estimation Based on Discrete-Return ALS Features

7.3.1   Principal Sources of Information in Discrete-Return ALS Data

The species-specific information extracted from typical discrete return ALS data is based on either the coordinate or intensity data. The vertical profiles of the point data describe the structural properties of the tree crowns and may therefore reflect species-specific differences in these structures. The structural information can be quantified in terms of height value distribution, distribution of different types of returns in different parts of the tree crown, and geometric measurements. Principally all these properties are related to the amount and allocation of foliage in different parts of the tree crown as seen by an ALS sensor. If the species have real-world differences in these properties, then this information may be extracted from the ALS data and used to discriminate between tree species.

In addition to the structural information, most ALS systems record properties of the backscattered laser signal (Wehr and Lohr 1999), which is commonly referred to as "intensity". For discrete return lasers, intensity often represents some measure of the strength of the returned echo, usually the peak amplitude (Wehr and Lohr 1999, see also Chaps. 2 and 3). A pioneer work of using intensity for vegetation studies is reported by Schreier et al. (1985). Today, the intensity is recorded for all echoes in discrete return lasers, which usually record from one to four echoes per emitted pulse. However, sensor algorithms for both echo triggering and intensity recordings are proprietary information, and accurate descriptions of how the intensity recordings are performed are normally not available for end users. Waveform recording sensors digitize the returned energy in consecutive narrow intervals giving more control of the physical meaning of these numbers.

The intensity captured by current commercial ALS systems offers a radiometric resolution of 8-bit, 12-bit or 16-bit (Höfle and Pfeifer 2007). The wavelength of the emitted pulse is 1,064 nm in most ALS systems used for forestry applications, but it could be located in other parts of the infrared spectrum (700–1,500 nm). Thus, intensity describes some properties related to the reflectivity in these wavelengths. However, intensity is not only related to the reflectivity of the target, but also the area of the target and the scattering from the target (e.g. Korpela et al. 2010b).

Since both the reflectivity in the near infrared wavelength and the sizes, shapes and arrangement of leaves and branches differ between tree species the intensity may be expected to indicate these differences in species classification.

7.3.2   Studies Based on Using the Structural and Intensity Features

Holmgren and Persson (2004) classified pine and spruce trees in southern Sweden using proportions of different types of echoes, and features based on height value distribution, point geometry, and intensity. Altogether 20 classification features were divided in 8 groups based on their mutual correlation, i.e. the different groups were assumed to describe different properties between the species. Proportion of vegetation hits, i.e. points above the estimated crown base height, and the standard deviation of the of the intensity values were the two most important features, producing classification accuracies of 88 % and 84 %, respectively. Accuracies of 93–95 % could be reached using combinations of 3–8 features. All of these combinations included the standard deviation of the intensity values and the proportion of the first echoes within the tree crown. Additional information was obtained using height distribution features and parameters describing a parabolic surface fitted to the tree-level point data.

In a later study, Holmgren et al. (2008) included deciduous trees in the analysis carried out in the same geographical area. They grouped the classification features into height distribution, canopy shape, proportions of echo types, and intensity features, with only slight differences to Holmgren and Persson (2004). In that study, the most important individual feature was the mean value of the shape parameters normalized by tree height, which resulted in about 72 % classification accuracy in the three species case. An accuracy of 87–88 % was obtained adding different combinations of ratio between ALS-based crown base height and tree height, proportion of echoes in the canopy, and mean intensity.

The height and intensity distribution features were further examined by Ørka et al. (2009a, 2010, 2012), who calculated the features closely corresponding to the area-based estimation (Næsset 2002, Chaps. 1 and 11) at individual tree level and separately for different echo types. Ørka et al. (2009a) found 17 features to explain a significant proportion of variability between spruce and birch trees in a Norwegian study area. Altogether 14 of these features were based on the height value distributions, but the maximum and mean intensity of first-of-many echoes and the mean intensity of the last-of-many echoes were found to be among the most important classification features, providing overall accuracies of 73 %, 70 % and 67 %, respectively. Both intensity and height distribution features differed considerably when calculated by different echo categories. Ørka et al. (2009a) further pointed out that the tree height significantly affected to the height distribution features and a better performance could be obtained by normalizing these features with respect to the tree height.

In a later analysis, Ørka et al. (2012) found the proportions of single, last and intermediate echoes more important than the height percentiles. These proportions combined with density features extracted from the highest and lowest vertical layers of the point cloud resulted in an accuracy 74–77 % in the classification of pine, spruce and deciduous trees in Norway. Correspondingly, the 90th percentile, kurtosis and coefficient of variation derived from range normalized intensities of first echoes were found to be a best combination of the intensity features, providing an accuracy of 63–75 % depending on the classifier used.

Vauhkonen et al. (2008, 2009) derived structural information applying a computational geometry concept called 3D alpha shapes (Edelsbrunner and Mücke 1994) for species discrimination in southern Finland. The proposed approach was based on triangulating the point clouds of the individual trees and deriving classification features describing the allocation of computational volume and connected structures at different relative height levels. This feature group was found to be comparable to the point distribution features in classifying coniferous trees, and could alone produce a 93 % accuracy of pine, spruce and deciduous trees. Although based on a limited set of trees, their results indicate that very detailed (branch level) differences can be extracted from the very high density (40 pulses m^{-2}) ALS data. In a simulation carried out by Vauhkonen et al. (2008), these features were however found to be more sensitive to the reduction of the point density than features based on height distribution, for example.

The studies of Vauhkonen et al. (2008, 2009) also include a comparison between the structural and intensity features, but the intensity data were given less priority as these were considered to contain noise between the species. The intensity features describing the differences in the distributions of the first echoes separated slightly between deciduous and conifer trees, but not between the coniferous species. When combining structural and intensity features to separate spruce, pine and deciduous trees, three out of nine features selected described the intensity distribution. In another study area in southern Finland Vauhkonen et al. (2010) used a non-parametric approach to estimate the species, stem diameter, tree height and stem volume simultaneously using ALS data in a more practical pulse density (6–8 pulses m^{-2}). Two thirds of the features considered were derived from the intensity distribution. The features selected were the higher percentiles (40th, 50th, 60th and 70th) computed from first echoes and the mean intensities in two crown layers corresponding to relative height layers 60–70 % and 70–80 %. The overall accuracy obtained ranged from 68 to 79 %.

7.3.3 Intensity Normalization for Tree Species Predictions

As opposed to the previous studies that used raw intensity values, the later studies by Korpela et al. (2010b) and Ørka et al. (2012) suggest a calibration or normalization of the intensity as a necessary step to produce improved species recognition accuracies. The recorded intensity values are dependent on many factors, such as the range from sensor to target, incidence angle, atmospheric transmittance and transmitted power (Ahokas et al. 2006; Wagner et al. 2006). Calibration methods based on both physical and more data-driven approaches are used (Ahokas et al. 2006; Coren and Sterzai 2006; Höfle and Pfeifer 2007). Among the effects for which the intensity data should be calibrated, a normalization based on the range from the sensor to the target (range normalization) is the most mature. The normalization methods and details on the improved accuracies can be found from Korpela et al. (2010b) and Ørka et al. (2012). Furthermore, calibration of the sensor specific effects related to the hardware and software, such as the Automatic Gain Control (AGC) of Leica sensors (Korpela 2008; Korpela et al. 2010a) and the differences in intensity between scan directions in Optech Airborne Laser Terrain Mappers (ALTM), which is referred to as banding (Ørka et al. 2012), may be required.

The reasons causing the different intensity values have been dealt with by Korpela et al. (2010b). For example trees with large and almost horizontal leaves (e.g. Norway maple) produce high intensities while trees with small leaves and narrow angles to the incoming pulse (birches, aspen) produce lower intensities. Generally, conifers produce lower average intensities than most of the deciduous trees. For the three most common boreal species, the intensities of birch, spruce and pine are usually in this order (Korpela et al. 2010b; Ørka et al. 2009a). Employing this information, Korpela et al. (2010b) obtained an accuracy of 88–90 % when classifying the aforementioned species with intensity features. The features selected were the mean and standard deviation, skewness, kurtosis and percentiles (20th, 40th, 60th, 80th and 90th) of crown echoes, mean intensity of four upper crown layers, where a crown layer was defined as one tenth of the tree height.

7.3.4 Considerations Regarding the Use of the Structural and Intensity Features

There are some important considerations regarding the use of ALS data in species classification. First, other forest properties than species may influence the recorded structural and intensity information. Such properties are tree height (Korpela et al. 2010b; Ørka et al. 2009a) and site fertility (Korpela et al. 2010b). Korpela et al. (2010b) found out that pine trees produced a lower average intensity of the first echoes than spruce in forested areas, while there was no difference between trees grown in open and fertile stands. Ørka et al. (2009a) showed that differences in tree heights between species could cause the incorrectly selection height features. Age is another property that might influence ALS measurements due to the change in crown architecture and reflectivity of trees with increasing age. However, Korpela et al. (2010b) did not find any influence of age in the classification accuracy based on intensity features.

Secondly, the intensity differs between echo categories (Ørka et al. 2009a, 2010). Thus, which categories to use is an important question. The 'first returns' (first-of-many and single echoes) are considered as an optimal combination as these echoes are less affected by transmission losses due to penetrating the crown (Korpela et al. 2010b). Single echoes have the highest intensity among the echo categories, which is natural as only one echo is recorded when there is not enough backscattered energy for additional echoes. However, the drawback of using only either one echo category compared to combining single and first of many is that the number of echoes per tree crown, and thus feature derivation possibilities, are reduced. The intensity of last of many echoes is related to the energy reflected by the preceding echoes, but also these have been found to support the separation of spruce and birch (Ørka et al. 2009a).

The third issue is that the ALS measurements and thus the accuracy of the tree species classification vary between sensors (Korpela et al. 2010b; Ørka et al. 2010). The classification accuracy obtained using a normalized intensity from two different sensors differed by 10 percentage points in a Finnish study (Korpela et al. 2010b) and 3 percentage points in a Norwegian study (Ørka et al. 2010).

Finally, the differences in the phenological status of the vegetation as observed by ALS sensors have been used to improve species classification accuracies. Ørka et al. (2010) studied the point height value distributions under leaf-on and leaf-off conditions and found campaign-specific properties that could be used in the species estimation task. The intensity of first of many echoes was lower in birch and aspen trees when using leaf-off ALS data compared to using leaf on data, for last of many echoes an opposite trend was found. Furthermore, they discovered a shift in the proportions of echoes in the first-of-many and single return categories between acquisitions performed under leaf-off and leaf-on canopy conditions. Specifically, the deciduous trees that had lost their foliage were found to produce more multiple (first-of-many) echoes under leaf-off than leaf-on canopy conditions, resulting in classification accuracies in the order of 10 percentage points higher in overall accuracy. Earlier, Liang et al. (2007) reported an accuracy of 89 % in coniferous-deciduous tree classification in Finland following a difference between the digital surface models constructed from either first and last echoes. In the study of Ørka et al. (2010), the use of different sensors also caused differences to the height value distributions, which did not affect remarkably on classification accuracies.

7.4 Estimation Based on Full-Waveform Features

Compared to discrete-return ALS, the parameters derived from full-waveform ALS data (Chap. 3) have been shown to produce more information especially for solving multiple species problems typical to North America and Central Europe, for example. The parameters of the reflected waveform have proved to be more sensible in the differentiation between similar species than other feature types (Heinzel and Koch 2012).

Reitberger et al. (2008), Hollaus et al. (2009), and Heinzel and Koch (2011) describe the use of waveform characteristics like width, amplitude, and intensity of the echo and total number of detectable targets as highly relevant for accurate species detection in temperate forests of Central Europe. Figure 7.2 illustrates major

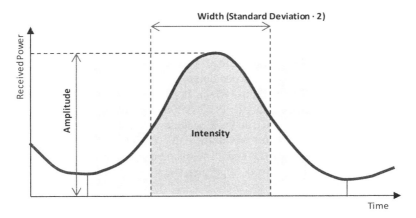

Fig. 7.2 Extract from a reflected laser signal depicting metrics of the waveform (schematic)

waveform parameters based on simply modeling a single echo with a Gaussian shape. The width of an echo is considered as the double standard deviation and the amplitude is the height of the Gaussian at its peak.

Related to the definition of an intensity value (see Sect. 7.2), the echo amplitude can be defined more precisely using full-waveform data (Fig. 7.2). From a physical point of view it describes the complete area below the Gaussian model of a reflector and therefore provides a measure for the reflected energy which is received by the system installed in the aircraft. Further explanations of the waveform parameters like the backscatter cross section are provided by Wagner et al. (2006) and Chap. 3 of this book.

Comprehensive explanations of the interaction between the ALS pulse and the surface of forest trees are currently missing. However, Hollaus et al. (2009) presents some hypotheses which explain the backscatter cross section of an echo with attributes from different species under leave-off conditions. For example they propose that the entity of thin spruce needles work as an extended target which leads to high backscatter cross sections. In contrast the leafless branches of a beech work as small scatterers with small cross sections.

As an example of the classification features and accuracies obtained in studies using the full-waveform features, Hollaus et al. (2009) used the mean backscatter cross section of all echoes above the 50th height percentile and a ratio of the number of echoes reflected above the 50th height percentile to classify coniferous and deciduous trees with an accuracy of 83 %. They further highlight the potential to separate beech, larch and spruce using the standard deviation of the echo widths per crown segment, which resulted in an accuracy of 75 %. Heinzel and Koch (2011) identified the intensity, the width and the total number of targets, corresponding to a measure of canopy density, as the most important features in discrimination between different species classes. In their study, up to six tree species were classified with an overall accuracy of 57 %, four main species with an accuracy of 78 %, and coniferous and deciduous tree groups with an accuracy of 91 %.

It is notable that in addition to the full-waveform features, similar structural features than outlined in previous sections are typically used. However, more detailed structural information may be obtained by calculating these features at an individual echo level (Heinzel and Koch 2011). Remarkably, very high classification accuracies between coniferous and deciduous trees have been obtained based on unsupervised classification methods, which depicts the clear difference between these classes. For example, Reitberger et al. (2008) reported an 85 % classification accuracy in a leaf-on situation and an 96 % accuracy in a leaf-off situation following a clustering approach of the full-waveform data. Yao et al. (2012), on the other hand, obtained accuracies of 93 % and 95 % by using unsupervised and supervised classification methods, respectively, reporting much less differences between leaf-on and leaf-off data sets.

Despite all benefits the disadvantage of using waveform parameters is that waveform information is not always available with the original ALS data and discrete ALS system are more commonly available. Since waveform systems are still more expensive and require higher data storage capacities such information is often not being recorded during survey. Most datasets are limited to the so called discrete ALS information which only contains the spatial coordinates of the individual echoes. According to Vaughn et al. (2012) the limitation to discrete ALS leads to a reduced overall accuracy. In their example of five tree species from Northern America classification rates reduce from 85 to 79 % even when completely leaving out the use of special waveform parameters in both classifications and only caused by the reduced number of extracted echoes per pulse.

7.5 Estimation Based on Combined ALS and Spectral Image Data

Complementary information for the species classification can be extracted from the combination of ALS and spectral image data, which may be provided by either multi- or hyper-spectral sensors. Multispectral sensors record spectral values usually covering three to four broad bands, whereas hyperspectral sensors are capable of acquiring data with a high spectral resolution that usually results in observations of tens to hundreds of spectral bands. The latter provide a detailed characterization of the spectral signature of the tree(s), which is useful in the classification process. The use of multispectral data has a longer tradition for the species classification whereas hyperspectral data has become more popular in the recent years due to their increased availability. The simultaneous acquisition of the two data types (e.g. Asner et al. 2007, 2012; Ørka et al. 2012) reduces the cost of data acquisition, but is potentially a limited alternative for practical forestry applications due to various technical, economic and weather reasons. Issues related to combining different data sources are presented in Chap. 4.

Compared to the species classification accuracies obtained by using ALS data alone, the improvement of adding image data is related to the spectral signatures differing between certain species. The spectral signature of tree species is usually characterized by a peak of reflectance in the green area of the spectrum (from about 520 to 570 nm). This peak is strictly connected with the chlorophyll content of vegetation. There is also a rapid increase of reflectance moving from the red region (from about 630 to 700 nm) to the infrared one (from about 700 to 1,500 nm). This area, between about 680 and 720 nm, is called red-edge region, and it is widely exploited in vegetation studies, as the exact location of it and its behavior are strictly correlated with the species, the age of the tree, and its health condition (Cho and Skidmore 2006). In the infrared area of the spectrum (after about 750 nm), there is a strong difference in reflectance between coniferous and broadleaves species. This is useful in the classification process, but the infrared area is usually also noisier, for the presence of the same absorption bands.

Using multispectral images, the available spectral bands describe the differences between the tree species at a limited interval. Most commercial sensors have four bands located in the blue, green, red and infrared parts of the spectrum. The choice of the spectral bands is made by the camera manufacturers and may not be optimized for tree species classification. For example, when an optical radiation model was used to examine the composition of the spectral bands of a Leica sensor, the conclusion was that the tree species classification could be improved if one additional band measuring the red edge region of the spectrum was included (Heikkinen et al. 2010).

One problem when using spectral values from aerial images is that the illumination conditions vary between images and within individual images. There are physical, empirical or semi-empirical models that can be used for radiometric correction (Collings et al. 2011). There are several factors influencing the radiometric values that one has to consider before operational classification of tree species. There are three main categories that need to be considered (Collings et al. 2011): (1) atmospheric composition that affects the scattering and absorption of light, (2) equipment noise and errors, and (3) the systematic variation in intensity across an image frame that is caused by different view and solar angles of the observations. This last effect is typically dependent on land-cover type. The data also vary according to seasonal differences of the vegetation. This variation, weather and sun conditions limit the time frame for a successful image acquisition to a short period especially in the high latitudes.

The majority of the studies in the literature present different ways to combine spectral and ALS data for individual tree crowns classification, but it is possible to define two main strategies: (A) an individual tree crown (ITC) approach and (B) a pixel approach. Approach A combines features extracted from both spectral and ALS data at an ITC level (Holmgren et al. 2008; Ørka et al. 2012). Conversely, approach B extracts the features from both data at a pixel level, and afterwards the pixel level classification map is aggregated for the ITCs extracted from ALS data. Some studies that follow approach B (e.g. Dalponte et al. 2008; Jones et al. 2010) do not perform the final pixel aggregation, but stop their analysis to the pixel-level classification map. This output can be an instrument for having an idea of the tree species distribution over a large area, which can be used for qualitative studies of tree species distribution.

The information extracted from the spectral images may be based on different statistics like spectral mean values. One could also use ratios between spectral bands because these could be less sensitive to different illumination conditions compared with individual spectral bands (Persson et al. 2004). Tree-level variations due to self-shading and shading from other trees may need to be considered since they have a high impact on the reflectance values. One way to reduce the errors due to self-shading is to only select pixels from the sunny side of the tree crowns as demonstrated by Olofsson et al. (2006) and Dalponte et al. (2014), for which there are also alternative techniques.

Regarding the classification accuracies obtained by fusing ALS and multispectral data, Holmgren et al. (2008) showed that classification accuracies of 87–88 % between pine, spruce and deciduous trees could be increased up to 95–96 % by including the spectral mean values calculated at per-pixel basis. In the study of Ørka et al. (2012), a benchmark accuracy level of 74–77 % obtained using structural information based on ALS data could be increased up to 87–89 %. The imagery was acquired simultaneously with the ALS data, which improves the operational possibilities to use a combination of these data sources. In both of these studies, the classification of the deciduous trees was the reason for the improved accuracies, whereas the information for the coniferous trees was obtained mainly from the ALS data.

Ørka et al. (2013) demonstrated a major improvement in the classification accuracy due to using a combination of hyperspectral and ALS data compared to only ALS or combined ALS and multispectral data. The authors applied a feature extraction to the hyperspectral images at pixel level and afterwards, the mean value of the principal components in each crown were considered. The results showed an improvement of kappa accuracy from 0.56 (ALS data only) to 0.78 (hyperspectral and ALS data).

On the same dataset, Dalponte et al. (2014, 2013) applied a classification system based on a pixel approach. The classification accuracy they obtained by combining hyperspectral and ALS data was marginal compared with only using hyperspectral data. This is mainly due to the fact that by considering a high spatial resolution of the spectral images, only few ALS echoes are contained in each pixel. Thus, it is not possible to extract high level statistics due to the low number of points per pixel, and many pixels of the ALS images are interpolated as they do not contain any ALS echo. However, the ITCs were extracted from the ALS data, which demonstrates their importance for the phase of pixel aggregation (Dalponte et al. 2013).

7.6 Species-Specific Estimation Using the Area-Based Approach

The previous sections focused the analyses on data with a rather high spatial resolution, being typically interpreted at an individual tree level. On the contrary, the area-based approaches use data in considerably lower densities (e.g. ALS data with a density <1 pulses m^{-2}; Maltamo et al. 2011) to produce forest attribute for plots, (micro-)stands or forest areas. Therefore, the structural information derived from the area-based data may not be so obviously related to species-specific properties, as the plot data may combine trees of several sizes and species. However, also the plot-level information can be generalized to form species-specific estimations in the Scandinavian conditions (cf. Maltamo et al. 2011). This approach is addressed in Chap. 12 of this book; therefore, this section only briefly reviews the methodological aspects related to the species-specific estimation.

Those area-based methods that produce species-specific estimates have been developed in Scandinavia especially by Packalén and Maltamo (2006, 2007, 2008) and these methods are currently used operationally in Finland (see Chap. 12). In these approaches, the species-specific predictions are produced using a plot-level imputation method with essential species-specific attributes (e.g. volume, basal area) as dependent variables and airborne features as independent variables. The estimation is based on searching for most similar neighbor plots in a feature space determined by the ALS and image features. The airborne features are typically based on height and the density features derived from the ALS data and due to the low density of the data, these features are assumed to be more related to the dimensions of the growing stock. The species-specific information is expected to be based on information derived from aerial images.

The image features used in the estimation include spectral mean values and the textural features introduced by Haralick et al. (1973). These features are derived from the locations corresponding to field plots; orthorectification techniques are used to co-register the image data with ALS. To develop the method further, Packalén et al. (2009) linked the ALS points to the pixel values of unrectified aerial photographs. The average spectral values were then used as predictors of the species proportions. The developed methods avoids the geometric errors related to the orthorectification process and furthermore has a potential to avoid radiometric correction due to fetching the spectral values from several images and averaging the images in that way (cf. Chap. 4).

The obtained results have indicated highly accurate results for the stand totals and also for main tree species whereas the accuracy is worse for minor tree species. Due to the properties of the airborne information, the estimation may be assumed to be based on the properties of the dominant canopy; therefore, it can be questioned how good the accuracy could be regarding the minor species, which are typically located below the dominant canopy (cf. Packalén and Maltamo 2007; Packalén et al. 2009). Also at the tree-level, Vauhkonen et al. (2011) found the accuracies of the species-specific estimates to be connected with the unequal tree detection probabilities of the different species, in that the species located in the dominant canopy were estimated more accurately.

In attempting to further develop the species-specific estimates, Niska et al. (2010) tested neural networks as an alternative estimation method to the nearest neighbor techniques, which resulted in only minor improvements. Villikka et al. (2012) found leaf-off data extracted at the plot level to provide more accurate estimates than leaf-on data and to discriminate between coniferous and deciduous trees even without the use of aerial images. Vauhkonen et al. (2013) found the plot-level intensity distributions useful for separating plots dominated by certain species from other types of forest (see also Donoghue et al. 2007). Furthermore, Vauhkonen et al. (2012) reported improvements by using the corresponding alpha shape metrics than calculated at the individual tree level (Vauhkonen et al. 2008, 2009). Ørka et al. (2013) estimated main species with an overall accuracy of 86 % and estimated species composition in deciles accurately for 83 % of the plots using only ALS data. Based on these observations, it seems to be possible to develop the feature space to produce further species-specific information also using area-based methods.

7.7 On the Classification of Multiple Species in Complex Forest Canopies

Although the previous text is mainly based on examples from studies carried out in Scandinavian forest conditions, it can be concluded that principally similar techniques in both feature extraction and classification are applied elsewhere (Brandtberg et al. 2003; Moffiet et al. 2005; Kim et al. 2009; Suratno et al. 2009; Jones et al. 2010; Heinzel and Koch 2012; Vaughn et al. 2012). The difference is the complexity of the classification problem, which can grow very high in the temperate forests and thus affect the obtainable species classification accuracies.

When comparing the results from temperate forests with those from boreal forests it is noticeable that for multiple species the accuracies may vary highly between the different classes. For example Jones et al. (2010) report a span from 7 to 97 % when classifying up to eleven species combining hyperspectral and ALS data. Very similar conclusions are reported by Heinzel and Koch (2011), who improved the initial 13–79 % accuracy when classifying six species to 78–86 % (4 main species) and up to 96 % (coniferous-deciduous classes) by limiting the classification depth to cover less species. On the other hand, when considering simple coniferous-deciduous case, the assignment to these classes can be done accurately at various scales using simple height distribution features (van Aardt et al. 2008).

The number of species classes considered in the boreal zone is usually lower, but also there the detailed classification of the deciduous species could cause considerably lower accuracies. Therefore, the detail of the classification problem poses an important factor affecting the potential accuracies obtained. However, even a seemingly complex classification problem may be solvable with a reasonable accuracy, but it calls for a detailed knowledge in species morphology, which can be reproduced to feature space (e.g. Brandtberg 2007; Niccolai et al. 2010).

Several authors have pointed out increased accuracies by combining ALS data with hyperspectral data especially in challenging, multiple species conditions. For example, Dalponte et al. (2012) report an increase from 74 to 83 % overall accuracy when classifying seven species by combining ALS features with spectral information. A similar increment of classification performance is confirmed by Heinzel and Koch (2012) who report an increase from 79 to 88 % when classifying the main species of a temperate forest. The main species of such a Central European forest comprise Scots pine, spruce, beech and oak.

The fusion of structural properties from ALS with hyperspectral data is expected to increase especially in challenging conditions such as tropics (cf. Asner 2013) and there already are examples of successful classifications of multiple savanna species (Naidoo et al. 2012; Sarrazin et al. 2012). However, the scale and complexity factors related to tropical forest canopies may also result to opposite decisions regarding the applied sensor(s) and scale(s). For example, Wolf et al. (2012) suggest tropical species richness to be related mainly to topography due to species-level associations with water and nutrient availability. They report a generalized least squares model based on measures of canopy and terrain structure (i.e. ALS features only) to be able to explain a significant proportion of the variation in tropical species richness at scales ranging from 0.01 to 1 ha, which could potentially be used as simplified *a priori* information for later species assessments.

7.8   Concluding Remarks and a Future Outlook

Regarding classification methods and applied classification features, certain conclusions can be made based on previous studies. Only minor differences were found between linear discriminant analysis and more complex classifiers such as Random Forests or SVMs when classifying the typical Scandinavian tree species (Korpela et al. 2010b; Ørka et al. 2012). Correspondingly, Niska et al. (2010) found only minor differences between NN and neural network based methods in an area-based imputation of stem volume per these species. However, in multiple species problems, non-linear or advanced classifiers may be preferred against simple linear classification methods (Heinzel et al. 2010). For example, Jones et al. (2010) manage to differentiate up to 11 species using SVMs. The results of Ørka et al. (2009b) and Heinzel et al. (2010) also indicate a better stability of the SVMs against linear discrimination in high dimensional feature spaces when discriminating multiple species classes. This can also be confirmed from a theoretically point of view (Gokcen and Peng 2002; Chen and Peter Ho 2008). On the other hand, a benefit of using NN techniques is the potential to simultaneously predict other attributes than species (Packalén and Maltamo 2007; Vauhkonen et al. 2010).

Regarding the combination of the classification features, i.e. the optimal combinations of intensity and structural data, the studies considered here report somewhat contradictory results even in forests with a rather similar structure (cf. Korpela et al. 2010b; Ørka et al. 2012). Consequently, a clear suggestion if either intensity or structural information or the combination is apparently difficult to give. As the majority of ALS sensors provide intensity recordings, the opportunity to utilize both intensity and structural measurements is present, but the use of them and the balance between the classification features likely changes between the test areas and needs to be considered using training or calibration data.

The development of new sensor technology could in the future make methods combining ALS and spectral data more efficient. First, higher geometrical precision of the data sources or integrated systems makes it easier to merge the data with high precision to allow analysis at an individual tree level. Also, each data source could in the future become more similar to the past combination of both data sources. For example, aerial images can already now be used to produce three dimensional measurements of the tree canopy with high precision.

One possibility is that multispectral laser systems will develop which could provide spectral data for tree species classification without above described problems with varying illumination conditions. There are results from experiments with TLS-based systems for which a high resolution laser scanner was combined with an active hyperspectral sensor to classify three tree species (Puttonen et al. 2010). The best classification results were achieved with the combined use of features derived from both spectral and geometrical data. By using a sensor in which the range and reflectance measurements were integrated, Vauhkonen et al. (2014) found the reflectance values of those pulses that had penetrated through the foliage to improve the discrimination between pine and spruce trees. Morsdorf et al. (2009) have simulated data from a multispectral full waveform laser and found that band ratios were useful for prediction of the amount of chlorophyll content. Systems have been tested that uses frequency multiplies of the laser pulse (Tan and Narayanan 2004). Also, new systems that use wide ranges of the reflectance spectrum are expected to be developed. However, the current tests are based on laboratories, and there are today limitations for operational use of these sensors that are difficult to combine with eye-safety restrictions (Puttonen et al. 2010).

The pulse densities of the instruments are likely to increase in the future, which enables the extraction of more detailed tree crown structures such as branch level information. The full waveform recording sensors becoming more commonly used supports this development. Analyses based on full waveform data are currently lacking in Scandinavian conditions, but expected to be reported in the near future. Besides increased point density, the waveform metrics likely produce additional information to area-based analyses carried out using a lower pulse density.

References

Ahokas E, Kaasalainen S, Hyyppä J, Suomalainen J (2006) Calibration of the Optech ALTM 3100 laser scanner intensity data using brightness targets. Int Arch Photogramm Remote Sens Spat Inf Sci XXXVI, 7 p

Asner GP (2013) Biological diversity mapping comes of age. Editorial to the special issue "Remote Sensing of Biological Diversity". Remote Sens 5:374–376

Asner GP, Knapp DE, Kennedy-Bowdoin T, Jones MO, Martin RE, Boardman J, Field CB (2007) Carnegie Airborne Observatory: in-flight fusion of hyperspectral imaging and waveform light detection and ranging for three-dimensional studies of ecosystems. J Appl Remote Sens 1:1–21

Asner GP, Knapp DE, Boardman J, Green RO, Kennedy-Bowdoin T, Eastwood M, Martin RE, Anderson C, Field CB (2012) Carnegie Airborne Observatory-2: increasing science data dimensionality via high-fidelity multi-sensor fusion. Remote Sens Environ 124:454–465

Barilotti A, Crosilla F, Sepic F (2009) Curvature analysis of lidar data for single tree species classification in alpine latitude forests. Laser scanning 2009. Int Arch Photogramm Remote Sens XXXVIII:129–134

Bishop CM (2006) Pattern recognition and machine learning. Springer, New York. ISBN 0-387-31073-8

Brandtberg T (2007) Classifying individual tree species under leaf-off and leaf-on conditions using airborne lidar. ISPRS J Photogramm Remote Sens 61:325–340

Brandtberg T, Warner TA, Landenberger RE, McGraw JB (2003) Detection and analysis of individual leaf-off tree crowns in small footprint, high sampling density lidar data from the eastern deciduous forest in North America. Remote Sens Environ 85:290–303

Chen CH, Peter Ho PG (2008) Statistical pattern recognition in remote sensing. Pattern Recognit 41:2731–2741

Cho MA, Skidmore AK (2006) A new technique for extracting the red edge position from hyperspectral data: the linear extrapolation method. Remote Sens Environ 101:181–193

Collings S, Caccetta P, Campbell N, Wu XL (2011) Empirical models for radiometric calibration of digital aerial frame mosaics. IEEE Trans Geosci Remote Sens 49:2573–2588

Coren F, Sterzai P (2006) Radiometric correction in laser scanning. Int J Remote Sens 27:3097–3104

Dalponte M, Bruzzone L, Gianelle D (2008) Fusion of hyperspectral and LIDAR remote sensing data for the classification of complex forest areas. IEEE Trans Geosci Remote Sens 46:1416–1427

Dalponte M, Bruzzone L, Gianelle D (2012) Tree species classification in the Southern Alps based on the fusion of very high geometrical resolution multispectral/hyperspectral images and LiDAR data. Remote Sens Environ 123:258–270

Dalponte M, Ørka HO, Gobakken T, Gianelle D, Næsset E (2013) Tree species classification in boreal forests with hyperspectral data. IEEE Trans Geosci Remote Sens 51:2632–2645

Dalponte M, Ørka HO, Ene LT, Gobakken T, Næsset E (2014) Tree crown delineation and tree species classification in boreal forests using hyperspectral and ALS data. Remote Sens Environ 140:306–317

Donoghue DNM, Watt PJ, Cox NJ, Wilson J (2007) Remote sensing of species mixtures in conifer plantations using LiDAR height and intensity data. Remote Sens Environ 110:509–522

Edelsbrunner H, Mücke EP (1994) Three-dimensional alpha shapes. ACM Trans Graph 13:43–72

Fukunaga K (1990) Introduction to statistical pattern recognition, 2nd edn. Academic, San Diego

Gokcen I, Peng J (2002) Comparing linear discriminant analysis and support vector machines. Lect Notes Comput Sci 2457:104–113

Haralick RM, Shanmugam K, Dinstein J (1973) Textural features for image classification. IEEE Trans Syst Man Cybern 3:610–621

Heikkinen V, Tokola T, Parkkinen J, Korpela I, Jääskelainen T (2010) Simulated multispectral imagery for tree species classification using support vector machines. IEEE Trans Geosci Remote Sens 48:1355–1364

Heinzel J, Koch B (2011) Exploring full-waveform LiDAR parameters for tree species classification. Int J Appl Earth Obs Geoinf 13:152–160

Heinzel J, Koch B (2012) Investigating multiple data sources for tree species classification in temperate forest and use for single tree delineation. Int J Appl Earth Obs Geoinf 18:101–110

Heinzel J, Ronneberger O, Koch B (2010) A comparison of support vector and linear classification of tree species. In: Silvilaser 2010, Freiburg, Freiburg, Germany, pp 377–384

Hill RA, Thomson AG (2005) Mapping woodland species composition and structure using airborne spectral and LiDAR data. Int J Remote Sens 26:3763–3779

Höfle B, Pfeifer N (2007) Correction of laser scanning intensity data: data and model-driven approaches. ISPRS J Photogramm Remote Sens 62:415–433

Hollaus M, Mücke W, Höfle B, Dorigo W, Pfeifer N, Wagner W, Bauerhansl C, Regner B (2009) Tree species classification based on full-waveform airborne laser scanning data. In: Silvilaser 2009, College Station, TX, USA, pp 54–62

Holmgren J, Persson Å (2004) Identifying species of individual trees using airborne laser scanner. Remote Sens Environ 90:415–423

Holmgren J, Persson Å, Söderman U (2008) Species identification of individual trees by combining high resolution LIDAR data with multi-spectral images. Int J Remote Sens 29:1537–1552

Hudak AT, Crookston NL, Evans JS, Hall DE, Falkowski MJ (2008) Nearest neighbor imputation of species-level, plot-scale forest structure attributes from LiDAR data. Remote Sens Environ 112:2232–2245

Hughes JS, Evans DL, Burns PY, Hill JM (1986) Identification of 2 southern pine species in high-resolution aerial MSS data. Photogramm Eng Remote Sens 52:1175–1180

Jones TG, Coops NC, Sharma T (2010) Assessing the utility of airborne hyperspectral and LiDAR data for species distribution mapping in the coastal Pacific Northwest, Canada. Remote Sens Environ 114:2841–2852

Kim S, McGaughey RJ, Andersen HE, Schreuder G (2009) Tree species differentiation using intensity data derived from leaf-on and leaf-off airborne laser scanner data. Remote Sens Environ 113:1575–1586

Korpela IS (2008) Mapping of understory lichens with airborne discrete-return LiDAR data. Remote Sens Environ 112:3891–3897

Korpela I, Tokola T (2006) Potential of aerial image-based monoscopic and multiview single-tree forest inventory: a simulation approach. For Sci 52:136–147

Korpela I, Ørka HO, Hyyppä J, Heikkinen V, Tokola T (2010a) Range and AGC normalization in airborne discrete-return LiDAR intensity data for forest canopies. ISPRS J Photogramm Remote Sens 65:369–379

Korpela I, Ørka HO, Maltamo M, Tokola T, Hyyppä J (2010b) Tree species classification using airborne LiDAR – effects of stand and tree parameters, downsizing of training set, intensity normalization, and sensor type. Silva Fenn 44:319–339

Koukoulas S, Blackburn GA (2005) Mapping individual tree location, height and species in broadleaved deciduous forest using airborne LIDAR and multi-spectral remotely sensed data. Int J Remote Sens 26:431–455

Liang X, Hyyppä J, Matikainen L (2007) Deciduous-coniferous tree classification using difference between first and last pulse laser signatures. Int Arch Photogramm Remote Sens XXXVI:253–257

Maltamo M, Packalén P, Kallio E, Kangas J, Uuttera J, Heikkilä J (2011) Airborne laser scanning based stand level management inventory in Finland. In: SilviLaser 2011, Hobart, Australia

Moffiet T, Mengersen K, Witte C, King R, Denham R (2005) Airborne laser scanning: exploratory data analysis indicates potential variables for classification of individual trees or forest stands according to species. ISPRS J Photogramm Remote Sens 59:289–309

Morsdorf F, Nichol C, Malthus T, Woodhouse IH (2009) Assessing forest structural and physiological information content of multi-spectral LiDAR waveforms by radiative transfer modelling. Remote Sens Environ 113:2152–2163

Næsset E (2002) Predicting forest stand characteristics with airborne scanning laser using a practical two-stage procedure and field data. Remote Sens Environ 80:88–99

Naidoo L, Cho M, Mathieu R, Asner G (2012) Classification of savanna tree species, in the Greater Kruger National Park region, by integrating hyperspectral and LiDAR data in a Random Forest data mining environment. ISPRS J Photogramm Remote Sens 69:167–179

Niccolai A, Niccolai M, Oliver CD (2010) Point set topology extraction for branch and crown-level species classification. Photogramm Eng Remote Sens 76:319–330

Niska H, Skön JP, Packalén P, Tokola T, Maltamo M, Kolehmainen M (2010) Neural networks for the prediction of species-specific stem volumes using airborne laser scanning and aerial photographs. IEEE Trans Geosci Remote Sens 48:1076–1085

Olofsson K, Wallerman J, Holmgren J, Olsson H (2006) Tree species discrimination using Z/I DMC imagery and template matching of single trees. Scand J For Res 21:106–110

Ørka HO, Næsset E, Bollandsås OM (2009a) Classifying species of individual trees by intensity and structure features derived from airborne laser scanner data. Remote Sens Environ 113:1163–1174

Ørka HO, Næsset E, Bollandsås OM (2009b) Comparing classification strategies for tree species recognition using airborne laser scanner data. In: Silvilaser 2009, College Station, TX, USA, pp 46–53

Ørka HO, Næsset E, Bollandsås OM (2010) Effects of different sensors and leaf-on and leaf-off canopy conditions on echo distributions and individual tree properties derived from airborne laser scanning. Remote Sens Environ 114:1445–1461

Ørka HO, Gobakken T, Næsset E, Ene L, Lien V (2012) Simultaneously acquired airborne laser scanning and multispectral imagery for individual tree species identification. Can J Remote Sens 38:125–138

Ørka HO, Dalponte M, Gobakken T, Næsset E, Ene LT (2013) Characterizing forest species composition using multiple remote sensing data sources and inventory approaches. Scand J For Res 28:677–688

Packalén P, Maltamo M (2006) Predicting the plot volume by tree species using airborne laser scanning and aerial photographs. For Sci 52:611–622

Packalén P, Maltamo M (2007) The k-MSN method for the prediction of species specific stand attributes using airborne laser scanning and aerial photographs. Remote Sens Environ 109:328–341

Packalén P, Maltamo M (2008) Estimation of species-specific diameter distributions using airborne laser scanning and aerial photographs. Can J For Res 38:1750–1760

Packalén P, Suvanto A, Maltamo M (2009) A two stage method to estimate species-specific growing stock. Photogramm Eng Remote Sens 75:1451–1460

Persson Å, Holmgren J, Söderman U, Olsson H (2004) Tree species classification of individual trees in Sweden by combining high resolution laser data with high resolution near-infrared digital images. ISPRS working group VIII/2, pp 204–207

Puttonen E, Suomalainen J, Hakala T, Räikkonen E, Kaartinen H, Kaasalainen S, Litkey P (2010) Tree species classification from fused active hyperspectral reflectance and LIDAR measurements. For Ecol Manage 260:1843–1852

Reitberger J, Krzystek P, Stilla U (2006) Analysis of full waveform lidar data for tree species classification. Int Arch Photogramm Remote Sens XXXVI/3, 6 p

Reitberger J, Krzystek P, Stilla U (2008) Analysis of full waveform LIDAR data for the classification of deciduous and coniferous trees. Int J Remote Sens 29:1407–1431

Sarrazin MJD, van Aardt JAN, Asner GP, McGlinchy J, Messinger DW, Wu J (2012) Fusing small-footprint waveform LiDAR and hyperspectral data for canopy-level species classification and herbaceous biomass modeling in savanna ecosystems. Can J Remote Sens 37:653–665

Sato M, Koike T (2003) Classification of tree types by polarimetric Pi-SAR. IEEE Int Geosci Remote Sens Symp 1:431–433

Säynäjoki R, Packalén P, Maltamo M, Vehmas M, Eerikäinen K (2008) Detection of aspens using high resolution aerial laser scanning data and digital aerial images. Sensors 8:5037–5054

Schreier H. Lougheed J, Tucker C, Leckie D (1985) Automated measurements of terrain reflection and height variations using an airborne infrared laser system. Int J Remote Sens 6:101–113

Suratno A, Seielstad C, Queen L (2009) Tree species identification in mixed coniferous forest using airborne laser scanning. ISPRS J Photogramm Remote Sens 64:683–693

Tan SX, Narayanan RM (2004) Design and performance of a multiwavelength airborne polarimetric lidar for vegetation remote sensing. Appl Opt 43:2360–2368

Theodoridis S, Koutroumbas K (2009) Pattern recognition, 4th edn. Academic, Burlington. ISBN 978-1-59749-272-0

Törmä M (2000) Estimation of tree species proportions of forest stands using laser scanning. Int Arch Photogramm Remote Sens XXXIII:1524–1531

van Aardt JAN, Wynne RH, Scrivani JA (2008) Lidar-based mapping of forest volume and biomass by taxonomic group using structurally homogenous segments. Photogramm Eng Remote Sens 74:1033–1044

Vaughn NR, Moskal LM, Turnblom EC (2012) Tree species detection accuracies using discrete point lidar and airborne waveform lidar. Remote Sens 4:377–403

Vauhkonen J, Tokola T, Maltamo M, Packalén P (2008) Effects of pulse density on predicting characteristics of individual trees of Scandinavian commercial species using alpha shape metrics based on airborne laser scanning data. Can J Remote Sens 34:S441–S459

Vauhkonen J, Tokola T, Packalén P, Maltamo M (2009) Identification of Scandinavian commercial species of individual trees from airborne laser scanning data using alpha shape metrics. For Sci 55:37–47

Vauhkonen J, Korpela I, Maltamo M, Tokola T (2010) Imputation of single-tree attributes using airborne laser scanning-based height, intensity, and alpha shape metrics. Remote Sens Environ 114:1263–1276

Vauhkonen J, Packalén P, Pitkänen J (2011) Airborne laser scanning-based stem volume imputation in a managed, boreal forest area: a comparison of estimation units. In: SilviLaser 2011, Hobart, Australia

Vauhkonen J, Seppänen A, Packalén P, Tokola T (2012) Improving species-specific plot volume estimates based on airborne laser scanning and image data using alpha shape metrics and balanced field data. Remote Sens Environ 124:534–541

Vauhkonen J, Hakala T, Suomalainen J, Kaasalainen S, Nevalainen O, Vastaranta M, Holopainen M, Hyyppä J (2013) Classification of spruce and pine trees using active hyperspectral LiDAR. IEEE Geosci Remote Sens Lett 10:1138–1141

Vauhkonen J, Packalen P, Malinen J, Pitkänen J, Maltamo M (2014) Airborne laser scanning based decision support for wood procurement planning. Scand J For Res. doi:10.1080/02827581.2013.813063

Villikka M, Packalén P, Maltamo M (2012) The suitability of leaf-off airborne laser scanning data in an area-based forest inventory of coniferous and deciduous trees. Silva Fenn 46:99–110

Wagner W, Ullrich A, Ducic V, Melzer T, Studnicka N (2006) Gaussian decomposition and calibration of a novel small-footprint full-waveform digitising airborne laser scanner. ISPRS J Photogramm Remote Sens 60:100–112

Wehr A, Lohr U (1999) Airborne laser scanning – an introduction and overview. ISPRS J Photogramm Remote Sens 54:68–82

Wolf JA, Fricker GA, Meyer V, Hubbell SP, Gillespie TW, Saatchi SS (2012) Plant species richness is associated with canopy height and topography in a neotropical forest. Remote Sens 4:4010–4021

Yao W, Krzystek P, Heurich M (2012) Tree species classification and estimation of stem volume and DBH based on single tree extraction by exploiting airborne full-waveform LiDAR data. Remote Sens Environ 123:368–380

Chapter 8
Estimation of Biomass Components by Airborne Laser Scanning

Sorin C. Popescu and Marius Hauglin

Abstract Airborne laser scanning (ALS) has evolved for the past three decades into becoming an established technology to accurately derive forest inventory parameters and assess aboveground biomass of forests. In addition to total above ground biomass, there is interest in estimating biomass of individual tree components, such as stem, branches, foliage, bark and even roots, for a better understanding of carbon sequestration by trees and their components, but also for better estimating tree biomass resources for bioenergy production utilizing various parts of forest trees. This chapter introduces the importance of forest biomass studies with airborne ALS remote sensing means and presents the various approaches for estimating above ground biomass of forests and tree components biomass. The chapter reviews the most common methodological approaches for estimating biomass, such as the area based approach (ABA) and the individual tree crown (ITC) approach, discusses advantages and disadvantages to both methods, presents the allometry involved, and includes a brief discussion on biomass change and multi-platforms ALS data used for estimating biomass.

8.1 Introduction

Traditionally, the tree stem and its volume have in many countries been the main subject of a forest inventory, and this is reflected in the methodological development of ALS-based estimation methods. Most of the methodological framework that form

S.C. Popescu (✉)
Department of Ecosystem Science and Management, Texas A & M University, College Station, TX, USA
e-mail: s-popescu@tamu.edu

M. Hauglin
Department of Ecology and Natural Resource Management, Norwegian University of Life Sciences, Ås, Norway

M. Maltamo et al. (eds.), *Forestry Applications of Airborne Laser Scanning: Concepts and Case Studies*, Managing Forest Ecosystems 27, DOI 10.1007/978-94-017-8663-8__8, © Springer Science+Business Media Dordrecht 2014

the basis for this chapter was first developed with a focus on estimation of timber volume. Interests towards a more extensive mapping of forest biomass have however been growing since the early 2000s, mainly caused by two interrelated factors: The inclusion of forest biomass in carbon accounting systems, and the possible utilization of forest biomass for energy purposes.

Man-made emissions of CO_2 are believed to be a major cause for global climate change. This has led to measures to control and reduce CO_2-emissions, including national or regional accounting of carbon flux. In this accounting, forests play a role since they absorb and store carbon from the atmosphere. Mapping of forest biomass – and how it changes over time – is in this context essential. In the last 10–15 years there has also been growing interest towards the use of forest biomass for energy purposes, and in addition to the environmental benefits of using renewable energy, also the expected decrease in supply of fossil fuels is held as a reason to utilize more of the forest biomass. The reserves of fossil fuels on earth are inevitably decreasing and even if the expected output from unconventional fossil fuels such as shale gas is growing, it is still suggested by many that alternative sources of energy should be searched for. Biomass from forests is thus likely to be one of several sources of energy that will replace conventional fossil fuels at some point in the future. Both carbon sequestration and utilization for bioenergy purposes raise the need for mapping of forest biomass. Whereas a large scale mapping is typically suggested for regional carbon sequestration, mapping at a finer scale– such as the stand, or even single-tree level – is more useful in the context of managing and harvesting from forests for bioenergy purposes. The two types of application can require different types of mapping and both will be discussed in the course of this chapter, which is organized into six sections: after this introductory section, the next section introduces terms and concepts related to the estimation of biomass. The third section, which is the main section of this chapter, is devoted to the procedures and concepts involved in estimating biomass and biomass components by ALS. Before some concluding remarks in the last section, we briefly touch upon topics related to biomass change, and the use of multi-platform laser remote sensing.

8.2   Estimating Biomass – Definitions and Terms

The biomass of a tree can be divided into fractions – or components. Descriptions of typical biomass components are given in Table 8.1, and some are illustrated in Fig. 8.1. Some components – such as foliage – are straight-forwardly defined, but others need more specifications to be unambiguously described. Biomass is typically quantified by dry weight, which is the total weight excluding the weight of the water contained within the matter in question (more on this in the next subsection). One will often encounter biomass figures given at a per area basis, typically in units of *Mg/ha*, i.e. metric ton per hectare.

Table 8.1 List of commonly used single-tree biomass components

Component	Remark
Total above-ground biomass	Stump height should be defined
Stem biomass	Can be divided into bark and stem wood biomass or may refer to merchantable stem biomass
Branch biomass	Can be divided into live and dead branches
Foliage/needle biomass	
Stump biomass	
Crown biomass	Typically the sum of the branch and foliage biomass
Total below-ground biomass	Inclusion of stump must be defined, also minimum thickness of roots
Root biomass	Can be subdivided according to root size
Total tree biomass	

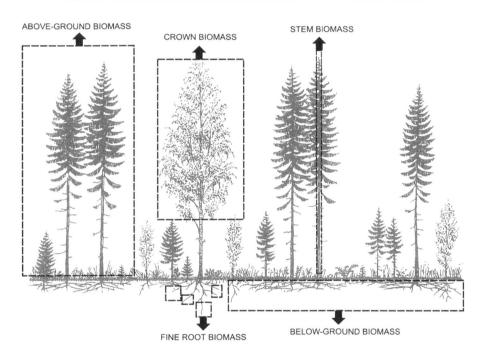

Fig. 8.1 Biomass components above- and below-ground

The carbon content of wood and bark of trees is about 50 % of dry biomass (Houghton et al. 1997), therefore, carbon estimates can be obtained by multiplying dry biomass estimates by 0.5. As such, most efforts of carbon estimation stored by forests focus on the estimation of dry biomass.

To derive information about these biomass fractions from the ALS data one must find a relationship – directly, or indirectly – between the remotely sensed data and the biomass component in question. What we here mean by direct relationship can be illustrated by an example: Consider a case where the height of a tree is reduced by cutting it halfway up the stem and removing the top part. This will clearly affect how the laser pulses will be reflected from this tree, so we can say that there is a direct relationship between the *tree height* and the ALS data. If one dug a tunnel and removed half of the roots of a tree, this would not have any effect on how the laser pulses are reflected from this tree (we are obviously not considering the long term effects of this treatment here). This means that there is no *direct* relationship between root biomass and ALS data. Fortunately, an *indirect* relationship between root biomass and ALS data does exist, and this will be discussed in the following. If we consider above-ground components such as stem biomass, when – contrary to the roots – the stem could actually be hit by laser pulses, it is reasonable to believe that an *indirect* relationship will be most useful when estimating stem biomass as well. After all, most of the information inherent in the ALS data will not be directly related to the stem, since few of the pulses will be reflected from the stem itself. So in such cases we must also look for an indirect relationship, which almost certainly will involve the concept of allometry, or the systematic relationship between biophysical properties – such as size or shapes of certain parts – found in many living organisms. The existence of these allometric relationships is a key factor in the methodological framework that form the basis for ALS-based estimations in forests, and it is especially visible in the process of estimating biomass components. Allometric relationships – that certain properties of the trees covary in a systematic fashion, allows for the development of allometric models, also known as allometric Equations. A typical example of an allometric model which most foresters will be familiar with is a single-tree diameter/height – volume model, i.e. a volume equation. By analyzing a large number of sampled trees with accurately measured diameter at breast height (DBH), tree height and stem volume one can derive a predictive model expressed as a function that takes DBH and height as input and yields stem volume as the output. A typical purpose of such a model is to use properties that are easy to measure – such as DBH and height in this example – to get a prediction of a property that is time-consuming or difficult to measure directly, like stem volume. These modeling principles can also be used for other properties of a tree, and we will in the next section discuss allometric *biomass* models.

The reason why allometry is important when estimating biomass with ALS is that there are only certain properties – in particular tree height – that can be directly related to the information in the ALS data. As described in Chap. 1, a crucial bit of information that is needed when estimating biomass with the ABA is the height above ground for the laser echoes. So when stand volume is derived using the ABA, it relies on the fact that volume is systematically related to the tree heights. With the ITC approach, estimation of tree properties such as stem volume also depends on allometric relationships.

8.2.1 *Allometric Biomass Models*

Single-tree biomass field measurements are often needed as reference data in estimation of biomass at various spatial scales. This is usually also true for remote sensing methods, thus even in large-scale regional mapping the reference data might consist of field measurements of biomass on single trees. The biomass of these trees is used – together with the remote sensing data – to build predictive models.

Since the moisture content of trees varies, dry weight is typically preferred when quantifying forest biomass, as we have mentioned earlier. Obtaining accurate measurements of the dry weight of biomass-components of a tree do however involve a time-consuming and work-intensive process, which include destructive sampling, drying and weighing. Consequently, the use of existing allometric models – often referred to as biomass equations – is common when obtaining single-tree forest biomass data from field measurements. These allometric models are typically species-specific models with diameter at breast height (DBH), and sometimes also tree height or other characteristics as explanatory variables. Allometric biomass models are usually derived using a set of destructively sampled trees, and there are studies that have collected and reviewed a large number of available regional and national allometric biomass models, such as Zianis et al. (2005) for Europe, Jenkins et al. (2003, 2004) and Ter-Mikaelian and Korzukhin (1997) for North America and Eamus et al. (2000) and Keith et al. (2000) for Australia. In addition to models for total above-ground biomass, also models for biomass of components such as branches, bark, roots, and foliage are available for certain species and in certain regions.

Note that the errors associated with predictions of some biomass components from these allometric models – such as branch or root biomass –can be larger than for predictions of other components such as stem wood. It means that e.g. branch biomass varies more than stem wood biomass for a given diameter and height, which for many tree species is intuitively reasonable. This implies however that stem wood biomass in many cases can be more accurately predicted than for example branch biomass, using these allometric models.

The choice of allometric models have been shown to affect the predictions from an ALS-based biomass model, and Zhao et al. (2012) conclude that the use of regional biomass models when deriving the reference data in most cases are better than using national models, provided the regional models were developed with a representative set of data (the reference data referred to here will be described in more detail later in this chapter, and it is also analogue to the reference data used in the ABA described in Chap. 1). Since collecting field reference data for allometric biomass models is costly and time-consuming, some regional models are based on only a limited number of sampled trees, and this could affect the applicability even at the regional level. It is thus advisable to thoroughly consider the choice of allometric models, if several are available.

8.2.2   Component Ratio Methods and Biomass
Expansion Factors

Another approach that was proposed for biomass estimates in the Forest Inventory
and Analysis (FIA) program of the United States Department of Agriculture
(USDA) Forest Service that would produce national-level biomass and carbon
estimates consistent with FIA volume estimates at the tree-level is described in
Heath et al. (2009). The approach is called the component ratio method (CRM)
and its steps can be summarized as follows: (1) convert wood volume in the stem to
biomass using a compiled set of wood specific gravities; (2) calculate the biomass of
bark on the stem using a compiled set of percent bark and bark specific gravities; (3)
calculate the biomass of tops and branches as a proportion of the stem biomass based
on component proportions from Jenkins et al. (2003); (4) calculate the biomass of
the stump based on equations in Raile (1982); and (5) sum the parts to obtain the
total aboveground biomass. Root biomass is calculated as a proportion of the stem
biomass based on component proportions described in Jenkins et al. (2003). There
are advantages to using such an approach, for example: (1) there is considerable
research and operational inventory experience in deriving individual tree volume
estimates; and (2) tying biomass to volume, and then converting biomass to carbon
by using the 0.5 conversion factor, provides not only consistent volume and biomass
estimates, but also carbon estimates that match volume and biomass predictions.
Based on the same logic of the relationship between biomass and carbon, there
is a relationship between biomass and volume. Somogyi et al. (2008) compiled
a database to calculate biomass or carbon for forests of the Eurasian region
from proxy variables derived by forest inventories such as, tree volume. This
database contains several types of expansion, conversion and combined factors,
by various tree species or species groups, to calculate individual tree biomass.
This approach is based on the following equation that involves biomass expansion
factors (BEF):

$$\text{Biomass } (t) = V \times D \times ExpF \times (1 + R) \tag{8.1}$$

where V is volume (m^3), D is wood density (t m^{-3}), $ExpF$ is biomass expansion
factor (dimensionless), and R is root-to-shoot ratio (dimensionless, the ratio between
root and total above ground biomass).

   The Intergovernmental Panel on Climate Change (IPCC) guidance for national
greenhouse gas inventories (Penman et al. 2003) lists the BEF approach as the
preferred method for some of the tiers involved in carbon estimates. The higher tier
methods demand for greater specificity, such as country-level factors and factors
specific to species. It is generally recognized that when individual tree data is
available, biomass estimates based on individual trees are preferred (Heath et al.
2009).

8.2.3 Model Type and ALS-Derived Metrics

There are several types of regression techniques in use when predicting biomass components from ALS data, some of which were referred to in Chap. 1. In general, the same types of statistical models are used for estimation of biomass components as those used for estimation of stem volume and other forest characteristics using ALS data. Linear regression models are common, and could involve variable reduction techniques such as stepwise selection if the ALS-derived variables are numerous, or highly correlated. Also non-parametric regression models based on machine learning techniques such as random forest and support vector machines have been successfully used to predict forest biomass components. With regard to the statistical features derived from the ALS data – i.e. the predictor variables in the models – many of the same features as those discussed in Chap. 1 are also used in predictive modeling of forest biomass with ALS data.

As described at the beginning of this section, it is necessary to find a relationship between the ALS data and the targeted biomass component in order to develop a predictive model. As we have already discussed, this relationship can be direct or indirect, but to distinguish between the two types of relationship is in practice not necessary, nor is it always possible. In the context of developing predictive models and finding suitable predictor variables it suffices to find that a relationship exists. In some modeling procedures – such as stepwise regression – the actual ALS-derived features that are used as predictor variables in the model will be determined through the process of fitting the model, and typically consists of a subset of a large range of ALS-derived features extracted prior to the model fitting. A different approach would be to *a-priori* select one or a few ALS-derived features that one believes will capture the relationship between the ALS data and the desired property. An example of this could be to use the maximum above-ground height of the echoes assigned to a single tree to model tree height or DBH. Another example could be the use of geometric features derived from the echoes assigned to the crown of a single tree as predictors for modeling properties such as crown or foliage biomass. More specific, this could mean to use the volume of a three-dimensional convex hull or the area of a two-dimensional projection of the laser echoes assigned to a tree crown as predictor variables, as was done by Hauglin et al. (2013) and Kankare et al. (2013b).

8.3 Estimating Biomass and Biomass Components by ALS

The use of ALS to estimate forest biomass has been investigated in a number of studies. Most of these studies follow the ABA described in Chap. 1, i.e. biomass is predicted at the plot, stand or regional level. Zolkos et al. (2013) and Koch (2010) summarize some of these studies and are good starting points for in-depth studies of biomass estimation with the ABA.

Lately there have also been studies investigating the estimation of forest biomass over large areas using ALS as a sampling tool, following the approach described in Chap. 14. See for example Andersen et al. (2011), Gobakken et al. (2012) or Stephens et al. (2012).

Some studies have proposed methods to estimate biomass at the single-tree level using ALS, following the ITC approach described in Chap. 1. Popescu (2007) estimated single-tree DBH from ALS data and used this ALS-derived DBH with allometric models to get total above-ground biomass, whereas Gleason and Im (2012), Hauglin et al. (2013) and Kankare et al. (2013b) related single-tree biomass directly to ALS-derived variables. An approach where elements from both ABA and ITC are combined is described in Chap. 6, and that chapter also contains a case study on estimation of biomass. In the rest of this section we will describe some of the concepts and procedures that are used for estimation of biomass and biomass components in forests by ALS, primarily focusing on the ABA and the ITC approach.

8.3.1 ALS Estimation of Biomass Components with the ABA

A typical methodological approach for estimation of biomass components by ALS is similar to the ABA described in Chap. 1, in which stem volume is predicted for area units.

The approach can be summarized as follows: The ALS data are first spatially assigned to area units, typically raster cells or sample plots. The ALS echoes in each area unit are then analyzed, and a range of statistical features are derived. Usually the features are computed from the height distribution of the laser echoes, and typically include measures such as moments and order statistics of the distributions. Also features derived from the returned intensity of the echoes have been used. These ALS features are then related to ground reference values through regression models. When computing biomass, the ground reference values are usually derived with existing allometric models. Since the biomass is derived using allometric models, the actual field measurements carried out on each plot are typically recordings of DBH, species and possibly also tree heights. These actual measurements are the inputs to the allometric models, which in turn will yield biomass as output. This modeled biomass will then be aggregated for all trees on a plot, and used as the reference value. Any allometric model taking the properties actually measured in the field as input can be applied in this step of the process, which means that for example biomass components such as below-ground biomass can be used. The underlying assumption here is that the amount of biomass of the selected component is related to the remote sensing data. This will usually be the case when allometric models which have DBH as one of the main inputs are used, since DBH is indirectly related to the ALS data. As described in the previous section the errors associated with the predicted biomass from allometric models varies according to biomass component, which means that for example below-ground biomass in most cases will be modeled

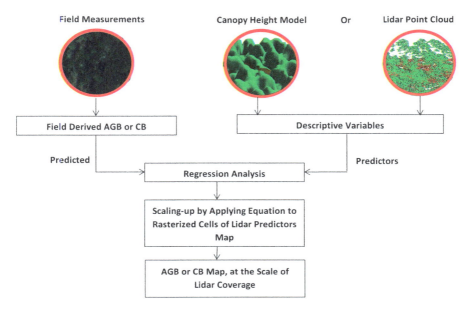

Fig. 8.2 Summary steps of the ABA. *AGB* refers to Above Ground Biomass and *CB* to Components Biomass

with a larger error than stem wood biomass. Due to the randomness of these errors they will decrease when aggregating up to a plot level.

In the next step predictive regression models are developed with the plot-level field measurements – in this case biomass – and ALS-derived metrics from the same plots, following the ABA described in Chap. 1.

The ALS regression models are in the next step used to predict biomass for new area units, typically raster cells. The raster cells can then be aggregated to get biomass figures for the desired spatial units, for example forest stands. Most commonly the total above ground biomass is modeled, but some studies also report separate estimates for biomass components, like foliage, branches or below-ground biomass (Hauglin et al. 2012; Kotamaa et al. 2010; Lim and Treitz 2004; Næsset and Gobakken 2008). This is achieved by simply applying an appropriate allometric model when calculating the ground reference biomass at the sample plots. Rather than using model variables only derived from the height distribution of the ALS-echoes some researcher have proposed methods where not only the above-ground heights of the laser echoes are taken into consideration but also the planar distribution of the ALS-echoes. This is typically done in studies identifying individual trees and where information related to these identified trees is extracted. The studies described by Bortolot and Wynne (2005) and Gleason and Im (2012) are examples of this latter approach, but predictions are in these cases still done at the plot level. The steps for the ABA, as it relates to deriving total aboveground biomass or components biomass, are summarized in Fig. 8.2.

A potential disadvantage of the area-based approach is that regression models used to link ALS-derived metrics to plot biomass have limited portability to other species or other study areas, or may change in time for the same area as the three-dimensional forest structure changes. More so, models may be dependent on instruments used to collect data because different ALS systems respond differently to the same forest (Næsset and Gobakken 2008). ALS metrics found to be significant in explaining plot-level biomass and regression models reported in the literature often lack commonalities. For example, ALS metrics commonly include mean, maximum and median canopy height, quadratic mean canopy height, or quantile heights. These metrics have been used alone or combined in linear models or nonlinear models, with or without transformations, such as logarithmic. In the context of airborne laser scanning of forest structural parameters, most estimation models described in existing literature are likely to be not only site- or species-specific but also scale-dependent, due to the fact that the models are fitted using data collected at a given plot size and should be applied at a scale or pixel (raster cell) size appropriate to the plot size used in the model fitting (Næsset 2002). As such, deriving biomass for tree components becomes more difficult for the area-based approach, whereas allometric equations for biomass components have already been developed at individual tree level for commonly occurring tree species, as previously mentioned in this chapter.

8.3.2 ALS Estimation of Biomass Components with the ITC Approach

As described in Chap. 1, an ITC approach is available as an alternative method when estimating forest attributes from ALS data, and in Popescu (2007) it is described how biomass components can be estimated using ALS at the single tree level. In the ITC approach described by Popescu (2007) DBH is estimated from the ALS data and then this ALS-derived DBH can be used with an allometric model to get a biomass estimate for the tree in question. A similar approach was used early on to derive single-tree stem volume from ALS data by Hyyppä et al. (2001). Although DBH is not directly measured by ALS, good estimates of DBH can be derived by using the information that *is* captured by ALS – particularly height and crown size related metrics. The allometric relationship between height and DBH are for many species strong, and is utilized in numerous allometric models. In the particular study described by Popescu, ALS-derived height and crown diameter was used to estimate DBH for Loblolly pine trees. When ALS is used to identify and map individual trees, dimensions and location of individual trees are derived with high accuracy, therefore a map of individual tree biomass can be generated at the extent of airborne ALS coverage. This approach was used by several studies, such as Popescu 2007, and Zhao et al. 2011, with the major steps illustrated in Fig. 8.3. To summarize their approach, the stem biomass was assigned to the pixel in the biomass map corresponding to the tree location and the foliage biomass was distributed uniformly

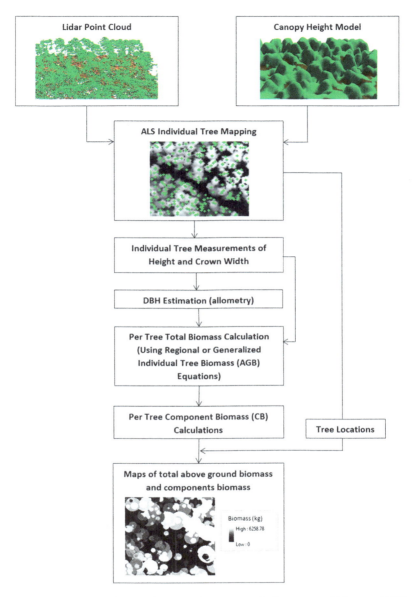

Fig. 8.3 Summary steps of the ITC approach to map total above ground biomass (*AGB*) and components biomass (*CB*)

over the pixels covered by the crown, as shown in Fig. 8.4. This biomass map was used for validating spaceborne ALS biomass estimates, such as those derived using the Geoscience Laser Altimeter System (GLAS) aboard the Ice Cloud and land Elevation Satellite (ICESat) (Popescu et al. 2011). Zhao et al. 2009, used a similar

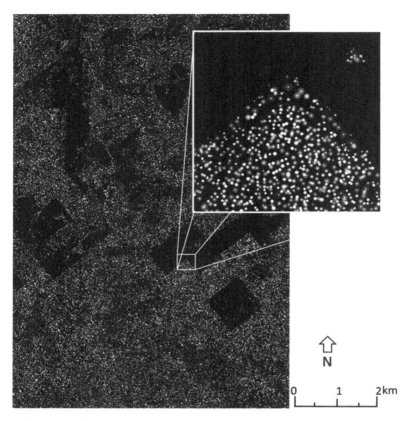

Fig. 8.4 Local-scale biomass map derived with the single-tree approach for forest conditions in East Texas, USA

biomass map to validate a scale-invariant biomass estimation approach. The map was used as reference data to synthesize training and test datasets at multiple scales for validating two scale-invariant biomass models, a linear functional model and an equivalent nonlinear model that uses ALS-derived canopy height distributions and canopy height quantile functions as predictors, respectively.

In the case of ITC estimation of biomass components, there are several advantages over ABA, most notably that, once trees are mapped with ALS, the ITC method can be applied to derive biomass components where an allometric model is available at tree level. In other words, this method takes advantage of current AGB allometric models from existing literature and ALS is only concerned with the accurate mapping and measurement of trees. In traditional forest biometry, allometric relationships are derived based on field measurements at individual tree level and not plot or area level. With the ABA, in and of itself the relationship between ALS metrics and biomass per some unit area has no practical meaning other than for scaling up biomass values to the extent of ALS data.

As with the ABA, also biomass components – such as roots – with no direct relationship to the ALS data can be estimated using the ITC approach. There are however some biomass components – in particular branches and foliage – where the ALS data might contain more direct information. In many cases, most of the laser echoes interacting with individual trees are reflected from the tree crowns, therefore from elements of tree crowns, such as foliage and branches. This suggests that the ALS data might contain direct information on the reflective surface, therefore on the branch- and foliage biomass. Hauglin et al. (2013) showed in a study with Norway spruce that ALS data in fact can contain more information regarding the branch biomass than the actual field measured DBH. An alternative approach could thus be to estimate the biomass of these components directly from the ALS data. Although it is appealing to use the remotely sensed data in such a direct fashion, there are some challenges that need to be resolved in order to fully utilize this information.

In any operational mapping of forest biomass one must make sure that the data material used to develop the model corresponds to the forest resources that are being mapped. In schemes such as those used in operational forest inventories (see Chap. 11) this is achieved by using a statistical sample from the target area as the model data. This means that a new set of model data will have to be collected for each area that is inventoried. In the direct modeling of branch biomass from ALS data Hauglin et al. (2013), Kankare et al. (2013b) and Räty et al. (2011) investigated the use of model data obtained through destructive sampling – that is – actually cutting down and weighing trees. In most operational settings this would be associated with prohibitively high costs, and would therefore not be an option. Further development of the method might however solve this by using model data obtained by other means, such as model data obtained using terrestrial laser scanning (TLS).

Mapping of biomass or biomass components at the individual tree level, following the general approach, has the advantage of using species-specific allometric equations extensively investigated and tested in existing literature or forest practice, whether regional or generalized equations. Another advantage is the fact that the ITC approach brings minimal scaling effects to inventorying forest biomass. Biomass maps at scales above tree levels will be immediately available by integrating tree-level results up to the desired scale. As long as the algorithms for processing ALS data and identifying individual trees perform accurately, one could expect the derivation of highly spatial specific and accurate biomass maps by following this approach. However, as reported in the literature, ITC segmentation and measurement algorithms vary in their performance of identifying all trees correctly and deriving their measurements. A possible concern is the inefficiency of many tree-segmentation algorithms to identify understory and suppressed trees, or to delineate trees under very dense canopy conditions. The systematic errors that will arise from this erroneous crown delineation were the motivation for the semi-ITC approach described in Chap. 6, which is an alternative to the general ITC approach, also for estimation of biomass components.

In some cases however, the ITC approach can be effective because the main contributors to above ground biomass are the dominant trees (Zavitkovski 1976) that could be identified more easily by tree-delineation algorithms on ALS data.

8.4   Biomass Change

As with traditional forest inventory methods, the use of ALS data will only provide a "snapshot" of the total above ground biomass or components biomass levels when the data were acquired. Changes in biomass due to temporal variation (e.g. disturbance, anthropogenic management activities, forest growth) could be monitored utilizing the pre-existing models and repeat data collection missions. As such, changes in total above ground biomass and components biomass can be analysed with spatially coincident ALS data acquired at multiple dates. However, published studies on using repeat or multi-temporal airborne ALS data sets to characterize biomass change are very few and none of these refer to components biomass change. Noting that the estimation of biomass change can be largely built upon the general inference on change (see Chap. 15), with a few most important published results on estimating biomass change in particular being listed below.

Hudak et al. 2012, used repeat ALS acquisition over conifer forests and concluded that ALS can provide accurate spatially explicit biomass maps to characterize C dynamics. Meyer et al. 2013, analysed biomass changes in a tropical forest from both ground measurements and ALS data and quantified uncertainty by propagating measurement and prediction errors across spatial scales. They found that errors associated with both the mean biomass stock and mean biomass change declined with increasing spatial scales. Naesset et al. 2013, used field data and a model based approach supported by repeat ALS data to analyze variance estimates and demonstrated the potential gain in terms of reduced uncertainties by adding ALS data to field measurements. They found that ALS data contributed to improved precision of the biomass change estimates. The standard errors for individual change categories, such as, forest degradation, deforestation, and untouched, were reduced by 18–84 %. The largest improvement in precision was experienced for degradation (73–84 %), which is a category that is difficult to assess with most other remote sensing techniques.

8.5   Biomass Assessment with Multi-platform ALS

While the focus of this text book is on airborne ALS applications in forest studies, ALS data are collected from multiple platforms, such as terrestrial, airborne, and spaceborne platforms. Each platform provides data over different spatial scales and enables C estimates with different levels of uncertainty and errors. TLS provide highly detailed scans for small areas, while data collected utilizing airborne and spaceborne ALS can estimate biomass from local to regional and continental scales. Utilizing multi-platform ALS data will allow us to better understand the variation in forest structure and biophysical parameters at multiple spatial scales, and to provide more accurate measurements of AGB with uncertainty estimates. Using TLS, Kankare et al. (2013a) developed single tree based biomass models from multiple-scan TLS data and reported improved accuracies for components biomass, such as branch biomass. An excellent meta-analysis of AGB estimation using airborne and spaceborne sensors is presented in Zolkos et al. 2013, which also discusses AGB estimation with various types of laser remote sensors, such as discrete return ALS or full-waveform sensors. This review paper identified the interest in mapping biomass so that carbon stocks and changes can be monitored consistently across a range of scales. Their findings are discussed in relation to monitoring, reporting and verification (MRV) in the context of reducing carbon emissions associated with deforestation and forest degradation (REDD). The components of MRV include measuring the extent and change in forest area, reporting carbon stocks and emissions, and verifying the findings and implementation of REDD activities (Houghton et al. 2009; Herold and Skutsch 2011). MRV guidelines do not explicitly state accuracy requirements, but previous studies assert that satellite remote sensing biomass estimates should meet biomass errors within 20 Mg/ha or 20 % of field estimates, whichever is greater, and should not exceed errors of 50 Mg/ha for a global biomass map at 1 ha resolution (Hall et al. 2011; Houghton et al. 2009).

8.6 Conclusions

ALS and multi-platform laser studies for assessing vegetation biomass and other biophysical parameters in many forest ecosystem types report reliable results with uncertainty estimates. Zolkos et al. (2013) found that the level of accuracy obtained when estimating biomass using ALS-data is dependent on forest type – among other factors. The mean errors from the studies reviewed by Zolkos et al. varied from less than 20 % to more than 40 % of the field measured biomass, and statistics for some selected published studies are given in Table 8.2. A general finding was that accuracy – that is, model error relative to the field measured biomass – was declining with increasing plot size. Zolkos et al. further compare ALS-based biomass models with models using data derived from other types of remote sensing sensors such as radar and passive optical sensors. From their results they draw the conclusion that biomass models developed from ALS are significantly better than those based on radar and passive optical sensors. Koch (2010) also reaches similar conclusions. An integration of ALS with other airborne and spaceborne sensors is suggested by Koch as a possible path of development, and the need for more knowledge on the interaction between the laser beams and the vegetation is emphasized.

We will end this chapter by highlighting a concluding remark from the meta-study by Koch (2010) where it is asserted that *laser scanning will play an important role in the future mapping of forest biomass.*

Table 8.2 Statistics for some published studies on estimation of biomass components by ALS (Conditions other than those listed in the table may vary between the studies)

Comp[a]	Forest type	Units	N	Mean field measured biomass	RMSE	Reference
ABG	Loblolly pine plantation (USA)	707 m² plots	25	96 Mg ha⁻¹	14 %	Bortolot and Wynne (2005)
ABG	Tropical forest (Costa Rica)	0.5 ha	18	160.5 Mg ha⁻¹	14 %	Drake et al. (2002)[b]
ABG	Mixed forest (USA)	380 m² plots	18	129 Mg ha⁻¹	32 %	Gleason and Im (2012)
ABG	Deciduous forest (USA)	0.1 ha plots	48	236 Mg ha⁻¹	19 %	Lefsky et al. (1999)[b]
ABG, BR	Old growth hardwood forest (Canada)	400 m² plots	36	ABG: 120 Mg ha⁻¹ BR: 34 Mg ha⁻¹	ABG: 47 % BR: 121 %	Lim and Treitz (2004)
ABG	Boreal – young forest (Norway)	200 m² plots	39	13.6 Mg ha⁻¹	28–36 %	Næsset (2011)
ABG	Mixed forest (USA)	Single trees	136	Coniferous: 605 kg Deciduous: 618 kg	Coniferous: 71 % Deciduous: 95 %	Gleason and Im (2012)
CR	Boreal coniferous forest (Norway)	Single trees – Norway spruce	50	65 kg	35 %	Hauglin et al. (2013)
ABG	Pine plantations and mixed pine forests (USA)	Single trees	43	486 kg	33 %	Popescu (2007)
ABG	Boreal coniferous forest (Finland)	Single trees	38	Scots pine: 157 kg Norway spruce: 235 kg	Scots pine: 26 % Norway spruce: 37 %	Kankare et al. (2013b)

[a]Biomass component: *ABG* above ground biomass, *BR* branch biomass, *CR* crown biomass (branches and foliage)
[b]Large footprint ALS (5–25 m)

References

Andersen HE, Strunk J, Temesgen H (2011) Using airborne light detection and ranging as a sampling tool for estimating forest biomass resources in the Upper Tanana Valley of Interior Alaska. West J Appl For 26:157–164

Bortolot ZJ, Wynne RH (2005) Estimating forest biomass using small footprint LiDAR data: an individual tree-based approach that incorporates training data. ISPRS J Photogramm Remote Sens 59:342–360

Drake JB, Dubayah RO, Clark DB, Knox RG, Blair JB, Hofton MA, Chazdon RL, Weishampel JF, Prince S (2002) Estimation of tropical forest structural characteristics using large-footprint lidar. Remote Sens Environ 79:305–319

Eamus D, Burrows W, McGuinness K (2000) Review of allometric relationships for estimating woody biomass for Queensland, the Northern Territory and Western Australia. Australian Greenhouse Office, Canberra

Gleason CJ, Im J (2012) Forest biomass estimation from airborne LiDAR data using machine learning approaches. Remote Sens Environ 125:80–91

Gobakken T, Næsset E, Nelson R, Bollandsås OM, Gregoire TG, Ståhl G, Holm S, Ørka HO, Astrup R (2012) Estimating biomass in Hedmark County, Norway using national forest inventory field plots and airborne laser scanning. Remote Sens Environ 123:443–456

Hall F, Bergen K, Blair JB, Dubayah R, Emanuel W, Houghton R, Hurtt G, Kellndorfer J, Ranson J, Saatchi S, Wickland D (2011) Characterizing 3-D vegetation structure from space: mission capabilities requirements. Remote Sens Environ 115:2753–2775

Hauglin M, Gobakken T, Lien V, Bollandsås OM, Næsset E (2012) Estimating potential logging residues in a boreal forest by airborne laser scanning. Biomass Bioenerg 36:356–365

Hauglin M, Dibdiakova J, Gobakken T, Næsset E (2013) Estimating single-tree branch biomass of Norway spruce by airborne laser scanning. ISPRS J Photogramm Remote Sens 79:147–156

Heath LS, Hansen MH, Smith JE, Smith WB, Miles PD (2009) Investigation into calculating tree biomass and carbon in the FIADB using a biomass expansion factor approach. Forest Inventory and Analysis (FIA) symposium 2008; 21–23 October 2008, Park City, UT. In: Proceedings RMRS-P-56CD. Department of Agriculture, Forest Service, Rocky Mountain Research Station. Fort Collins, CO, USA

Herold M, Skutsch M (2011) Monitoring, reporting and verification for national REDD + programmes: two proposals. Environ Res Lett 6:014002

Houghton JT, Meira Filho LG, Lim B, Treanton K, Mamaty I, Bonduki Y, Griggs DJ, Callander BA (1997) Revised 1996 guidelines for national greenhouse gas inventories, vol 3. Intergovernmental Panel on Climate Change, Meteorological Office, Bracknell, United Kingdom

Houghton RA, Hall F, Goetz SJ (2009) Importance of biomass in the global carbon cycle. J Geophys Res 114:G00E03. doi:10.1029/2009JG000935

Hudak AT, Strand EK, Vierling LA, Byrne JC, Eitel JUH, Martinuzzi S, Falkowski MJ (2012) Quantifying aboveground forest carbon pools and fluxes from repeat LiDAR surveys. Remote Sens Environ 123:25–40

Hyyppä J, Kelle O, Lehikoinen M, Inkinen M (2001) A segmentation-based method to retrieve stem volume estimates from 3-D tree height models produced by laser scanners. IEEE Trans Geosci Remote Sens 39:969–975

Jenkins J, Chojnacky D, Heath L, Birdsey R (2003) National-scale biomass estimators for United States tree species. For Sci 49:12–35

Jenkins JC, Chojnacky DC, Heath LS, Birdsey RA (2004) Comprehensive database of diameter-based biomass regressions for North American tree species. General technical report, Department of Agriculture, Forest Service, Newtown Square, PA, USA

Kankare V, Holopainen M, Vastaranta M, Puttonen E, Yu X, Hyyppä J, Vaaja M, Hyyppä H, Alho P (2013a) Individual tree biomass estimation using terrestrial laser scanning. ISPRS J Photogramm Remote Sens 75:64–75

Kankare V, Räty M, Yu X, Holopainen M, Vastaranta M, Kantola T, Hyyppä J, Hyyppä H, Alho P, Viitala R (2013b) Single tree biomass modelling using airborne laser scanning. ISPRS J Photogramm Remote Sens 85:66–73

Keith H, Barrett D, Keenan R (2000) Review of allometric relationships for estimating woody biomass for New South Wales, the Australian Capital Territory, Victoria, Tasmania and South Australia. Australian Greenhouse Office, Canberra

Koch B (2010) Status and future of laser scanning, synthetic aperture radar and hyperspectral remote sensing data for forest biomass assessment. ISPRS J Photogramm Remote Sens 65:581–590

Kotamaa E, Tokola T, Maltamo M, Packalén P, Kurttila M, Mäkinen A (2010) Integration of remote sensing-based bioenergy inventory data and optimal bucking for stand-level decision making. Eur J For Res 129:875–886

Lefsky MA, Harding D, Cohen W, Parker G, Shugart H (1999) Surface Lidar remote sensing of basal area and biomass in deciduous forests of Eastern Maryland, USA. Remote Sens Environ 67:83–98

Lim KS, Treitz PM (2004) Estimation of above ground forest biomass from airborne discrete return laser scanner data using canopy-based quantile estimators. Scand J For Res 19:558–570

Meyer V, Saatchi SS, Chave J, Dalling J, Bohlman S, Fricker GA, Robinson C, Neumann M (2013) Detecting tropical forest biomass dynamics from repeated airborne Lidar measurements. Biogeosci Discuss 10:1957–1992

Næsset E (2002) Predicting forest stand characteristics with airborne scanning laser using a practical two-stage procedure and field data. Remote Sens Environ 80:88–99

Næsset E (2011) Estimating above-ground biomass in young forests with airborne laser scanning. Int J Remote Sens 32:473–501

Næsset E, Gobakken T (2008) Estimation of above- and below-ground biomass across regions of the boreal forest zone using airborne laser. Remote Sens Environ 112:3079–3090

Næsset E, Bollandsås OM, Gobakken T, Gregoire TG, Ståhl G (2013) Model-assisted estimation of change in forest biomass over an 11 year period in a sample survey supported by airborne LiDAR: a case study with post-stratification to provide "activity data". Remote Sens Environ 128:299–314

Penman J, Gytarsky M, Hiraishi T, Krug T, Kruger D, Pipatti R, Buendia L, Miwa K, Ngara T, Tanabe K, Wagner F (2003) Good practice guidance for land use, land-use change and forestry. IPCC national greenhouse gas inventories programme and Institute for Global Environmental Strategies, Kanagawa, Japan. Available from: http://www.ipcc-nggip.iges.or.jp/public/gpglulucf/gpglulucf_contents

Popescu SC (2007) Estimating biomass of individual pine trees using airborne lidar. Biomass Bioenerg 31:646–655

Popescu SC, Zhao K, Neuenschwander A, Lin C (2011) Satellite lidar vs. small footprint airborne lidar: comparing the accuracy of aboveground biomass estimates and forest structure metrics at footprint level. Remote Sens Environ 115:2786–2797

Raile GK (1982) Estimating stump volume. Research paper NC-224. Department of Agriculture, Forest Service, North Central Forest Experiment Station, St. Paul, MN, USA

Räty M, Kankare V, Yu X, Holopainen M, Vastaranta M, Kantola T, Hyyppä J, Viitala R (2011) Tree biomass estimation using ALS features. In: Proceedings of Silvilaser 2011, Hobart, Australia

Somogyi Z, Teobaldelli M, Federici S, Matteucci G, Pagliari V, Grassi G, Seufert G (2008) Allometric biomass and carbon factors database. iForest 1:107–113

Stephens PR, Kimberley MO, Beets PN, Paul TSH, Searles N, Bell A, Brack C, Broadley J (2012) Airborne scanning LiDAR in a double sampling forest carbon inventory. Remote Sens Environ 117:348–357

Ter-Mikaelian MT, Korzukhin MD (1997) Biomass equations for sixty-five North American tree species. For Ecol Manage 97:1–24

Zavitkovski J (1976) Ground vegetation biomass, production, and efficiency of energy utilization in some Northern Wisconsin forest ecosystems. Ecology 57:694–706

Zhao K, Popescu SC, Nelson RF (2009) Lidar remote sensing of forest biomass: a scale-invariant estimation approach using airborne lasers. Remote Sens Environ 113:182–196

Zhao KG. Popescu SC, Meng X, Agca M (2011) Characterizing forest canopy structure with composite lidar metrics and machine learning. Remote Sens Environ 115:1978–1996

Zhao F, Guo Q, Kelly M (2012) Allometric equation choice impacts lidar-based forest biomass estimates: a case study from the Sierra National Forest, CA. Agric For Meteorol 165:64–72

Zianis D, Muukkonen P, Mäkipää R, Mencuccini M (2005) Biomass and stem volume equations for tree species in Europe. Silva Fenn 4:1–63

Zolkos SG, Goetz SJ, Dubayah R (2013) A meta-analysis of terrestrial aboveground biomass estimation using lidar remote sensing. Remote Sens Environ 128:289–298

Chapter 9
Predicting Tree Diameter Distributions

Matti Maltamo and Terje Gobakken

Abstract Diameter distribution of trees is an important stand attribute that describes stand structure in terms of volume, biomass, value, growth and biodiversity factors. Diameter distribution can be characterized using different approaches such as probability density functions, percentile-based distributions or nearest neighbour applications. We review the research related to airborne laser scanning (ALS)-based predictions of diameter distributions. This includes the above-mentioned plot level approaches, as well as predicting the diameter of individual trees and combinations of different approaches. Although ALS does not directly measure tree diameter, there is a strong statistical relationship between ALS metrics and the characteristics of a diameter distribution. The capability of ALS to reproduce different shapes of diameter distribution is the most notable feature of these applications.

9.1 Introduction

Tree diameter at breast height (DBH) is, together with species, the most common attribute to be measured or registered for an individual tree. The DBH is easy to measure in field and due to strong allometric relationships within a tree, it allows for rather accurate prediction of other attributes, such as basal area, height and volume. Measurement of DBH for trees within a certain area produces the diameter

M. Maltamo (✉)
School of Forest Sciences, University of Eastern Finland, Joensuu, Finland
e-mail: Matti.maltamo@uef.fi

T. Gobakken
Department of Ecology and Natural Resource Management, Norwegian University
of Life Sciences, Ås, Norway

M. Maltamo et al. (eds.), *Forestry Applications of Airborne Laser Scanning: Concepts and Case Studies*, Managing Forest Ecosystems 27, DOI 10.1007/978-94-017-8663-8_9, © Springer Science+Business Media Dordrecht 2014

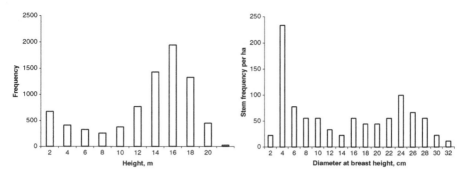

Fig. 9.1 An example of ALS height distribution (*left*) and underlying diameter distribution (*right*)

distribution, which is an important descriptor of stand structure. For example, together with height and tree taper models, diameter distribution allows for flexible calculation of timber assortments. Diameter distribution is also a highly important attribute characterizing the economic value, growth and structural biodiversity characteristics of forests. The shape of the diameter distribution is unimodal in most managed stands. Thus, because of this feature, several studies have relied on theoretical probability density functions, such as the Weibull distribution (e.g. Bailey and Dell 1973), to model diameter distributions. Some other modelling approaches include percentile-based distributions or non-parametric imputation, which also allows for characterization of multimodal diameter distributions. The modelling of diameter distributions is usually related to applications where the actual DBH's are not known but the distribution is predicted by applying some stand level attributes as independent variables. This is the usual case in stand-level management inventories. Estimation of diameter distributions is also done for analyses of forest structure, for example by comparing the parameter values of the underlying distribution in different areas. The criteria applied to compare different distributions estimates include e.g., root-mean-square error (RMSE) of the derived volume values, error indices and statistical tests (Reynolds et al. 1988; Haara et al. 1997; Siipilehto 1999).

The role of diameter distribution modelling in the airborne laser scanning (ALS) context is different from that based on field measurements. Due to the close correlation between forest canopy and ALS metrics there is also a relationship between the latter and diameter distribution (Fig. 9.1). However, ALS data are mainly affected by the vertical distribution of the canopy elements, making tree height the primary attribute that can be obtained from a tree. Tree height cannot, however, replace the role of DBH in all applications. While tree height is an excellent descriptor of the vertical component of stand structure, all tree attribute models are traditionally based on DBH. This means that information on DBH is still a primary requirement. At tree level, the prediction of DBH from height-based variables is problematic because DBH for a given tree height varies considerably, especially for the tallest trees (e.g. Maltamo et al. 2004). An alternative method is diameter distribution modelling, using similar approaches that have been used with ground-based measurements (e.g. Gobakken and Næsset 2005).

9.2 Predicting Theoretical Diameter Distributions

9.2.1 Parametric Prediction of Diameter Distributions

The diameter distribution can be constructed by assuming that it corresponds to some probability density function. Several density functions have been used for modelling the diameter distribution (Eriksson and Sallnäs 1987). To give more weight to larger and more valuable trees, diameter distributions weighted by basal area (called hereafter weighted distributions) are often used (Päivinen 1980; Van Deusen 1986). The objective of most of the studies considering various distribution functions has been to assess the accuracy of different methods and models to estimate probability density function parameters and to compare different distributions. The Weibull density function is the most frequently used density function for describing diameter distribution (Poudel and Cao 2013).

For practical applications in forest planning and management it is important that the procedures used to derive the required output data are robust. Therefore, the two-parameter Weibull has often been selected for use in forestry contexts rather than the three-parameter formulation. The advantages of the two-parameter Weibull include simplicity of mathematical derivation, the low number of parameters to be estimated and its flexibility in describing different shapes of unimodal distributions (Bailey and Dell 1973). The probability density function of the two-parameter Weibull distribution for a random variable x is (Dubey 1967):

$$f(x) = \frac{c}{b}\left(\frac{x}{b}\right)^{c-1} \exp\left[-\left(\frac{x}{b}\right)^{c}\right], \quad x \geq 0; b, c > 0 \tag{9.1}$$

where b is a scale parameter and c is a shape parameter. The two-parameter Weibull distribution parameters of Eq. 9.1 are derived for each field plot or stand by fitting the two-parameter Weibull distribution to the discrete ground reference diameter or basal area distributions (empirical trees).

After the estimation of the Weibull distribution is done regression models are constructed for the parameters using stand characteristics (e.g. basal area, stem number, basal area median diameter, basal area mean diameter, stand age, mean height, and site index) as explanatory variables. When the parameters of Weibull distribution are predicted the cumulative distributions can be calculated. The total number of trees in each diameter class is found by scaling the relative number of trees and relative basal area for the weighted distributions to ground truth, i.e. stem number and basal area, respectively.

It has been argued that statistics, such as percentiles computed directly from empirical diameter distributions, may have stronger correlation to stand inventory characteristics compared to the distribution parameters themselves. Thus, if reliable regressions for predicting such statistics could be obtained from stand inventory characteristics, they could be equated to theoretical parameters through analytical relationships (e.g., Bailey et al. 1981). Dubey (1967) derived percentile estimators for parameter recovery of the two-parameter Weibull and it is possible to use several combinations of two percentiles to estimate b and c. However, Dubey (1967) showed that the 24 and 93 percentiles were jointly best for estimating b and c in a two-parameter Weibull.

9.2.2 ALS-Based Prediction of Diameter Distributions

Gobakken and Næsset (2004) were the first to derive diameter distributions from laser scanner data. They compared the accuracy of diameter and weighted distributions obtained using parameter prediction and parameter recovery methods. Distribution parameters and the 24 and 93 percentiles for parameter recovery of a two-parameter Weibull were derived for empirical diameter and weighted distributions. Regression analysis was used to relate the distribution parameters and percentiles to various metrics for canopy height and canopy density that were derived from ALS data. Stem number and basal area were also predicted from the laser data. On average, the distance between transmitted laser pulses was 1.0 m on the ground. Aerial photo-interpretation was used to divide the plots into three strata according to age class and site quality. Stratum-specific regressions modelling the observed percentiles and total plot volume predicted from the estimated distributions were used to assess the accuracy of the regressions. The precision was slightly better for the predictions based on parameter recovery using the 24 and 93 percentiles compared to predictions where the cumulative distributions were defined directly by the predicted Weibull parameters (b, c). An example of a field-measured ground reference distribution for a sample plot and its predicted distribution is presented in Fig. 9.2.

A satisfactory characterization of the diameter distribution does not require a mathematical probability distribution known a priori. Borders et al. (1987) developed a percentile-based diameter prediction method using a system of percentiles defined across the range of observed diameters. This approach is more flexible as can be applied to irregularly shaped distributions and makes no assumption about the diameter distribution. Gobakken and Næsset (2005) predicted the weighted distributions of coniferous plots derived from ALS data using the parameter recovery method and a method based on a system of 10 percentiles defined across the range of observed diameters and compared the accuracies. Regression analysis was used to relate the percentiles to the metrics of various canopy heights and canopy density derived from the laser data. The methods were evaluated in typical one-layered forests and in forests with large diameter variability, such as those

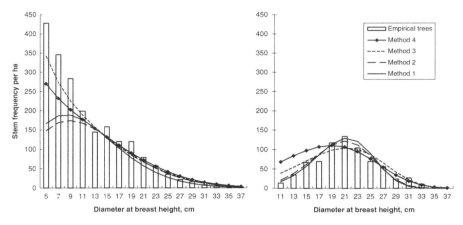

Fig. 9.2 Ground reference diameter distributions (Empirical trees) and corresponding ALS-based distributions using parameter prediction (Method 1), parameter recovery (Method 2), weighted distribution using parameter prediction (Method 3), and weighted distribution using parameter recovery (Method 4) for a young field plot (*left*) and a mature field plot (*right*) (Modified from Gobakken and Næsset 2004)

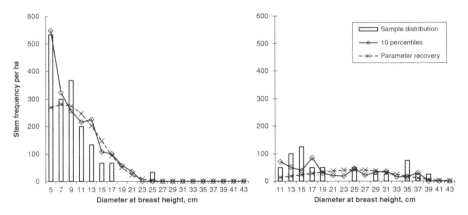

Fig. 9.3 An example of an inverse J-shaped diameter distribution (*left*) and an irregular diameter distribution (*right*), modified from Gobakken and Næsset (2005). The figure depicts the sample distributions and predicted distributions based on methods with parameter recovery and 10 percentiles, respectively. N number of trees, SD_d standard deviation of callipered tree

with multimodal and inverse J-shaped distributions. The total plot volume predicted from the estimated distributions was used to assess the accuracy of the regressions. Neither bias nor standard deviation differed significantly between the two validated methods. An example of the field-measured ground reference distribution for a sample plot and the predicted distributions is presented in Fig. 9.3.

Bollandsås and Næsset (2007) made regression models for uneven-sized Norway spruce stands predicting 10 percentiles and stand basal area. An independent validation showed non-significant mean differences in 20 of 21 diameter classes for data corresponding to the model calibration data. The model reproduced diameter distributions that corresponded well with the model calibration data (uneven-sized forest). However, the model was not flexible enough to reproduce normal and uniform diameter distributions. Volume estimates derived from predicted diameter distributions were generally well-determined, irrespective of the observed distribution.

9.2.3   Specific Cases to Utilize Theoretical Diameter Distributions

When theoretical diameter distributions are predicted, they are usually scaled according to either stem number or basal area. However, it would be beneficial if the resulting distribution was compatible with all known (measured or estimated) stand attributes. Kangas and Maltamo (2000) applied a calibration estimation technique developed by Deville and Särndal (1992) to predict distributions with such properties. In general this means that trees are either added to or removed from size classes in order to modify the distribution so that it becomes compatible with all stand attributes of interest. In the ALS context, Maltamo et al. (2007) predicted parameters of a Weibull distribution as well as stem number, basal area and stand volume. The Weibull distribution was first predicted and scaled according to the predicted stem number. Then, it was modified to be compatible also with estimated basal area and stand volume. The novel point in this study was that the calibration estimation technique was applied using stand volume estimates. In field context tree volumes are not measurable without destructive sampling, but the predicted stand volume estimate is available using the area-based ALS approach corresponding to other stand attributes. The other finding was that volume estimates obtained from diameter distributions were as accurate as those obtained from weighted distributions. This is contrary to what has been found in entirely field based studies.

Breidenbach et al. (2008) applied a generalized linear model (GLM) to estimate the shape and scale parameters of the Weibull distribution by using ALS metrics as predictors. The benefit of the GLM approach is that it is a one-step procedure, so there is no need to fit the Weibull distribution separately and then predict the parameter values (Cao 2004). The specific point of this study was that trees with different DBH values were measured from four differently sized concentric circle plots. Only trees with DBH larger or equal than 30 cm were measured from the largest plot with radius of 16 m. The change in plot size was taken into consideration by applying left- and right-truncated Weibull distributions, conditional on the DBH.

Instead of predicting Weibull parameters with ALS it is also possible to utilize the stand attributes predicted using an area-based approach in diameter distribution modelling. These estimates can be used in existing parameter prediction models. For example, in Nordic countries there exist several such field information-based models (e.g. Päivinen 1980; Tham 1988; Holte 1993; Maltamo 1997). Such general models for predicting diameter distributions have been used in ALS-related studies by, e.g. Maltamo et al. (2006) and Holopainen et al. (2010). In general, this approach may not be optimal since no local information on diameter distribution is utilized, although the area-based approach always includes field reference measurements for DBHs. Conversely, this approach can be applied to all ALS inventory areas where stand attribute models have been constructed by simply applying existing models without the need to go to revert to the modelling data. A more advanced but similar type of approach was applied by Mehtätalo et al. (2007) and Peuhkurinen et al. (2011), in which both diameter distribution and height curve were recovered from ALS-based stand attributes. An advantage of this approach is that no parameter model for Weibull is needed, but distributions are instead recovered according to mathematical relationships between distribution parameters and stand attributes. The recovered diameter distribution and height curve combination is also compatible with the utilized stand attributes.

Finally, Thomas et al. (2008) applied ALS-based Weibull parameter prediction in different types of forests including coniferous, hardwoods and mixed-woods. A special case in their study was a bimodal stand structure with a few large old trees and a large number of smaller trees. For this forest structure, they applied a finite mixture approach (e.g. Zhang et al. 2001) in which different modes of the multimodal distribution are characterized using separate Weibull functions and then combined to yield the final estimate. The study successfully predicted the parameters of two separate modes of the distribution, and ALS-based finite mixture distribution was able to characterize bimodal distributions. The drawback of the study was that there was no ALS-based separation of the different forest types that would be required for the application phase.

9.3 Non-parametric Prediction of Tree Lists

An alternative to parametric diameter distribution prediction is the utilization of non-parametric approaches. Nearest neighbour (k-nn) imputation methods in particular provide excellent possibilities for predicting diameter distributions. The basic idea of this approach is that diameter distributions of field-measured reference plots are imputed to target plots based on a measure of similarity between the reference and target plots. The measures of similarity are independent variables such as stand attributes or ALS metrics. The characteristics of diameter distribution, or some other defined parameters, are used as dependent variables in the imputation. Since the diameter distribution estimate is based on actual trees it is called a tree list. A benefit of using actual trees is that the resulting estimate can then also be a multimodal or descending distribution.

In the context of field data the first non-parametric nearest neighbour application was made by Haara et al. (1997). However, field-based k-nn diameter distribution models have not been very successful in practical applications. This is due to the fact that there is only a low number of stand attributes that can be used as independent variables. Additionally, the correlation between stand attributes, such as number of stems and mean height, and the diameter distribution can be low or at least have a strong averaging effect on the estimates. A large field data set is also required, and in addition to the plot data used in the non-parametric estimation, tree data are also needed. This might be problematic due to data rights since, in the field context, the data sets used are collected for other purposes and have usually been at very large geographical scales, including forest areas of numerous forest owners.

In the case of ALS-based forest inventories, the above-mentioned problems can mostly be avoided. A number of different area-based ALS metrics can be used in imputation (e.g. Vauhkonen et al. 2010). Additionally, if other data sources (e.g. image data) are utilized, the number of predictor candidates is further increased. On the other hand, this leads to problems related to selecting the optimal independent variables and the threat of model over-fitting (Packalén et al. 2012). In the case of ALS, the field data are also usually available because area-based methods require local sampling of field plots. Although only sum and mean attributes are modelled in operational ALS forest inventories, the underlying diameter distributions always exist since field measurements are carried out at tree level. The main benefit of this is that local variability in diameter distributions is included in the reference data. Additionally, there is no need or even the possibility of applying geographically wider data sets, due to differences in the technical settings of a laser scanner between flights and the differences in properties between laser scanners.

Research related to ALS-based k-nn diameter distributions is still rather rare, however (Packalén and Maltamo 2008; Peuhkurinen et al. 2008; Maltamo et al. 2009a). There has also been some earlier work related to tree list imputation using aerial images (Temesgen 2003), but in general the level of accuracy associated with the use of independent variables derived from aerial images has not been good enough for wider application. Packalén and Maltamo (2008) applied the non-parametric most similar neighbour (k-MSN) method for the prediction of species-specific diameter distributions, and in this case the situation was even more complex, since diameter distribution had to be predicted simultaneously for three different tree species (pine, spruce and deciduous species). Metrics calculated from both ALS data and aerial images were used as independent variables and 15 species-specific stand attributes were used as dependent variables in the canonical correlation analysis. In this case, species-specific distributions were imputed from the reference data as a whole, i.e. all species-specific distributions were imputed from the same chosen neighbour and averaged over all k-neighbours. This means that not all of the species considered may exist in each of the chosen neighbour plots. Correspondingly, some species that are not growing in the target plot may be imputed.

Studies by Packalén and Maltamo (2008) and Peuhkurinen et al. (2008) have shown that species-specific diameter distributions can be imputed by applying remote sensing data. The accuracy, especially for main tree species, was found to be good according to the RMSE figures and error indices. As a comparison, k-MSN based species-specific stand attribute estimates were used as an input to existing Weibull distribution models. The accuracy of the predictions using the Weibull

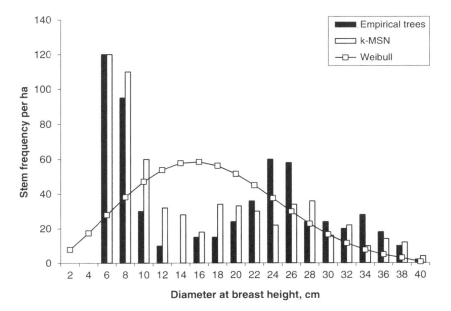

Fig. 9.4 Comparison of Weibull and k-MSN predictions (Modified from Packalén and Maltamo 2008)

models was close to the results obtained using the k-MSN application, but the benefit of k-MSN arises from the possibility of predicting multimodal distributions, as seen in Fig. 9.4. The diameter distributions presented were averaged to stand-level while original predictions were made at plot level.

9.4 Prediction of Diameter Distribution in the Case of Single Tree Detection

The prediction of diameter distribution differs fundamentally from the area based approach in the single tree detection case. Single tree detection provides estimates of tree height and canopy characteristics which can be applied in the prediction of DBH at tree level. The diameter distribution is then summed from single trees. This approach is rather straightforward but there are some issues related to tree detection and DBH prediction.

The challenge of the prediction of DBH from tree height is mainly related to the largest trees. When a tree is maturing, DBH growth continues but the increase in tree height gradually diminishes. Consequently, DBH for a given tree height varies considerably. There is additionally a considerable variation in the DBH/height relationship between different stands. This might be caused by factors such as site fertility, soil class, stand density and stand management history. Additionally, in the ALS context, the trees that are recognized do not usually include the smallest trees (e.g. Persson et al. 2002). This does not have a major effect on the stand-level volume estimates, but the resulting diameter distribution might be left-truncated. Such a distribution can be problematic when predicting future growth and yield and the need for silvicultural operations. Furthermore, single trees that are detected must be classified to the correct tree species to be counted towards the correct species-specific diameter distribution.

The models between DBH and ALS-based tree height and crown characteristics (diameter or area) were firstly presented by Hyyppä and Inkinen (1999) and Persson et al. (2002). The models between ALS metrics and tree attributes are usually ALS campaign specific due to the properties of ALS scanners. Kalliovirta and Tokola (2005) constructed species-specific DBH- models based on national forest inventory (NFI) data from Finland. The idea behind developing such models is that they can be applied to derive DBH information from remotely sensed height and crown characteristics data. Since the models are based on NFI data they can be applied in the area of whole country. The drawback of using field data in diameter modelling is that crown characteristics derived in the field and from remote sensing data may not correspond to each other.

In addition to utilizing height and crown area/diameter in modelling DBH from ALS data it is also possible to utilize ALS point cloud metrics (e.g. Villikka et al. 2007). Thus, height and density metrics corresponding to those used in the area-based approach are used, but calculated from the area of the detected tree crown rather than from an entire field-plot. Instead of using only tree-level metrics, a circular area around the tree can also be used to calculate the area-based metrics to also describe tree social status compared to surrounding trees (Vauhkonen et al. 2010).

The data are often hierarchical when predicting DBH in a stand, i.e. several trees are measured from one stand. In such cases, it is expected that the biophysical properties of trees growing in the same stand are more similar than when compared to trees growing in other stands, and this knowledge should be utilized. Mixed-effects modelling provides a means of predicting hierarchically structured variables. Maltamo et al. (2004) applied this to DBH prediction in the ALS context and, later, Salas et al. (2010) compared different spatial models constructed for this purpose. Recently, Maltamo et al. (2012) applied mixed-effects modelling and stand-level field calibration of mixed-effects models to improve model applicability. Most of the DBH prediction studies are based on regression modeling but k-nn imputation has been used as well (Maltamo et al. 2009b; Vauhkonen et al. 2010).

In conclusion, it can be stated that the use of k-nn imputation, mixed-effects modelling and tree-level ALS point cloud data have improved the accuracy of diameter prediction. Both k-nn and mixed-effects models also allow simultaneous prediction of different tree attributes. In the best cases so far, the RMSE values of DBH-estimates have been slightly over 1 cm. However, the accuracy is usually still, say, about 3 cm, which yields rather high errors in estimating tree volume.

9.5   Combining the Area-Based Approach and Single Tree Detection to Improve Diameter Distribution Estimates

Single tree detection usually results to distribution estimates where the largest trees are detected but smaller are not. In this context Mehtätalo (2006) presented theoretical framework to recover smaller trees based on the spatial pattern and detectability of trees. The problem of not detecting small trees has also led to some experiments in which single tree detection and the area-based approach are combined (e.g., Maltamo et al. 2004; Lindberg et al. 2010; Ene et al. 2012). The idea generally is to utilize detected trees and some area-based sum variable, such as stem number or volume, to calibrate the estimates so that the detected single trees sum up to the area-based variable. While Ene et al. (2012) only considers detection rates, other above-mentioned studies examined combined diameter or height distributions.

Maltamo et al. (2004) first performed single tree detection. According to shortest identified tree a truncation point was defined. The height distribution of the detected trees was then characterized by a Weibull distribution applying both the two-parameter approach and the truncated form of the two-parameter approach. According to the chosen truncation point, the height estimates for smaller trees were then derived with the Weibull function (see Fig. 9.5). For application purposes, the Weibull parameters were regressed using stand-level independent variables derived from the individual trees detected. Then the height distribution was converted to DBH's using a model based on tree height and crown area. This kind of approach

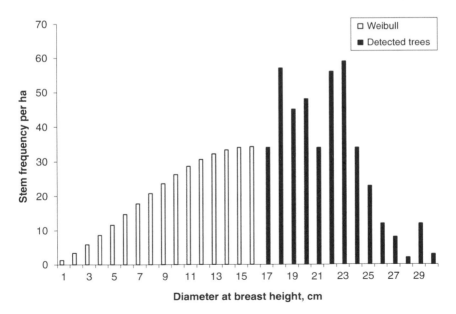

Fig. 9.5 The principle of combining detected trees and theoretical Weibull distribution

led to considerably improved estimates of stand level volume and stem numbers when compared to approaches relying only on single tree detection. However, it involves a risk of unrealistic estimates of diameter distribution and, in general, this kind of approach is sensitive to stem number errors since it does not include any ALS-based information about forest structure for small trees.

Lindberg et al. (2010) first detected individual trees and estimated their height and diameter. In the next step they imputed area based target distributions at plot level by utilizing k-nn. These target distributions included estimates of stem volume and number of stems as well as percentiles of tree height and DBH. These two approaches were then compared and calibrated. The number of trees based on single tree detection was first rescaled to fulfil the estimated area based stem volume at plot level. Trees included in the tree lists from the single tree detection were then either removed or added into the percentiles according to the information concerning the area based target distribution. The result of this procedure was a tree list that was consistent with unbiased estimates at the area level.

The accuracy of the single tree detection and area-based approaches has been compared in a few studies (Packalén et al. 2008; Breidenbach et al. 2010; Yu et al. 2010, Peuhkurinen et al. 2011). These studies found that the accuracies of predicted stand attributes were similar, although it can be very difficult to make fair comparisons with respect to e.g., data acquisition costs and how to handle tree species recognition. Nevertheless, Peuhkurinen et al. (2011) found that the diameter distribution of a stand was more accurately predicted using the area-based approach utilizing the parameter recovery of the Weibull function compared to individual tree detection. However, the saw log size proportion of the diameter distribution was more accurately predicted using single tree detection. It is also notable that an ALS-based Weibull estimate was able to characterize J-shaped descending distribution forms (Peuhkurinen et al. 2011). Diameter distributions provide a favourable combination of accuracy and costs when evaluating the value of information and comparing different inventory approaches (Chap. 16).

Different approaches to combine single tree detection and area based approach have been developed. Usually the idea has been to calibrate the result of single tree detection by taking into account also small trees. In principle this improves accuracy and removes bias but these approaches need high density ALS data and the estimation procedure is usually complex. It is also not guaranteed that the accuracy will increase in all application stands.

References

Bailey RL, Dell TR (1973) Quantifying diameter distributions with the Weibull function. For Sci 19:97–104
Bailey RL, Abernthy NC, Jones EP (1981) Diameter distributions models for repeatedly thinned slash pine plantations. In: Barnett JP (ed) Proceedings of the 1st Biennial Southern Silvicultural Research Conference, Atlanta, Georgia

Bollandsås OM, Næsset E (2007) Estimating percentile-based diameter distributions in uneven-aged Norway spruce stands using airborne laser scanner data. Scand J For Res 22: 33–47

Borders BE, Souter RA, Bailey RL, Ware KD (1987) Percentile-based distributions characterize forest stand tables. For Sci 33:570–576

Breidenbach J, Gläser C, Schmidt M (2008) Estimation of diameter distributions by means of airborne laser scanner data. Can J For Res 38:1611–1620

Breidenbach J, Næsset E, Lien V, Gobakken T, Solberg S (2010) Prediction of species-specific forest inventory attributes using a nonparametric semi-individual tree crown approach based on fused airborne laser scanning and multispectral data. Remote Sens Environ 114:911–924

Cao QV (2004) Predicting parameters of a Weibull function for modelling diameter distribution. For Sci 50:682–685

Deville J-C, Särndal C-E (1992) Calibration estimators in survey sampling. J Am Stat Assoc 87:376–382

Dubey SD (1967) Some percentile estimators for Weibull parameters. Technometrics 9:119–129

Ene L, Næsset E, Gobakken T (2012) Single tree detection in heterogeneous boreal forests using airborne laser scanning and area-based stem number estimates. Int J Remote Sens 33:5171–5193

Eriksson LO, Sallnäs O (1987) A model for predicting log yield from stand characteristics. Scand J For Res 2:253–261

Gobakken T, Næsset E (2004) Estimation of diameter and basal area distributions in coniferous forest by means of airborne laser scanner data. Scand J For Res 19:529–542

Gobakken T, Næsset E (2005) Weibull and percentile models for LIDAR-based estimation of basal area distribution. Scand J For Res 20:490–502

Haara A, Maltamo M, Tokola T (1997) The k-nearest-neighbour method for estimating basal-area diameter distribution. Scand J For Res 12:200–208

Holopainen M, Vastaranta M, Rasinmäki J, Kalliovirta J, Mäkinen A, Haapanen R, Melkas T, Yu X, Hyyppä J (2010) Uncertainty in timber assortment predicted from forest inventory data. Eur J For Res 129:1131–1142

Holte A (1993) Diameter distribution functions for even-aged (*Picea abies*) stands. Norsk Institutt for Skogforskning, Ås

Hyyppä J, Inkinen M (1999) Detecting and estimating attributes for single trees using laser scanner. Photogramm J Finl 16:27–42

Kalliovirta J, Tokola T (2005) Functions for estimating stem diameter and tree age using tree height, crown width and existing stand data bank information. Silva Fenn 39:227–248

Kangas A, Maltamo M (2000) Calibrating predicted diameter distribution with additional information. For Sci 46:390–396

Lindberg E, Holmgren J, Olofsson K, Wallerman J, Olsson H (2010) Estimation of tree lists from airborne laser scanning by combining single-tree and area-based methods. Int J Remote Sens 31:1175–1192

Maltamo M (1997) Comparing basal area diameter distributions estimated by tree species and for the entire growing stock in a mixed stand. Silva Fenn 31:53–65

Maltamo M, Eerikäinen K, Pitkänen J, Hyyppä J, Vehmas M (2004) Estimation of timber volume and stem density based on scanning laser altimetry and expected tree size distribution functions. Remote Sens Environ 90:319–330

Maltamo M, Eerikäinen K, Packalén P, Hyyppä J (2006) Estimation of stem volume using laser scanning based canopy height metrics. Forestry 79:217–229

Maltamo M, Suvanto A, Packalén P (2007) Comparison of basal area and stem frequency diameter distribution modelling using airborne laser scanner data and calibration estimation. For Ecol Manage 247:26–34

Maltamo M, Næsset E, Bollandsås OM, Gobakken T, Packalén P (2009a) Non-parametric estimation of diameter distributions by using ALS data. Scand J For Res 24:541–553

Maltamo M, Peuhkurinen J, Malinen J, Vauhkonen J, Packalén P, Tokola T (2009b) Predicting tree attributes and quality characteristics of Scots pine using airborne laser scanning data. Silva Fenn 43:507–521

Maltamo M, Mehtätalo L, Vauhkonen J, Packalén P (2012) Predicting and calibrating tree size and quality attributes by means of airborne laser scanning and field measurements. Can J For Res 42:1896–1907

Mehtätalo L (2006) Eliminating the effect of overlapping crowns from aerial inventory estimates. Can J For Res 36:1649–1660

Mehtätalo L, Maltamo M, Packalén P (2007) Recovering plot-specific diameter distribution and height-diameter curve using ALS based stand characteristics. In: Proceedings of ISPRS workshop laser scanning 2007 and Silvilaser 2007, Finland, September 12–14, 2007. Int Arch Photogramm Remote Sens Spat Inf Sci XXXVI:288–293

Packalén P, Maltamo M (2008) The estimation of species-specific diameter distributions using airborne laser scanning and aerial photographs. Can J For Res 38:1750–1760

Packalén P, Pitkänen J, Maltamo M (2008) Comparison of individual tree detection and canopy height distribution approaches: a case study in Finland. In: Proceedings of SilviLaser 2008, 8th international conference on LiDAR applications in forest assessment and inventory. Heriot-Watt University, Edinburgh, UK, 17–19 September 2008

Packalén P, Temesgen H, Maltamo M (2012) Variable selection for nearest neighbor imputation in remote sensing based forest inventory. Can J Remote Sens 38:557–569

Päivinen R (1980) Puiden läpimittajakauman estimointi ja siihen perustuva puustotunnusten laskenta. (On the estimation of stem-diameter distribution and stand characteristics). Folia For 442:1–28 (In Finnish with English summary)

Persson A, Holmgren J, Söderman U (2002) Detecting and measuring individual trees using an airborne laser scanner. Photogramm Eng Remote Sens 68:925–932

Peuhkurinen J, Maltamo M, Malinen J (2008) Estimating species-specific diameter distributions and saw log recoveries of boreal forests from airborne laser scanning data and aerial photographs: a distribution-based approach. Silva Fenn 42(4):625–641

Peuhkurinen J, Mehtätalo L, Maltamo M (2011) Comparing individual tree detection and the area-based statistical approach for the retrieval of forest stand characteristics using airborne laser scanning in Scots pine stands. Can J For Res 41:583–598

Poudel KP, Cao QV (2013) Evaluation of methods to predict Weibull parameters for characterizing diameter distributions. For Sci 59:243–252

Reynolds MR Jr, Burk TE, Huang W-C (1988) Goodness-of-fit tests and model selection procedures for diameter distribution models. For Sci 34:373–399

Salas C, Ene L, Gregoire TG, Næsset E, Gobakken T (2010) Modelling tree diameter from airborne laser scanning derived variables: a comparison of spatial statistical models. Remote Sens Environ 114:1277–1285

Siipilehto J (1999) Improving the accuracy of predicted basal-area diameter distribution in advanced stands by determining stem number. Silva Fenn 33:281–301

Temesgen H (2003) Estimating tree-lists from aerial information: a comparison of a parametric and most similar neighbor approaches. Scand J For Res 18:279–288

Tham Å (1988) Structure of mixed *Picea abies* (L.) Karst. and *Betula pendula* Roth and *Betula pubescens* Ehrh. stands in South and Middle Sweden. Scand J For Res 3:355–369

Thomas V, Oliver RD, Lim K, Woods M (2008) Lidar and Weibull modeling of diameter and basal area. For Chron 84(6):866–875

Van Deusen PC (1986) Fitting assumed distributions to horizontal point sample diameters. For Sci 32:146–148

Vauhkonen J, Korpela I, Maltamo M, Tokola T (2010) Imputation of single-tree attributes using airborne laser scanning-based height, intensity, and alpha shape metrics. Remote Sens Environ 114:1263–1276

Villikka M, Maltamo M, Packalén P, Vehmas M, Hyyppä J (2007) Alternatives for predicting tree-level stem volume of Norway spruce using airborne laser scanner data. Photogramm J Finl 20:33–42

Yu X, Hyyppä J, Holopainen M, Vastaranta M (2010) Comparison of area based and individual tree based methods for predicting plot level attributes. Remote Sens 2:1481–1495

Zhang L, Gove JH, Liu C, Leak WB (2001) A finite mixture distribution for modeling the diameter distribution of rotated-sigmoid, uneven-aged stands. Can J For Res 31:1654–1659

Chapter 10
A Model-Based Approach for the Recovery of Forest Attributes Using Airborne Laser Scanning Data

Lauri Mehtätalo, Jukka Nyblom, and Anni Virolainen

Abstract As three-dimensional wall-to-wall information on forest structure, ALS echoes provide information on the growing stock and canopy structure. Even though the ALS echo heights are associated with the dimensions of trees, a theoretical model to relate ALS data with interesting forest attributes is missing. The recorded observation of echo height can be viewed as an outcome of a complex process mixing several random sub-processes related to the forest and the atmosphere in a non-trivial way. The forest-related processes include those generating stand density, tree heights, tree locations, tree crown shapes, and the internal structure of tree crowns. This chapter presents our recent work on development of a theoretical model for ALS echo heights. Furthermore, extensions are presented to take into account randomness in tree crown shape, to incorporate the penetration of laser pulses into tree crowns, and to develop the model for mixed stands.

10.1 Introduction

The possibilities of airborne laser scanning (ALS) as a forest inventory tool has been recognized for decades (e.g. Solodukhin et al. 1977; Nelson et al. 1984). Decreased costs of data collection in the 2000s have made it an alternative to the traditional field surveys (e.g., Næsset et al. 2004). However, as Lim et al. (2003) and Junttila et al. (2008) have pointed out, theoretical understanding of the relationship between forest

L. Mehtätalo (✉)
School of Computing, University of Eastern Finland, Joensuu, Finland
e-mail: lauri.mehtatalo@uef.fi

J. Nyblom
Department of Mathematics and Statistics, University of Jyväskylä, Jyväskylä, Finland

A. Virolainen
School of Forest Sciences, University of Eastern Finland, Joensuu, Finland

M. Maltamo et al. (eds.), *Forestry Applications of Airborne Laser Scanning: Concepts and Case Studies*, Managing Forest Ecosystems 27, DOI 10.1007/978-94-017-8663-8__10,
© Springer Science+Business Media Dordrecht 2014

structure and laser data is still incomplete. Theoretical models on the behavior of ALS pulses in tree canopies have been developed (Sun and Ranson 2000; Ni-Meister et al. 2001), but these models focus on the modeling of the emitted energy in forest, not in the recovery of the forest attributes from recorded echoes. A model for the forest stand would be useful in estimation of forest attributes of interest, such as the stand density and distribution of tree heights. Such a model was recently presented by Mehtätalo and Nyblom (2009, 2012) and Mehtätalo et al. (2010).

The connection between the canopy heights (Z) and tree heights (H) is evident, but the distributions of these two random variables are not identical (Magnussen and Boudewyn 1998). There are several reasons for this difference. We start by considering laser echoes from one single tree. The mean of these echo heights do not equal the tree height (i) because observed canopy height for a given tree equals tree height only at the tree apex (Nelson 1997; Magnussen et al. 1999). Therefore, most echo heights are from lower heights than the total tree height. A common strategy to overcome this property of *missed treetop* in the individual tree detection approaches (cf. Chap. 1) is to take the maximum echo height as the estimate of the tree height and by using high-density laser data. However, (ii) the echo height at any location within a tree canopy underestimates the height of the surface of the canopy at the given point. This underestimation results from the *pulse penetration* into the crown before producing a detectable echo. Gaveau and Hill (2003) observed the mean penetration of 1.27 m (range -0.14, 3.06) in a broadleaf forest. The degree of penetration depends on the structure of the canopy (e.g., Gaveau and Hill (2003) found a smaller penetration for shrub canopies than for forest canopies) and the properties of the laser pulse itself (e.g. wavelength, footprint size, pulse discretization method, etc.). Some authors include also the tendency of a pulse to miss the treetops in the penetration component (e.g. Magnussen et al. 1999), even though they are caused by different processes.

When one switches from tree level to stand level, the following three additional issues can be recognized that make the distribution of echo heights different from the distribution of tree heights: (iii) uniformly spread laser pulses hit more likely a large individual tree crown than a small one, which means that large trees are over-represented in the data of ALS echoes. This *sampling probability proportional to size (PPS)* property (e.g. Næsset 1997; Magnussen et al. 1999) is further emphasized by the effect of (iv) *canopy overlap*: considering two overlapping crowns at the location of the echo, the pulse echoes from the one that has the canopy at a higher level. The overlap may have important implications in stands with large size variation, high stand density and clustered pattern of tree locations, whereas it may be less important in even-sized plantations with low density and systematic spatial pattern of tree locations (Mehtätalo 2006). Finally, (v) the ALS data includes *ground echoes*, i.e., echoes from the openings between individual trees. To summarize, the laser echo heights include information on the tree heights, but the relationship between ALS echo heights and tree heights is rather complex. Especially, the five properties listed above demonstrate that the distributions of these two variables differ not only in mean and standard deviation but also in shape.

In this chapter, we assume that the marginal distribution of ALS echo heights (z) has been observed for the forest area of interest using e.g. airborne laser scanner.

However, we also assume that the latitude (x) and longitude (y) of the echoes are not available or, if available, they are not utilized. Such a restriction is justified in situations where the x-y coordinates of the pulses relative to each other do not provide meaningful information on the forest, for example, if the area for which the distribution of z is observed (e.g. a sample plot) is rather homogeneous and the pulse density is so low that individual crowns cannot be recognized from the point cloud. The widely-used area-based approach (cf. Chap. 1) is also based on the marginal distribution of the echo heights, whereas the individual-tree recognition approaches utilize also the x-y coordinates of the echo.

The recorded observation of ALS echo height can be seen as an outcome of a process that is a complex mixture of several random sub-processes, including those generating stand density, tree heights, tree locations, tree crown shapes, the internal structure of tree crowns, and the properties of the laser pulse. In this chapter, the aim is to present a general model of forest canopy height. The model integrates the above-mentioned sub-processes, parameterizing the distribution of echo heights Z by meaningful parameters of them. The aim is to develop a model that can be used in estimation and inference from a marginal distribution of canopy heights collected by an airborne laser scanner. We will present the theoretical basis of the models, an overview of the previous publications, as well as some theoretical developments that substantially relax the quite restrictive assumptions of the previously published models.

10.2   Models for ALS Echo Heights in Forest Stands

10.2.1   The Tree-Specific Crown Envelope

The *tree specific crown envelope* is defined as a smooth solid surface, which covers the whole tree crown like a hood and has a minimum volume under it, as illustrated by Fig. 10.1. The crown envelope makes two important simplifications compared to true trees. First, the envelope does not take into account the high-level structure of tree crowns caused by branches and leaves, but treats an individual tree crown as a three-dimensional object with solid surface on the top. Second, the area of the cross-section of the crown envelope decreases from the bottom to the top, even though the true cross-section may increase. This is justified because the true crown width of the tree at a given height is not observable from above if the crown has larger width at any height above. We introduce this simplifying concept as the starting point for our mathematical derivations.

There is a trade-off between the smoothness of the top surface of the envelope and the envelope volume: the higher the requested degree of smoothing, the larger the volume. In practice, the crown envelope is defined by a parametric function. In this case, the crown envelope is of the assumed mathematical form, covers all branches (or at least a large majority of them), and has the minimum volume.

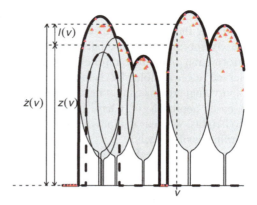

Fig. 10.1 Illustration of the concept of crown envelope. The *gray* objects with thin borderlines demonstrate six trees. The *dashed thick line* shows the tree-specific crown envelope for one tree. The *thick black lines* show the area-specific crown envelope and the triangles demonstrate laser echoes form pulses that have penetrated into the crowns envelope according to an exponential distribution. Height $\dot{z}(v)$ is the height of the area-specific crown envelope at location v, $z(v)$ and $l(v)$ are the corresponding echo height and realized penetration, respectively

In what follows, we assume for simplicity that the horizontal cross-section of a tree crown is circular. Then it is possible to specify the crown shape in terms of crown radius at a given height above the ground. However, some of the results presented in this chapter are valid also for cross-sections of any closed shape of the cross-section (Stoyan et al. 1995).

The parametric function that is used specifies the crown radius r at height $\dot{z}$ above the ground for a tree with total height h. A commonly used simple function for tree crowns is the ellipsoid (Nelson 1997), which was used also by Mehtätalo and Nyblom (2009),

$$r\left(\dot{z}, h \,|\, \boldsymbol{\varphi}\right) = \begin{cases} a(h) & 0 < \dot{z} \le h_{max} \\ a(h)\sqrt{1 - \frac{[\dot{z}-b(h)]^2}{[h-b(h)]^2}} & h_{max} < \dot{z} \le h \end{cases} \qquad (10.1)$$

where the half axes $a(h)$ and $b(h)$ specify the shape of the tree as a function of tree height. The function is constant from ground level to the height of maximum crown width h_{max}. Thereafter, it is a decreasing function of $\dot{z}$ until the height of the tree top, h. Mehtätalo and Nyblom (2009) further specified the half axes to be proportional to the total height as $a(h) = ph$ and $b(h) = qh$, therefore the parameters specifying the crown shape were $\boldsymbol{\varphi} = (p,q)'$.

Mehtätalo and Nyblom (2012) found the simple ellipsoid insufficient for modeling large Scots pine trees, which had the shape between a cone and an ellipsoid. Therefore, they extended the model by allowing the center of the ellipsoid to move away from the x-axis, which is at the tree stem. The resulting three-parameter model allows the shape to vary between conical and ellipsoidal forms

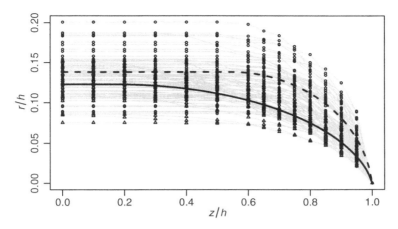

Fig. 10.2 Empirical crown shape functions for Norway spruce (*solid*) and Scots Pine (*dashed*) based on Eq. 10.2. The individual symbols and *gray lines* show the original observations (triangle = spruce, circle = pine)

$$r\left(\dot{z}, h \mid \boldsymbol{\varphi}\right) = \begin{cases} h\left(y_0 + b\right) & 0 < \dot{z} \le h_{max} \\ h\left(y_0 + b\sqrt{1 - \frac{(\dot{z}/h - x_0)^2(b^2 - y_0^2)}{b^2(1 - x_0)^2}}\right) & h_{max} < \dot{z} \le h \end{cases} \quad (10.2)$$

where $\boldsymbol{\varphi} = (y_0, x_0, b)'$ specify the location of the ellipsoid center and the vertical half axis. Figure 10.2 shows examples of empirical crown shape functions based on Eq. 10.2.

A more flexible three-parameter function is provided by the Lame function (Rautiainen and Stenberg 2005; Mehtätalo et al. forthcoming)

$$r\left(\dot{z}, h \mid \boldsymbol{\varphi}\right) = \begin{cases} R & 0 < \dot{z} \le h_{max} \\ R\left(1 - \left(\frac{\dot{z} - qh}{h - qh}\right)^t\right)^{1/t} & h_{max} < \dot{z} \le h \end{cases} \quad (10.3)$$

where the parameter vector is $\boldsymbol{\varphi} = (R, q, t)'$. Parameter R specifies the maximum crown radius, which is reached at the height qh. Parameter t ($t > 0$) specifies the shape so that $t < 1$ provides convex shapes, $t = 1$ produces a cone, $1 < t < 2$ produces shapes between cone and ellipsoid, $t = 2$ produces an ellipsoid and $t > 2$ produces even more concave shapes.

Consider a tree at location u. Assume that the height of the tree is generated by a random process specified by a cumulative distribution function $F(h)$, such as the widely used Weibull distribution (Bailey and Dell 1973)

$$F\left(h \mid \boldsymbol{\xi}\right) = P\left(H < h\right) = 1 - \exp\left(-\left(\frac{h}{\beta}\right)^{\alpha}\right) \quad (10.4)$$

where the parameter vector $\boldsymbol{\xi} = (\alpha, \beta)'$ includes the shape and scale parameters of the distribution, respectively. Because of the randomness of tree height, the crown envelope radius (and hence the area of the cross-sectional area of the crown envelope) at height $\dot{z}$ is also random. From a technical point of view, the crown envelope shape defines a transformation of the random tree height (Mehtätalo and Nyblom 2009, 2012). Therefore, the probability distribution of the crown radius has the following relationship with that of tree height (Eq. 10.4):

$$P(R \le r) = F\{h(\dot{z}, r \,|\, \boldsymbol{\varphi}) \,|\, \boldsymbol{\xi}\}, \tag{10.5}$$

where $h(z, r|\boldsymbol{\varphi})$ expresses the height of a tree that has crown envelope radius r at height $\dot{z}$ above the ground. This function results from solving the crown shape function (e.g. one of Eqs. 10.1, 10.2 and 10.3) for h.

However, Virolainen et al. (forthcoming) observed that field-measured crown profiles of Scots pine vary considerably among individual trees, and even the profile of same tree can change with the direction of view. Therefore, additional randomness to the crown radius is introduced by the among-tree and within-tree variability in crown envelope shape. This variation can be taken into account by allowing the parameters of the crown shape functions Eqs. 10.1, 10.2 and 10.3 to vary between trees. In this situation, the crown radius function is defined as $r(\dot{z}, h \,|\, \boldsymbol{\varphi}_i)$, where the additional subscript i indicates that the parameter vector $\boldsymbol{\varphi}$ is specific for each tree. Allowing separate parameters for all trees is, however, not an option in reality. This problem can be overcome by treating the tree-specific parameters as realizations from a common distribution of parameters, as is commonly done in the mixed-effects models (Pinheiro and Bates 2000).

10.2.2 The Area-Specific Crown Envelope

In area-based approaches, the echoes cannot be assigned to certain trees. Instead, only the sample plot (or stand) of origin is known, yielding an observed marginal distribution of echo heights for the plot in question. For analysis of such data, the concept of crown envelope is extended to the area level.

Similarly to the tree-level crown envelope, the *area-specific crown envelope* is such a smooth solid surface, which covers all crowns of the plot like a hood and has minimum volume under it. Furthermore, to connect the plot-specific envelope with the tree specific one, we define that the height of the area-specific envelope at a given point in the horizontal plane is the maximum over all tree-specific envelopes that extend to the point in question (Fig. 10.1, see also Fig. 10.3).

Mehtätalo and Nyblom (2009, 2012) interpreted the ALS echo heights as observed heights of the area-specific crown envelope at the points of laser echoes. This approach implicitly assumes that the footprint size is 0, the pulses are exactly vertical, and the pulses do not penetrate to the area-specific envelope but return

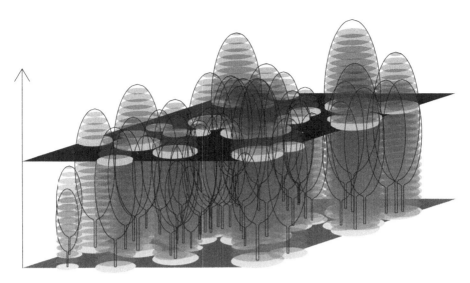

Fig. 10.3 An illustration of the theoretical model of Mehtätalo and Nyblom (2009). The probability for a randomly located ALS pulse to echo from below the level of the black plane z equals the ratio of the plane area to the total area

immediately when hitting the top surface of the envelope. Even though these assumptions are simplifying and unrealistic to some degree, the concept provides a sound approach that implicitly takes into account the previously discussed properties of *missed treetops, canopy overlap, PPS sampling* and *ground echoes*. An approach to take into account the *laser pulse penetration* will be presented later in this chapter.

The laser echo height is interpreted as the observed height of the area-specific crown envelope. Because the locations of trees and laser returns are unrelated, we think that the laser pulse hits the area-specific envelope at a random point within the plot, with uniform probability over the plot area. Therefore, the cumulative distribution function (c.d.f.) of ALS echo height, $G(\dot{z})$, can be specified as the probability that the area-specific envelope at an randomly selected point v (in x-y plane) within the area is below the height $\dot{z}$, i.e.,

$$G(\dot{z}) = P\left(\dot{Z}(v) \leq \dot{z}\right).$$

Furthermore, this probability is equal to the probability that the (randomly placed, zero-footprint, vertical) pulse misses the union of crown discs at height $\dot{z}$, i.e., it hits the black area of Fig. 10.3. Here the crown discs are the cross-sections of the tree-specific crown envelopes. Assuming that the cross-sectional areas of the crown envelopes of neighboring trees are independent, the probability of the pulse at v to miss all cross-sections at height $\dot{z}$ is the product of probabilities to miss all tree-specific cross-sections. Furthermore, point v misses the cross-section of tree i at height z if the crown envelope radius at height z is smaller than the distance between the pulse location v and the tree location u_i. We get

$$G\left(\dot{z}\,|\boldsymbol{\varphi}\right) = \prod_{i=1}^{N} P\left(r\left(\dot{z}, h_i\,|\boldsymbol{\varphi}\right) \leq \|u_i - v\|\right) \tag{10.6}$$

where N is the total number of trees in the stand and $r\left(\dot{z}, h_i\right)$ is a function specifying the crown radius at height $\dot{z}$ above the ground for tree i that has the total height of h_i.

Equation 10.6 now specifies in very general terms the distribution of ALS echoes from the area-specific envelope at point v within the area of interest, where tree locations are given by u_i, $i = 1, \ldots, N$. However, the equation is not useful if the tree heights and locations are unknown as they usually are. This problem can be overcome by making assumptions on the processes generating tree heights and locations.

To include the distribution of tree heights into the model, recall that we can treat the crown radius as a transformation of tree height (Eq. 10.5). Each term of the product in Eq. 10.6 is actually the value of the c.d.f. of crown radius for $r = \|u_i - v\|$ (Eq. 10.5). Therefore, the c.d.f. of ALS echo heights can be expressed by writing Eq. 10.5 into Eq. 10.6 to get

$$G\left(\dot{z}\,|\boldsymbol{\varphi}, \boldsymbol{\xi}\right) = \prod_{i=1}^{N} F\left\{h\left(\dot{z}, \|v - u_i\|\,|\boldsymbol{\varphi}\right)\,|\boldsymbol{\xi}\right\}.$$

The expression includes a product over all trees of the area of interest. However, the probability of the crown i to not extend to point v, $F\left\{h\left(\dot{z}, \|v - u_i\|\,|\boldsymbol{\varphi}\right)\,|\boldsymbol{\xi}\right\}$, is 1 for trees sufficiently far from the point v. Therefore, it suffices to take the product over the nearby trees only:

$$G\left(\dot{z}\,|\boldsymbol{\varphi}, \boldsymbol{\xi}\right) = \prod_{i;\|v - u_i\| < R_{max}} F\left\{h\left(\dot{z}, \|v - u_i\|\,|\boldsymbol{\varphi}\right)\,|\boldsymbol{\xi}\right\}, \tag{10.7}$$

where R_{max} is the upper limit for the crown radius of a tree (Mehtätalo et al. 2010; Mehtätalo and Nyblom 2012).

The results of Virolainen et al. (forthcoming) indicated a need for tree-specific vectors $\boldsymbol{\varphi}_i$ to account for the tree-to-tree variation in crown shape. In order to properly take into account the tree-to-tree variation in crown shape, we specify the probability density associated with $\boldsymbol{\varphi}_i$ as $p(\boldsymbol{\varphi}_i)$. However, we assume that all the trees are realizations of the same model. For example, if the crown shape is described by appropriate re-parameterizations of Eqs. 10.2 or 10.3, $p(\boldsymbol{\varphi}_i)$ might be the tri-variate normal density with mean $\boldsymbol{\mu}$ and variance-covariance matrix $\boldsymbol{\Sigma}$. Then the total number of parameters in this model would be nine. For each tree, we may define the probability to have the crown radius below distance $\|v - u_i\|$ as the mean probability over the distribution of $\boldsymbol{\varphi}_i$. Therefore, we get

$$G\left(\dot{z}\,|\boldsymbol{\mu}, \boldsymbol{\Sigma}, \boldsymbol{\xi}\right) = \prod_{i;\|v - u_i\| < R_{max}} \int_{\Omega_\varphi} F\left\{h\left(\dot{z}, \|v - u_i\|\,|\boldsymbol{\varphi}_i\right)\,|\boldsymbol{\xi}\right\} p\left(\boldsymbol{\varphi}_i\,|\boldsymbol{\mu}, \boldsymbol{\Sigma}\right) d\boldsymbol{\varphi}_i.$$

$$\tag{10.8}$$

10.2.3   Distribution of Echo Heights Under a Grid Pattern of Tree Locations

Mehtätalo et al. (2010) applied Eq. 10.6 to a Eucalyptus plantation where trees are planted according to a grid pattern. In this particular situation, the value of Eq. 10.6 varies according to the location v within a rectangle defined by four consecutive trees from two adjacent lines. On the other hand, all rectangles of the forest stand are identical with respect to the probabilistic properties of canopy. Therefore, to compute the distribution of echo heights for the stand, it is enough to average Eq. 10.6 over one rectangle. Denoting the distance between rows by l and the distance between trees of a row *by* m, the mean of Eq. 10.6 over the cell becomes

$$G\left(\dot{z}\,|\,\boldsymbol{\varphi}\right) = \frac{1}{lm}\int_0^m\int_0^l\prod_{i\,;\,\|v-u_i\|<R_{max}}P\left(r\left(\dot{z},h_i\,|\,\boldsymbol{\varphi}\right)\leq\|u_i-v\|\right)\mathrm{d}v_1\mathrm{d}v_2 \quad (10.9)$$

where v_1 and v_2 are the x and y coordinates related to location v.

Mehtätalo et al. (2010) assumed that the randomness in crown envelope radius arises only through the randomness of tree heights. In this case, Eq. 10.9 becomes

$$G\left(\dot{z},\boldsymbol{\varphi},\boldsymbol{\xi}\right) = \frac{1}{lm}\int_0^m\int_0^l\prod_{i\,;\,\|v-u_i\|<R_{max}}F\left\{h\left(\dot{z},\|v-u_i\|\,|\,\boldsymbol{\varphi}\right)\,|\,\boldsymbol{\xi}\right\}\mathrm{d}v_1\mathrm{d}v_2. \quad (10.10)$$

Equations 10.9 and 10.10 provide a model for a situation where the stand density and spacing of a plantation are known. Mehtätalo and Nyblom (2012) presented a model for a special case of a square grid pattern with unknown grid-spacing. In a stand with λ trees per m^2 and square grid pattern of tree locations, inter-tree distances are $l = m = \frac{1}{\sqrt{\lambda}}$ (meters). Hence,

$$G\left(\dot{z}\,|\,\boldsymbol{\varphi},\boldsymbol{\xi},\lambda\right) = \lambda\int_0^{1/\sqrt{\lambda}}\int_0^{1/\sqrt{\lambda}}\prod_{i\,;\,\|v-u_i\|<R_{max}}F\left\{h\left(\dot{z},\|v-u_i\|\,|\,\boldsymbol{\varphi}\right)\,|\,\boldsymbol{\xi}\right\}\mathrm{d}v_1\mathrm{d}v_2$$

$$(10.11)$$

In this case, the c.d.f. of echo heights is expressed using three vectors of parameters: that specifying the tree crown envelope shape as a function of tree height, that specifying the plot-specific distribution of tree heights, and the stand density.

10.2.4   Distribution of Echo Heights Under a Random Pattern of Tree Locations

A common starting point for any analysis related to spatial pattern of tree locations is to assume complete spatial randomness, where tree locations are generated

independently from a uniform distribution over the area of interest. In such a
situation, the c.d.f. of echo heights (Eq. 10.7) simplifies to

$$G\left(\dot{z}\,|\boldsymbol{\varphi},\boldsymbol{\xi},\lambda\right) = \exp\left\{-\lambda \int_0^{\infty} \pi\,r(\dot{z}, h_i\,|\boldsymbol{\varphi})^2 f\left(h\,|\boldsymbol{\xi}\right) dh\right\} \tag{10.12}$$

where $f(h|\boldsymbol{\xi}) = F'(h|\boldsymbol{\xi})$ is the probability density function of tree heights
(Mehtätalo and Nyblom 2009).

10.2.5   The Density Function of Echo Heights

Sections 10.2.3 and 10.2.4 provided expressions for the c.d.f. of echo heights
from the area-specific envelope in the case of three different patterns of tree
locations: for a rectangular grid with known spacing (Eq. 10.10), for a square
grid pattern with unknown stand density (Eq. 10.11), and for a random pattern
of tree locations with unknown stand density (Eq. 10.12). The c.d.f. was parameterized using parameters for the plot-specific distribution of tree heights, for
the crown envelope shape for a given tree height, and for stand density ($\boldsymbol{\varphi}$, $\boldsymbol{\xi}$
and λ). For clarity of presentation, we now define a single parameter vector $\boldsymbol{\theta}$
which includes all those parameters that are unknown. The exact contents of this
vector may differ according to the specific application: for example, the stand
density of a plantation may be known in advance (Mehtätalo et al. 2010, see also
Chap. 13), or the parameters specifying the average crown shape for the region may
have been estimated beforehand using a training dataset (Mehtätalo and Nyblom
2012).

A specific property of the c.d.f. of echo heights, $G\left(\dot{z}\,|\boldsymbol{\theta}\right)$, is that the distribution
consists of two components: a discrete component for the echoes from the ground
and a continuous component for the echoes from the canopy (Fig. 10.4, left). The
discrete component is visible as a jump at $\dot{z} = 0$ to the level that quantifies the
proportion of ground echoes. As $\dot{z}$ increases from 0, the c.d.f. is first relatively
flat until $\dot{z}$ gets values that correspond to the upper parts of the canopy, where the
majority of individual crowns have a fast decrease in the cross-sectional area.

A common way to explore ALS echo heights is through a histogram, which is
consistent with the probability density function (p.d.f.). The p.d.f. corresponding to
$G\left(\dot{z}\,|\boldsymbol{\theta}\right)$ has also discrete and continuous parts. More specifically, the value of the
p.d.f. at $\dot{z} = 0$ is equal to the jump of the c.d.f. at $\dot{z} = 0$; the proportion of ground
echoes. For strictly positive values of $\dot{z}$, the p.d.f. is the first derivative of $G\left(\dot{z}\,|\boldsymbol{\theta}\right)$
(Fig. 10.4, right)

$$g\left(\dot{z}\,|\boldsymbol{\theta}\right) = \begin{cases} G\left(0\,|\boldsymbol{\theta}\right) & \dot{z} = 0 \\ G'\left(\dot{z}\,|\boldsymbol{\theta}\right) & \dot{z} > 0 \end{cases}. \tag{10.13}$$

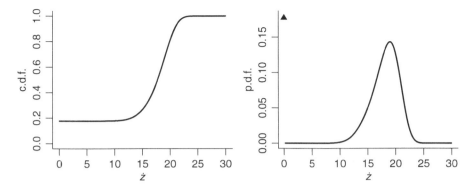

Fig. 10.4 Illustration of the cumulative distribution function of echo heights from an area-specific envelope (*left*) and the corresponding probability density function (*right*). The discrete part of the p.d.f. is demonstrated by the triangle

10.2.6   Including a Component for Penetration Within the Crown

The models in Sects. 10.2.3 and 10.2.4 take into account the effects and factors related to *missing treetops, PPS sampling, canopy overlap,* and *ground echoes.* However, the penetration of the laser pulses into tree crowns was not addressed by Mehtätalo and Nyblom (2009, 2012) nor by Mehtätalo et al. (2010). Mehtätalo et al. (forthcoming) provides a crude solution to the problem by introducing an additional fixed penetration parameter. However, the results were not very good due to the fact that penetration is better modeled as a random variable. For example, Gaveau and Hill (2003) observed penetrations ranging from 0 to 3 m in mixtures of broad-leaved trees, the mean penetration being around 1 m.

We hypothesize that the penetration of the pulse is related to the canopy gap fraction of the area under study. Therefore, the higher number of gaps, the higher the value of the mean penetration. A generally accepted starting point for modeling the canopy gap fraction is the Beer-Lamberts law (e.g. Grover et al. 1999), which states that the proportion of openings in a canopy of height l can be expressed as

$$P(\Delta) = \exp(-K(\Delta) \Omega \Delta l / \cos(\Delta))$$

where parameters Δ, $K(\Delta)$, and $\Omega \Delta$ are the beam direction, canopy extinction coefficient for beam direction Δ, and non-randomness factor, respectively. Let random variable L be the penetration of the pulse into a tree crown. The probability for a laser pulse to penetrate less than a fixed value l into a tree crown is therefore the complement of the canopy gap fraction. Furthermore, assuming parallel beams and homogeneous canopies with randomly oriented convex leaves and substituting $K(\Delta) = \delta$ yields

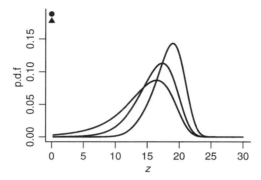

Fig. 10.5 The p.d.f. of ALS echo heights under the model that assumes random tree locations. The stand density is 1,500 trees per ha, tree height follows the Weibull (10,20) distribution, crowns are elliptic (Eq. 10.1) with p = 0.1 and q = 0.4, and penetration follows an exponential distribution with a mean of 0 (the most peaked curve), 2 and 4 m (the flattest curve)

$$P\,(L \le l) = B(l) = 1 - \exp\,(-\delta l)$$

which is the c.d.f. of the exponential distribution with rate parameter δ and density $b(l) = \delta\exp(-\delta l)$. The triangles in Fig. 10.1 demonstrate echo heights under the exponential distribution of penetration of laser pulses into tree crowns.

To integrate the penetration into the models of Sects. 10.2.3 and 10.2.4, we define the echo height at point v as the difference between the crown envelope height $\dot{Z}$ and pulse penetration L as $Z = \dot{Z} - L$. The p.d.f. of the resulting random variable is the convolution of the two distributions, yielding

$$g_l\,(z|\,\theta) = \begin{cases} G\,(\dot{z}) + \displaystyle\int_0^\infty g\,(\dot{z})\,[1 - B\,(\dot{z})]\,d\dot{z} & z = 0 \\ \displaystyle\int_0^\infty g\,(\dot{z})\,b\,(\dot{z} - z)\,d\dot{z} & z > 0 \end{cases} \qquad (10.14)$$

where $g\,(\dot{z})$ is the probability density of the laser pulses without the penetration and $g_l(z|\theta)$ is the density with penetration. The additional parameter (the rate of penetration) is now included in the parameter vector θ. A graphical illustration of the effect of integrated penetration on the probability density of echo heights is shown in Fig. 10.5.

Unfortunately, empirical results of Virolainen et al. (forthcoming) indicate that the penetration is not necessarily distributed according to the exponential distribution. A natural explanation is that the gaps in the crowns are not distributed uniformly. A natural extension could be a distribution that a has two parameters and provides the exponential distribution as a special case. Such alternatives are, for example, the gamma and Weibull distributions.

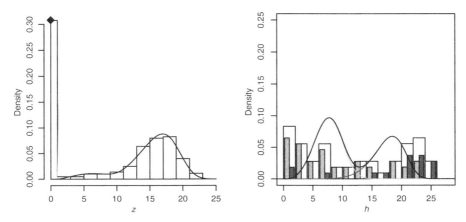

Fig. 10.6 The empirical histogram of 400 echo heights from a mixed Scots pine – Norway spruce sample plot in southern Finland (*left*), and the true histogram of tree heights (*right*, wide bars). The narrow bars show spruces (*light*) and pines (*dark*) separately. The *black smooth lines* show the fitted density function (1st derivative of Eq. 10.15, left), and the corresponding height distribution for whole tree stock (*right*). The *gray lines* on the *right* show the species-specific components of the Weibull mixture distribution

10.2.7 Extension to Mixed Stands

The model can also be extended to mixed species stands. In this extension, it is assumed that the stand density and the functions for the individual crown shape and the height distribution are specific for each tree species. For example, assuming a mixed stand with two tree species and a random spatial pattern of tree locations yields (c.f. Eq. 10.12):

$$
\begin{aligned}
G\left(\dot{z}\,|\boldsymbol{\varphi}_1,\boldsymbol{\xi}_1,\boldsymbol{\varphi}_2,\boldsymbol{\xi}_2,\lambda,\rho,\right) = \exp\Bigg\{&-\lambda\pi\Bigg[\rho\int_0^\infty r(\dot{z},h\,|\boldsymbol{\varphi}_1)^2 f\left(h\,|\boldsymbol{\xi}_1\right)\mathrm{d}h \\
&+ (1-\rho)\int_0^\infty r(\dot{z},h\,|\boldsymbol{\varphi}_2)^2 f\left(h\,|\boldsymbol{\xi}_2\right)\mathrm{d}h\Bigg]\Bigg\},
\end{aligned}
\tag{10.15}
$$

where the parameter ρ specifies the proportion of tree species 1 of the total stand density λ, parameters $\boldsymbol{\xi}_1$ and $\boldsymbol{\varphi}_1$ specify the height distribution and crown shape for tree species 1 and correspondingly $\boldsymbol{\xi}_2$ and $\boldsymbol{\varphi}_2$ for tree species 2. Therefore, the height distribution of the stand is a mixture of two densities with weight specified by ρ. Figure 10.6 shows an example application of the model in Eq. 10.15 in a forest stand with dominated Norway spruces trees growing under a dominant crown layer of Scots pine; the shape of individual crowns is described by the models shown in Fig. 10.2.

The echo height is not the only characteristic recorded by laser scanners. Especially, the proportion of energy reflected by the target (echo intensity) might provide information about the reflectance of the forest. Extension of the model-based approach to bivariate ALS data of echo height and intensity, $Z = (\dot{Z}, I)'$, arises by defining the joint distribution of Z as

$$G(z) = P(Z \leq z).$$

This model might be useful especially in mixed stands where two tree species are rather similar in crown shape but have different distributions of echo intensity (for definition, see Chaps. 2 and 3). Additional parameters specifying the species-specific distribution of echo intensity would be introduced by this extension.

10.3 Applications of the Approach

10.3.1 Parameter Estimation

The models presented in Eqs. 10.9, 10.10, 10.11, 10.12, 10.13 and 10.14 define the distribution of ALS echo heights under the different assumptions on the process and forest stand. The estimation of the parameters of these models is naturally done using conventional estimation methods. Studies applying a model-based approach have used the method of Maximum Likelihood (ML) due to its strong theoretical basis (Casella and Berger 2002). The method searches for such estimates for the parameter vector θ that produce the maximum value of the log-likelihood function. Assuming independent, identically distributed echo heights, the log-likelihood as a function of parameters is defined as

$$ll(\theta) = \sum_{i \leq N} \ln g(z_i \mid \theta) \qquad (10.16)$$

where z_i is the observed echo height for pulse i, $i = 1, \ldots, N$ and g is the density corresponding to the assumed model (Eq. 10.13 if no penetration is assumed or Eq. 10.14 if penetration is assumed). Due to the structure of the density function in these equations, the log likelihood can be written alternatively as a sum of two components corresponding to ground echoes and echoes from the crown, respectively, as

$$ll(\theta) = M_0 \ln G(0 \mid \theta) + \sum_{i \leq M} \ln g(z_i \mid \theta), \qquad (10.17)$$

where M_0 is the total number of ground echoes, and the latter sum is taken over the non-zero echoes to tree crowns ($i = 1, \ldots, M$). The ML estimators are asymptotically unbiased and efficient. The method also provides estimators of the standard errors of the parameters. However, possible lack of independence among observations (e.g. nearby echo heights likely stem from the same tree) may lead to lack of efficiency and underestimation of the standard errors of the estimators.

However, the point estimators of θ remain asymptotically unbiased even with dependent data. All properties of the ML-estimators are valid only if the assumed model is correct.

With our model, evaluating of the p.d.f. for the likelihood may be computationally highly intensive, especially under the grid-based spatial models where the distribution function itself includes numerical integrals over an area (Eqs. 10.10 and 10.11). Therefore, some efforts have been made to approximate the likelihood using methods where the p.d.f. is evaluated only for a small set of regularly spaced values of z. Virolainen (2010) and Mehtätalo et al. (2010) used an approximation based on splines, whereas Mehtätalo et al. (forthcoming) used classified ALS echo heights when computing the likelihood.

10.3.2   Applications to Actual Data

Mehtätalo and Nyblom (2009) evaluated the model for random pattern of tree locations (Eq. 10.12) in a simulated dataset, which exactly followed the assumptions behind the model. Mehtätalo and Nyblom (2012) reported on a similar simulation study for a square grid pattern of tree locations (Eq. 10.11). Results were encouraging, but they do not demonstrate the performance of the model in practice.

Mehtätalo and Nyblom (2012) reported on an application to a real dataset from an old-growth Norway spruce sample plot. Models for crown shape were first estimated from empirical data where individual trees had been recognized and extracted from an ALS point cloud (Fig. 10.2). The example showed good potential, but the evaluation still left many questions unanswered. Especially, 'true' tree heights were based on ALS data and only one sample plot was used in the demonstration. Furthermore, the feasibility of fitting a generic crown-shape function remains an open question.

Mehtätalo et al. (2010, forthcoming) presented a more pragmatic application of the approach. They proposed a two-stage fitting approach that has similar principal data needs as the area-based approaches. The first step involves 'model' training using ground-measured sample plots. Because all trees of these plots were known for total height, accurate estimates for the parameters specifying the distribution of heights were available. Therefore, the training stage provided estimates of the parameters related to tree crown shape, conditional to the known height distribution. In the evaluation stage, the same model was used for estimation of the height distribution of the evaluation plots, but with the previously estimated parameters of crown shape treated as fixed constants. The empirical part of the study used Eucalyptus plantations where trees were planted according to a known rectangular grid pattern. Therefore, Eq. 10.10 was used with the known stand density and distances between rows and trees within rows. The quality of the estimated height distribution was evaluated using the mean and dominant height based on that distribution. The procedure resulted in RMSEs of 1.4 and 0.8 m for mean and dominant height, respectively. The corresponding empirical biases were -0.35 and

0.22 m. These values include also the errors related to the determining of the true tree heights using measurements and locally calibrated height-diameter models.

An application utilizing no field data was implemented and evaluated in Mehtätalo et al. (forthcoming). The likelihood based on the model of Eq. 10.10 was maximized with respect to all of its parameters simultaneously, and no field data were used to aid the estimation. The resulting estimated distributions of tree heights showed an RMSE of 2.9 m and 0.9 m for mean and dominant height, the biases being 0.03 m and −0.23 m, respectively, when the estimated dominant and mean heights were compared to the assumed true heights.

10.4 Discussion and Conclusions

This chapter gave an overview of the model-based approach proposed by Mehtätalo and Nyblom (2009) and further developed by Mehtätalo et al. (2010), Mehtätalo and Nyblom (2012), and Mehtätalo et al. (forthcoming). The approach is based on a model for the marginal distribution of the ALS echo heights from forest. The echo heights are modeled as the outcomes of a stochastic process, which integrates the underlying forest-related sub-processes in a theoretically justified way.

The derivations presented in this chapter demonstrate that even though the task is challenging, it is soluble at least in the simplest cases. However, the current versions of the model is too limited for practical use, and mathematically and computationally much more demanding than, e.g., the currently used area-based approach. However, if these issues can be solved, the model provides several benefits compared to the empirical approaches. The major benefits would be:

(i) The approach implicitly and simultaneously takes into account properties related to *missed treetops, pulse penetration, PPS sampling, canopy overlap* and *ground echoes* through modeling the processes behind these in a theoretically justified manner.

(ii) The model is parameterized using the stand attributes of interest. Therefore, the problem of stand attribute estimation with the model-based approach is a problem of parameter estimation.

(iii) Estimation of characteristics related to stand density, stand structure, and tree canopy is possible under the same general approach.

(iv) The approach provides a possibility to estimate forest characteristics without campaign-specific calibration data using low-density ALS data.

(v) There are apparent ways to extend the approach to more complex situations, as demonstrated by the bivariate model for echo height and intensity.

The model assumes specific random processes for stand density, tree locations, and heights of individual trees. In contrast, the studies reported so far have assumed that the process generating tree crown envelope shapes as a function of tree heights is fixed. In addition, penetration of ALS pulses into tree crowns has not been assumed or it has been assumed fixed. The model has been published only for single-species stands. This chapter provided extensions of the model to relax these restrictions. Equation 10.8 now provides a starting point for areas with random tree shapes. The need for this extension was demonstrated by Virolainen et al. (forthcoming). We also presented a model that allows penetration of pulses into tree crowns (Eq. 10.14) according to a given distribution function, such as the exponential distribution justified by the Beer-Lambert's law. Finally, an extension to a mixed stand with two tree species with different crown shape was presented with a real-data example in Sect. 10.2.7. An approach to include the echo intensity into the model by assuming bivariate observations was outlined, too.

In forest inventory, the parameters of primary interest are those specifying the stand density and the distribution of tree heights. The other parameters, i.e. those specifying the crown shape, the randomness in them, or the penetration of pulses into the tree crowns are not of primary interest. They can therefore be called nuisance parameters in this context. They are necessary to make the model as realistic as possible, and are therefore included for more efficient and less biased estimation of the parameters of primary interest. However, a parameter that is regarded as a nuisance parameter in the context of forest inventory can be of primary interest in other situations. For example, the estimated distribution of penetration could be related with the canopy gap fraction. Therefore, it could be the parameter of primary interest if the model is used for estimation of the leaf area index.

The model-based approach provides means to analyze the marginal distribution of ALS echo heights for a given area. The data are marginal with respect to locations of the pulses. For this reason, the information on the location of pulses within the area is not utilized. Such an approach is justified and may be efficient in a situation where (i) the tree locations do not provide any information on the underlying forest or (ii) the locations of laser echoes with in relation to each other are unknown.

The previous studies (e.g. Mehtätalo et al. 2010) have considered the model-based approach only for situations where the pulse density is low. In that situation, analysis of the marginal distribution of z is justified because the x-y coordinates do not include any essential information about the stand structure beyond the echo heights z. The situation is similar to the area-based approach, where predictors are usually quantities that are calculated from the marginal distribution of echo heights (e.g., percentiles, or moments of it). The individual tree detection (ITD) approaches have potential for higher accuracy than the area-based and model-based approaches because ITD is able to utilize the information of the relative locations of the echoes, too. On the other hand, ITD can be used only with high-density data where the x-y data includes essential information.

The locations may be unknown if the data are full-waveform data from scanners with a large footprint (e.g. Sun and Ranson 2000; Ni-Meister et al. 2001). In such case the information provided by the sensor is the marginal distribution of the canopy heights within the single footprint. The model-based approach could provide means to analyze such data for estimation of interesting forest attributes for the footprint area.

The model has been developed only for a limited set of spatial patterns (random and strictly regular). For practical applications, an extension for a general spatial model, such as the Gibb's process (Illian et al. 2008) is needed. However, nice solutions as the ones of Eqs. 10.12 and 10.15 do not necessarily exist for the more general model.

There is still an open question about whether all necessary nuisance parameters can be satisfactorily included in the model, still keeping the model simple enough for estimation. The estimation is already quite demanding using the models presented in this chapter. Especially, the models for grid pattern (Eqs. 10.10 and 10.11) are computationally demanding because the cumulative distribution function already includes a numerically evaluated integral over a two-dimensional space. For estimation, this c.d.f. needs still to be differentiated to compute the likelihood to be repeatedly evaluated in the estimation. We have done some work to find approximations of the likelihood, and these approximations have shown potential with the models used until now. However, the extensions of the model introduce more complexity of the model. Especially, inclusion of random penetration and random crown shape introduce additional integrals to the models which most likely need to be evaluated numerically. Switching to the Bayesian framework in estimation could provide a step towards a good solution, but the need to evaluate the likelihood still remains.

References

Bailey R, Dell T (1973) Quantifying diameter distributions with the Weibull function. For Sci 19:97–104

Casella G, Berger RL (2002) Statistical inference, 2nd edn. Duxbury, Pacific Grove

Gaveau DLA, Hill RA (2003) Quantifying canopy height underestimation by laser pulse penetration in small-footprint airborne laser scanning data. Can J Remote Sens 29:650–657

Grover ST, Kucharick CJ, Norman JM (1999) Direct and indirect estimation of leaf area index fAPAR and net primary production of terrestrial ecosystems. Remote Sens Environ 70:29–51

Illian J, Penttinen A, Stoyan D, Stoyan H (2008) Analysis and modelling of spatial point patterns: from spatial data to knowledge. Wiley, New York

Junttila V, Maltamo M, Kauranne T (2008) Sparse Bayesian estimation of forest stand characteristics from airborne laser scanning. For Sci 54:543–552

Lim K, Treitz P, Wulder M, St-Onge B, Flood M (2003) Lidar remote sensing of forest structure. Prog Phys Geogr 27:88–106

Magnussen S, Boudewyn P (1998) Derivations of stand heights from airborne laser scanner data with canopy-based quantile estimators. Can J For Res 28:1016–1031

Magnussen S, Eggermont P, LaRicca VN (1999) Recovering tree heights from airborne laser scanner data. For Sci 45:407–422

Mehtätalo L (2006) Eliminating the effect of overlapping crowns from aerial inventory estimates. Can J For Res 36:1649–1660

Mehtätalo L, Nyblom J (2009) Estimating forest attributes using observations of canopy height: a model-based approach. For Sci 55:411–422

Mehtätalo L, Nyblom J (2012) A model-based approach for ALS inventory: application for square grid spatial pattern. For Sci 58:106–118

Mehtätalo L, Virolainen A, Tuomela J, Nyblom J (2010) A model-based approach for estimating the height distribution of eucalyptus plantations using low-density ALS data. In: Proceedings of SilviLaser 2010, the 10th annual conference on lidar applications for assessing forest ecosystems. Freiburg, Germany

Mehtätalo L, Virolainen A, Packalen P, Tuomela J (forthcoming) A model-based approach for estimating tree height distribution using airborne laser scanning data with and without field measurement data (Manuscript)

Næsset E (1997) Determination of mean tree height of forest stands using airborne laser scanner data. ISPRS J Photogramm Eng Remote Sens 52(2):49–56

Næsset E. Gobakken T, Holmgren J, Hyyppä H, Hyyppä J, Maltamo M, Nilsson M, Olsson H, Persson Å, Söderman U (2004) Laser scanning of forest resources: the Nordic experience. Scand J For Res 19:482–499

Nelson RF (1997) Modeling forest canopy heights: the effects of canopy shape. Remote Sens Environ 60:327–334

Nelson RF, Krabill WB, Maclean GA (1984) Determining forest canopy characteristics using airborne laser data. Remote Sens Environ 15:201–212

Ni-Meister W, Jupp DLB, Dubayah R (2001) Modeling lidar waveforms in heterogeneous and discrete canopies. IEEE Trans Geosci Remote Sens 39:1943–1958

Pinheiro JC, Bates DM (2000) Mixed-effects models in S and S-Plus. Springer, New York

Rautiainen M, Stenberg P (2005) Simplified tree crown model using standard forest mensuration data for Scots pine. Agric For Meteorol 128:123–129

Solodukhin VI, Zhukov Y, Mazhugin IN, Bokova TK, Polezhai VM (1977) Vozmozhosti lazernoi aeros emka profilei lesa (Possibilities of laser arial photography of forest profiles). Lesn Khoz 10:53–58

Stoyan D, Kendall WS, Mecke J (1995) Stochastic geometry and its applications, 2nd edn. Wiley, Chichester, 436 p

Sun G, Ranson KJ (2000) Modeling lidar returns from forest canopies. IEEE Trans Geosci Remote Sens 38:2617–2626

Virolainen A (2010) Estimation of forest characteristics using mathematical models. MSc thesis, University of Eastern Finland, Joensuu, Finland

Virolainen A, Mehtätalo L, Korhonen L, Korpela I (forthcoming) Modelling crown envelope shape and penetration of laser pulse into tree crown (Manuscript)

Part II
Forest Inventory Applications

Chapter 11
Area-Based Inventory in Norway – From Innovation to an Operational Reality

Erik Næsset

Abstract The aim of this chapter is to give an overview of the development of ALS as an operational tool for forest management inventories in Norway. The chapter will shed light on some of the technical and institutional challenges that were faced. Interaction between the scientific community and private sector was seen as a critical factor for successful adoption of the new technology for practical purposes and it will briefly be described. A description of local adoptions of the methods and of research conducted to improve the technical and economic performance will be given. Finally, some future needs and directions will be discussed. It is believed that the lessons learned in Norway may be found useful for similar efforts in other countries.

11.1 Introduction

Like in many other western countries, forest inventories in Norway are traditionally conducted at two very different geographical scales serving two distinct purposes, namely (1) the National Forest Inventory (NFI) providing data for national and regional policy-making and (2) forest management inventories (FMI) primarily serving the need for local data for management of individual forest properties. The NFI which dates back to 1919, is basically a sample survey based on a probability sample of field plots used to estimate important parameters such as total timber volume for the entire country and individual regions, such as counties with a typical size of around 2,000–50,000 km^2. The management inventories, which are spatially explicit, aim at providing data for every treatment unit (forest stand) for tactical and strategic planning. Thus, FMIs will in most cases have a wall-to-wall coverage.

E. Næsset (✉)
Department of Ecology and Natural Resource Management, Norwegian University
of Life Sciences, Ås, Norway
e-mail: Erik.naesset@nmbu.no

M. Maltamo et al. (eds.), *Forestry Applications of Airborne Laser Scanning: Concepts and Case Studies*, Managing Forest Ecosystems 27, DOI 10.1007/978-94-017-8663-8__11, © Springer Science+Business Media Dordrecht 2014

Although the first management inventories date back to around 1870–1880, a systematic development of methodologies and implementation of inventory programs started in the 1950s and early 1960s. Like in many other European countries, the inventories relied on field visits to every stand that was identified and delineated using aerial photography and topographic maps. Point sampling taking advantage of the recent developments by Bitterlich in 1948 and onwards (the relascope; see e.g. Bitterlich 1984) was commonly performed in every stand to produce volume estimates, whereas tree species distribution, site quality, treatment proposals and other properties required for forecasts and analyses were subjectively assessed during the field visit.

An important shift in technology and methodological approaches took place in the late 1970s, which paved the road for introduction of airborne laser scanning (ALS) in management inventories 20 years later (around year 2000, see below). During a period of 10–15 years (~1975–1990), stereo photogrammetry supported by geographical information systems became the basic method for stand-based inventories. Analogue stereo plotters were used for stand delineation, while measurement and interpretation of stand tree height and crown closure were used to predict volume using established volume models. Other parameters such as tree species and site quality were interpreted by manual interpretation. Field visits were only occasionally performed, and mainly to determine specific silvicultural treatment needs and to verify the photo interpretation.

An important political decision was made which later had a great impact on how the management inventories were conducted. During this period it was decided to offer the forest owners direct economic support from the state (subsidies) if they could collaborate on inventories across property boundaries. The motivating factor for the support was an expectation of higher activity in the forestry sector when the economical values of the forests became evident to the owners through more complete and reliable information about the resources. Since the support was offered directly to the private owners, it stimulated competition in the market between providers of inventory services who looked for better products than their competitors and it promoted large-scale solutions (economy of scale). Wall-to-wall inventories covering large tracts of land, say, 5,000–10,000 ha, suddenly became the rule rather than the exception. The average size of private forest properties in Norway is around 50 ha. Field-based sample surveys covering the entire geographical area of interest were performed in conjunction with the wall-to-wall inventories to provide overall estimates of wood volumes, and the sample surveys were also used to adjust for the systematic errors that frequently occur in manual photo interpretation. Thus, by the end of the 1990s there was a strong tradition and acceptance for large-area wall-to-wall management inventories based on photo interpretation and supported by field-based sample surveys. Another interesting observation from the last epoch of the 1990s, was a close collaboration between the service providers (inventory firms, the forest owners associations) and the research community. Digital photogrammetry was introduced in the forestry sector in 1997 and soon after the so-called area-based method to produce stand-wise volume and other estimates, which was proposed in 1997 (Næsset 1997a, b), was applied to three dimensional (3D) point clouds from digital photogrammetry (Næsset 2002a).

The aim of this chapter is to give an overview of the development of ALS as an operational tool for FMI in Norway. The chapter will shed light on some of the technical and institutional challenges that were faced. The interaction between the scientific community and private sector that took place and was seen as a critical factor for successful adoption of the new technology for practical purposes will briefly be described. A description of local adoptions of the methods and of research conducted to improve the technical and economic performance will be given. Finally, some future needs and directions will be discussed. It is believed that the lessons learned in Norway may be found useful for similar efforts in other countries.

11.2 Early Tests and Method Development

The first test with ALS with the aim of estimating properties such as mean height and volume of individual stands was conducted in 1995 (Næsset 1997a, b). That was only a couple of years after the first commercial ALS instruments were introduced for topographic mapping (Ussyshkin and Theriault 2010). In 1995, some state-of-the-art proprietary instruments were capable of recording two echoes per emitted pulse, some could only collect either first or last echo (Baltsavias 1999). In our experiments we used the ALTM 1020 sensor produced by Optech Inc, Canada. The instrument could collect up to two echoes per pulse, but one would have to set the operating mode (first only, last only, or first and last) prior to acquisition. In our first dataset we collected last echoes only simply because there was very little awareness in those early days that the first echo data could be useful for vegetation studies. It was recognized though that last echoes were the most appropriate ones for derivation of the digital terrain surface. In our first acquisition the applied pulse repetition frequency was 2 kHz, although ALTM 1020 could be operated up to 5 kHz (Baltsavias 1999). The pulse density was around 0.09–0.12 pulses m^{-2}. This illustrates an intriguing technological development. Over a time period of around 15 years the capacity has increased by a factor of around 100, representing a doubling of the capacity about every second year.

Although this first test was conducted without any previous experience about relevant point densities and flight and instrument parameters, several useful results were documented:

1. *Gridding:* The concept of gridding the data into regular grid cells was introduced (Næsset 1997a) and proved to be useful for estimation of mean height and volume of stands.
2. *Mean height estimation:* ALS data was shown to provide accurate results of mean height (Næsset 1997a). It was also shown empirically that the relationship

between height of the ALS echoes and the mean height of the trees depends on the size of the grid cells. This latter result was formalized in a more theoretical manner by Magnussen and Boudewyn (1998).

3. *Volume estimation:* Stand volume could be estimated from ALS data following the grid approach (Næsset 1997b). *Canopy cover density* and *volume below the canopy surface* were defined and introduced as variables complementary to the height in volume estimation. The use of ALS-derived height and cover is a fundamental approach to estimation of volume as well as biomass.

4. *The area-based approach:* The so-called area-based approach was briefly outlined (Næsset 1997a, p. 55, b, p. 252) and defined the research agenda for the following 5 years.

However, the identification of accurate geo-positioning of the field plots as a prerequisite for application of ALS in operational forest inventory (Næsset 1997b, p. 252) called for intensive testing of equipment, methods, and procedures for accurate positioning under forest canopies. In the mid 1990s state-of-the-art differential Global Positioning System (GPS) positioning in forests indicated an expected accuracy of around 3–4 m (e.g. Deckert and Bolstad 1996) using C/A (course/acquisition) code observations. A series of experiments was carried out in order to improve the positioning accuracy to better match with that of the ALS echoes (30–40 cm). Carrier phase observations were used; the number of available satellites was increased by also observing and recording data from the Russian Global Navigation Satellite System (GLONASS); different types of receivers were applied, including survey grade receivers; and different processing algorithms and software packages were tested, which even included use of precise satellite ephemeris and thus avoiding the need for differential correction against a base station with known coordinates. The conclusion of this series of tests (Næsset 1999, 2001a; Næsset et al. 2000; Næsset and Jonmeister 2002; Næsset and Gjevestad 2008) indicated that an average accuracy of better than 0.5 m should be achievable for plot positioning under forest canopies, at least in boreal forests.

In 1998, the various steps of the area-based method were detailed and about 1,600 field plots were measured within a 1,000-ha study to empirically test the performance of the methodology with point clouds derived from ALS (Næsset and Bjerknes 2001; Næsset 2002b) or image matching (Næsset 2002a).

The basic steps of the area-based method are as follows:

1. *Stand delineation and interpretation:* The forest stands in the area of interest (AOI) are delineated. Delineation using orthophotos or stereo photogrammery is commonly adopted. Manual interpretation is conducted to derive properties like for example tree species, site quality, and age which are required either as primary variables that are sought for every stand in the inventory or used for stratification of the inventory.

2. *Sample survey:* A field sample survey is conducted across the entire AOI. Such surveys are commonly performed according to stratified sampling schemes applying systematic sampling with a random start or systematic cluster sampling. Thus, the collected sample of field plots can be considered as a (stratified)

probability sample. Conventional field data are collected on the plots (e.g. tree diameter and height) and the plots are accurately geo-referenced.

3. *ALS acquisition and processing:* ALS data are acquired for the AOI. The entire AOI is partitioned into regular grid cells with cell size equal to the sample plot size. Various metrics are derived from the 3D ALS point cloud for each cell and field plot. These metrics basically belong to two distinct "families" of variables, namely (1) canopy height-related variables like order statistics, mean height, and variability among the heights of the ALS echoes and (2) canopy density metrics which characterize the frequency of echoes above a certain height threshold relative to the total number of recorded echoes. A wide range of other metrics or slightly modified metrics can be found in the literature, see e.g. Junttila et al. (2008).

4. *Model fitting:* Based on the sample plots regression models with biophysical properties such as mean height, basal area and volume as dependent variables are fitted with ALS metrics used as independent variables.

5. *Stand-wise estimation:* Finally, the fitted models are used to predict the biophysical properties of interest for every grid cell. The individual cell estimates are aggregated to stand estimates.

In the early description in Næsset (1997a, b) this methodology was characterized as "two-phase sampling" while Næsset and Bjerknes (2001) and Næsset (2002b) labeled it "a two-step procedure". According to sampling terminology a more appropriate characterization would be "synthetic regression estimation for small areas" (Särndal et al. 1992). The term "synthetic estimation" refers to a situation where a sample for a large area is used to provide estimates for a smaller area based on predictions. This terminology was introduced in the ALS literature by Andersen and Breidenbach (2007).

The first full-scale tests reported results based on independent validation for entire forest stands following every step in the procedure outlined above. Forest stands are the relevant units in operational forest management. Thus, the tests evaluated the products that were obtained in the last step (step 5) described above. In young planted forest stands (tree heights typically below 8–10 m) with an observed average mean height of 6.64 m it was found that stand mean height could be predicted with a precision (standard deviate of the difference between observed and predicted mean height) of 0.56 m (Næsset and Bjerknes 2001). The mean difference was 0.23 m ($p > 0.05$). It is expected that biomass at stand level in such very young stands may be predicted with an accuracy of 20–30 % of the observed mean value (Næsset 2011). In young forest stands (i.e., stands that have reached an age where commercial thinning is a relevant treatment option) and mature stands quantitative biophysical properties of relevance for forest planning are amongst others mean tree height, dominant height, mean stem diameter, basal area, stem number, and volume. The first full-scale testing revealed a precision for these properties of around 4–8 % (mean and dominant height), 6–12 % (mean diameter), 9–12 % (basal area), 17–22 % (stem number), and 11–14 % (volume) (Næsset 2002b). Subsequent full-scale testing (Næsset 2004a, 2007), including a testing of the first operational inventory (Næsset 2004b), and full-scale tests conducted by colleagues in Finland (e.g. Maltamo et al. 2006a, b, Chapter 12) and Sweden (e.g. Holmgren 2004) confirm these results. A rule of thumb which seems to apply to boreal forest stands, at least under Nordic conditions, indicates that volume estimates at stand level are expected to have a precision of around 10–15 % of the mean value. A summary of the results from some of the Nordic validation studies can be found in Næsset (2007). Similar studies have been conducted in other countries and continents as well (Hollaus et al. 2006; Hudak et al. 2006; Jensen et al. 2006; Thomas et al. 2006; Rombouts et al. 2008; Latifi et al. 2010).

11.3 From Research to Commercial Applications: Marketing and Adoption to Local Practices

Forestry is sometimes considered a conservative branch. The extremely long time horizons under which many decisions are made call for continuity rather than frequently shifting policies and practices. For example, investment decisions in planting and establishment of new forests can have up to a 100 year horizon or sometimes even longer. Introducing new methods and techniques in forest inventory can therefore sometimes be challenging.

Once it became evident around 1999 that ALS-based inventories could provide more accurate estimates of important biophysical properties than conventional methods (see also Chap. 12), a strategy consisting of five pillars was adopted by the research community with the aim of introducing the area-based method for operational use in the commercial market within "reasonable time". The five pillars were:

1. *Scientific evidence:* It was anticipated that careful scientific documentation and acceptance in the scientific community for the promising results that were provided were the best "selling arguments" for the proposed method. In particular, independent validation of the new method under practical conditions and in full scale was given priority and resulted in four full-scale tests with independent validation (Næsset 2002b, 2004a, b, 2007).
2. *Utility and economical benefit:* Acquisition of ALS data was considered expensive around the turn of the millennium. However, the promising results showing improved accuracies indicated a potential for increased utility of the data and improved management decisions resulting from better data. Many practitioners tend to focus on immediate costs of forest inventory and ignore the benefits of improved data. Emphasis was therefore put on showing through scientific studies what benefits in monetary terms one potentially could expect by the new method. As an example, it was shown that under Norwegian conditions one could expect a net utility (net present value) over conventional but cheaper inventory methods of at least 25 €/ha (Eid et al. 2004).

3. *Commercial engagement in concepts and ideas:* Ownership to the new method among commercial service providers was considered a key to adoption of the new method in operational inventory. Once the proof of concept and the first results from the full-scale testing and validation (Næsset 2002b) were presented to the users in a user workshop in Norway in year 2000, private service providers showed an immediate interest in the concept. However, it was considered important from both sides – the scientific community and private forest industry – to repeat previous tests on full scale in a practical setting with strong involvement from the service providers (1) to get a confirmation of previous results and (2) to give the private sector an opportunity to gain "hands-on" experience with the method. A second full-scale demonstration (6,000 ha) was therefore carried out in 2001–2002 in conjunction with an operational FMI in a joint research project between the scientific community and the inventory company Prevista AS, Norway (Næsset 2004a). The demonstration project also addressed new scientific questions of great relevance to the forest industry in Norway, namely what the effects of steep and rough terrain would be on the accuracy of the inventory. The study did not reveal any serious degradation of the accuracy (Næsset 2004a). In 2001, the results based on the two first studies and a description of how the method could be incorporated in operational inventory was presented in a professional journal in Norway (Næsset 2001b). The article received a lot of attention from the forest industry.

4. *Continuity of accepted practices:* To gain ownership to the new concept among the service providers and to receive acceptance and maintain confidence in the forest inventory products from the customers (the forest owners) it was considered to be of great importance to maintain as many elements of the conventional inventory routines as possible. For example, the stand map of a forest property is one of the most intuitive products of a management inventory which most forest owners are familiar with. Maintaining the conventional stand map and how it traditionally had been produced – by manual stand delineation using stereo photogrammetry – was essential. Stereo photogrammetry was therefore kept as an initial step of the production chain and even the map products remained unchanged, although ALS obviously offer some advantages in terms of automatic stand delineation (Koch et al. 2009, Chapter 5) and even a map representation at finer geographical scales, like at an individual cell level (typically 200–400 m^2). Instead of replacing existing routines, emphasis was put on improving them. One such improvement was to present the ALS echoes of the vegetation visually as points in 3D in digital photogrammetry in the same view as the 3D representation of the aerial image. Thus, the operator could easily see the height variation in the canopy to support the manual stand delineation. This is still a preferred method for stand delineation in many of the Norwegian forest inventory companies.

5. *Marketing and communication:* After a major strategic decision was made by Prevista AS, the company started as the first one to market ALS-assisted management inventories in Norway in 2001. The concept was also presented to relevant companies and agencies in Finland (e.g. Metsähallitus and Metsämannut) and

Sweden (e.g. Skogssällskapet and Västra Skogsägarna). These efforts were supported by the research community and researchers often participated with scientific presentations on some of these occasions.

Prevista AS was awarded the first commercial contract for an operational ALS-assisted management inventory in 2002. The contract comprised an area of 46,000 ha in Nordre Land municipality in southeastern Norway (Næsset et al. 2004). There is no doubt that the public subsidies stimulating collaboration among forest owners in an area and favoring inventory methods which can offer economic scale effects were of crucial importance for Prevista's success. Another "success factor" was the model for cost-sharing between different sectors and actors that had been adopted by public agencies in Norway in 1992. Through the so-called Geovekst initiative, a public-private partnership coordinated by the National Mapping Authority, the forestry sector was given access to the public geo-data infrastructure, including ALS data. This partnership also stimulated coordinated ALS acquisitions and cost-sharing between the National Mapping Authority (topographic mapping) and private forestry, which the ALS-assisted forest inventories have profited from. The inventory in Nordre Land municipality was also conducted in close collaboration with the research community. An independent validation of the accuracy of the stand-based estimates was carried out in order to document the performance of the method in a full-scale environment. It was also deemed essential for future marketing of the method that the results were published even according to scientific standards. The results of the validation were in line with previous findings (Næsset 2004b).

Once this first operational project had started and served as a reference for future work, similar contracts were signed with forest owners in other parts of the country. Three years later, in 2005, seven more projects were under contract and the same year Prevista AS entered into the first commercial contract in Sweden based on the same methodology. That was just 2 years after research colleagues in Sweden had done their first full-scale tests (Holmgren 2004; Holmgren and Jonsson 2004). Even 5 years later, around 2010, most FMIs in Norway were performed with support of ALS data and every inventory company in the market offered products based on ALS-assisted inventory methods.

Before the end of the decade, the first full-scale operational projects had been carried out in several other countries as well, like for example Finland (2008; see Chap. 12), Canada, Sweden and Spain (Turunen et al. 2012). Forest inventory companies, as well as surveying companies which traditionally have not been engaged in forest inventory, therefore saw an opportunity for developing an international market for services in forest inventory based on ALS. For example, the Norwegian surveying companies Blom ASA and TerraTec AS through their respective daughter companies Blom Geomatics AS (Norway), Blom Sweden AB (Sweden), Blom Kartta Oy (Finland), TerraTec OY (Finland) and TerraPro Oü (Estonia) started to market ALS-assisted FMI in the respective countries, covering the entire chain from ALS data acquisition to final data delivery of refined forest inventory products. In a similar way, the Danish company COWI through its Norwegian branch COWI AS also offered full solutions for the clients – from ALS acquisition to forest inventory end products for forest management. COWI has been active in especially the Swedish market. The Finnish company Arbonaut Oy, which is one of the major actors in the Finnish market for forest inventory services, has engaged in countries throughout the world, and in particular in testing and demonstration of ALS-assisted carbon inventories under the UN REDD process in developing countries.

It is interesting to notice how the various commercial actors have adjusted their services to the inventory traditions and expectations in the various countries. It has been a common practice to apply the ALS-assisted, area-based method as it is described in this chapter in e.g. Norway and Sweden. In those two countries both Blom Geomatics AS, Blom Sweden AB and COWI AS have offered more or less the same products. In Finland, however, information about tree species and species-specific biophysical parameters is a primary requirement in FMI. This has led to the development of inventory procedures relying on non-parametric methods (see Chap. 12). In the Finnish market, Blom ASA through Blom Kartta Oy has adopted to this tradition by offering even such products, in competition with amongst others Arbonaut Oy and TerraTec Oy.

Finally, it should be mentioned that inventories based on single-tree segmentation also have been marketed. The Swedish company Foran Remote Sensing AB in collaboration with other companies in the Foran Group (Foran Sverige AB, Sweden; Foran Norge AS, Norway) seems to have taken a leading role to offer such products – especially in Norway and Sweden. The number of signed contracts based on single-tree segmentation is limited though, but companies in the Foran Group also offer – and have been actively involved – in numerous area-based, ALS-assisted inventories.

In the commercial efforts commenced in all the Nordic countries there have been strong alliances between the scientific research community and the commercial actors, at least in the early phase of method development. Individuals from the research community have either collaborated directly with the various companies or been recruited by them. As opposed to what seems to have been the tradition in previous decades when new methodological achievements were implemented in practical operations, the introduction of ALS in operational management inventory had a strong scientific component with a desire to conduct controlled and independent testing and validation of the new procedures and products. Numerous scientific publications resulting from this work were published in Finland (see Chap. 12), Norway and Sweden (see above). The forest industry even conducted its own inquiries and communicated the results to a broader audience (e.g. Lindgren 2006, 2012).

11.4   Technical and Economic Improvements

When the first full-scale testing of the area-based method was conducted there was hardly any experience with the influence on the accuracy of the biophysical stand properties of technical specifications of the ALS acquisitions, sensor effects as such and design parameters of the field sample surveys. Thus, no attempt had been made to quantify how the relationship between costs and accuracy were affected by parameters that the analyst potentially could control in order to design the surveys in a manner that reflected individual clients' specific requirements.

A number of factors that will have an influence on the costs and potentially also may have an impact on the accuracy were identified and gave rise to a series of studies exploring how these factors and design parameters affected the accuracy of the forest stand estimates. Factors that were considered can be divided into five distinct groups, which will be discussed below.

11.4.1 ALS Acquisition Parameters

11.4.1.1 Point Density Per Unit Area

Reduced point density per unit area may allow greater areal coverage per over-flight and thus reduced costs. In order to address this issue, Gobakken and Næsset (2008) thinned data with an initial point density of 1.13 points m^{-2} to densities of 0.25, 0.13 and 0.06 points m^{-2}, respectively. They found that the values of most laser-derived metrics (canopy height-related densities such as percentiles, and density-related metrics such as canopy density for different vertical layers) to some extent were affected by point density, although the effect was rather small in most cases. One important exception was noted though: the maximum recorded height for a given target area (grid cell, plot or stand) was strongly influenced by point density, with significantly lower maximum height being recorded at low densities. These effects will also be influenced by forest type, with larger effects of point density for open forests with a highly variable canopy surface. Further, it was reported that the variability of most metrics was influenced by point density, with increasing variability for lower densities. Point density is thus expected to influence the accuracy of biophysical properties predicted from ALS data.

Gobakken and Næsset (2008) conducted a Monte Carlo simulation by which they assessed the influence of point density on stand predictions of biophysical properties in an independent validation. They found that the accuracy of stand-based predictions of mean tree height, basal area and volume was almost unchanged from 1.13 to 0.25 points m^{-2}. Thus, the current practice in Norway to collect ALS data for forest inventory purposes with a density of around 0.7 points m^{-2} was confirmed to be relatively robust. The study confirmed previous results obtained under similar forest conditions in other Nordic countries (e.g. Maltamo et al. 2006a).

11.4.1.2 Flying Altitude

Increasing flying altitude may allow greater performance and areal coverage per over-flight if the instrument has capacity for higher point repetition frequencies under the constraint that a requested point density per unit area is met. Næsset (2004c, 2009) conducted controlled empirical experiments with repeated over-flights over two study areas where field data for sample plots and independent validation stands were collected. Within each of the two studies the same sensor was used, i.e., the Optech ALTM 1210 sensor in Næsset (2004c) and the Optech ALTM 3100 in Næsset (2009). In order to isolate the effect of flying altitude, all other parameters were kept constant. It was considered of particular importance to keep the point repetition frequency constant within each of the experiments. The flying altitudes were 530–540 and 840–850 m above ground level (Næsset 2004c), i.e., an increase in altitude by approximately 60 %, and 1,100 and 2,000 m (Næsset 2009), an increase of around 80 %. ALS-derived canopy height and canopy density metrics were compared across acquisitions. Both studies reported significant differences for some of the metrics typically used in forest inventory, but basically the canopy height distributions derived from the various flying altitudes were fairly stable. It was noted though that there was a weak tendency of an upwards shift in the canopy height distributions by increasing flying altitude, which explained why some of the metrics deviated between the acquisitions. Næsset (2009) also reported that the distribution on echo categories differed between the flying altitudes, with fewer multiple echoes at the higher flying altitude. Thus, the consequence of changing altitude will differ between echo categories. Næsset (2004c) indicated that metrics derived from the first echoes seemed to be more stable than last echo metrics.

Since flying altitude above ground may vary considerably within a given project if there are large differences in the terrain elevation, which is very common in Norway, Næsset (2004c) conducted a Monte Carlo simulation in which ALS data from the two altitudes were mixed randomly for different field plots and the area-based method was followed by fitting regression models and predicting biophysical properties for the independent validation stands. The results revealed that the precision of the stand estimates of mean height, basal area and volume was robust against variability in flying altitude within a project. The same conclusion was reached by Næsset (2004a). Næsset (2009) fitted regression models for the two acquisitions separately and showed in a cross validation that the accuracy of mean height and volume predictions was approximately the same for the two flying altitudes. However, fitting models with data from one flying altitude and applying the models to ALS data from another altitude revealed that some systematic differences in predicted biophysical properties may be expected.

A general conclusion from these experiments is that variability in flying altitude within an operational project is less of a concern if field data for model fitting is collected across all terrain elevations. However, fitting models with data from one flying altitude and applying the models to data from another flying altitude for prediction purposes may introduce systematic errors in the predictions. Also these results confirmed general recommendations for ALS-assisted forest inventory in Norway, namely (1) that terrain variability within a project area is not a critical factor and (2) it is not recommended to use regression models across inventories if there are large differences in flying altitude between the inventories.

11.4.1.3 Pulse Repetition Frequency (PRF)

Higher PRF may, in combination with increased flying altitude, permit a greater areal coverage per over-flight and thus reduced costs. However, higher PRF may generate more noise in the data and also lower measurement precision. Næsset (2009) compared two ALS acquisitions with different PRFs (50 kHz and 100 kHz) from repeated over-flights over 40 field-measured plots (1,000 m^2 in size) using the same instrument (Optech ALTM 3100) under otherwise the same settings (flying altitude, scan angle). At these PRFs the instrument at hand had a pulse width of 10 and 16 ns, respectively, pulse energy of 121 and 66 μJ, and peak power of 12.0 and 4.1 kW. Significant effects were found for the ALS-derived metrics as well as for predicted biophysical properties. There was a general upwards shift in the canopy height distributions with increasing PRF. The proportion of multiple echoes decreased with increasing PRF and there were significant differences for many of the laser-derived metrics between the two PRFs, especially for metrics derived from the last echoes. The effects of doubling the PRF were generally stronger than of increasing the flying altitude by almost the same factor (80 %, see above). It was noted though that forest inventories based on ALS data acquired with different PRFs do not necessarily provide differences in the precision of the stand estimates if the PRF is kept constant within a project.

Due to shifts in the entire point clouds acquired at different PRFs, fitting regression models for biophysical properties with data from one PRF will tend to produce systematic errors in predictions with the models when applied to data acquired with another PRF. The general recommendation in Norway is to avoid using different PRFs within an inventory and also avoid transferring regression models developed with data acquired at different PRFs across inventories.

11.4.2 Sensor Effects

Individual ALS instruments have unique specifications (e.g. characteristics of the emitted laser beam, i.e., peak power, pulse width etc.). These characteristics may influence on the vegetation measurements as illustrated above. If different instruments can be used interchangeably in an inventory, ALS acquisitions may be conducted more effectively by using several aircrafts and instruments in an area to take advantage of favourable weather conditions and expensive infrastructure. Sensor effects were assessed by Næsset (2005, 2009) in empirical studies in two different areas where field data for sample plots and independent validation stands had been collected. The instrument used in Næsset (2005) was an Optech ALTM 1210 sensor with a PRF of 10 kHz which in fact was upgraded to 33 kHz. Thus, it is actually the same instrument that was used but with very different specifications before and after upgrade. Some labelled this specific instrument ALTM 2033 after upgrade (e.g. Maltamo et al. 2006b). Some key specifications before and after upgrade were pulse widths of 7 and 11 ns, respectively, pulse energy of 138 and

84 μJ, and peak power of 20.0 and 7.6 kW. It should be noted that the flying altitude on average was 26 % higher in the latter acquisition, but it was adjusted for the effects of different flying altitudes based on the experiences gained by Næsset (2004c). In Næsset (2009), the two instruments that were used were the Optech ALTM 1210 and Optech ALTM 3100, with pulse widths of 11 and 10 ns, respectively, pulse energy of 84 and 121 μJ, and peak power of 7.6 and 12.0 kW. The two instruments used by Næsset (2009) were operated at a fairly similar PRF (33 and 50 kHz) and flying altitude (1,200 and 1,100 m above ground level). Næsset (2009) also corrected carefully for the effect which different point densities may have on the results by thinning the ALS data to the same point densities. It is important to recognize that instrument differences to a large extent can be related to the same instrument properties that are important to consider when using a given instrument with different operational settings, like flying altitude and PRF (resulting in alteration of important properties like pulse width, pulse energy and peak power).

The two studies were presented as examples and demonstrations of what sensor effects one must expect when using standard instruments available in the commercial market. Thus, the results in Næsset (2005) showed a tendency of an upwards shift in the canopy height distributions for the data acquired after instrument upgrade – an effect one would expect as a consequence of a higher PRF (see above). The instrument effects were much more pronounced in the study by Næsset (2009). A pronounced shift in the point clouds resulting in statistically significant shifts in canopy height-related metrics as well as in metrics related to canopy density was reported. For some of the height metrics (e.g. the 90th percentile) a shift of around 0.3 m was found.

By fitting regression models for mean height and volume using the field plots and ALS-metrics derived from the ALTM 1210 acquisition and applying those models for prediction of the same biophysical properties using the ALS data from the ALTM 3100 acquisition statistically significant under-predictions of mean height of 2.2 % and volume of 7.5 % were reported. When this exercise was repeated with prediction based on ALS-data acquired by the ALTM 3100 instrument with a PRF of 100 kHz rather than 50 kHz, the under-prediction for mean height and volume increased to 2.5 and 10.7 %, respectively. Thus, this study demonstrated that in cases where the sensor or operating properties that influence the resulting point clouds from tree canopies deviate much between the model training data and the data used for prediction, systematic errors in for example volume of, say, 10 % or more must be expected. Therefore, it has been recommended in Norway to avoid usage of different instruments in the same inventory or to use fitted models across ALS datasets acquired with different sensors.

An important sensor-specific factor that to date has not been assessed is the algorithm (and its parameters) used to trigger an echo in the sensor. Simulations have shown that the algorithms used to trigger an echo may have a great impact on the resulting height values of discrete return ALS systems (Wagner et al. 2004). Unfortunately, the commercial system providers like for example Optech, do not offer detailed information concerning their detection method.

11.4.3   Field Sample Survey

11.4.3.1   Stratification

An effective stratification of the sample survey may improve accuracy and/or reduce the field sampling effort to reach a given accuracy requirement. When the first full-scale tests (Næsset 2002b, 2004a) as well as the first operational inventory (Næsset 2004b) were conducted, there was not much scientific evidence to build upon as far as stratification was concerned. Two considerations were important though, namely (1) that it was considered useful for reasons given above (Sect. 11.3) to follow a general stratification that already had been practiced for years in FMI using conventional methods, and (2) that the stratification should aim at identifying unique forest classes for which it was reasonable to expect similar relationships within each class between ALS-derived canopy distributions and relevant biophysical properties. Thus, since the ALS point cloud will reflect the size and shape of the tree crowns, the composition of biological material within the crowns (branches, foliage) and the structure of the forest, similarities in the forest structure and these crown properties on one hand and their relation with biophysical properties like height, basal area and volume on the other hand were sought. A reasonable compromise between the two motivations was to use age class and site quality as the basic criteria for stratification. In a boreal forest dominated by spruce, pine and a few deciduous species, the crown shape will tend to be more rounded with increasing age. Similarly, poorer sites will tend to have shorter trees with more rounded crowns and with an open forest structure with more scattered trees. Poorer sites will also tend to be dominated by pines rather than spruce. Thus, site quality is also a fairly good species discriminator. Age class and site quality are properties that for a long time have been derived from existing maps and by stereo photo interpretation.

However, a first comprehensive study of the discriminating power of these properties in modeling of biophysical properties from ALS data was not conducted before Næsset and Gobakken (2008) analyzed a large dataset of 1,395 field plots with above-ground biomass as response variable. In their regression model they included indicator variables for age class and tree species in addition to site index and relevant ALS-derived canopy height and canopy density metrics. It appeared that the indicator variables for tree species were the strongest predictor variables among the non-ALS variables. When the tree species indicator variables were included in the model, site index did not make any meaningful contribution to the model. Similarly, site index was not a statistically significant variable. However, it was shown that site index and tree species were highly inter-correlated. Thus, if tree species is not at hand for stratification, a combination of age class and site index, which sometimes is available in existing maps, is a good compromise. This is also in accordance with current practice in operational ALS-assisted forest inventories in Norway. It should be noted though that since tree species as well as age class and site quality at a stand level commonly are derived on the basis of photo interpretation (see details above) the assignment of individual stands to strata will be subject to errors. This may induce errors also in the final stand estimates because less appropriate regression models may be used for prediction of the stand properties. In some projects with large variation in altitude, stratification has also been based on altitude since the crown shape of especially spruce trees tends to become narrower closer to the alpine tree line.

11.4.3.2 Sample Size

Reducing the number of field sample plots will have a direct impact on costs and the issue of determining an appropriate sample size has been a major concern in the scientific literature as well as in practical inventories. Gobakken and Næsset (2008) conducted a Monte Carlo study in which they by simulation assessed the effects of reducing the plot number on the accuracy of stand-based predictions in an independent validation dataset following the area-based method. Starting with 50, 34 and 48 plots, respectively, in each of three pre-defined strata, they reduced the number of plots to 75 and 50 % of the initial plot numbers. The results showed a moderate reduction in accuracy when reducing the sample size from the full sample to the 75 % sample and even more when the sample size was only 50 % of the initial size. However, effective ways of selecting the sample can compensate for smaller sample sizes. Maltamo et al. (2011) analyzed the effects of sampling strategy for selection of sample plots to be used as training data. Using a common validation dataset for multiple sampling designs, Maltamo et al. (2011) used nearest neighbors methods and found that with ALS-based stratification, the sample size could be substantially reduced with no adverse effects. Further details can be found in Chap. 14.

What can be inferred from these and other international studies is that there is a potential for reduced sample sizes without any serious reduction in accuracy of the predictions if the field resources are used in a "smart" way. Using ALS data as prior information in the design phase of the field sampling is one important means to maintain accuracy with reduced sample sizes. It has been used as a rule of thumb in ALS-assisted FMIs in Norway to allocate around 50 plots to each stratum in pre-stratified sampling. All full-scale tests with independent validation data (see details above) have used around 50 plots per stratum as an initial sample size. The recent studies reviewed here indicate that this general recommendation can be somewhat relaxed when using ALS data as prior information in the design phase.

11.4.3.3 Plot Size

Plot size has a direct impact on costs. In the study by Gobakken and Næsset (2008) in which ALS data were thinned to analyze the effects of point density, the effects of plot size were also assessed. The analysis was based on plots distributed on three predefined strata and each plot consisted of tree measurements within two concentric circles on each plot. For young and fairly dense and even forest the plots (n = 50) around a common plot center had a size of 200 and 300 m^2, respectively. For plots in mature forest on poor sites (more open and low-biomass forest) (n = 34) the plot sizes were 200 and 400 m^2, respectively, while in mature forest on good sites (more closed and high-biomass forest) (n = 48) the plot sizes were 200 and 400 m^2. When regression models for mean height, basal area and volume were fitted to ALS-derived metrics for these plots it was revealed that the RMSE of the models in most cases was lower for the larger plots while the coefficient of correlation tended to be higher. There were some differences between the forest strata though, indicating that the influence of plot size on the accuracy depends on the properties of the forest. Open forests with scattered trees will have a canopy surface with higher variability. Thus, when ALS is used to obtain a sample of points of the canopy from which the ALS-metrics are derived, a highly variable surface will need more observations in terms of higher point density and/or larger plots to maintain accuracy relative to plots with larger tree stocking and a more even and closed canopy surface.

The Monte Carlo simulation mentioned in Sect. 11.4.1.1 was even applied to assess the influence of plot size on stand predictions of biophysical properties in an independent validation. It was found that the precision of the stand predictions of mean height, basal area and volume in most cases was improved by increasing plot size.

Næsset et al. (2011) applied design-based, model-assisted estimators to estimate the biomass in a forest inventory of a 960 km^2 area where a pre-stratified probability sample of field plots was used in the estimation along with a complete wall-to-wall coverage of ALS data. Standard error estimates of the estimated biomass were also provided. Biomass estimates and associated standard error estimates were provided on the basis of only field plots as well (i.e., a so-called direct estimate). Even in this study the field plots consisted of tree measurements within two concentric circles (200 and 400 m^2, respectively). Both the ALS-assisted estimation and the pure field-based estimation relied primarily on the 200 m^2 plots. However, when the plot size was increased to 400 m^2, the precision (standard error) improved relatively more for the ALS-assisted estimate than for the pure field-based estimate. This finding indicates that larger field plots may be favorable in forest inventories assisted by ALS. It is therefore important that design and planning of forest inventories jointly considers the field survey and the ALS acquisition. Numerous causes for the seemingly higher relative efficiency of ALS-assisted surveys with larger plots can be given. Some of them are mentioned above. Other factors are the so-called plot boundary effects (Mascaro et al. 2011; Næsset et al. 2013). Plot boundary effects have the potential to cause discrepancies between ground and remote sensing-based assessments, which will favor larger plots, see further details in Chap. 14.

Because there is a trade-off between point density and plot size on one hand, and plot size and sample size (number of plots) on the other, these three important design properties must be considered jointly when planning ALS-assisted forest inventories.

11.4.3.4   Field Plot Positioning Accuracy

Accurate positioning of field plots is expensive because it requires use of advanced GPS and GLONASS receivers and also time-consuming and demanding procedures for data processing (post-processing with data from base-station). Relaxed requirements for positioning accuracy may reduce costs but may also have a negative impact on accuracy of the biophysical estimates.

In a Monte Carlo simulation Gobakken and Næsset (2009) assessed the influence of positional errors of the field plots on ALS-derived metrics. The plot positions were randomly altered from the true positions by horizontal shifts in steps from 0.5 m and up to 20 m. They also used different plot sizes (200 m^2 versus 300 or 400 m^2) since it is likely that the relative impact of positional errors will be smaller for larger plots. The difference in ALS-derived canopy height-related metrics and canopy density metrics between the correct positions and the altered positions increased gradually with increasing positional error and the differences were more pronounced for the smallest plot size. The differences were also more pronounced for uneven forest with scattered trees than for dense forests. The impact on the ALS-derived metrics varied somewhat between the various metrics, as one would expect. For example, in a stand with even tree heights but uneven spacing of the trees, it is reasonable that the density-related metrics are more influenced by positional errors than the canopy height-related metrics.

In a simulation where regression models were fitted using the field plots with altered positions and applied to predict biophysical properties of independent validation stands it was revealed that the precision of predictions was fairly stable up to a positional error of around 5 m for mean height. For basal area and volume the precision degraded faster than for mean height, most likely due to uneven stand density which will have a limited impact on the mean height estimate. The precision degraded faster for smaller training plots than for large plots. For some of the larger positional errors a sudden drop in precision was observed, caused by erroneous location of the plots in neighboring stands with entirely different characteristics.

Although the work by Gobakken and Næsset (2009) can only be considered an individual case study, it demonstrated that plot size is an issue also when it comes to the impact of positional accuracy. Further, it can be concluded that different ALS-metrics and biophysical properties will be influenced differently by positional errors, depending on what properties of the forest that varies the most over space. The study site is quite typical for the boreal forest in Norway. Based on plot sizes commonly used in operational ALS-assisted management inventories in Norway (250–400 m^2), it has been recommended to keep the positional errors below 2–3 m on average. All inventory companies now use GPS receivers and data recording and processing routines that comply with this recommendation.

11.4.4 Seasonal Effects

If season (leaf-on, leaf-off) affects the accuracy for otherwise equal acquisition parameters, then there is a potential for cost-savings by relaxing other design parameters by choosing the most favorable season.

Næsset (2005) demonstrated how ALS-derived canopy height and canopy density metrics in a mixed conifer-deciduous forest may be affected by canopy conditions by comparing the metrics derived under leaf-on and leaf-off conditions, respectively. It was revealed that the canopy density was reduced significantly under leaf-off conditions while the relative height distribution within the canopy was less influenced. However, a significant downwards shift was noted also for the canopy height distribution.

When regression models for biophysical properties (mean height, basal area, volume) were fitted to ALS-derived metrics for mixed forest sample plots under leaf-on and leaf-off conditions, respectively, the correlation coefficient was higher and the RMSE was lower for the leaf-off data. When the regression models were used for independent validation for large plots (~0.34 ha in average size) following the area-based approach it appeared that the precision was equal or higher under leaf-off conditions. Similar findings have been reported in Finland, see Chap. 12. The reason for the similar or superior performance of the leaf-off data seems to be that in a mixed forest, the relationship between ALS-metrics and the biophysical properties are more similar for coniferous and deciduous trees.

The study demonstrated that there is a significant impact of canopy conditions on relationships between ALS metrics and biophysical properties. Thus, it is essential to avoid mixing leaf-on and leaf-off data in the same inventory. Further, leaf-off data may offer some opportunities for better models and more precise predictions in mixed forests. However, the leaf-off periods after snow-melting in the spring and before snow-fall in the autumn are rather short in Norway. Therefore, there is a high risk associated with acquisition of ALS data in these two narrow time windows, given that the data need to be collected under either leaf-on or leaf-off conditions in a given project. Thus, for planning of ALS-assisted forest inventories in Norway it is generally recommended to collect data in the leaf-on season. However, if leaf-off data already have been collected for other purposes and can be proven to have been collected under stable conditions, they will be well suited for forest inventory purposes.

11.4.5 Effects of Geographical Region

If the relationship between ALS-metrics and essential biophysical stand properties such as mean height, basal area and volume are similar in different geographical regions, then geographically distant inventories may be treated as one project, sharing some of the same infrastructure (e.g. field sample plots used for regression model construction). Field plots and associated ALS metrics from a pool of data may even be used across inventories allowing great savings in field inventory.

Næsset et al. (2005) and Næsset (2007) assessed the effects of geographical origin of the ALS and field data by combining ALS and field data from two different districts in each of the two studies for regression model construction. Within each of the two comparisons the same sensor was used for both districts and the sensor was operated with roughly the same settings (flying altitude, PRF, scan angle). The plot sizes used to fit the individual models across districts were identical or almost identical. Models were constructed for six different biophysical properties in each study, and both studies used field data from pre-stratified field surveys. For the 30 models that were fitted, an indicator variable for district was statistically significant in only five of the models. The five models were either for mean height or dominant height. No significant effect was found in any of the models for basal area and volume. Similar stability in regression relationships has been reported across a wide range of geographical areas and forest conditions in other parts of the world (see Chap. 14 for further details).

However, it is worth noting that in the most comprehensive study conducted to date (1,395 observations collected in 10 different geographical areas in the boreal forest in Norway), Næsset and Gobakken (2008) reported significant effects in biomass models of tree species as well as of geographical district. To the extent that tree species composition and other properties of the forest that influence on tree crown shapes, sizes and composition (branches, foliage) vary between regions it is reasonable to expect regional effects. Differences in relationships between crown properties and biophysical properties of interest, like for example biomass and volume, will necessarily have an impact on regression models because the information inherent in an ALS point cloud from a tree canopy reflects the properties of the crowns and structure of the canopy but not the properties of the stems, which total tree height, basal area, stem volume and total tree biomass relate to. In the study by Næsset and Gobakken (2008) most of the areas were located in parts of the country strongly influenced by a dry continental climate with low winter temperatures and high summer temperatures. However, one among the ten areas was located on the Atlantic coast of Norway with an oceanic climate with high annual precipitation and mild winters. An indicator variable for this particular area was highly significant in the model, indicating that factors not captured by other variables that were included in the regression model (e.g. tree species and age class) had an influence on the relationship between the ALS-metrics and biomass. It is reasonable to assume that differences in stand and crown structure and stem forms caused by differences in climatic conditions are among the basic factors captured by the indicator variable.

In conclusion, these studies suggest that field plots and regression models may be shared between geographical regions if it can be justified that the relationships between crown and canopy properties on one hand and stem properties (especially taper) on the other are fairly similar. A general recommendation for operational inventories has been to exercise some caution if data from different regions are combined. One option might be to use data across regions and include indicator variables in the models to account for regional effects. Thus, one may take advantage of a larger overall dataset in the form of potentially positive effects on the uncertainty of the parameter estimates and/or lower inventory costs due to reduced overall sample sizes.

11.5   Future Research and Development – New Products and Opportunities

Forest management planning in Norway has traditionally applied an approach where mean values for essential biophysical stand properties like mean height and total volume distributed on species have been required for management decisions and forecasting. One important direction of research has been to exploit the relationship between the ALS height distribution of a stand and the tree size distribution (e.g. the diameter distribution) for predicting size distribution from the ALS point cloud. Studies in this field (Gobakken and Næsset 2004, 2005; Bollandsås and Næsset 2007; Maltamo et al. 2009; Bollandsås et al. 2013b; Magnussen et al. 2013) have demonstrated that diameter distributions can be predicted at stand level with a precision that is accepted by the customers, especially by using percentiles of the size distributions as target variables (see further details in Chap. 9). Although diameter distributions have been delivered as part of the management plan in a couple of operational ALS-assisted inventories lately, extraction of size distributions still seems to remain very much within the research community. However, because there is a growing awareness of the value of more detailed information on species and size distributions to better communicate the wood qualities and quantities available in a stand prior to harvest decisions, it is assumed that there will be a growing interest in this kind of information in the future.

Likewise, distribution of wood quantities on qualities (assortments) prior to harvest has been seen as a means to increase the value of the wood by indicating how the wood potentially could be utilized (e.g. use of sawn wood for construction and panels). Since essential wood quality properties like amount, size and properties of the knots in the wood are related to tree crown properties, there is definitely opportunities for providing some of the needed information from ALS-based inventories. Recent research confirm that relevant crown properties may be predicted at plot or stand level in the area-based framework using parametric or non-parametric estimation techniques (Maltamo et al. 2010; Bollandsås et al. 2011). It is also expected that harvesters through improvements of positioning techniques in the future can provide single-tree information on taper and crown properties with geo-location accuracy for individual trees at sub-meter level. This may open up for more detailed modeling and estimation of wood quality-related properties.

When the first operational ALS-assisted inventory was conducted in 2002, it was decided to restrict the inventory to areas covering the age classes from young to mature forest. Thus, areas of recently planted forest and young forest up to an age corresponding to a tree height of around 8–10 m were left out. For these areas, which constitute around 30 % of the productive forest area in Norway, traditional methods based on photo interpretation and field visits were maintained. The traditional methods are costly and also not very precise, given the small amount of resources available for field work in most management inventories. The data requirement for such very young stands also differs a lot from older stands. The main decision in such stands is often related to tending operations. The most valuable information is considered to be height, stem numbers and tree species distribution – especially the amount of decisions trees, since conifer trees usually are considered more valuable for wood production and a high proportion of deciduous trees may indicate a need for tending. Although estimation of height and stem number in such very young stands were among the first topics that were addressed (Næsset and Bjerknes 2001), little research has been conducted during the last decade. There is obviously a potential for extraction of relevant information also for the very young forest (see Chap. 12), especially by fusing ALS data with digital aerial images for tree species classification.

The scientific studies conducted in Norway from the very beginning in 1995 and until today have taken place in the southeastern part of the country with typical boreal forest conditions and a continental climate which is very similar to the vast tracts of managed forests in Finland and Sweden. This has facilitated easy and relevant comparison between studies conducted in the different countries. Most of the forest resources are also found in the southeastern part of the country. However, Norway's geography and climate with large and rapid changes in altitude and weather conditions, with steep gradients in terrain altitude, precipitation, temperature and other factors influencing the forest structure and therefore also the performance of forest inventory, suggest that more experience is needed on the performance of ALS-assisted forest inventories under other conditions than those considered so far.

There is a potential for entirely new information products which previously have not been part of the management inventories. A forest owner is obliged to maintain cultural remains predating the Reformation in 1537. Such remains are automatically protected by law. Information on cultural remains is thus required in planning of e.g. harvest operations. Promising results have been achieved for detection of such remains from ALS data in forested areas (Bollandsås et al. 2012; Risbøl et al. 2013). There is obviously a great potential for extraction of properties related to biodiversity and even assessment of habitat qualities for birds and mammals (see Chaps. 17 and 18). The forest biomass is becoming ever more important as a resource for renewable energy production while the forests also may become an important instrument for mitigation of climate change. The first calls for methods to accurately quantify biomass in the management plans in a similar way as the plans provide information about timber volumes and other biophysical properties. Promising results have been reported in this field (Hauglin et al. 2012). The latter points to development of new products and markets for carbon trade, by which carbon-offsets need to be quantified. Thus, estimation of change in biomass and carbon stocks becomes a necessity. Recent studies have shown great opportunities for ALS even in this area (Bollandsås et al. 2013a; Næsset et al. 2013).

As indicated above, the traditions and data requirements in management planning differ somewhat between countries. Current practice in Norway does not assume that biophysical stand properties like mean height, basal area and volume are estimated separately for each tree species. Instead, mean tree height is assumed to be the same for all species while total basal area and volume are distributed on species according to the tree species distribution derived by photo interpretation. Future studies may look closer at for example the Finnish practice to estimate species-wise information directly (see further details in Chap. 12).

Finally, it should be mentioned that extraction of information at an individual tree level has always been an intriguing research challenge. Although the ALS point density that is required to extract single-tree information in most cases is considered too expensive for operational use, previous efforts to improve algorithms for single-tree delineation (Solberg et al. 2006; Ene et al. 2012), to classify tree species (Ørka et al. 2009, 2012, 2013), to extract properties of the trees that can characterize the wood quality in the interior of the stems prior to harvest, and to find ways to mitigate the serious underestimation of tree heights and total stand volume (Breidenbach et al. 2010) will continue, using ALS data alone as well as in combination with optical data, such as multispectral (Ørka et al. 2012, 2013) and hyperspectral (Dalponte et al. 2013; Ørka et al. 2013) aerial images.

11.6 Conclusions

In retrospect, the development and introduction of ALS-based forest management inventory in Norway went surprisingly fast and the new procedures and products gained confidence and acceptance in the market during a period of only 2–3 years. Some of the "success factors" have been presented in this chapter. It is also relevant to point at the experience that has been gained later on through scientific studies that have tried to improve quality and reduce costs of the inventories. The technical results from these studies to a large degree suggest designs that the first operational inventories already followed. Thus, one of the reasons for the fast and successful implementation of the new inventory concept may be that they were quite well designed from the very beginning, despite lack of in-depth experience with ALS at that time. Although the practices that were established when the first inventories were conducted around 2002 have been somewhat modified over the years, there is obviously a potential for continuous improvements with positive effects on the products (improved accuracy, new types of information, lower costs). There is obviously some good opportunities for improvements through continued research but also by learning from practical experiences in other countries (see e.g. Chap. 12).

References

Andersen H-E, Breidenbach J (2007) Statistical properties of mean stand biomass estimators in a LIDAR-bases double sampling forest survey design. In: Proceedings of the ISPRS workshop laser scanning 2007 and SilviLaser 2007, Espoo, Finland, 12–14 September 2007. IAPRS, vol XXXVI, Part 3/W52, 2007, pp 8–13

Baltsavias EP (1999) Airborne laser scanning: existing systems and firms and other resources. ISPRS J Photogramm Remote Sens 54:164–198

Bitterlich W (1984) The relascope idea: relative measurements in forestry. Commonwealth Agricultural Bureaux, Slough

Bollandsås OM, Næsset E (2007) Estimating percentile-based diameter distributions in uneven-sized Norway spruce stands using airborne laser scanner data. Scand J For Res 22:33–47

Bollandsås OM, Maltamo M, Gobakken T, Lien V, Næsset E (2011) Prediction of timber quality parameters of forest stands by means of small footprint airborne laser scanner data. Int J For Eng 22:14–23

Bollandsås OM, Risbøl O, Ene LT, Nesbakken A, Gobakken T, Næsset E (2012) Using airborne small-footprint laser scanner data for detection of cultural remains in forests: an experimental study of the effects of pulse density and DTM smoothing. J Archaeol Sci 39:2733–2743

Bollandsås OM, Gregoire TG, Næsset E, Øyen B-H (2013a) Detection of biomass change in a Norwegian mountain forest area using small footprint airborne laser scanner data. Stat Methods Appl 22:113–129

Bollandsås OM, Maltamo M, Næsset E, Gobakken T (2013b) Comparing parametric and non-parametric modeling of diameter distributions on independent data using airborne laser scanning. Forestry 86:493–501

Breidenbach J, Næsset E, Lien V, Gobakken T, Solberg S (2010) Prediction of species specific forest inventory attributes using a nonparametric semi-individual tree crown approach based on fused airborne laser scanning and multispectral data. Remote Sens Environ 114:911–924

Dalponte M, Ørka HO, Gobakken T, Gianelle D, Næsset E (2013) Tree species classification in boreal forests with hyperspectral data. IEEE Trans Geosci Remote Sens 51:2632–2645

Deckert C, Bolstad PV (1996) Forest canopy, terrain, and distance effects on global positioning system point accuracy. Photogramm Eng Remote Sens 62:317–321

Eid T, Gobakken T, Næsset E (2004) Comparing stand inventories based on photo interpretation and laser scanning by means of cost-plus-loss analyses. Scand J For Res 19:512–523

Ene L, Næsset E, Gobakken T (2012) Single tree detection in heterogeneous boreal forests using airborne laser scanning and area based stem number estimates. Int J Remote Sens 33:5171–5519

Gobakken T, Næsset E (2004) Estimation of diameter and basal area distributions in coniferous forest by means of airborne laser scanner data. Scand J For Res 19:529–542

Gobakken T, Næsset E (2005) Weibull and percentile models for lidar-based estimation of basal area distribution. Scand J For Res 20:490–502

Gobakken T, Næsset E (2008) Assessing effects of laser point density, ground sampling intensity, and field sample plot size on biophysical stand properties derived from airborne laser scanner data. Can J For Res 38:1095–1109

Gobakken T, Næsset E (2009) Assessing effects of positioning errors and sample plot size in biophysical stand properties derived from airborne laser scanner data. Can J For Res 39:1036–1052

Hauglin M, Gobakken T, Lien V, Bollandsås OM, Næsset E (2012) Estimating potential logging residues in a boreal forest by airborne laser scanning. Biomass Bioenerg 36:356–365

Hollaus M, Wagner W, Eberhöfer C, Karel W (2006) Accuracy of large-scale canopy heights derived from LiDAR data under operational constraints in a complex alpine environment. ISPRS J Photogramm Remote Sens 60:323–338

Holmgren J (2004) Prediction of tree height, basal area and stem volume using airborne laser scanning. Scand J For Res 19:543–553

Holmgren J, Jonsson T (2004) Large scale airborne laser scanning of forest resources in Sweden. In: Thies M, Koch B, Spiecker H, Weinacker H (eds) Proceedings of the ISPRS working group VIII/2. Laser-scanners for forest and landscape assessment, Freiburg, Germany, 3–6 October, 2004. International Archives of Photogrammetry, Remote Sensing and Spatial Information Sciences, vol XXXVI, Part 8/W2. ISSN 1682–1750, pp 157–160

Hudak AT, Crookston NL, Evans JS, Falkowski MJ, Smith AMS, Gessler PE, Morgan P (2006) Regression modeling and mapping of coniferous forest basal area and tree density from discrete-return lidar and multispectral satellite data. Can J Remote Sens 32:126–138

Jensen JLR, Humes KS, Conner T, Williams CJ, DeGroot J (2006) Estimation of biophysical characteristics for highly variable mixed-conifer stands using small-footprint lidar. Can J For Res 36:1129–1138

Junttila V, Maltamo M, Kauranne T (2008) Sparse Bayesian estimation of forest stand characteristics from ALS. For Sci 54:543–552

Koch B, Straub C, Dees M, Wang Y, Weinacker H (2009) Airborne laser data for stand delineation and information extraction. Int J Remote Sens 30:935–963

Latifi H, Nothdurft A, Koch B (2010) Non-parametric prediction and mapping of standing timber volume and biomass in a temperate forest: application of multiple optical/LiDAR-derived predictors. Forestry 83:395–407

Lindgren O (2006) Validering av laserdata och flygbildstolkade data från Storådalen [Validation of laser data and photointerpreted data from Storådalen]. Internal report to Sveaskog AB, 11 pp (in Swedish)

Lindgren O (2012) Validation of stand-wise forest data based on ALS. In: Proceeding SilviLaser 2012, Vancouver, Canada, 16–19 September 2012, 8 pp

Magnussen S, Boudewyn P (1998) Derivations of stand heights from airborne laser scanner data with canopy-based quantile estimators. Can J For Res 28:1016–1031

Magnussen S, Næsset E, Gobakken T (2013) Prediction of tree-size distributions and inventory variables from cumulants of canopy height distributions. Forestry 86:583–595

Maltamo M, Eerikäinen K, Packalén P, Hyyppä J (2006a) Estimation of stem volume using laser scanning based canopy height metrics. Forestry 79:217–229

Maltamo M, Malinen J, Packalén P, Suvanto A, Kangas J (2006b) Nonparametric estimation of stem volume using airborne laser scanning, aerial photography, and stand-register data. Can J For Res 36:426–436

Maltamo M, Næsset E, Bollandsås OM, Gobakken T, Packalen P (2009) Non-parametric prediction of diameter distributions using airborne laser scanner data. Scand J For Res 24:541–553

Maltamo M, Bollandsås OM, Vauhkonen J, Breidenbach J, Gobakken T, Næsset E (2010) Comparing different methods for prediction of mean crown height in Norway spruce stands using airborne laser scanner data. Forestry 83:257–268

Maltamo M, Bollandsås OM, Næsset E, Gobakken T, Packalén P (2011) Different plot selection strategies for field training data in ALS-assisted forest inventory. Forestry 84:23–31

Mascaro J, Detto M, Asner GP, Muller-Landau HC (2011) Evaluating uncertainty in mapping forest carbon with airborne LiDAR. Remote Sens Environ 115:3770–3774

Næsset E (1997a) Determination of mean tree height of forest stands using airborne laser scanner data. ISPRS J Photogramm Remote Sens 52:49–56

Næsset E (1997b) Estimating timber volume of forest stands using airborne laser scanner data. Remote Sens Environ 51:246–253

Næsset E (1999) Point accuracy of combined pseudorange and carrier phase differential GPS under forest canopy. Can J For Res 29:547–553

Næsset E (2001a) Effects of differential single- and dual-frequency GPS and GLONASS observations on point accuracy under forest canopies. Photogramm Eng Remote Sens 67:1021–1026

Næsset E (2001b) Ressursregistrering med flybåren laser-scanner: snart virkelighet (Resource inventory with airborne laser scanner: soon a reality) (in Norwegian). Norsk Skogbruk 47(5):20–23

Næsset E (2002a) Determination of mean tree height of forest stands by means of digital photogrammetry. Scand J For Res 17:446–459

Næsset E (2002b) Predicting forest stand characteristics with airborne scanning laser using a practical two-stage procedure and field data. Remote Sens Environ 80:88–99

Næsset E (2004a) Practical large-scale forest stand inventory using a small airborne scanning laser. Scand J For Res 19:164–179

Næsset E (2004b) Accuracy of forest inventory using airborne laser-scanning: evaluating the first Nordic full-scale operational project. Scand J For Res 19:554–557

Næsset E (2004c) Effects of different flying altitudes on biophysical stand properties estimated from canopy height and density measured with a small-footprint airborne scanning laser. Remote Sens Environ 91:243–255

Næsset E (2005) Assessing sensor effects and effects of leaf-off and leaf-on canopy conditions on biophysical stand properties derived from small-footprint airborne laser data. Remote Sens Environ 98:356–370

Næsset E (2007) Airborne laser scanning as a method in operational forest inventory: status of accuracy assessments accomplished in Scandinavia. Scand J For Res 22:433–442

Næsset E (2009) Effects of different sensors, flying altitudes, and pulse repetition frequencies on forest canopy metrics and biophysical stand properties derived from small-footprint airborne laser data. Remote Sens Environ 113:148–159

Næsset E (2011) Estimating above-ground biomass in young forests with airborne laser scanning. Int J Remote Sens 32:473–501

Næsset E, Bjerknes K-O (2001) Estimating tree heights and number of stems in young forest stands using airborne laser scanner data. Remote Sens Environ 78:328–340

Næsset E, Gjevestad JG (2008) Performance of GPS precise point positioning under forest canopies. Photogramm Eng Remote Sens 74:661–668

Næsset E, Gobakken T (2008) Estimation of above- and below-ground biomass across regions of the boreal forest zone using airborne laser. Remote Sens Environ 112:3079–3090

Næsset E, Jonmeister T (2002) Assessing point accuracy of DGPS under forest canopy before data acquisition, in the field and after postprocessing. Scand J For Res 17:351–358

Næsset E, Bjerke T, Øvstedal O, Ryan LH (2000) Contributions of differential GPS and GLONASS observations to point accuracy under forest canopies. Photogramm Eng Remote Sens 66:403–407

Næsset E, Bollandsås OM, Gobakken T (2005) Comparing regression methods in estimation of biophysical properties of forest stands from two different inventories using laser scanner data. Remote Sens Environ 94:541–553

Næsset E, Gobakken T, Holmgren J, Hyyppä H, Hyypää J, Maltamo M, Nilsson M, Olsson H, Persson Å, Söderman U (2004) Laser scanning of forest resources: the Nordic experience. Scand J For Res 19:482–499

Næsset E, Gobakken T, Solberg S, Gregoire TG, Nelson R, Ståhl G, Weydahl D (2011) Model-assisted regional forest biomass estimation using LiDAR and InSAR as auxiliary data: a case study from a boreal forest area. Remote Sens Environ 115:3599–3614

Næsset E, Bollandsås OM, Gobakken T, Gregoire TG, Ståhl G (2013) Model-assisted estimation of change in forest biomass over an 11 year period in a sample survey supported by airborne LiDAR: a case study with post-stratification to provide "activity data". Remote Sens Environ 128:299–314

Ørka HO, Næsset E, Bollandsås OM (2009) Classifying species of individual trees by intensity and structure features derived from airborne laser scanner data. Remote Sens Environ 113:1163–1174

Ørka HO, Gobakken T, Næsset E, Ene L, Lien V (2012) Simultaneously acquired airborne laser scanning and multispectral imagery for individual tree species identification. Can J Remote Sens 38:125–138

Ørka HO, Dalponte M, Gobakken T, Næsset E, Ene L (2013) Characterizing forest species composition using multiple remote sensing data sources and inventory approaches. Scand J For Res 28:677–688

Risbøl O, Bollandsås OM, Nesbakken A, Ørka HO, Næsset E, Gobakken T (2013) Interpreting cultural remains in airborne laser scanning generated digital terrain models: effects of size and shape on detection success rates. J Archaeol Sci 40:4688–4700

Rombouts J, Ferguson IS, Leech JW (2008) Variability of LiDAR volume prediction models for productivity assessment of radiata pine plantations in South Australia. In: Hill R, Rosette J, Suárez J (eds) Proceedings of SilviLaser 2008, 8th international conference on LiDAR applications in forest assessment and inventory, 17–19 September 2008. Heriot-Watt University, Edinburgh, UK, pp 39–49

Särndal C-E, Swensson B, Wretman J (1992) Model assisted survey sampling. Springer, New York

Solberg S, Næsset E, Bollandsås OM (2006) Single tree segmentation using airborne laser scanner data in a structurally heterogeneous spruce forest. Photogramm Eng Remote Sens 72:1369–1378

Thomas V, Treitz P, McCaughey JH, Morrison I (2006) Mapping stand-level forest biophysical variables for a mixedwood boreal forest using lidar: an examination of scanning density. Can J For Res 36:34–47

Turunen L, Pesonen A, Suvanto A (2012) Fjernanalysebasert skogregistrering i Finland (Remote sensing based forest inventory in Finland). Kart og Plan 72:184–187

Ussyshkin V, Theriault L (2010) ALTM ORION: bridging conventional lidar and full waveform digitizer technology. In: Wagner W, Székely B (eds) ISPRS TC VII symposium – 100 years ISPRS, Vienna, Austria, 5–7 July 2010, IAPRS, vol XXXVIII, Part 7B, 6 pp

Wagner W, Ullrich A, Melzer T, Briese C, Kraus K (2004) From single pulse to full-waveform airborne laser scanners: potential and practical challenges. In: Proceedings of ISPRS XXth congress, Vol XXXV, part B/3, Istanbul, Turkey, 12–23 July 2004, pp 201–206

Chapter 12
Species-Specific Management Inventory in Finland

Matti Maltamo and Petteri Packalen

Abstract A new remote sensing based stand management inventory system was developed and adopted to operational forestry in Finland during the years 2005–2010. The inventory is based on wall-to-wall mapping of the inventory area. The outcome of the inventory is species-specific stand attributes which are estimated with the help of ALS, aerial images and field sample plots. The new inventory system has been successful and within a few years all the actors of the practical forestry have updated their inventory and planning systems to support the new method. The new inventory system is now applied for almost 3,000,000 ha annually. This chapter presents the main properties of the system.

12.1 Introduction

In Finland two main types of forest inventories are (1) the sampling based National Forest Inventory (NFI) primarily serving national and regional purposes and (2) the inventory by compartments for stand level management. In this context, a stand is considered the smallest management unit. When NFI data are combined with remote sensing data and other auxiliary information in multi-source NFI, reliable estimates for small areas are also obtained (Tomppo 2006). However, the areas considered in multi-source NFI are still considerably larger than a typical stand used in operational forest management planning and the information needs for stand level management are also different. From the perspective of practical forestry the accuracy requirement for the stand level inventory is about 15–30 % RMSE for stand volume (Uuttera et al. 2002).

M. Maltamo (✉) • P. Packalen
School of Forest Sciences, University of Eastern Finland, Joensuu, Finland
e-mail: Matti.maltamo@uef.fi

M. Maltamo et al. (eds.), *Forestry Applications of Airborne Laser Scanning: Concepts and Case Studies*, Managing Forest Ecosystems 27, DOI 10.1007/978-94-017-8663-8__12,
© Springer Science+Business Media Dordrecht 2014

Traditionally, the information for stand level management was collected with a stand-wise field inventory method, that is, inventory by compartments, in which species-specific forest attributes were estimated using subjective angle count sampling and partly visual assessment (e.g. Koivuniemi and Korhonen 2006). This method included stand delineation from aerial photographs, field visits to each stand and calculation of stand attributes of interest, mainly volume by tree species and timber assortments. The stand attributes assessed in the field included age, basal area, mean diameter and height. These data were used in forest planning, for example to determine the need for silvicultural treatments. In practical forestry the data acquisition costs were about 10 euros per hectare and the costs of the whole forest planning process were over 20 euros. However, in private forestry this process is highly subsidized by the state, which covers about 70 % of the costs. Annually this method was applied to over 1,000,000 million hectares of private forests and additionally also to considerable areas of state forests and forests owned by large companies.

Inventory by compartments has been applied since the 1950s, which means that practically the whole country has been inventoried several times. During the last decades the development was related to field measurements and calculation routines (e.g. Kilkki and Päivinen 1986; Kangas et al. 2004). During the last 15 years there had, however, been a strong emphasis on modernizing this methodology completely. The main reasons for this development were the high costs, subjectivity and inaccuracy of the existing method. In the 1990s there was already a lot of research concerning the development towards remote sensing applications (e.g. Päivinen et al. 1993; Varjo 2002). However, the accuracy demands and usability for operational purposes were not fulfilled by available remote sensing materials, i.e. by satellite and aerial images. Usually, the RMSE of the total volume at stand level exceeded 30 % (Hyyppä et al. 2000; Uuttera et al. 2006). It was also difficult to separate tree species and the heterogeneity between images was an issue.

The usability of remote sensing data in forest management inventory changed considerably when airborne laser scanning (ALS) data became available. The first studies conducted in Norway indicated both good accuracy and good usability from an operational point of view (Næsset 1997, 2004). The method developed in Norway is based on height and density variables calculated from low density ALS data over a certain area and is, therefore, called an area-based approach (see Næsset 2004, Chapter 11). In Finland, the research on ALS data related to forest applications began with the single tree detection approach, in which individual trees were recognized from the canopy height model constructed from high density ALS data (Hyyppä and Inkinen 1999). The accuracy of this approach was also promising in the first studies (Hyyppä and Inkinen 1999; Maltamo et al. 2004) but high data acquisition costs, a lack of algorithms to detect tree species and the feature of exhibiting downward bias of stand sum attributes (e.g. total stand volume) have so far restricted the development towards an operational use.

As in other Nordic countries (Næsset 2002; Holmgren 2004) and later elsewhere (Hudak et al. 2006; Jensen et al. 2006; Hollaus et al. 2007; Rombouts et al. 2008; Latifi et al. 2010), the area-based approach was tested experimentally in

Finland starting in 2004 (Suvanto et al. 2005; Maltamo et al. 2006). The accuracy obtained for stand total volume was superior compared to the previous field-based inventory method (Haara and Korhonen 2004) or other remote sensing based inventory methods (see e.g. Uuttera et al. 2006). However, the requirement for species-specific stand attributes was not fulfilled. As a means to provide species-specific information, Packalén and Maltamo (2006, 2007, 2008) combined ALS data with aerial images and applied non-parametric nearest-neighbor imputation to simultaneously estimate several stand attributes. This approach has now been adopted as a basis for new stand level management inventories in Finland. In the following we will go through the basis and implementation of the method.

12.2 Airborne Laser Scanning Based Stand Level Management Inventory System in Finland

12.2.1 Field and Aerial Data

Accurate and georeferenced field sample plot measurements are a keystone in the new stand level management inventory (see e.g. Gobakken and Næsset 2009). Georeferencing enables the extraction of the aerial data from exactly the same location as where the field measurements were carried out. The ALS-based inventory concerns sapling, young, maturing and mature forests but quite often seedling stands (height <1.3 m) have been left out. The sample plot measurements should represent the variation within the inventory area and normally the number of field sample plots within an inventory area is about 500 plots in young, maturing and mature stands and 150 in sapling stands. In nearest neighbor (NN) methods the limitation is that it is impossible to extrapolate. If the data in the sample do not cover the entire range of values in the population, then the lowest values of any distribution will be overestimated, while the highest values will be underestimated. The multivariate modelling (i.e. simultaneously predicting the required forest attributes) makes this even more complex since some of the attributes are rare. To capture the true variation within the forest area, the placement of the sample plots should be considered carefully. Usually, the field sample plot data do not constitute a probability sample in Finland.

The low density ALS data that commonly are used in forest inventory usually consist of less than one transmitted laser pulse per square meter. Mainly for economical reasons, modern Leica and Optech laser scanners and fixed wing aircrafts are used in data collection. The National Land Survey of Finland is using ALS data for national DTM production (Ahokas et al. 2008). The technical requirements for the ALS data in national terrain modelling are quite similar to the requirements in forest inventory but in terrain modelling leaf-off data are preferred. In forest inventory both leaf-off and leaf-on ALS data are used but they cannot be mixed in the same project (Villikka et al. 2012). Leaf-off data is collected in April or

May immediately when the snow has melted. It is often difficult to cover the whole inventory area by leaf-off data because the narrow time window between the snow has melted and bud breaks. Leaf-off data is collectively used in forest inventories and in national terrain modeling in order to share costs and efforts (Ahokas et al. 2008; Villikka et al. 2012).

Aerial images are captured with large format digital aerial cameras such as Vexcel UltraCam or Z/I DMC. The Leica ADS pushbroom scanner is rarely used. Often multispectral bands without pan-sharpening are used in the prediction of forest attributes whereas in visual stand delineation pan-sharpened orthophotos are common. The ground sample distance of the panchromatic band is usually about 35 cm and the images are captured during the leaf-on season from mid-June till the end of August. The aim is to use only one ALS and one imaging instrument in one inventory area but every now and then this requirement has to be relaxed. In the best case ALS data, aerial images and field measurements are all collected in the same year but often there is a discrepancy of 1 year between the acquisitions.

12.2.2   Estimation Methods

To predict species-specific stand variables, Packalén and Maltamo (2006, 2007, 2008) combined ALS data with aerial images. As in other variants of the area-based method, predictor variables are those calculated from height and density distributions of low density ALS data. Additionally, the spectral statistics and the texture metrics of the aerial images are utilized in order to improve the separation of the tree species. The fusion of ALS data and aerial images was further developed in Packalén et al. (2009) in order to improve the accuracy of species-specific predictions. The response variables are the most essential stand sum and mean attributes, namely volume, basal area, number of stems, mean diameter and mean height estimated separately for Scots pine, Norway spruce and deciduous species as a species group.

The most common method to predict stand attributes is NN imputation (Packalén and Maltamo 2007) but Bayesian regression is also used (Junttila et al. 2008). In NN imputation species-specific stand attributes, and as a sum of these the total attributes, are predicted simultaneously using a combination of ALS and image predictors. Simultaneous imputation of several response variables guarantees more logical results. In general the predictions are accurate for stand totals and also for main tree species whereas the accuracy is worse for minor tree species. The chosen NN approach also allows the prediction of species-specific diameter distributions that are consistent with stand attributes (Packalén and Maltamo 2008). NN based diameter distribution are imputed from tree diameter measurements of sample plots, thus, they describe the local variability in an inventory area. Alternatively in Finland, it is possible to utilize predicted species-specific stand attributes and existing parameter models of theoretical diameter distribution functions (see Chap. 9 for more details).

Fig. 12.1 An example of a grid with cell size of 16 × 16 m overlaid with stand boundaries (Modified from Maltamo et al. 2009. The predicted volume (m³ha⁻¹) is depicted by *gray level*)

A specific feature to take into consideration is that seedling and sapling stands cover usually about 25 % of an inventory area. Considerable cost savings can be obtained if even a part of the sapling stands could be inventoried reliably by remote sensing. A lot of field checks are made in sapling stands which is a bottleneck of the current system. The information need is different for saplings. Instead of species-specific stand attributes it is more important to know the timing of the next silvicultural treatment. Therefore, in sapling stands the need for the next silvicultural treatment is predicted directly from ALS metrics (Korhonen et al. 2013). Another special application which is quite common is to predict the need for the first thinning in young stands but sometimes the need for thinning is predicted directly in maturing stands too. The site fertility class based on Cajander's (1909) theory on forest site types has also been predicted in some projects.

12.2.3   *Stand Level Attributes*

When applying the constructed model the stand attributes are predicted in a wall-to-wall manner in the whole inventory area. The predictor variables are calculated for the field sample plots and for grid cells with a size of about 16 × 16 m, which is laid over the inventory area (Fig. 12.1). Thus, the cells of the grid are used as units, for which the forest attributes are predicted by means of aerial data. The use of a grid corresponds with the Norwegian application (Næsset 2004) and is more or less similar to how systematic approaches have been proposed for remote sensing applications in general for stand level inventories (e.g. Poso 1994). Stand level predictions are aggregated from the grid cells that fall inside stand boundaries entirely or at least partly. There are several practices to avoid the effect of mixed cells in stand borders but often about 70 % of the cell area must be located inside the stand before the cell is used when computing stand level attributes. Alternatively grid cells may be intersected by stand borders and then the formed small cell parts are merged to the neighboring cells within the stand. This is done to avoid the effect of mixed cells along stand borders.

Semi-automated stand delineation is also an established part of an inventory. Stand delineation is done by means of segmentation in which ALS-based canopy height models and aerial images are used (Mustonen et al. 2008). Instead of creating the final stand delineation automatically by segmentation, the segmentation is usually used to create so-called micro-stands. Micro-stands are homogenous patches from which final stands used in silviculture are composed. ALS-based canopy height models are also used in manual stand delineation in parallel with orthorectified aerial images.

12.2.4   Reliability

The reliability of stand attribute models can be tested in many phases. When constructing models, the goodness of fit of the developed models is assessed. Predictions are assessed by means of RMSE and bias for continuous stand attributes and overall accuracy and kappa rate for categorical variables (e.g. treatment needs). When applying separate validation data the validation is usually done at the stand level, either by aggregating several validation plots within a stand or at least applying larger plots than in the case of modeling. This variation in validation plot sizes as well as variation in tree stocks makes the comparison of reliability figures between different studies difficult. However, the RMSE figures obtained in Finland seem to correspond very well with those reported in other studies conducted in the boreal region, both for stand totals (e.g. Næsset 2007) and for species-specific attributes (Breidenbach et al. 2010).

Service providers always give some kind of quality report of the inventory project comprising accuracies of the sample plot data. Every now and then end users also make their own field control by laying a systematic network of 5–20 field sample plots for each stand selected for the validation. Mainly circular sample plots are used but if the interest is only related to volume also angle count plots can be used. The choice of the control stands can be random or stratified according to, for example, tree species, site fertility, stand development stage or silvicultural treatment needs. In general, the reliability has been good in the eastern and northern parts of the country where pine dominates whereas in south in the mixed and fertile stands the reliability of some species-specific stand properties has been worse.

The results of one control inventory were reported by Wallenius et al. (2012). The study area was located in eastern Finland and the control was done at micro–stand level locating 5–14 sample plots in each chosen micro-stand. The average size of the micro-stands was 0.74 ha. The segment-level RMSEs for volume attributes were 33.1, 63.3, 69.4 and 14.5 % for pine, spruce, deciduous tree species and stand totals, respectively. These figures can be regarded as very good for completely independent field control. Correspondingly, there are some results concerning the accuracy of direct classification of sapling stands according to treatment needs (Korhonen et al. 2013). In independent stand level validation data kappa values of the classification were about 0.4 and overall accuracy about 70 %.

12.2.5 Practical Arrangements

Since scientific studies have shown that the species-specific accuracy obtained with the ALS based inventory is comparable with the traditional field inventory method (Haara and Korhonen 2004; Maltamo et al. 2009; Packalén and Maltamo 2007), the actors in the Finnish forest sector were willing to adopt the new and less fieldwork-intensive inventory procedure in the hope of reduced costs. In the case of private forestry it is assumed that a 60 % cost saving would occur when compared to conventional field inventory. The scientific basis for the new system was developed in the research projects where the actors of practical forestry were also involved.

In Finnish forestry there is no tradition of using remote sensing elsewhere than in the visual stand delineation. Therefore, the transition from the old inventory system to the new one happened surprisingly fast and smoothly. Normally there is a very conservative attitude towards changes. The causes of the smooth and rapid adoption of the new methodology might be twofold. Firstly, there was a long and profound history of finding new and more effective inventory solutions. Secondly, there was a close co-operation between researchers and actors of the practical forestry sector when the new inventory system was developed.

In practical applications, processing of the ALS data and aerial images into the predictor variables, stand attribute modelling and wall-to-wall inventory are done by service providers but the process differs between the actors. In the following the practices of three different actors are presented.

The forest company UPM Kymmene Oyj started its pilot projects in 2004 and the first fully operational project was in 2008. Currently about 85 % of its forest area in Finland has been inventoried by using the new method. In UPM Kymmene Oyj micro-stands are applied when planning silvicultural treatments. The service provider also takes care of the field data collection. The situation is almost the same for the state forests managed by Metsähallitus. ALS-based forest inventory is already in the operational stage in forest planning and so far about 2,000,000 ha have been inventoried. Metsähallitus is also using ALS data for planning cutting of the marked stands and assessing the need for ditch cleaning.

In the private forests the aim is to inventory annually about 1.5 million hectares with the new method. This means that all the private forests will be inventoried during less than one decade. These inventories are organized collectively by the Finnish Forest Centre which was established in 2012 by merging the Regional Forestry Centres. After several pilot projects in 2006–2009 the first operational inventory projects were implemented in 2010. In the case of privately owned forests, the stand attributes are usually further processed to holding-specific forest plans and treatment schedules. The forest planning process takes 2 years. During the first year data acquisition and data processing are accomplished. This is followed by the interpretation of stand attributes during winter. In the summer of the second year there are field checks and after that construction of a forest plan. The results of the inventory are also used for guidance of forest owners (so-called Metsaan.fi internet service).

The difference between other actors and private forestry is that the Finnish Forest Centre measures field sample plots by themselves. The field measurements are carried out independently in each inventory area of about 100,000–500,000 ha in size and the intention is to measure about 500–800 circular sample plots. The forest planning experts who previously assessed stand level forest attributes by means of angle count sampling now measure georeferenced field sample plots. They also concentrate more on planning, guidance and service for forest owners. Besides the field sample plot measurements, fieldwork is still needed for checking species proportions and the treatment proposals for seedling stands and saplings that cannot be estimated accurately enough with the new method. There is also a need for recording of some categorical attributes, such as site fertility class and soil class, and age and biodiversity attributes which must be taken from the old stand register data or field visit might be required.

In private forests the sample plot placement mimics the Finnish NFI design with subjectively allocated additional measurements. The stratification is performed on the basis of the old stand register data in the inventory area. The strata are forest site type, dominating tree species, basal area and mean diameter. This design aims to obtain a good non-probability sample of the forest of the inventory area that includes the true variations and also the extremes of all the variable distributions.

The ALS grid cells location is fixed in inventories organized by the Finnish Forest Centre. Service providers provide only cell level attributes to the Finnish Forestry Centre. Cell level attributes are held in a database independently from the stand delineation. This approach has certain advantages, such as easy recalculation of stand level attributes when the stand delineation is changed and the possibility to maintain coherent temporal database.

12.3   Discussion

The new inventory system has been successful in Finland. During just a few years almost all actors of the practical forestry have modified their practices and information systems to support the new inventory system. The new inventory methodology is applied to almost 3,000,000 ha annually. Nowadays the area-based approach is applied but due to the possibility to use diameter distribution models the characterization of tree stock can be transformed to tree level. Although this inventory system is still quite new there is already a lot of research work related to the development of this system or completely new systems.

In relation to the current inventory system, one of the most serious drawbacks is the separation of tree species. Although the main tree species is usually predicted well, the error of the minor tree species can be very high. It is a severe error for a holding-specific forest plan if the main tree species of a stand is not correctly predicted. Another serious issue is the ability to detect needs for silvicultural operations, especially in saplings and young stands.

There are still some shortcomings in the operational process. For example, co-operation between actors (e.g. when remote sensing data are collected and field measurements are done by different actors), rapid changes in established practices and different local conditions in different parts of the country cause difficulties in operational use. The compatibility of information systems used in forest planning with the new inventory data must also be improved. Weather restrictions are common too. Aerial data cannot always be acquired within a short time period using just one aircraft and a single instrument. That is why a stratification of an inventory area by aerial data sets is quite often required. The drawback of this stratification is that increased number of sample plots is needed which means increased costs.

Besides the basic work in the development of the new inventory system many other issues have been piloted as well. Since the NFI provides a systematic network of field plots across the country, the question of whether these data could be utilized in stand level management inventory was studied by Maltamo et al. (2009). It must be remembered that NFI uses angle count plots, which means that perfect correspondence with remote sensing data is impossible. According to the results by Maltamo et al. (2009) the effect of using angle count samples on accuracy is, however, minor but the current georeferencing of the plots is not adequate and also the sampling design is not optimal for stand level inventory purposes. NFI and remote sensing based management inventory also require very different kinds of tree and plot measurements. Thus, from a practical point of view it may be impractical to unite these field measurements. However, due to great potentials for cost savings there is an interest in using the same field sample plots in both inventories.

The usability of national terrain modelling based leaf-off ALS data in stand level forest inventory was examined by Villikka et al. (2012). The overall conclusion was that leaf-off ALS data are suitable for area-based forest inventory in which deciduous and coniferous trees need to be separated. However, the narrow time window when the leaf-off ALS data can be collected restricts the applicability. Correspondingly, in DTM production the laser scanner used may change during one campaign, but in forest inventory applications a scanner should not be changed (Næsset 2009). In general, the results of Villikka et al. (2012) were better with leaf-off than leaf-on data, a result which is also supported by similar findings in Norway (Næsset 2005). Leaf-off ALS data per se also had the ability to discriminate between deciduous and coniferous trees (Villikka et al. 2012). This may decrease the inventory costs if the acquisition of aerial images is, therefore, avoided entirely. The inventory costs are also decreased if the same data can be used in forest inventory and DTM production. Nowadays leaf-off ALS data are already widely used in projects of private forests. So far about 1,000,000 ha have been inventoried using leaf-off data and the annual amount will increase.

To sum up, a new remote sensing based inventory system was developed and adopted to operational forestry in Finland during the period 2005–2010. A continuous and further development is ongoing At the moment there is a lot of research under way on how to improve species discrimination, e.g. by improved algorithms, utilizing old stand register data or using new data sources. Other notable issues under development are the detection of young stands that should be thinned immediately, the accuracy of the inventory in sapling stands, minimum diameter limits in the field measurements, the characterization of multilayered forests as well as information on biodiversity aspects. The second remote sensing based inventory cycle will also start soon and it has not been decided whether the method should be the same or revised. Any major changes of the inventory system must be motivated and justified by potentials for cost savings and considerable improvement of accuracy.

References

Ahokas E, Kaartinen H, Hyyppä J (2008) On the quality checking of the airborne laser scanning-based nationwide elevation model in Finland. 21st ISPRS Congress Beijing 2008. Int Arch Photogramm Remote Sens Spat Inf Sci 37(B1/I):267–270

Breidenbach J, Næsset E, Lien V, Gobakken T, Solberg S (2010) Prediction of species specific forest inventory attributes using a nonparametric semi-individual tree crown approach based on fused airborne laser scanning and multispectral data. Remote Sens Environ 114:911–924

Cajander AK (1909) Ueber die Waldtypen. Acta For Fenn 1:1–175

Gobakken T, Næsset E (2009) Assessing effects of positioning errors and sample plot size on biophysical stand properties derived from airborne laser scanner data. Can J For Res 39:1036–1052

Haara A, Korhonen KT (2004) Kuvioittaisen arvioinnin luotettavuus. Metsätieteen aikakauskirja 4/2004:489–508 (in Finnish)

Hollaus M, Wagner W, Maier B, Schadauer K (2007) Airborne laser scanning of forest stem volume in a mountainous environment. Sensors 7:1559–1577

Holmgren J (2004) Prediction of tree height, basal area and stem volume using airborne laser scanning. Scand J For Res 19:543–553

Hudak AT, Crookston NL, Evans JS, Falkowski MJ, Smith AMS, Gessler PE, Morgan P (2006) Regression modeling and mapping of coniferous forest basal area and tree density from discrete-return lidar and multispectral satellite data. Can J Remote Sens 32:126–138

Hyyppä J, Inkinen M (1999) Detecting and estimating attributes for single trees using laser scanner. Photogramm J Finl 16:27–42

Hyyppä J, Hyyppä H, Inkinen M, Engdahl M, Linko S, Zhu YH (2000) Accuracy comparison of various remote sensing data sources in the retrieval of forest stand attributes. For Ecol Manage 128:109–120

Jensen JLR, Humes KS, Conner T, Williams CJ, DeGroot J (2006) Estimation of biophysical characteristics for highly variable mixed-conifer stands using small-footprint lidar. Can J For Res 36:1129–1138

Junttila V, Maltamo M, Kauranne T (2008) Sparse Bayesian estimation of forest stand characteristics from ALS. For Sci 54:543–552

Kangas A, Heikkinen E, Maltamo M (2004) Accuracy of partially visually assessed stand characteristics – a case study of Finnish forest inventory by compartments. Can J For Res 34:916–930

Kilkki P, Päivinen R (1986) Weibull function in the estimation of the basal area DBH-distribution. Silva Fenn 20:149–156

Koivuniemi J, Korhonen KT (2006) Inventory by compartments. In: Kangas A, Maltamo M (eds) Forest inventory. Methodology and applications, vol 10, Managing forest ecosystems. Springer, Dordrecht

Korhonen L, Pippuri I, Packalen P, Heikkinen V, Maltamo M, Heikkilä J (2013) Detection of the need for seedling stand tending using high-resolution remote sensing data. Silva Fenn 47. Article 952. http://dx.doi.org/10.14214/sf.952

Latifi H, Nothdurft A, Koch B (2010) Non-parametric prediction and mapping of standing timber volume and biomass in a temperate forest: application of multiple optical/LiDAR-derived predictors. Forestry 83:395–407

Maltamo M, Eerikäinen K, Pitkänen J, Hyyppä J, Vehmas M (2004) Estimation of timber volume and stem density based on scanning laser altimetry and expected tree size distribution functions. Remote Sens Environ 90:319–330

Maltamo M, Eerikäinen K, Packalén P, Hyyppä J (2006) Estimation of stem volume using laser scanning based canopy height metrics. Forestry 79:217–229

Maltamo M, Packalén P, Suvanto A, Korhonen KT, Mehtätalo L, Hyvönen P (2009) Combining ALS and NFI training data for forest management planning – a case study in Kuortane, Western Finland. Eur J For Res 128:305–317

Mustonen J, Packalén P, Kangas A (2008) Automatic segmentation of forest stands using canopy height model and aerial photography. Scand J For Res 23:534–545

Næsset E (1997) Estimating timber volume of forest stands using airborne laser scanner data. Remote Sens Environ 51:246–253

Næsset E (2002) Predicting forest stand characteristics with airborne scanning laser using a practical two-stage procedure and field data. Remote Sens Environ 80:88–99

Næsset E (2004) Practical large-scale forest stand inventory using a small airborne scanning laser. Scand J For Res 19:164–179

Næsset E (2005) Assessing sensor effects and effects of leaf-off and leaf-on canopy conditions on biophysical stand properties derived from small-footprint airborne laser data. Remote Sens Environ 98:356–370

Næsset E (2007) Airborne laser scanning as a method in operational forest inventory: status of accuracy assessments accomplished in Scandinavia. Scand J For Res 22:433–442

Næsset E (2009) Effects of different sensors, flying altitudes, and pulse repetition frequencies on forest canopy metrics and biophysical stand properties derived from small-footprint airborne laser data. Remote Sens Environ 113:148–159

Packalén P, Maltamo M (2006) Predicting the volume by tree species using airborne laser scanning and aerial photographs. For Sci 52:611–622

Packalén P, Maltamo M (2007) The k-MSN method in the prediction of species specific stand attributes using airborne laser scanning and aerial photographs. Remote Sens Environ 109:328–341

Packalén P, Maltamo M (2008) The estimation of species-specific diameter distributions using airborne laser scanning and aerial photographs. Can J For Res 38:1750–1760

Packalén P, Suvanto A, Maltamo M (2009) A two stage method to estimate species-specific growing stock by combining ALS data and aerial photographs of known orientation parameters. Photogramm Eng Remote Sens 75:1451–1460

Päivinen R, Pussinen R, Tomppo E (1993) Assessment of boreal forest stands using field assessment and remote sensing. In: Operalization of remote sensing. Proceedings of Earsel 1993 Conference, 19–23 April 1993. ITC Enshedene, the Netherlands, 8 p

Poso S (1994) Metsätalouden suunnittelu uusiin puihin. Voidaanko silmävaraisesta kuvioittaisesta arvioinnista luopua? Metsätieteeen Aikakauskirja 1/1994:85–89 (in Finnish)

Rombouts J, Ferguson IS, Leech JW (2008) Variability of LiDAR volume prediction models for productivity assessment of radiata pine plantations in South Australia. In: Hill R, Rosette J, Suárez J (eds) Proceedings of SilviLaser 2008, 8th international conference on LiDAR applications in forest assessment and inventory, 17–19 September 2008. Heriot-Watt University, Edinburgh, UK, pp 39–49

Suvanto A, Maltamo M, Packalén P, Kangas J (2005) Kuviokohtaisten puustotunnusten ennustaminen laserkeilauksella. Metsätieteen aikakauskirja 4/2005:413–428 (in Finnish)

Tomppo E (2006) The Finnish multi-source National Forest Inventory – small area estimation and map production. In: Kangas A, Maltamo M (eds) Forest inventory. Methodology and applications, vol 10, Managing forest ecosystems. Springer, Dordrecht

Uuttera J, Hiltunen J, Rissanen P, Anttila P, Hyvönen P (2002) Uudet kuvioittaisen arvioinnin menetelmät – arvio soveltuvuudesta yksityismaiden metsäsuunnitteluun. Metsätieteen aikakauskirja 3/2002:523–531 (in Finnish)

Uuttera J, Anttila P, Suvanto A, Maltamo M (2006) Yksityismetsien metsävaratiedon keruuseen soveltuvilla kaukokartoitusmenetelmillä estimoitujen puustotunnusten luotettavuus. Metsätieteen aikakauskirja 4/2006:507–519 (in Finnish)

Varjo J (2002) Metsäsuunnittelun tietohuollon järjestäminen tulevaisuudessa. Metsätieteen aikakauskirja 3/2002:537–540 (in Finnish)

Villikka M, Packalén P, Maltamo M (2012) The suitability of leaf-off airborne laser scanner data for forest inventory. Silva Fenn 46:99–110

Wallenius T, Laamanen R, Peuhkurinen J, Mehtätalo L, Kangas A (2012) Analysing the agreement between an airborne laser scanning based forest inventory and a control inventory – a case study in the state owned forests in Finland. Silva Fenn 46:111–129

Chapter 13
Inventory of Forest Plantations

Jari Vauhkonen, Jan Rombouts, and Matti Maltamo

Abstract We review the research related to the application of inventory techniques utilizing airborne laser scanning (ALS) in intensively managed forest plantations. The single tree detection and area based methods originally developed in more complex forest types were found to be no less effective in plantation forests for the estimation of important stand parameters such as stocking, height, basal area, volume and leaf area. Further accuracy gains have been achieved by leveraging ancillary information commonly available in plantation databases, such as plantation age, planting distance and genetics of planting stock. A case study is presented describing a site productivity inventory system implemented in South Australian *Pinus radiata* plantations. It demonstrates that ALS based solutions can be effectively introduced in existing resource information systems without disrupting information continuity and consistency, while achieving accuracy and cost efficiencies. It is concluded that airborne (and terrestrial) laser scanning based inventories are an effective option for forest plantations. The spatially explicit nature of ALS based information can support intensive site specific management and can assist in optimizing recovery of the value inherent in trees and stands.

J. Vauhkonen (✉)
Department of Forest Sciences, University of Helsinki, Helsinki, Finland
e-mail: jari.vauhkonen@helsinki.fi

J. Rombouts
ForestrySA, Mt Gambier, Australia

M. Maltamo
School of Forest Sciences, University of Eastern Finland, Joensuu, Finland
e-mail: Matti.maltamo@uef.fi

M. Maltamo et al. (eds.), *Forestry Applications of Airborne Laser Scanning: Concepts and Case Studies*, Managing Forest Ecosystems 27, DOI 10.1007/978-94-017-8663-8__13,
© Springer Science+Business Media Dordrecht 2014

13.1 Introduction

An increasing part of the world's forests is managed in the form of plantations, and forest plantations have become more significant in terms of timber supply, but also as providers of other non-wood forest products (Evans 2009). Like natural and semi-natural forests, plantations can serve multiple purposes including wood production, carbon sequestration, watershed or soil management, recreation and as a habitat for wildlife, but often there is a difference in the intensity and objectives of silviculture and management practices in planted forests (Carle and Holmgren 2003).

This chapter focuses on intensively managed, mostly single species plantations, grown in short to mid length rotations (6–35 years) with the primary objective of producing wood products. The planted species may be introduced or indigenous. The planting stock may be genetically improved for desired traits such as wood quality, volume growth and resistance to pests and diseases. Specific genotypes may be selected to produce the best outcome on a specific site. Trees are planted in rows with a fixed spacing, often in compartments with a regular shape. Silvicultural practice may include fertilization, selective thinning and pruning. Plantation forestry of this type leads to uniformity of end product (Savill et al. 1997).

Thus far the majority of research into application of ALS in forest assessment has occurred in multi-species, structurally more complex (semi-)natural forests. Evans et al. (2001) cautioned that the regular structure of plantation forests, in particular the occurrence of trees in parallel rows, may have implications for the performance of ALS based forest assessment. There is however considerable evidence that the basic ALS inventory techniques (see Chap. 1) are also applicable to plantations (e.g. Popescu et al. 2002, 2003; Koukoulas and Blackburn 2005; Nelson et al. 2007). Furthermore, similar conclusions regarding, for example, signal formation apply (cf. Lovell et al. 2005; Chasmer et al. 2006; Donoghue and Watt 2006). In fact, one would expect that the reduced complexity (structure and species composition) of plantations should be conducive to the success of ALS assessment methods. Moreover, the geometric patterns created by planting trees in rows in easily identifiable stands, and the availability of ancillary plantation records on age, genetics, planting spacing, and silvicultural history present opportunities to improve ALS inventory methods.

In this chapter we first review the research related to the use of basic ALS inventory techniques in various types of plantations and possibilities to take into account the special features of the plantation forests (e.g. tree spacing, information from plantation database). Then, we describe site quality assessment of radiata pine (*Pinus radiata*) plantations in South Australia as an example of an operational ALS based inventory system. Finally, we present conclusions on the potential and applicability of different ALS-based inventories in plantation forests.

13.2   The Applicability of Basic ALS Techniques
for the Inventory of Forest Plantations

Resource managers need accurate and timely information to make decisions wisely (Schreuder et al. 1993). The information also needs to be relevant to management objectives. Often information is needed about both current and future state of the resource. ALS techniques can assist in describing the current state of the resource. Growth models are needed to predict future states starting from the current state.

The growth and yield of plantations is typically predicted using a system of models that require a range of tree or stand parameters as input (e.g., Sullivan and Clutter 1972; Clutter et al. 1983; see Skovsgaard and Vanclay 2008 for a review). The application of existing modeling systems usually results in compatible current and projected yield predictions (Packalén et al. 2011a). A suitable strategy of a remote sensing based approach is to attempt to generate the required input parameters for these models. Examples of such parameters are diameter, height, basal area and some measure of site productivity (such as dominant height or total volume production at reference age).

Both basic approaches of ALS inventory, i.e. single tree detection and area based estimation, have been tested in a range of broadleaved and coniferous plantations. A pioneer study applying the **single tree detection** approach was carried out in a *Eucalyptus* plantation in Portugal (Wack et al. 2003). For a small reference field data set (only 4 plots were available) and fairly low stand density (less than 400 stems/ha) the study achieved a tree detection success rate of 93 %. The accuracy of tree height estimates was comparable to that found in studies in other forest types, with a mean underestimation of 0.69 m. Stem volumes were predicted using existing tree-level models. The required tree diameter was iteratively solved from a dominant height model. Volume predictions were verified using harvesting records in six stands, which showed that predicted wood volume was 7 % lower than harvested.

The study by Wack et al. (2003) used models that did not require any new field calibration measurements. Similar models were used by Tesfamichael et al. (2010b), who derived the number of trees and dominant height using a single tree detection approach. Combining these observations with the known stand age in existing models (Pienaar and Harrison 1988) allowed estimating basal area, quadratic mean DBH, and finally plot volumes. ALS data have also been proven effective for inventories of loblolly pine (*Pinus taeda*) plantations (McCombs et al. 2003; Roberts et al. 2005). The latter studies also reported stand density and height characteristics based on single tree detection, yet the main focus was on the estimation of very specific attributes, such as leaf area.

Vauhkonen et al. (2012) performed an extensive comparison of previously published single tree detection algorithms, including test sites from various boreal and broadleaved forests in Europe but also one data set from *Eucalyptus* plantations in Brazil. The comparison indicated that tree detection rates achievable in *Eucalyptus* plantations were higher than in European semi-natural stand structures. This was notwithstanding the fact that the Brazilian ALS data set had the lowest point density in the study (1.5 m^{-2}). It must be noted however that tree coordinates were not measured in the Brazilian field dataset, so that it was impossible to compute some of the accuracy criteria (omission and commission errors) that were available for the European studies.

These research results indicate that individual tree algorithms work well in plantation forests. This is not surprising given that the stand characteristics that complicate individual tree detection such as overshadowed, merging or flattened crowns (e.g. Chap. 5) are less likely to be encountered in plantations, especially if they are young, have been planted at wide spacing or have been thinned.

The application of the **area based approach** has also been tested in plantations. As in other forest types the estimation of forest attributes relies on plot-level ALS metrics as predictors (e.g. Zonete et al. 2010). The application of the area-based approach has been found to be straightforward and to result in higher accuracies compared to applications in other regions such as boreal forests. For example, Rombouts et al. (2010) and Packalén et al. (2011a) obtained root mean squared errors (RMSEs) in the order of 10 % of the total plot-level volume. These studies took place in unthinned plantations with a narrow age range. They also made use of field plots that were larger (500–530 m^2) than those typically used in other studies, which tends to reduce the RMSE of the results. Stephens et al. (2007) applied area based techniques to estimate carbon stocks in radiata pine plantations as part of the New Zealand National Carbon Inventory. They achieved a RMSE of 19 % of total plot carbon.

Donoghue et al. (2007) successfully separated species using ALS intensity values in two-species conifer plantations. Tree species recognition can be expected to be considerably simpler in plantations than in other forests (cf. Chap. 7), for example due to the absence of a suppressed tree layer consisting of minor species. Tesfamichael et al. (2010a) examined the effect of ALS point density on the plot level regressed mean height and dominant height characteristics. Their results agreed with studies conducted in other regions and forest types, i.e. point density did not have a major impact on the accuracies.

13.3 Improvements by Explicitly Considering the Specific Attributes of Plantation Forests

Stand structure tends to be closely related to age, genetics, site properties and silvicultural history. Data on planting dates, planting material, planting distances and silvicultural operations such as thinning, fertilizer and pruning, when recorded in plantation databases, can therefore assist plantation assessment, including ALS based assessments. Despite the already high accuracies obtainable using basic ALS techniques, a further improvement can be achieved by exploiting this information.

The regular stand structure and the availability of precise silvicultural management data in particular produce a special situation, in that a high degree of information is available in addition to the ALS data.

In fact, the regular stand structure is already implicitly considered by several studies mentioned in the previous section. Especially the single-tree studies by Wack et al. (2003), Roberts et al. (2005), Tesfamichael et al. (2009, 2010b) and Vauhkonen et al. (2012) benefit from the fact that it is possible to **make strong assumptions on the trees' spatial arrangement**. For example, the studies by Roberts et al. (2005) and Tesfamichael et al. (2009) specified the maximum search window for the tree detection algorithm based on the nominal planting distance, whereas in less regular conditions the proper parameter values need to be obtained by experimenting with a training data set (e.g. Popescu et al. 2003). Furthermore, the *a priori* knowledge of the expected stem number may boost the algorithm performance, whereas this information is considerably less accurate in semi-natural forests (Vauhkonen et al. 2012). However, the reported best-case omission error rates of 7–18 % still leave room for improvements, especially when the objective is timber volume estimation (cf. Tesfamichael et al. 2010b).

Planting distance data from the stand register were explicitly integrated in the individual tree detection method developed by Vauhkonen et al. (2011). They found tree rows to be clearly detectable in their data, but expected difficulties in detecting individual trees within the rows. Therefore their approach was to first detect the tree rows by means of the ALS data and then to sample the detected rows by considering the planting distance along the row. This approach produced an unbiased estimate of stem number with an RMSE of only 6 %, which was a clear improvement to the best-case RMSE and bias of about 20 % and 3 %, respectively, obtained by Vauhkonen et al. (2012) using general single tree detection algorithms in a subset of these data.

The use of plantation records further allows **site and clone effects to be taken into account in the estimation**. This approach requires the use of more advanced statistical methods such as mixed-effects models (Searle 1971). There are two fundamental reasons for using the mixed-effect models. First, the data often exhibit a hierarchical structure (e.g. sample plot within stand within clone) and the correlation of residuals between the hierarchical levels needs to be taken into account for statistical inference to be valid (ordinary least squares regression techniques assume independence of the residuals). Second, clone data are available in the plantation database and their inclusion in the modeling process imposes no additional cost. Mixed-effect models are flexible in that they may be applied at "local" or "global" level, depending on whether random effects are known at the location where the model is applied. In "local" prediction mode mixed-effect models enable improved small-area prediction.

The application of mixed-effects models in the context of ALS-based estimation was demonstrated by Packalén et al. (2011b) and Vauhkonen et al. (2011) in a *Eucalyptus* plantation. Of these studies, Packalén et al. (2011b) formulated non-linear mixed-effects models for site index and plot volume using the typical area-based ALS features as predictors, whereas Vauhkonen et al. (2011) modeled stem volume using height values extracted by the previously described individual tree detection approach. In both cases, the fixed parts of the models were augmented with random effects for categorical variables obtained from the stand database. The introduction of these random effects resulted in improved prediction accuracy. Especially the information on the applied clone was found to be of value in improving the predictions compared to the global population level model.

An advantage of the mixed-effects modeling approach from the practical point of view is that at least some part of the model can be applied at any situation. The population-level part of the model can be used in all cases corresponding to an ordinary regression model. Clone-level predictions may be made provided that the clones were represented in the modeling data. If there is a clone not included in the modeling data, the random effects for it can be predicted from a minimum of one clone-specific observation of the modeled attribute. The latter is explicitly shown by Wang et al. (2007), who included annual rainfall and daily maximum temperatures in mixed-effects models of growth and yield for *Eucalyptus* plantations in Australia.

The fast growth of plantation forests presents a challenge to **the collection of the required field data**. Rombouts (2011) found that in radiata pine plantations (with current annual increment often in excess of 30 m^{-3} ha^{-1}) the predictive relationships between field and ALS data change rapidly. Analysis of re-measured field plots showed that the parameter values of linear regression models relating stand volume and the ALS predictor variable changed by one to two times the standard error of the parameter estimate over a 12 month period. It was concluded that field data collection in fast growing plantations cannot be spread out over lengthy periods of time and preferably should take place during the season of lowest growth. Due to the potential to use existing models, the amount of field data required by the tree-level approaches is likely considerably less compared to area based methods (Vauhkonen et al. 2011).

Besides inventorying timber within fixed stand boundaries, the ALS data can be employed to **generate optimized treatment units**. Packalén et al. (2011a) formed such dynamic treatment units by aggregating small inventory units by means of spatial optimization. The idea was to first predict the growing stock characteristics for hexagons of 300 m^2, which acted as the initial inventory units. These units were simulated for treatment schedules by varying the rotation length, and units with similar treatments were aggregated according to varying degrees of cutting area targets. The use of the dynamic treatments units was in all cases found to be more optimal than fixed stand boundaries with respect to total volume production goals. Cutting areas formed by spatial optimization were also found to often deviate from compartment boundaries due to variations in productivity within the stands.

Besides goals related to log quantities, information on **log quality in individual stands** may also be of interest. Some researchers have considered terrestrial laser scanning (TLS) data to augment ALS-based inventories. TLS data provide information on stem straightness and branch thickness which help to determine how a stem may be cut into products of different quality and value. TLS works best when undergrowth is sparse and stems are clearly visible. Those conditions are more likely to be encountered in plantations. Murphy et al. (2010) compared predicted product recovery in 33 plots in six radiata pine stands in Australia for three assessment methods: (1) TLS based product assessment, (2) a product oriented field inventory system that includes measurement of sweep and branching and (3) product recovery measured using harvester sensors. They found a difference between the field inventory and harvester measurement of 4 %, compared to a difference between TLS and harvester measurement of 7–8 %. Murphy et al. (2010) describe the inventory workflow and propose certain improvements to increase the accuracy of the estimates. In addition to TLS-based estimates, information on branch characteristics, for example, can be obtained by ALS (Dean et al. 2009).

It may also be possible to apply ALS data towards **fertilizer decision support**. In the southeastern United States, Peduzzi et al. (2012) successfully assessed leaf area index (LAI) of intensively managed loblolly plantations. Research has shown that differences between a stand's current leaf area and its potential leaf area can be used to estimate responsiveness and the appropriate timing and rates of fertilization (Fox et al. 2007). By mapping LAI using ALS techniques it may therefore be possible to optimize fertilizer application based on spatial variations in LAI.

13.4 Case Study: Operational ALS Based Site Quality Assessment of Radiata Pine Plantations in South Australia

13.4.1 Background

Radiata pine is the principal softwood plantation species in the Southern Hemisphere. In South Australia some 123,400 ha have been planted with this species (Anon 2010). First plantings date back from 1876 and intensive management practices have developed in which site quality information plays a prominent role.

Site Quality, written in capital letters to denote the specific information as implemented in South Australia, is a measure of the site productivity realised under a specified management regime and given genotypes. Site Quality information provides the basis for growth modeling, is a source of stratification for pre-harvesting inventory, drives thinning and fertiliser regimes and is used to monitor productivity of successive rotations. The South Australian Site Quality system was introduced around 1931 and developed over a period of 20 years. By 1949 a Site Quality assessment method had been developed which was then consistently applied for the next 60 years. The criterion of Site Quality is total volume production, under bark, to 10 cm top diameter (V10), by reference age. Site Quality assessment typically takes place at age 9 or 10, i.e. after differentiation of seven Site Quality classes and before any volumes have been harvested.

Conventional Site Quality assessment involved both objective (plot based) and subjective (ocular) methods. The end-product of site quality assessment was a map showing seven Site Quality classes and areas classified as unproductive plantation. The smallest polygons mapped were a tenth of a hectare. The approach was labour intensive and the quality of outcomes depended on the aptitude and experience of assessors.

Site Quality mapping is essentially a problem of assessing the spatial variation of standing volume across the area of interest at age 9 or 10. In 2005 a research program was initiated to evaluate ALS for Site Quality assessment. It was established that area based methods, with calibration of predictive regression models linking standing volume to ALS variables, and the application of such models across the area of interest, provided a robust solution. By the end of 2007 the South Australian Forestry Corporation (trading as ForestrySA) decided to adopt ALS based Site Quality assessment. Conventional Site Quality assessment was discontinued and in early 2009 the first operational ALS based Site Quality survey was conducted. This was followed by a second operational survey in early 2012.

The following paragraphs discuss key decisions with regard to design and operational parameters associated with the 2012 survey. These paragraphs are organised around the four functional stages in an operational Site Quality survey: (1) ALS data acquisition, (2) field sampling, (3) model calibration and (4) generation of Site Quality maps.

13.4.2 ALS Data Acquisition

Given that plantations need to be assessed at age 9 or 10, and given that this age group on average represents only one twentieth of the plantation estate, survey areas are relatively small and highly fragmented, driving up ALS data acquisition costs. Research has therefore focused on providing practical means of minimising ALS data acquisition costs.

One way to reduce costs is to acquire data at the lowest possible density the application will permit. Studies comparing site quality assessment using high and low density ALS datasets (Rombouts et al. 2010) showed that there were no material differences between site quality maps based on dense ALS data (2.6 pulses m^{-2}, flown at 1,100 m altitude) and sparse data (0.3 pulses m^{-2}, flown at 2,600 m altitude). It was therefore decided to acquire the data at that lower density.

To achieve economies of scale 8 year old plantations were also included in the survey. However, measurement of field plots was delayed until those plantations reached the minimum site quality assessment age of 9 years. This was deemed acceptable because prior research involving re-measurement of field plots showed that plots could be measured 1 or 2 years after scanning without unacceptable detrimental effect on the correlation between ALS and forest variables (Rombouts 2011). Adding age eight plantations increased the survey area by 50 % to 6,726 ha. To further increase the survey area data acquisition was coordinated with three other forest growers.

Table 13.1 Data specifications in data acquisition contract		
Scanning swath overlap	$\geq 25\,\%$	
Max scanning angle ($^{\circ}$)	15.0	
Pulse density (m^{-2})	≥ 0.4, evenly spaced	
Horizontal accuracy (m, 1σ)	0.5 m	
Vertical accuracy (m, 1σ, relative to reference DEM)	0.25 m	
Classification of returns to ground, non-ground and overage points		

Further cost savings were made by allowing the contractor to reference their dataset to a pre-existing regional Digital Elevation Model (DEM) developed in 2007 during a large scale ALS terrain modeling project. The arrangement permitted the ALS contractor to align the new DEM to the pre-existing DEM using an empirically determined and site specific adjustment term for the z-coordinates of the ALS point cloud. As a result the contractor did not need to establish ground control points other than for calibration of the scanner, reducing costs by some 30 % or more.

Table 13.1 shows the full set of data specifications included in the contract. It was left to the contractor to optimise flying altitude, speed and pulse repetition rates to achieve the required data specifications. Data were acquired January 2012 using an Optech ALTM Orion scanner.

13.4.3 Field Sampling

Site Quality maps need to accurately reflect the spatial variation in site productivity. It follows that small area prediction (at sub-compartment level) is very important and that the field sampling strategy must aim to provide adequate field data to calibrate a model that is unbiased across the various plantations comprising the survey area.

The first decision made was to calibrate a new model, and not to re-use the model developed for the 2009 survey. This decision was guided by prior research (Rombouts et al. 2010) which had shown that predictive models are influenced by "campaign effects" (instrumentation, flight parameters, seasonal influences) indicating that they should not be re-used.

Non-productive plantation was excised from the survey area using aerial photography and ALS derived crown cover data. A stratified random sampling design was adopted at site level based on the ALS predictor variable. Twenty-two sites were identified with each site comprising of one or more 9 or 10 year old plantations. At each site, depending on site area, three to five plots were randomly selected after stratifying the site based on the ALS predictor variable calculated in a 20×20 m grid. The number of strata was made to match the number of plots to be established, and one plot location was selected per stratum. A further ten plots were established to sample the very extremes of the predictor variable range.

This sampling strategy ensured that both the geographical and ALS predictor variable range were sampled, ensuring that any important site effects could be detected. In total a hundred plots were measured. This was double the sample size suggested by prior research.

Rectangular 0.05 ha plot locations were established using planting row maps and measuring tape. In some cases plantation maps and ALS data were not perfectly aligned due to legacy surveying errors. In those cases the ALS data were moved to improve alignment. Field measurement took place in April-May at a time when plantation growth was minimal. It was completed by one field crew within 3 weeks. Plot diameters and predominant height were measured enabling estimation of plot volume using the Keeves Stand Volume Tables (Lewis et al. 1976).

13.4.4 Model Calibration

Past experience had consistently shown that a linear model with a single ALS predictor variable (the mean quadratic height of first returns) was the most effective predictor model for V10. This was confirmed to also be the case for the 2012 calibration dataset. Site effects were statistically significant but confined to three sites. The three sites were unremarkable in terms of soil, geographical location and site effects were further ignored.

Figure 13.1 shows the regression line fitted to the calibration dataset. The prediction model fitted to the calibration dataset of the 2009 is shown for comparison. While structurally the same and of comparable precision (%RMSE), the 2009 and 2012 models had significantly different slopes. Yet the operational parameters applied during the two surveys were very similar: pulse density (0.4 vs. 0.6 m^{-2}), flying altitude (2,500 vs. 2,250 m), pulse repetition rate (50 kHz in both cases), footprint size (50 vs. 55 cm), maximum scan angle (15° in both cases). Data acquisition took place at the same time of the year (December and January respectively). The only marked differences were instrumentation (Optech ALTM Gemini versus Optech ALTM Orion) and contractors. The model differences, even if they cannot be explained, do justify the decision to calibrate a new predictive model (see also Næsset 2009, Chap. 11).

13.4.5 Site Quality Mapping

Mapping is the final step in the assessment process and involves model application, spatial interpolation and stand volume to site quality conversion. The objective of the process is to produce an accurate map with appearance similar to that of conventional Site Quality maps. Consistency with existing maps was deemed desirable because conventional and ALS Site Quality maps have to co-exist in

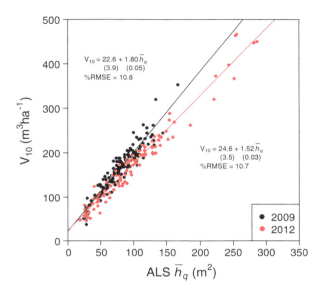

Fig. 13.1 Models fitted to the calibration datasets of the 2009 and 2012 surveys. h_q – the mean quadratic height of the first returns

corporate datasets for the foreseeable future. Moreover, successive rotation site quality comparisons have to remain feasible indefinitely.

The main decision with regard to model application is the choice of the grid dimensions by which to subdivide the area of interest. Prior research (Rombouts 2011) indicated that the choice of the grid dimensions, for the specific model and predictor variable used, did not significantly influence the accuracy of mapped volume surfaces, as long as the grid cell area was not substantially larger than the area of the calibration plots and there is a large enough laser point density. Pixel size did however influence map appearance. Grid dimensions of 20×20 m were applied producing grid cells with areas 20 % smaller than the calibration plots. Grid cells were made to overlap by 50 % to improve mapping near compartment edges and corners. Models were applied at the scale of these grid cells producing a point dataset of volume predictions spaced on a 10×10 m grid. Natural neighbour spatial interpolation was applied to convert this point dataset to a 1×1 m raster dataset. Rombouts (2011) demonstrated that this step does not affect accuracy but it does produce a smoothened volume surface that can easily be converted to a polygon dataset with an appearance very similar to conventional site quality maps.

The final step in the production process is the determination of Site Quality based on standing volume and age using a simple look-up table. Figure 13.2 shows a section of the ForestrySA Site Quality spatial dataset selected because it shows both conventionally (1981, 1983, 1998) and ALS assessed plantations (2001, 2002, 2003). It shows the consistency of Site Quality patterns of conventionally and ALS assessed maps, and of successive ALS surveys (compare 2001–2002, 2003 plantations).

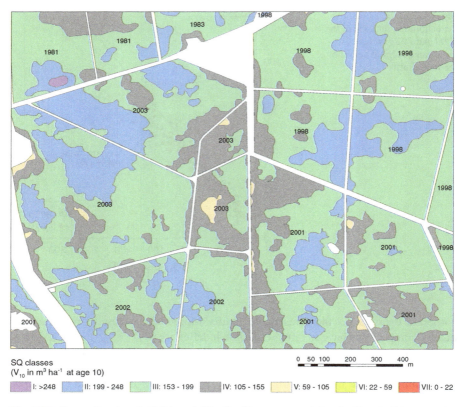

SQ classes
$(V_{10}$ in m^3 ha^{-1} at age 10)

0 50 100   200   300   400 m

▮ I: >248   ▮ II: 199 - 248   ▮ III: 153 - 199   ▮ IV: 105 - 155   ▮ V: 59 - 105   ▮ VI: 22 - 59   ▮ VII: 0 - 22

Fig. 13.2 Conventional and ALS based Site Quality maps coexisting in ForestrySA corporate datasets. All compartments planted post-1998 were mapped using ALS data

13.4.6 Evaluation

ALS based site quality assessment offers multiple benefits compared to conventional approaches. ALS based site quality maps tend to be more accurate (Rombouts 2011), are objective and less dependent on the experience and aptitude of staff. The ALS based approach offers flexibility in presenting site quality information as a familiar map of polygons or as a raster surface; in treating it as a categorical variable or as a continuous variable.

Field based workload decreased by a factor of four to five compared to conventional methods. Office work decreased significantly from the first to the second survey owing to re-use of contract templates, continued automation of data processing, improvement of in-house software tools and adoption of commercial software such as LAStools. The technical staff that used to walk the plantations has embraced ALS technology and has taken ownership of all survey stages with the exception of model calibration.

Overall site quality assessment costs have decreased by 50–70 % (depending on the price of ALS data). The success of ALS based site quality assessment has generated corporate interest in exploring the feasibility of other applications such as ALS-based pre-harvesting inventory and fertilizer decision support.

13.5 Conclusions

Both single tree and area based approaches have been tested under a range of plantation forest conditions, and have been found to be as effective, if not more effective, as in other forest types. Estimation techniques developed under other forest types, such as basic regression analysis based methods are usable in plantation forest inventories to assess a range of stand parameters including stocking, height, basal area, volume and leaf area. However, techniques have been developed that leverage ancillary information that is typical of plantation forests, such as age, planting distance and the genetic origin of planting stock. Wherever such information is available it is recommended that it be used to improve inventory accuracy using some of the techniques listed in Sect. 13.3.

In their commentary paper of 2001, Evans et al. raised concerns about the possible interaction of the ALS point pattern and density with the spatial pattern of trees planted in rows, suggesting that scan line design could have unwanted effects on inventory results in plantation forest. However, none of the ALS studies in plantation forests have reported such effects. The work by Rombouts et al. (2010) and Vauhkonen et al. (2011, 2012) indicate that reliable results can be achieved with data acquired using similar designs and methods but seemingly at lower data densities than those typically applied in (semi-)natural forest types.

Many of the techniques covered by Sects. 13.2 and 13.3 were applied on high-growth plantations aiming at intensive pulp wood production rates. Not all plantation forests have such high growth rates, however, and depending on the intended use of wood products, some plantations may have higher planting densities, for example. The plantation density will have an impact on success rates of individual tree detection similar to semi-natural forests (cf. Vauhkonen et al. 2012), and in this sense the plantation forests with longer rotation lengths do not fundamentally differ from semi-natural forests, in terms of applying the principles for ALS-based forest inventories described elsewhere in this book.

Several studies were discussed that demonstrate how ALS based approaches could assist intense plantation management through optimization of fertilizer regimes, creation and scheduling of harvesting units and product recovery. Applications on optimizing wood supply and improved value recovery exist and could be further extended to timber logistics in terms of ALS data based forest road planning, for example. Compared to conventional inventories ALS-based inventories can produce information at a high spatial resolution, showing variation of the parameter of interest at a stand or sub-stand level. Such information is invaluable if the objective is to make management more site specific and precise.

The South Australian case study demonstrates that ALS based methods can provide effective alternatives to long-established conventional approaches while maintaining information consistency, can be operationally introduced in existing corporate structures and offer significant cost and other benefits.

References

Anon (2010) Australia's plantations – 2010 inventory update. Department of Agriculture, Fishery and Forestry, Bureau of Rural Sciences, Canberra

Carle J, Holmgren P (2003) Definitions related to planted forests. Working paper 79, FAO, Rome, Italy

Chasmer L, Hopkinson C, Treitz P (2006) Investigating laser pulse penetration through a conifer canopy by integrating airborne and terrestrial lidar. Can J Remote Sens 32:116–125

Clutter JL, Fortson JC, Pienaar LV, Brister GH, Bailey RL (1983) Timber management: a quantitative approach. Wiley, New York

Dean TJ, Cao QV, Roberts SD, Evans DL (2009) Measuring heights to crown base and crown median with LiDAR in a mature, even-aged loblolly pine stand. For Ecol Manage 257:126–133

Donoghue DNM, Watt PJ (2006) Using LiDAR to compare forest height estimates from IKONOS and Landsat ETM+ data in Sitka spruce plantation forests. Int J Remote Sens 27:2161–2175

Donoghue DNM, Watt PJ, Cox NJ, Wilson J (2007) Remote sensing of species mixtures in conifer plantations using LiDAR height and intensity data. Remote Sens Environ 110:509–522

Evans J (2009) Planted forests – uses, impacts and sustainability. Food and Agriculture Organization of the United Nations (FAO) and CAB International, Rome, Italy. ISBN 978-1-84593-564-1. 203 p

Evans DL, Roberts SD, McCombs JW, Hatrington RL (2001) Detection of regularly spaced targets in small-footprint LIDAR data: research issues for consideration. Photogramm Eng Remote Sens 67:1133–1136

Fox TR, Allen HL, Albaugh TJ, Rubilar R, Carlson CA (2007) Tree nutrition and forest fertilization of pine plantations in the Southern United States. South J Appl For 31:5–11

Koukoulas S, Blackburn GA (2005) Mapping individual tree location, height and species in broadleaved deciduous forest using airborne LIDAR and multi-spectral remotely sensed data. Int J Remote Sens 26:431–455

Lewis NB, Keeves A, Leech JW (1976) Yield regulation in South Australian *Pinus radiata* plantations. Bulletin No 23, Woods and Forests Department, South Australia

Lovell JL, Jupp DLB, Newham GJ, Coops NC, Culvenor DS (2005) Simulation study for finding optimal lidar acquisition parameters for forest height retrieval. For Ecol Manage 214:398–412

McCombs JW, Roberts SD, Evans DL (2003) Influence of fusing LiDAR and multispectral imagery on remotely sensed estimates of stand density and mean tree height in a managed loblolly pine plantation. For Sci 49:457–466

Murphy GE, Acuna MA, Dumbrell I (2010) Tree value and log product yield determination in radiata pine (*Pinus radiata*) plantations in Australia: comparisons of terrestrial laser scanning with a forest inventory system and manual measurements. Can J For Res 40:2223–2233

Næsset E (2009) Effects of different sensors, flying altitudes, and pulse repetition frequencies on forest canopy metrics and biophysical stand properties derived from small-footprint airborne laser data. Remote Sens Environ 113:148–159

Nelson R, Hyde P, Johnson P, Emesienne B, Imhoff ML, Campbell R, Edwards R (2007) Investigating RaDAR-LiDAR synergy in a North Carolina pine forest. Remote Sens Environ 110:98–108

Packalén P, Heinonen T, Pukkala T, Vauhkonen J, Maltamo M (2011a) Dynamic treatment units in *Eucalyptus* plantation. For Sci 57:416–426

Packalén P, Mehtätalo L, Maltamo M (2011b) ALS-based estimation of plot volume and site index in a *Eucalyptus* plantation with a nonlinear mixed-effect model that accounts for the clone effect. Ann For Sci 68:1085–1092

Peduzzi A, Wynne RH, Fox TR, Nelson RF, Thomas VA (2012) Estimating leaf area index in intensively managed pine plantations using airborne laser scanner data. For Ecol Manage 270:54–65

Pienaar B, Harrison WM (1988) A stand table projection approach to yield prediction in unthinned even-aged stands. For Sci 34:804–808

Popescu SC, Wynne RH, Nelson RF (2002) Estimating plot-level tree heights with lidar: local filtering with a canopy-height based variable window size. Comput Electron Agric 37: 71–95

Popescu SC, Wynne RH, Nelson RF (2003) Measuring individual tree crown diameter with lidar and assessing its influence on estimating forest volume and biomass. Can J Remote Sens 29:564–577

Roberts SD, Dean TJ, Evans DL, McCombs JW, Harrington RL, Glass PA (2005) Estimating individual tree leaf area in loblolly pine plantations using LiDAR-derived measurements of height and crown dimensions. For Ecol Manage 213:54–70

Rombouts J (2011) Assessing site quality of South Australian radiata pine plantations using airborne LiDAR data. PhD dissertation, Department of Forest and Ecosystem Science, University of Melbourne, Australia

Rombouts JH, Ferguson IS, Leech JW (2010) Campaign and site effects in LiDAR prediction models for site quality assessment of radiata pine plantations in South Australia. Int J Remote Sens 31:1155–1173

Savill P, Evans JS, Auclair D, Falck J (1997) Plantation silviculture in Europe. Oxford University Press, Oxford/New York/Tokio

Schreuder HT, Gregoire TG, Wood GB (1993) Sampling methods for multiresource forest inventory. Wiley, New York

Searle SR (1971) Linear models. Wiley, New York

Skovsgaard JP, Vanclay JK (2008) Forest site productivity: a review of the evolution of dendrometric concepts for even-aged stands. Forestry 81:13–31

Stephens PR, Watt PJ, Loubser D, Haywood A, Kimberley MO (2007) Estimation of carbon stocks in New Zealand planted forests using airborne scanning LiDAR. ISPRS workshop on laser scanning 2007 and SilviLaser 2007, Espoo, Finland, 12–14 September 2007

Sullivan AD, Clutter JL (1972) A simultaneous growth and yield model for loblolly pine. For Sci 18:76–86

Tesfamichael SG, Ahmed F, van Aardt J, Blakeway F (2009) A semi-variogram approach for estimating stems per hectare in *Eucalyptus grandis* plantations using discrete-return lidar height data. For Ecol Manage 258:1188–1199

Tesfamichael SG, Ahmed FB, van Aardt J (2010a) Investigating the impact of discrete-return lidar point density on estimations of mean and dominant plot-level tree height in *Eucalyptus grandis* plantations. Int J Remote Sens 31:2925–2940

Tesfamichael SG, van Aardt J, Ahmed F (2010b) Estimating plot-level tree height and volume of *Eucalyptus grandis* plantations using small-footprint, discrete return lidar data. Prog Phys Geogr 34:515–540

Vauhkonen J, Mehtätalo L, Packalén P (2011) Combining tree height samples produced by airborne laser scanning and stand management records to estimate plot volume in *Eucalyptus* plantations. Can J For Res 41:1649–1658

Vauhkonen J, Ene L, Gupta S, Heinzel J, Holmgren J, Pitkänen J, Solberg S, Wang Y, Weinacker H, Hauglin KM, Lien V, Packalén P, Gobakken T, Koch B, Næsset E, Tokola T, Maltamo M (2012) Comparative testing of single-tree detection algorithms under different types of forest. Forestry 85:27–40

Wack R, Schardt M, Lohr U, Barrucho L, Oliveira T (2003) Forest inventory for *Eucalyptus* plantations based on airborne laser scanner data. Int Arch Photogramm Remote Sens Spat Info Sci XXXIV:40–46

Wang Y, LeMay VM, Baker TG (2007) Modelling and prediction of dominant height and site index of *Eucalyptus globulus* plantations using a nonlinear mixed-effects model approach. Can J For Res 37:1390–1403

Zonete M, Rodriguez L, Packalén P (2010) Estimação de parâmetros biométricos de aerotrans-portada. Sci For 86:225–235 (In Portuguese with an English abstract)

Chapter 14
Using Airborne Laser Scanning Data to Support Forest Sample Surveys

Ronald E. McRoberts, Hans-Erik Andersen, and Erik Næsset

Abstract Forest surveys, in the form of both stand management and strategic inventories, have a long history of using remotely sensed data to support and enhance their design and estimation processes. By the use of airborne laser scanning data this capacity has emerged as one of its most important and prominent applications. The chapter includes a brief overview of forest inventory uses of remotely sensed data, a section on aspects of ground sampling that can be managed to optimize estimation of relationships between ground and airborne laser scanning (ALS) data, and a section on stand management inventories. The latter section reviews underlying and motivating factors crucial to the primary focus of the chapter, formal statistical inference for ALS-assisted forest inventories. Inferential methods are described for two primary cases, full and partial ALS coverage. Within each case, estimators for both design-based and model-based inference are presented.

14.1 Introduction

Sample surveys have a long history and have played a fundamental role in forest inventory at multiple geographical scales, from national levels where the aim is to provide strategic estimates and to inform national policies, to local inventories

R.E. McRoberts (✉)
Northern Research Station, U.S. Forest Service, Saint Paul, MN, USA
e-mail: rmcrobert@fs.fed.us

H.-E. Andersen
Pacific Northwest Research Station, U.S. Forest Service, Seattle, WA, USA

E. Næsset
Department of Ecology and Natural Resource Management, Norwegian University
of Life Sciences, Ås, Norway
e-mail: Erik.naesset@umb.no

M. Maltamo et al. (eds.), *Forestry Applications of Airborne Laser Scanning: Concepts and Case Studies*, Managing Forest Ecosystems 27, DOI 10.1007/978-94-017-8663-8__14,
© Springer Science+Business Media Dordrecht 2014

aimed at management decisions for individual properties, and even to individual treatment units such as forest stands. National-level forest surveys in the form of national forest inventories were first established in Norway in 1919 and in Finland, Sweden, and the United States of America in the early 1920s. These inventories were based on probability samples consisting of strips or belts laid out across vast tracts of land and individual trees were measured in the field along those belts (Tomppo et al. 2010). Common practice has been to augment field measurements with information obtained from auxiliary sources such as maps to improve the precision of the estimates of population parameters. For example, timber volume may reasonably be expected to be correlated with other land characteristics and attributes such as land use, site productivity, stand age and soil properties that are commonly mapped for managed forests. Remotely sensed data from aerial images and satellite sensors such as Landsat with partial or complete coverage of the population have been shown to be useful sources of such auxiliary data.

Auxiliary variables whose observations are correlated with observations of the response variable can be used to enhance inventory estimates. First, a map based on auxiliary information can be used to facilitate unequal probability sampling. For example, an existing model based on auxiliary variables can be used to construct a map of a variable such as growing stock volume. Unequal probability sampling can then be implemented by selecting sample plot locations with greater predicted volume with greater probability, thereby including more volume in the sample.

Second, auxiliary information can be used in the design-phase of a survey to stratify the population into strata that are more homogenous with respect to the response variable than the population as a whole. A stratification allowing separate samples with different sampling intensities for each stratum can greatly improve the precision of the estimates within given budget constraints.

Third, even after a sample has been collected, auxiliary information may be used to enhance the estimation process. With post-stratification, auxiliary data are used to construct strata in the same way as for stratified sampling. Although stratified sampling is not used, the precision of estimates may still be increased by aggregating plots into homogeneous strata and using stratified estimation. Further, the precision of estimates can be increased using models that relate response variables to one or more auxiliary variables. If observations of the auxiliary variables are available for the entire population, then the model can be used to predict the variable of interest for all population units such as Landsat pixels. Multiple estimators, including ratio and regression estimators, have been derived for these applications.

When acquisition of auxiliary data is difficult or expensive, data for only part of the population, albeit a larger portion than the ground sample, can also increase precision. Designs for these applications, characterized as multi-phase designs, consist of a large sample of observations of only the auxiliary variables in the first-phase followed by a smaller second-phase sample of observations of the response variable for a subset of the first-phase sample. Multiple estimators, all of which are design-specific, have been derived to estimate parameters related to the variable of interest for the entire population using only samples of auxiliary and response variables.

Airborne laser scanning (ALS) is a source of auxiliary data that can be combined with or replace other auxiliary data such as maps and satellite-based spectral data. Because ALS data are often strongly correlated with forest attributes such as volume and biomass, they have frequently been found more useful for enhancing estimates of parameters related to those variables than other remotely sensed data. Although acquisition of ALS data is expensive, the utility of the information for purposes such as forest management decisions may be greater than many alternative sources of information. For situations for which full ALS coverage is economically favorable (Eid et al. 2004), appropriate survey designs and corresponding estimators have been developed. In other situations such as surveys of larger regions covering thousands of square kilometers where only an overall estimate is required, full ALS coverage may be prohibitively costly. For these situations, two-phase designs that require only partial ALS coverage may be viable options. Of importance, the increase in uncertainty resulting from partial ALS versus complete coverage of the population may be marginal, whereas cost reductions may be substantial.

The chronological maturation of scientific disciplines, particularly disciplines that are inherently spatial, typically follows a path characterized by multiple steps: (i) exploratory analyses featuring comparisons of means and correlation analyses, (ii) development of models of relationships between response and predictor variables, followed by investigations of methods for enhancing the models, (iii) mapping including map-based estimation, and (iv) statistical inference. The first two steps and part of the third are addressed in detail in Chap. 1 which includes accompanying literature references. This chapter focuses more narrowly on issues and the literature related to ALS support for and enhancement of sample-based forest inventories. Section 14.2 focuses on issues related to improving models used for inventory applications and on a specific model application, stand management inventories. Section 14.3 focuses on more formal statistical inference for inventory applications and is organized into subsections on full and partial ALS coverage.

14.2   ALS-Assisted Estimation

Remote sensing-assisted inventory estimation relies heavily on the development and application of models that characterize relationships between ground observations of forest attributes and the remotely sensed data. Construction of models, and their application to support formal statistical inference, is a natural step in the maturation of nearly all scientific disciplines including remote sensing. This section focuses on two primary model-related topics: (i) sampling design and plot configuration issues that affect the quality of models of relationships between forest attributes and ALS data, and (ii) use of ALS-based models for stand management inventory applications.

14.2.1 Sampling Designs

When ground data for calibrating remote sensing-based models are acquired from existing sampling programs such as national forest inventories, few opportunities are possible to optimize the ground sampling design to accommodate features of the remotely sensed data. However, the remotely sensed data could be considered when selecting plot configurations and sampling designs for acquisition of ground data (McRoberts et al. 2013c). For example, if ALS data are to be acquired in strips covering ground plots, then systematically positioning the plots at the intersections of perpendicular grids facilitates acquisition of ALS data from airborne platforms via straight flight lines. Separating the grid lines by greater distances in one of the directions permits intensification of the sample plots under fewer flight lines, thereby facilitating ALS data acquisition and reducing costs.

Optimization of ground sampling designs for the specific purpose of acquiring calibration data for remote sensing-based models has also been considered. Næsset (2002, 2004a) stratified on age class and site quality and then constructed stratum-specific models. Hawbaker et al. (2009) compared a simple random and a stratified sampling design. The stratified design featured sampling within strata consisting of population units with similar means and standard deviations of the ALS height distributions. The latter design distributed sampling locations more uniformly with respect to the ALS height distributions, produced more observations in the tails of the ALS height distributions, produced smaller root mean square errors for linear models of relationships between biomass and ALS metrics, and required fewer extrapolations beyond the range of the ALS sample data when the model was applied to the entire population. Using a common validation dataset for multiple sampling designs, Maltamo et al. (2011a) found similar results when using nearest neighbors methods, although the superiority of the ALS-based stratified sampling design dissipated as the model calibration sample size increased. Maltamo et al. (2011a) also found that with the ALS-based stratified sampling design, the sample size could be substantially reduced with no adverse effects. The explanation for the results is that the stratified sampling designs produce better predictions for population units corresponding to the extremes of the parameters of the ALS height distributions used as model predictor variables; conversely, for smaller model calibration sample sizes, simple random sampling designs force the models to extrapolate beyond the range of the data in the model calibration dataset.

When spatial data from independent sources such as field plots and remote sensors are combined, co-registration to a common coordinate system is crucial. Multiple reports of the effects of forest conditions on GPS plot location accuracy have been published. In general, GPS location accuracy increases with decreasing forest density and height and increasing observation period, number of satellites, and satellite distribution. For a positional error of a specified magnitude, the adverse effects are less for more homogeneous forest conditions, less fragmented forests, circular plots relative to rectangular plots of the same area, and larger plots. Gobakken and Næsset (2008) reported that the detrimental effects of location error on estimates of forest attribute parameters were greater for stand basal area and volume than for mean tree height. Næsset (Chap. 11) concluded that accuracies on the order of 0.5 m or less are possible under boreal forest canopies using 2013-era technology. For more details, see Chap. 4.

Plot boundary effects have the potential to cause discrepancies between ground and remote sensing-based assessments. In particular, portions of trees whose stems are outside plots may extend into the vertical extensions of plot boundaries and, similarly, portions of trees whose stems are inside plots may extend beyond the vertical extensions of plot boundaries (Mascaro et al. 2011; Næsset et al. 2013a). Discrepancies arise because ground assessments of biophysical variables such as volume and biomass are generally based on stem locations, whereas ALS assessments are based on the vertical extensions of the plot boundaries. These discrepancies decrease the quality of fit of models of relationships between field measurements and ALS metrics. Several steps can be taken to reduce these detrimental effects. First, circular plots have smaller ratios of circumference to area than rectangular plots; second, larger plots have smaller ratios than smaller plots, regardless of shape; and third, larger plots relative to the sizes of tree crowns contribute to resolving the problem.

14.2.2 Stand Inventories

With the recognition that multiple approaches to modeling can be used to characterize relationships between forest attributes and ALS metrics, current research emphases have shifted to spatial applications of the models. The simplest such application is map construction with stand inventories as an important extension. In essence, mapping and stand inventory applications share two components, construction of models and application of the models to predict forest attributes for grid-based tessellations of areas of interest. However, stand inventory applications include a third component, calculation of the mean of the grid cell predictions as a stand-level estimate.

The modeling and mapping components of stand inventories typically feature much greater complexity and sophistication than general mapping applications. In particular, because stand inventories are the basis for important management decisions, greater emphasis is placed on the quality of the model predictions. This emphasis is manifested in the recognition that relationships between stand attributes such as volume and ALS metrics often vary with respect to factors such as species, age, site properties, and topography, and that stands are typically fairly homogeneous with respect to these factors. Thus, stand inventories as conducted in many countries incorporate three pre-modeling steps: (i) stand boundaries are carefully delineated, often using digital stereo photogrammetry; (ii) the entire area of interest is stratified by stand with respect to the relevant factors; and (iii) stratum-specific model training or calibration data are acquired to ensure adequate data for construction of stratum-specific models. If stratification is not used because it is not necessary or because information sufficient for constructing efficient stratifications is lacking, then common rather than stratum-specific models must be used. Once the models are constructed, they are used to predict the attribute of interest by grid cell within stands. The cell predictions are aggregated by stand to produce estimates of totals or means; see also Chaps. 11 and 12 for further details and examples.

The chronological development of this approach to stand inventories is illustrated via review of several seminal papers. Næsset and Bjerknes (2001) constructed regression models of the relationships between mean height of dominant trees and various ALS-based canopy height metrics, used the models to predict height for 200-m^2 cells, and calculated the mean over cells within stands. The mean deviation between ALS-based means of stand dominant height and corresponding plot measurement means was 0.23 m with standard deviation of 0.56 m. Næsset (2002) further developed this approach by stratifying a Norwegian study area by age class and site quality, acquiring stratum-specific model calibration data, constructing separate ALS-based models for each of six response variables separately by stratum, and then predicting the response variables for grid cells within stands. Stand-level means estimated by averaging over cell predictions were not statistically significantly different from means estimated from ground data obtained for plots within the stands. Næsset (2004a, b, 2007) replicated the previous study for three additional Norwegian study areas with approximately similar results. Holmgren and Jonsson (2004) used the same approach in Sweden, except models were constructed for different species rather than different strata. They reported that under limiting uniformity conditions, an ALS-based approach produced more precise estimates than means of plots within stands. Maltamo et al. (2011b) described a similar approach using a non-parametric nearest neighbors method for multivariate species-specific prediction but without stratification. Finally, a review of validation results for stand inventories conducted in the Nordic countries can be found in Næsset (2007). Operational ALS-based stand inventories have been conducted in Norway, Finland, and Sweden since 2002–2004 (Chaps. 11 and 12, Næsset et al. 2004; Turunen et al. 2012) and later in Australia (Turner et al. 2011), Canada (White et al. 2013) and Spain.

Several statistical features of ALS-based stand inventories merit noting. First, the models are developed using data from a large area and then applied to smaller individual stands within the larger area, even though the stands are not necessarily represented in the model calibration data. In this sense, the estimators corresponding to such models are characterized as *synthetic* (Särndal et al. 1992). Second, although considerable care is typically exercised to ensure that the models adequately represent relationships exhibited in the calibration data, no attempt is typically made to adjust estimates of stand-level means for systematic model prediction error. Third, no attempt is typically made to assess the uncertainty of the estimated stand-level means. Of importance, the variance of the within-stand cell predictions does not fully quantify the uncertainty of the estimates of stand-level means, because no accommodation is made for the fact that the predictions themselves have uncertainty. Nevertheless, although the stand-level estimates do not qualify as formal statistical inference, this approach to stand inventories represents a seminal step forward in the use of ALS data to support forest inventories. Additional information on ALS-based stand inventories can be found in Chaps. 11, 12 and 13.

14.3 Inference for ALS-Assisted Inventories

By definition, the term *survey* refers to a population, and the term *sample survey* refers to acquisition of a sample of the population for purposes of constructing an inference for a parameter, μ, of the population. In a scientific context, the term *inference* is understood to mean expressing the population parameter in a probabilistic manner, most commonly in the form of a confidence interval such as $\widehat{\mu} \pm t \cdot \sqrt{\widehat{\mathrm{V}}\mathrm{ar}\,(\widehat{\mu})}$ where $\widehat{\mu}$ and $\widehat{\mathrm{V}}\mathrm{ar}\,(\widehat{\mu})$ designate the sample-based estimates of the population parameter and its variance, respectively, and t corresponds to a probability level. Thus, the ultimate objective of a sample survey, of which a forest inventory is one example, is to produce an inference for a parameter of a population. Two forms of inference have been used for forest inventories. Design-based inference is more common and familiar, requires probability samples, and includes multiple estimators for use with auxiliary data including stratified estimators and model-assisted ratio and regression estimators. Model-based (model-dependent) inference is less familiar and often more complex and computationally intensive, but it has the important advantage of not requiring a probability sample.

14.3.1 Modes of Inference

Design-based inference, also characterized as probability-based inference (Hansen et al. 1983), is based on three assumptions: (1) population units are selected for the sample using a probability-based randomization scheme; (2) the probability of selection for each population unit into the sample is positive and known; and (3) the observation of the response variable for each population unit is a constant. Estimators are derived to correspond to sampling designs and are typically unbiased or nearly unbiased, meaning that the expectation of the estimator, $\widehat{\mu}$, over all samples that could be obtained with the sampling design is the population parameter; i.e., $\mathrm{E}\,(\widehat{\mu}) = \mu$ where $\mathrm{E}(\cdot)$ denotes statistical expectation. However, even with unbiased estimators, the estimate of a population parameter obtained from any particular sample may deviate considerably from the true value. The latter phenomenon motivates construction of confidence intervals as a means of mitigating some of the uncertainty concerning the degree to which an estimate obtained from a particular sample deviates from the true value.

Satisfying the design-based requirement for a probability sample may be challenging when the population includes remote areas with limited access or where security issues inhibit field work. Such conditions are frequent in some developing countries. Further, estimates may be sought for small sub-units of a larger population for which samples are too small to produce an acceptably precise estimate or perhaps samples are lacking entirely. An alternative that is attracting increasing attention is model-based (model-dependent) inference. The assumptions underlying model-based inference differ considerably from the assumptions underlying design-based inference: (1) the observation for a population unit is a random variable whose value is considered a realization from a distribution of possible values, rather than a constant as is the case for probability-based inference; (2) the basis for inference is the model, not the probabilistic nature of the sample as is the case for probability-based inference; and (3) randomization for model-based inference enters through the random realizations from the distributions for population units, rather than through the random selection of population units into the sample. Multiple important consequences derive from these assumptions: first, probability samples may be used with model-based inference although they are not required; second, model-based inference may be used for small areas for which sample sizes are insufficient for design-based inference; third, uncertainty may be estimated for all population units, not just those selected for the sample.

Current approaches to model-based inference originated in the context of survey sampling and can be attributed to Mátern (1960), Brewer (1963), and Royall (1970). Given the origins of model-based inference in survey sampling, it is not surprising that forestry applications have often been in the context of forest inventory (Rennolls 1982; Gregoire 1998; Kangas and Maltamo 2006; Mandallaz 2008).

Many of the exploratory studies that have led to development of ALS-based techniques used wall-to-wall data, albeit mostly for small areas. For large areas, the cost of acquiring wall-to-wall ALS data is currently extremely expensive, perhaps prohibitively so. The primary means of circumventing the cost issue is by acquiring samples of ALS data, primarily in strips corresponding to aircraft flight lines. As described in the following sections, both design-based and model-based inferential methods can be used, regardless of whether full or only partial ALS coverage is achieved.

14.3.2 Inference for Full ALS Coverage

14.3.2.1 Design-Based Inference

Design-based estimators that use auxiliary data take multiple forms including stratified estimators and model-assisted ratio and regression estimators. However, all forms require probability samples for estimation and inference, although not for model calibration.

The essence of stratified estimation is to assign population units to groups or strata that are more homogeneous than the population as a whole, calculate within stratum sample plot means and variances, and then calculate the population estimate as a weighted average of the within stratum estimates where the weights are proportional to the stratum sizes. Stratified estimation requires accomplishment of two tasks: (1) calculation of the stratum weights as the relative proportions of the population area corresponding to strata, and (2) assignment of each sample unit to a single stratum. The first task is accomplished by calculating the stratum weights as proportions of population units in strata. The second task is accomplished by assigning plots to strata on the basis of the stratum assignments of the population units containing the plot centers.

Sampling programs such as national forest inventories often use equal probability sampling designs with systematic components to accommodate the large number and variety of variables measured. Although stratified sampling for purposes of reducing variance is generally not possible when the same systematic sample of plots is measured at regular intervals, post-stratification has been used successfully for this purpose. Post-stratification is essentially stratified estimation but without stratified sampling. McRoberts et al. (2012, 2013b) predicted volume per unit area using ALS-based models and used aggregations of population units with similar predictions to form strata for post-stratified estimation. Relative to the simple random sampling (SRS) variance, post-stratification reduced the variance of the estimate of mean volume per unit area by a factor greater than 3.

When stratified estimation is used with stratified sampling, as opposed to post-stratification, variances may be reduced by even greater factors. For stratified sampling, the strata could be constructed by aggregating sample units with similar ALS-based predictions obtained from an existing model for the area of interest or from a different area if a common model may be assumed. For many natural resource variables, variability among sample units with large observed values is greater than among sample units with small observed values. Thus, overall variances may be reduced by using unequal probability sampling whereby strata with larger expected observations as indicated by larger ALS-based predictions are sampled more intensely than strata with smaller ALS-based predictions. Again, the ALS-based predictions could be obtained from either an existing model for the area of interest or from a model for a different area if a common model may be assumed.

Model-assisted estimators use models based on auxiliary data to enhance inferences but rely on probability samples for validity. Model-assisted estimators take several forms including the ratio estimator (Corona and Fattorini 2008), the difference estimator (Steinmann et al. 2013), and the regression estimator which is more general and is emerging as particularly useful. Model-assisted regression estimators begin with a model of the relationship between the forest attribute, Y, and a vector X, of auxiliary variables, formulated as,

$$y_i = f(X_i; \beta) + \varepsilon_i \qquad (14.1)$$

where $f(X_i; \beta)$ expresses the statistical expectation, the βs are parameters to be estimated, and ε is a random residual term. Although $f(X_i; \beta)$ is often formulated as a linear model, nonlinear models may also be used (Firth and Bennett 1998).

For a population of size N denoted U and an equal probability sample of size n denoted S, a general formulation of the regression estimator for the mean is,

$$\widehat{\mu}_y^{reg} = \frac{1}{N}\sum_{i \in U}\widehat{y}_i + \frac{1}{n}\sum_{i \in S}(y_i - \widehat{y}_i), \qquad (14.2)$$

where $\widehat{y}_i$ is obtained from Eq. 14.1 (Section 6.5 in Särndal et al. 1992). The first right-hand term in Eq. 14.2 is the sum of the model predictions for the entire population, and the second right-hand term is a bias estimate which compensates for systematic model prediction error. Although different forms of this estimator are also used (Chapter 7 in Cochran 1977; Section 6.4 in Särndal et al. 1992), the form corresponding to Eq. 14.2 is more intuitive when the auxiliary variables are remotely sensed or map sources, because the first right-hand term may be interpreted as a map-based estimate. For N much greater than n, an approximately unbiased estimator of the corresponding variance is formulated as,

$$\widehat{Var}\left(\widehat{\mu}_y^{reg}\right) = \frac{1}{n(n-p)}\sum_{i \in S}(\varepsilon_i - \overline{\varepsilon})^2, \qquad (14.3)$$

where $\varepsilon_i = y_i - \widehat{y}_i$ and p is the number of model parameters. The advantage of the regression estimator over the simple random sampling (SRS) estimator is that the variance estimator is based on residuals, $y_i - \widehat{y}_i$ rather than differences, $y_i - \overline{y}$, between observations and their mean. The degree to which Eq. 14.1 characterizes the relationship between Y and X reflects the degree to which the variance estimated using Eq. 14.3 is smaller than the SRS variance estimate.

Strunk et al. (2012) reported that the regression estimator of mean volume per unit area using a model based on ALS height deciles and other variables produced variances that were smaller by a factor greater than 2.3 than the SRS estimator for a study in Washington, USA. For a similar study in Brazil, d'Oliveira et al. (2012) reported variance reduction by a factor greater than 3.6 for an estimate of mean biomass per unit area. McRoberts et al. (2013b) used a logistic regression model to characterize the relationship between volume and ALS height deciles and estimates of mean volume per unit area for SRS, stratified, and regression estimators. For the estimate of mean volume per unit area, the model-assisted regression variance estimate was smaller by a factor of 2.0 than the stratified variance estimate and smaller by a factor of 6.3 than the SRS variance estimate. Næsset et al. (2011) combined stratified and model-assisted estimation. Models of relationships between ground-based observations of biomass and ALS height and canopy density metrics were constructed separately for strata based on forest maturity and site productivity factors. The stratified variance estimate when using model-assisted estimation within strata was smaller by a factor of 5.3 than the corresponding variance estimate when using SRS estimation within strata.

14.3.2.2 Model-Based Inference

Although model-based inference has been used since the 1980s for inventory applications (Rennolls 1982), McRoberts (2006) was among the first to use it with remotely sensed auxiliary data. The primary advantage of model-based inference is that a probability sampling design is not necessary which makes it attractive for small areas with insufficient sample sizes for design-based inference and for remote and inaccessible regions such as tropical forests where ground sampling may be prohibitively expensive or logistically impossible.

With model-based inference, an observation, y_i, for the ith population unit is assumed to have been obtained randomly from a distribution with mean, μ_i, and standard deviation, σ_i. A model of the relationship between the population unit means and auxiliary variables, X, is expressed as,

$$\mu_i = f(X_i; \beta) + \varepsilon_i, \tag{14.4}$$

where $f(X_i;\beta)$ expresses the mathematical relationship between the auxiliary variables and the parameters and ε_i is a random deviation with mean zero and standard deviation σ_i. The model parameters are estimated using data for a sample which is not required to be a probability sample, although a probability sample does provide moderate assurance of capturing the range of data in the population.

The model-based estimator of the population mean is based on the set of estimates, $\{\hat{\mu}_i, i = 1, 2, \ldots, N\}$, of the means for individual population units,

$$\hat{\mu} = \frac{1}{N} \sum_{i=1}^{N} \hat{\mu}_i. \tag{14.5}$$

With model-based inference, the difference between an observation and a corresponding model prediction is not considered to be a prediction error as for design-based inference, but rather is simply a random deviation between a particular observation and the mean of its distribution. Thus, the concept of bias must be addressed in a different manner. Whereas for model-assisted methods bias is defined in terms of prediction errors, for model-based methods bias is defined in terms of model lack of fit exhibited by systematic mischaracterization of the means for individual population units. Importantly, when the model is correctly specified, the estimator is unbiased (Lohr 1999), but when the model is incorrectly specified, the adverse effects on inference may be substantial (Royall and Herson 1973; Hansen et al. 1983). Therefore, because model-based inference includes no bias adjustment term to compensate for model lack of fit, additional analyses must be conducted to ensure that the model adequately characterizes the relationship between the response and predictor variables.

The variance of the model-based estimator of the estimated population mean is,

$$\widehat{V}ar(\hat{\mu}) = \frac{1}{N^2} \sum_{i=1}^{N} \sum_{j=1}^{N} \widehat{C}ov(\hat{\mu}_i, \hat{\mu}_j), \tag{14.6}$$

where

$$C\widehat{ov}\left(\widehat{\mu}_i, \widehat{\mu}_j\right) = Z_i' \ \widehat{V}_{\widehat{\beta}} \ Z_i, \qquad (14.7)$$

$$z_{ij} = \frac{\partial f\left(X_i; \widehat{\beta}\right)}{\partial \beta_j}, \qquad (14.8)$$

$$\widehat{V}_{\widehat{\beta}} = \left(Z_i' \ W \ Z_i\right)^{-1} \qquad (14.9)$$

is a first-order Taylor series approximation of the parameter covariance matrix, and W is a diagonal matrix with $w_{ii} = \sigma_i^{-2}$ (McRoberts et al. 2013b).

A disadvantage of model-based inference is the computational intensity involved in estimation of the variance using Eq. 14.6. However, McRoberts et al. (2013a, b) demonstrated that use of Eq. 14.6 with only a systematic sample of the population can be used to reduce computational intensity with no adverse effects on the variance estimate.

Although model-based inference can be used for estimation for small areas with small sample sizes, sufficient data are still necessary to construct and calibrate the models. An appealing feature of model-based inference is that it facilitates *synthetic estimation* (Särndal et al. 1992, p. 399) whereby a model is constructed and calibrated using data for a large area and then is applied to a smaller area, often within the larger area, under the assumption that the model is still appropriate. Thus, the greater amount of data available for the larger area is used to construct and calibrate the model, the same as for design-based inference, but the uncertainty of estimates is evaluated only for the smaller area (McRoberts 2006; 2013a, b; Ståhl et al. 2011). Global models may be used also to provide local estimates even with model-assisted estimators under the design-based approach provided that the local sample is sufficiently large. One such example is provided by Næsset et al. (2011) who used sample plots from a large area to fit a model for biomass with ALS as auxiliary data and provided separate model-assisted estimates of biomass for each of three smaller districts within the larger area. McRoberts (2010, 2013b, Table 1) noted though that whereas the variance of design-based, model-assisted regression estimates decrease for larger areas as the result of larger sample sizes, such is not the case for model-based variance estimation using synthetic model estimators. The reason is twofold. First, as per Eq. 14.6, the variance estimator of the mean is a two-dimensional mean over population units, not just sample units, and is scaled to a per unit area basis so that the size of an area for which the estimate is calculated is mostly irrelevant. Second, as per Eq. 14.7 the primary component of the variance estimator is the covariance matrix of the model parameters which is estimated for the large area and does not change, regardless of the size of the area to which the model is applied.

14.3.3   Inference for Partial ALS Coverage

In lieu of satellite platforms for acquiring ALS data, acquisition via airborne platforms is the current default. However, acquisition of full airborne ALS coverage may be prohibitively expensive for the large areas that are of interest for operational use. An alternative is to acquire full coverage for only a sample of units consisting of geographically contiguous areas, an approach characterized as partial coverage. To minimize logistical and cost constraints, these contiguous areas are most often strips corresponding to aircraft flight lines.

When used specifically as an inventory tool, ALS data usually fulfills an intermediate role – in terms of both cost per unit area and quality/detail of information provided – between field measurements and satellite imagery. Therefore ALS is a natural source of intermediate-level sampled data – where ALS data only partially covers the area of interest – in two-phase sampling designs for forest inventory. Two-phase sampling involves collection of measurements for two different samples. First, a relatively large first-phase sample is collected from the population using a probability sampling design, and information such as ALS data is collected for each first-phase sample element. The second-phase sample consists of a subsample of the first-phase sample selected using an arbitrary probability sampling design. The observations or measurements of the variable of interest, in addition to all information acquired for the first-phase sample, are then obtained for each second-phase element (Särndal et al. 1992, p. 344). If the inventory variable measured for the second-phase sample is strongly correlated with the information collected in the first phase, two-phase sampling designs can be efficient and economical alternatives to single-phase sampling. In this section, multiple approaches, including both model-assisted and model-based estimation, that incorporate partial ALS coverage are described.

14.3.3.1   Double Sampling

Use of information acquired from aerial photography in the first phase of two-phase sampling designs has a long history for forest inventory applications. The first-phase sample is essentially a simple random or systematic sample of aerial photo plots distributed across the area of interest. On each aerial photo plot, measurements of volume, stand size, stand density, and forest type can be acquired relatively quickly and inexpensively via manual photointerpretation. A subsample of photo plots is then visited in the field, where individual tree measurements are used to estimate plot-level inventory attributes such as volume, biomass, and basal area (Fig. 14.1). Regression models are then constructed to represent the relationship between the field and photo measurements and used in a double-sampling for regression design to increase the precision of the inventory population parameter estimates. Another

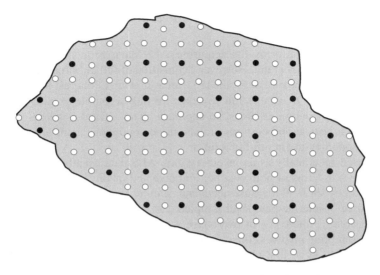

Fig. 14.1 Double sampling design. *White circles* represent first-phase plots measured using remotely sensed data; *black circles* represent second-phase plots with both remotely sensed and field measurements

desirable quality of double-sampling estimators is that they are unbiased in the sense that their mean over repeated sampling equals the population parameter of interest.

Under the SRS assumption, the estimator of the population mean using a double-sampling for regression (dsr) design is,

$$\widehat{\mu}^{dsr} = \frac{1}{m}\sum_{i\in S_1} \widehat{y}_i + \frac{1}{n}\sum_{i\in S_2} (y_i - \widehat{y}_i), \qquad (14.10)$$

where m is the size of the first-phase sample, S_1, of remote sensing plots (in this case every grid cell covering the area) and n is the size of the second phase sample, S_2, of field plots. The expression $\frac{1}{m}\sum_{i\in S_1} \widehat{y}_i$ in Eq. 14.10 is an estimator in its own right, the *synthetic* estimator, but in contrast to $\widehat{\mu}^{dsr}$ is not unbiased because no compensation is made for systematic model prediction error. When m is much greater than n, a variance estimator for $\widehat{\mu}^{dsr}$ is given by:

$$\widehat{\text{V}}\text{ar}\left(\widehat{\mu}^{dsr}\right) \approx \frac{s_\varepsilon^2}{n} + \frac{s_y^2}{m} \qquad (14.11)$$

where s_ε^2 is the variance of the regression residuals and s_y^2 is the variance of the second phase sample S_2. Under the SRS assumption and a linear model, and ignoring finite population correction factors, the variance estimator for the double-sampling estimator can be expressed as,

$$\widehat{V}\widehat{\mathrm{ar}}\left(\widehat{\mu}^{\mathrm{dsr}}\right) \approx s_y^2 \cdot \left(1 - R^2\right), \tag{14.12}$$

where R^2 is the coefficient of determination of the regression model. This form of the double-sampling variance estimator illustrates that if the first-phase sample measurements (remote sensing data) have a strong relationship with the variable of interest measured only on the second-phase units as indicated by R^2, substantial reduction in the variance of the two-phase estimator can be realized.

Multiple reports of the use of ALS data as auxiliary information in double-sampling forest inventory designs have been published where ALS plots are used in the same manner as aerial photo plots. For example, Parker and Evans (2004) demonstrated use of a double-sampling design where ALS data were acquired in first-phase strips, and 15 second-phase ALS plots were established in each strip. Tree diameter at-breast-height (dbh) and height were measured on every fifth second-phase ALS plot, and ALS-based individual tree estimates of basal area and volume were aggregated at the plot level. A linear model was constructed for the relationship between the field-based plot-level estimates and the corresponding ALS-based estimates of basal area and volume, and the authors report that mean volume was estimated with a standard error of 5.9 %. In another study (Parker and Evans 2009), the same authors used a double-sampling approach with stratified sampling and ALS individual tree measurements. A 9:1 ratio of first-phase ALS plots to second-phase field plots was used and yielded relative standard errors for volume estimates in the range of 2.1–11.4 %. Stephens et al. (2012) used a double-sampling forest carbon inventory in New Zealand with ALS data as auxiliary information. Andersen and Breidenbach (2007) used a simulation approach to investigate the statistical properties of both synthetic and double-sampling estimators of mean stand biomass and confirmed that double-sampling estimators were unbiased at the stand-level, although the variance could be quite large in cases where the number of field plots per stand was small. Næsset et al. (2013b) used double-sampling with a model-assisted generalized regression estimator to estimate biomass for Hedmark County (27,000 km^2), Norway and found that the standard error of the biomass estimate using double sampling was reduced to 35 % of that obtained using field plots alone.

14.3.3.2   Two-Phase ALS Strip Sampling

Strip sampling has a long history in forest inventory, largely due to the efficiencies gained from collecting measurements along a fixed line such as a compass bearing. Schreuder et al. (1993) mention a type of multi-level sampling design termed *line plot sampling* whereby topographical and forest type data are collected along the

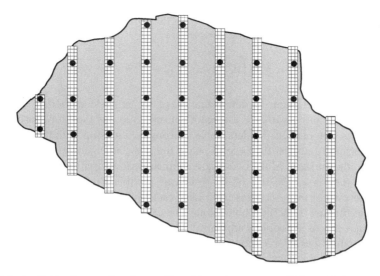

Fig. 14.2 An ALS strip sampling design. The *white strips* represent ALS acquisition corresponding to aircraft flight lines; the squares within the strips represent ALS grid cells; the *black circles* represent plots with both ALS and field measurements

strips, and forest measurements are acquired at inventory plots established at fixed intervals along the same strip lines. In this sense, the information collected along the lines serves as auxiliary information to increase the precision of estimates obtained using the plot data.

Acquisition of ALS data in strips corresponding to straight aircraft flight lines is currently regarded as the most efficient approach (Wulder et al. 2012). The rationale is twofold: first, fixed-wing aircraft are currently the most economical acquisition platform, and second, acquisition costs are a linear function of flight time, apart from fixed mobilization costs. When strips corresponding to flight lines are systematically distributed over an area of interest and field plots are distributed along the ALS strips, statistical estimators corresponding to two-phase sampling designs are appropriate (Fig. 14.2).

Approaches to estimation and inference based on this ALS strip sampling have been proposed that use both model-assisted (Andersen et al. 2009; Gregoire et al. 2011) and model-based two-phase (Ståhl et al. 2011; Andersen et al. 2011a) statistical frameworks.

One approach for model-assisted estimation using ALS data acquired in strips consists of viewing the design as a cluster sample for which auxiliary information is available for entire clusters (Särndal et al. 1992, p. 323). Each ALS strip is an individual cluster for which auxiliary ALS-derived covariate information is obtained for every element in the cluster. A sample of field plots is then established within each ALS strip or cluster, and the response variable of interest such as biomass is measured for every plot. A plot-level regression model is constructed and used to predict the variable for each element in each sampled cluster. The model-assisted estimator of the mean is,

$$\hat{\mu} = \frac{1}{m}\sum_{i\in S_1}\hat{y}_i + \frac{1}{mn}\sum_{i\in S_1}\sum_{k\in S_{i2}}\varepsilon_{ik} \qquad (14.13)$$

where S_1 is the first-phase sample of clusters with sample size m, S_{i2} is the second-phase sample for the ith first-phase cluster with common sample size n, and $\varepsilon_{ik} = y_{ik} - \hat{y}_{ik}$. Under the assumption of equal size clusters and equal size samples of plots within clusters, and ignoring finite population correction factors, the corresponding variance estimator can be formulated as,

$$\widehat{Var}(\hat{\mu}) \approx \left(\frac{1}{m}\right)\left[s_{\hat{\mu}}^2 - \frac{1}{m}\sum_{i\in S_1}\left(\frac{s_{yi}^2}{n}\right)\right] + \frac{1}{m^2}\sum_{i\in S_1}\left(\frac{s_{\varepsilon i}^2}{n}\right), \qquad (14.14)$$

where

$$s_{\hat{\mu}}^2 = \frac{1}{m-1}\sum_{i\in S_1}(\bar{y}_i - \bar{y})^2,$$

$$\bar{y} = \frac{1}{m}\sum_{i\in S_1}\bar{y}_i,$$

$$s_{yi}^2 = \frac{1}{n-1}\sum_{k\in S_{i2}}(y_{ik} - \bar{y}_i)^2,$$

$$s_{\varepsilon i}^2 = \frac{1}{n-1}\sum_{k\in S_{i2}}(\varepsilon_{ik} - \bar{\varepsilon}_i)^2.$$

Gregoire et al. (2011) provides a more complete formulation of the estimator for more general assumptions. In Eq. 14.14, the term with $s_{\hat{\mu}}^2$ represents the portion of the variance contributed by the variability among all possible samples of clusters (flight lines), the term with s_{yp}^2 represents the portion contributed by variability for all possible samples of elements within clusters and the term with $s_{\varepsilon p}^2$ represents the variability around the estimated model predictions.

This model-assisted inferential approach was demonstrated by Andersen et al. (2009) for the Kenai Peninsula of Alaska, USA and by Gregoire et al. (2011) for Hedmark County, Norway. For both studies, aboveground biomass was the inventory variable of interest, and ALS data were acquired in strips over a systematic grid of national forest inventory (NFI) plots. Gregoire et al. (2011) investigated the statistical properties of model-assisted biomass estimators at both the population level as well as subsets of the population such as different forest types. As previously, the primary advantage of using a model-assisted approach is that estimators are asymptotically unbiased, even in cases for which the model is misspecified or exhibits lack of fit. Unbiasedness is a very desirable property for forest inventory applications for which estimates of carbon-based parameters and their stated level of precision can have important economic or policy implications. The design requirements for implementing a model-assisted cluster sampling approach to ALS sampling can be challenging, however. In particular, because the field plots within strips are assumed to represent a probability sample, every unit in the population must have a non-zero probability of being selected in the second-stage sample. This requirement is not problematic in cases of NFI plot grids such as for the Kenai and Hedmark studies. However, where the area of interest is large, remote, and inaccessible, establishment of true probability samples of field plots can be prohibitively expensive. In the latter case, a model-based approach may be the only feasible option (Sect. 14.3.2.2). Both Andersen et al. (2009) and Gregoire et al. (2011), with subsequent analyses by Gobakken et al. (2012), reported that the estimates obtained from a model-assisted approach using ALS sampling tended to be close to those obtained from the ground NFI data. However, both studies also reported that the standard errors for the model-assisted estimator of Eq. (13) tended to be greater than the standard errors for the estimates obtained from the NFI plots alone, even after accounting for the greater number of plots used for the NFI estimates. Simulation analyses based on the Hedmark data (Ene et al. 2012) indicated that the model-assisted variance estimators for cluster sampling performed well when the ALS strips were collected as a simple random sample but substantially overestimated the true variance when the strips were actually collected as a systematic sample (strips are implicitly assumed to be collected via simple random sampling in the variance estimator). This result may largely explain the overestimation reported in both Gregoire et al. (2011) and Andersen et al. (2009). By treating the NFI field data in Hedmark according to a cluster design with plots along the ALS flight lines as individual clusters and thus making the assumed design for the field survey more similar to the design of the ALS-assisted sample survey, it was shown that the ALS-assisted variance estimates were only 40 % of the field-based variance estimates (Næsset et al. 2013b).

Although an unbiased design-based estimator of variance is not available under systematic sampling, approximate solutions such as obtained using successive differences estimators, have been suggested and could provide more satisfactory estimates of precision under systematic sampling (Wolter 2007). Ene et al. (2013) found that successive differences variance estimators produced estimates that were consistently closer to the true value than the model-assisted estimators that implicitly assumed SRS.

Model-based approaches to estimation and inference have also been proposed for data collected using ALS strip sampling designs. As per Sect. 14.3.2.2, the assumptions underlying model-based inference are fundamentally different than those underlying model-assisted inference with the consequence that the statistical properties of the two kinds of estimators are difficult to compare directly. For example, in contrast to model-assisted inference, model-based inferences are conditional on the models, so that variance estimators do not incorporate the effects of any model lack of fit. Ståhl et al. (2011) described the statistical properties of a model-based estimator of total and stratum-level biomass using the same data set as Gregoire et al. (2011). A seminal contribution of this paper is a detailed description of the uncertainty budget in terms of the variability due to both model prediction uncertainty for individual population elements and population variability of the model predictions. Understanding how various components of the uncertainty budget contribute to the overall variance can be helpful in planning future ALS-assisted inventories. The authors reported that model-based ALS estimates over all sites were about 6 % greater than ground-based biomass estimates and that for the productive forest stratum, the component of the variance due to model parameter uncertainty ranged from 58 to 85 % of the total variance.

Andersen et al. (2011a) used a model-based approach to estimate forest biomass using field plots and ALS strip sampling over a 201,226 ha area within the upper Tanana valley of interior Alaska, USA. Because of the remoteness of the study area, field plots could only be established in areas relatively close to roads, trails, or rivers, whereas the ALS data were collected as a systematic sample of strips over the entire study area. Although the field plots were collected across a range of representative forest conditions and were randomly located within the accessible area, it was not a probability sample over the entire area and therefore a model-based approach to estimating biomass was necessary. The sampling design was considered a single-stage cluster sample with model-based estimation of biomass within each ALS grid cell. A bootstrap resampling approach was used to estimate the uncertainty of the total biomass estimates where the resampling was carried out at multiple levels to account for both sampling error due to cluster sampling and model error (analogous to the first and second terms, respectively, in the analytical variance estimator of Ståhl et al. (2011)). A standard error of 8 % for the estimate of total biomass was reported.

14.3.3.3 Multi-level Sampling Designs Using ALS and Other Auxiliary Data Sources

ALS strip sampling and field plots can be used in combination with wall-to-wall satellite remotely sensed data to further improve the precision of inventory parameter estimates (Fig. 14.3).

Andersen et al. (2011b) used a model-based approach to estimate biomass in the upper Tanana valley of interior Alaska using field plot data, ALS data acquired in strips, and both satellite multispectral (Landsat TM) imagery and satellite L-band (ALOS PALSAR) radar imagery for the entire study area. First, a regression model of the relationship between plot-level biomass and corresponding ALS data was constructed; second, the model was used to predict biomass for all grid cells within the ALS strips; and third, biomass was predicted for each grid cell in the entire study area using nearest-neighbor imputation with predictions for the ALS

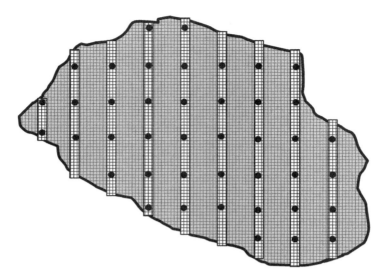

Fig. 14.3 Sampling design with multiple sources of remotely sensed data. The *white strips* represent ALS acquisition corresponding to aircraft flight lines; the squares within the strips represent ALS grid cells; the grid cells over the entire area represent satellite image pixels; the *black circles* represent plots with both ALS, satellite, and field measurements

grid cells as reference data and the Landsat TM spectral reflectance and dual-pole radar backscatter data as feature or predictor variables. The results indicate that augmenting ALS data with wall-to-wall satellite data might substantially improve the precision of total biomass estimates.

14.4 Summary

The use of ALS data has been convincingly demonstrated to enhance both stand management and strategic forest inventories. Both approaches rely heavily on models of relationships between inventory attributes of interest and ALS metrics. Models that better fit the reference data and that exhibit less residual variability produce spatial estimates of greater quality. The design-based estimators presented include an adjustment for estimated bias induced by model lack of fit, whereas the model-based estimators assume no lack of model fit. For stand management inventories, no adjustments for estimated bias are typically included because less exact approximations can be tolerated.

Of crucial inferential importance, estimators must conform to sampling designs. For full ALS coverage, the estimators are considerably less complex because greater flexibility, and hence greater simplicity, is possible for ground sampling designs. However, because costs associated with full ALS coverage are prohibitive for large areas, ALS is increasingly used in a sampling mode. Further, in the absence of space borne lasers, ALS data are typically, but not exclusively, acquired by fixed wing aircraft. For these acquisitions, cost efficiencies argue in favor of strip samples corresponding to straight aircraft flight lines. The results are greater cost efficiencies but also greater computational complexities. Thus, considerable care must be exercised in both the design of ALS-assisted inventories and the analysis of the resulting data.

Designing cost-effective, multi-phase ALS-inventories is a complex task. To assist with this task, simulation tools may be used to investigate the statistical properties of estimators corresponding to different sampling designs for a variety of populations (Ene et al. 2012). Currently, however, such tools are not available to practitioners for most application. Nevertheless, as the results of sampling simulations across spatial scales and forest types accumulate over time, a common international basis of knowledge for designing ALS-assisted inventories will emerge.

References

Andersen H-E, Breidenbach J (2007) Statistical properties of mean stand biomass estimators in a LIDAR-bases double sampling forest survey design. In: Proceedings of the ISPRS Workshop Laser Scanning 2007 and SilviLaser 2007. 12–14 September 2007, Espoo, Finland. IAPRS, Vol XXXVI, Part 3/W52, 2007, pp 8–13

Andersen H-E, Barrett T, Winterberger K, Strunk J, Temesgen H (2009) Estimating forest biomass on the western lowlands of the Kenai Peninsula of Alaska using airborne lidar and field plot data in a model-assisted sampling design. In: Proceedings of the IUFRO Division 4 conference, extending forest inventory and monitoring over time and space, Quebec City, Canada, 19–22 May 2009. Available at: http://blue.for.msu.edu/meeting/proceed.php

Andersen H-E, Strunk J, Temesgen H (2011a) Using airborne light detection and ranging as a sampling tool for estimating forest biomass resources in the Upper Tanana Valley of interior Alaska. West J Appl For 26:157–164

Andersen H-E, Strunk J, Temesgen H, Atwood D, Winterberger K (2011b) Using multilevel remote sensing and ground data to estimate forest biomass resources in remote regions: a case study in the boreal forests of interior Alaska. Can J Remote Sens 37:596–611

Brewer KR (1963) Ratio estimation in finite populations: some results deductible from the assumption of an underlying stochastic process. Aust J Stat 5:93–105

Cochran WG (1977) Sampling techniques, 3rd edn. Wiley, New York, p 428

Corona P, Fattorini L (2008) Area-based lidar-assisted estimation of forest standing volume. Can J For Res 38:2911–2916

d'Oliviera MVN, Reutebuch S, McGaughey R, Andersen H-E (2012) Estimating forest biomass and identifying low-intensity logging areas using airborne scanning lidar in Antimary State Forest, Acre State, Western Brazilian Amazon. Remote Sens Environ 124:479–491

Eid T, Gobakken T, Næsset E (2004) Comparing stand inventories based on photo interpretation and laser scanning by means of cost-plus-loss analyses. Scand J For Res 19:512–523

Ene LT, Næsset E, Gobakken T, Gregoire TG, Ståhl G, Nelson R (2012) Assessing the accuracy of regional LiDAR-based biomass estimation using a simulation approach. Remote Sens Environ 123:579–592

Ene LT, Næsset E, Gobakken T, Gregoire TG, Ståhl G, Holm S (2013) A simulation approach for accuracy assessment of two-phase post-stratified estimation in large-area LiDAR biomass surveys. Remote Sens Environ 133:210–224

Firth D, Bennett KE (1998) Robust models in probability sampling. J R Stat Soc B 60:3–21

Gobakken T, Næsset E (2008) Assessing effects of laser point density, ground sampling intensity, and field sample plot size on biophysical stand properties derived from airborne laser scanner data. Can J For Res 38:1095–1109

Gobakken T, Næsset E, Nelson R, Bollandsås OM, Gregoire TG, Ståhl G, Holm S, Økra HO, Astrup R (2012) Estimating biomass in Hedmark, County, Norway using national forest inventory field plots and airborne laser scanning. Remote Sens Environ 123:443–456

Gregoire TG (1998) Design-based and model-based inference: appreciating the difference. Can J For Res 28:1429–1447

Gregoire TG, Ståhl G, Næsset E, Gobakken T, Nelson R, Holm S (2011) Model-assisted estimation of biomass in a LiDAR sample survey in Hedmark county, Norway. Can J For Res 41:83–95

Hansen MH, Madow WG, Tepping BJ (1983) An evaluation of model-depending and probability-sampling inferences in sample surveys. J Am Stat Assoc 78:776–793

Hawbaker TJ, Keuler NS, Lesak AA, Gobakken T, Contrucci K, Radeloff VC (2009) Improved estimates of forest vegetation structure and biomass with a LiDAR-optimized sampling design. J Geophys Res 114:G00E04

Holmgren J, Jonsson T (2004) Large scale airborne laser scanning of forest resources in Sweden. In: Thies M, Koch B, Spiecker H, Weinacker H (eds) Laser-scanners for forest and landscape assessment. International Society of Photogrammetry and Remote Sensing. International archives of photogrammetry, remote sensing and spatial information sciences, Freiburg, Germany, pp 157–160

Kangas A, Maltamo M (eds) (2006) Forest inventory: methodology and applications. Springer, Dordrecht, 363 pp

Lohr S (1999) Sampling: design and analysis. Duxbury, Pacific Grove, p 494

Maltamo M, Bollandsås OM, Næsset E, Gobakken T, Packalén P (2011a) Different plot selection strategies for field training data in ALS-assisted forest inventory. Forestry 84:23–31

Maltamo M, Packalén P, Kallio E, Kangas J, Uuterra J, Heikkilä (2011b) Airborne laser scanning based stand level management inventory in Finland. In: Proceedings of SilviLaser 1011, 11th international conference on LiDAR applications for assessing forest ecosystems, University of Tasmania Australia, 16–20 October 2011, pp 1–10

Mandallaz D (2008) Sampling techniques for forest inventories. Chapman & Hall, New York, p 256

Mascaro J, Detto M, Asner GP, Muller-Landau HC (2011) Evaluating uncertainty in mapping forest carbon with airborne LiDAR. Remote Sens Environ 115:3770–3774

Mátern B (1960). Spatial variation. Medd. Statens Skogsforskningsinst. Band 49, No. 5. (Reprinted as volume 36 of the series Lecture notes in statistics. 1986. Springer, New York, 151 p)

McRoberts RE (2006) A model-based approach to estimating forest area. Remote Sens Environ 128:268–275

McRoberts RE (2010) Probability- and model-based approaches to inference for proportion forest using satellite imagery as ancillary data. Remote Sens Environ 114:1017–1025

McRoberts RE, Gobakken T, Næsset E (2012) Post-stratified estimation of forest area and growing stock volume using lidar-based stratifications. Remote Sens Environ 125:157–166

McRoberts RE, Næsset E, Gobakken T (2013a) Accuracy and precision for remote sensing applications of nonlinear model-based inference. IEEE J Sel Top Appl Earth Obs Remote Sens 6:27–34

McRoberts RE, Næsset E, Gobakken T (2013b) Inference for lidar-assisted estimation of forest growing stock volume. Remote Sens Environ 128:268–275

McRoberts RE, Tomppo EO, Freitas J, Vibrans AC (2013c) Design considerations for tropical forest inventories. Braz J For Res 33:1–14

Næsset E (2002) Predicting forest stand characteristics with airborne scanning laser using a practical two-stage procedure and field data. Remote Sens Environ 80:88–99

Næsset E (2004a) Practical large-scale forest stand inventory using a small airborne scanning laser. Scand J For Res 19:164–179

Næsset E (2004b) Accuracy of forest inventory using airborne laser-scanning: evaluating the first Nordic full-scale operational project. Scand J For Res 19:554–557

Næsset E (2007) Airborne laser scanning as a method in operational forest inventory: status of accuracy assessments accomplished in Scandinavia. Scand J For Res 22:433–442

Næsset E, Bjerknes K-O (2001) Estimating tree heights and number of stems in young forest stands using airborne laser scanner data. Remote Sens Environ 78:328–340

Næsset E, Gobakken T, Holmgren J, Hyyppä H, Hyyppä J, Maltamo M, Nilsson M, Olsson H, Persson Å, Söderman U (2004) Laser scanning of forest resources: the Nordic experience. Scand J For Res 19:482–499

Næsset E, Gobakken T, Solberg S, Gregoire TG, Nelson R, Ståhl G, Weydahl D (2011) Model-assisted regional forest biomass estimation using LiDAR and InSAR as auxiliary data: a case study from a boreal forest area. Remote Sens Environ 115:3599–3614

Næsset E, Bollandsås OM, Gobakken T, Gregoire TG, Ståhl G (2013a) Model-assisted estimation of change in forest biomass over an 11 year period in a sample survey supported by airborne LiDAR: a case study with post-stratification to provide "activity data". Remote Sens Environ 128:299–314

Næsset E, Gobakken T, Bollandsås OM, Gregoire TG, Nelson R, Ståhl G (2013b) Comparison of precision of biomass estimates in regional field sample surveys and airborne LiDAR-assisted surveys in Hedmark County, Norway. Remote Sens Environ 130:108–120

Parker RC, Evans DL (2004) An application of LiDAR in a double-sampling forest inventory. West J Appl For 19:95–101

Parker R, Evans D (2009) LiDAR forest inventory with single-tree, double-, and single-phase procedures. Int J For Res 2009:864108

Rennolls K (1982) The use of superpopulation-prediction methods in survey analysis, with application to the British National Census of Woodlands and Trees. In: Lund HG (ed) In place resource inventories: principles and practices. 9–14 Aug 1981. Society of American Foresters, Orono/Bethesda, pp 395–401

Royall RM (1970) On finite population sampling theory under certain linear regression models. Biometrika 57(2):377–387

Royall RM, Herson J (1973) Robust estimation in finite populations II. J Am Stat Assoc 68:890–893

Särndal C-E, Swensson B, Wretman J (1992) Model assisted survey sampling. Springer, New York

Schreuder H, Gregoire T, Wood G (1993) Sampling methods for multiresource forest inventory. Wiley, New York

Ståhl G, Holm S, Gregoire TG, Gobakken T, Næsset E, Nelson R (2011) Model-based inference for biomass estimation in a LiDAR sample survey in Hedmark county, Norway. Can J For Res 41:96–107

Steinmann K, Mandallaz D, Ginzler C, Lanz A (2013) Small area estimations of proportion of forest and timber volume combining Lidar data and stereo aerial images with terrestrial data. Scand J For Res 28(4):373–385

Stephens PR, Kimberley MO, Beets PN, Paul TSH, Searls N, Bell A, Brack C, Broadly J (2012) Airborne scanning LiDAR in a double sampling forest carbon inventory. Remote Sens Environ 117:348–357

Strunk J, Reutebuch S, Andersen H-E, Gould P, McGaughey R (2012) Model-assisted forest yield estimation with light detection and ranging. West J Appl For 27(2):53–59

Tomppo E, Gschwantner T, Lawrence M, McRoberts RE (eds) (2010) National forest inventories – pathways for common reporting. Springer, Dordrecht, 612 pp

Turner R, Goodwin N, Friend J, Mannes D, Rombouts J, Haywood A (2011) A national overview of airborne lidar applications in Australian forest agencies. In: Proceedings SilviLaser 2011. 16–19 October 2011, Hobart, Tasmania, Australia, 13 pp

Turunen L, Pesonen A, Suvanto A (2012) Fjernanalysebasert skogregistrering i Finland (Remote sensing based forest inventory in Finland). Kart og Plan 72:184–187

White JC, Wulder MA, Varhola A, Vastaranta M, Coops NC, Cook BD, Pitt D, Woods M (2013) A best practice guide for generating forest inventory attributes from airborne laser scanning data using an area-based approach (Version 2.0). Canadian Forest Service, Information report FI-X-010, 39 pp

Wolter K (2007) Introduction to variance estimation, 2nd edn. Springer, New York

Wulder M, White J, Nelson R, Naesset E, Orka H, Coops N, Hilker T, Bater C, Gobakken T (2012) Lidar sampling for large-area forest characterization: a review. Remote Sens Environ 121:196–209

Chapter 15
Modeling and Estimating Change

Ronald E. McRoberts, Ole Martin Bollandsås, and Erik Næsset

Abstract Airborne laser scanning (ALS) data are a source of spatial information that can be used to assist in the efficient and precise estimation of forest attributes such as biomass and biomass change, particularly for remote and inaccessible forests. This chapter includes an introduction to the use of ALS data for estimating change and a detailed review with tabular summary of the small number of known published reports on the topic. The review proceeds chronologically, noting the progression from exploratory and correlation studies to modeling and mapping and finally to statistically rigorous inference for population parameters. Both direct and indirect approaches for constructing models and mapping change are summarized. Although maps can be used to assist in the estimation procedure, systematic model prediction errors as reflected in maps induce bias into estimators. Thus, if the objective is rigorous inference for population parameters such as mean biomass change per unit area, rather than just maps of the populations, then bias must be estimated and incorporated into the parameter estimates, and uncertainty must be estimated. The design-based, model-assisted estimators that are presented for both independent estimation samples and single samples with repeated observations satisfy these criteria and produce inferences in the form of confidence intervals.

R.E. McRoberts (✉)
US Department of Agriculture Forest Service, Northern Research Station,
1992 Folwell Avenue, St Paul, Minnesota, MN, USA
e-mail: mcrob001@umn.edu

O.M. Bollandsås • E. Næsset
Department of Ecology and Natural Resource Management, Norwegian
University of Life Sciences, Ås, Norway
e-mail: Erik.naesset@umb.no

M. Maltamo et al. (eds.), *Forestry Applications of Airborne Laser Scanning: Concepts and Case Studies*, Managing Forest Ecosystems 27, DOI 10.1007/978-94-017-8663-8_15,
© Springer Science+Business Media Dordrecht 2014

15.1   Introduction

Estimates of changes in the amounts and distributions of forest biomass are important for both general monitoring purposes and for assessment in the context of climate change. For example, countries that ratified the Kyoto protocol have also committed to reporting estimates of their emissions and removals of atmospheric carbon for land use, land use change and forestry (LULUCF) activities (Höhne et al. 2007) such as afforestation, reforestation, deforestation, and forest management. To be credible, the reported estimates must be based on objective data. The most common and comprehensive sources of objective forest data are national forest inventories (NFI) (e.g. Rypdal et al. 2005; Tomppo et al. 2010). The large range of variables observed and measured on the extensive and systematic NFI networks of field plots produce estimates of multiple parameters including many related to biomass and carbon. Change for NFI variables such as biomass between times t_1 and t_2 is typically estimated as the difference between the estimate for t_2 and the estimate for t_1. However, because the aggregate coverage of NFI plots is relatively small, even with intense sampling, the spatial resolution of pure NFI-based estimates is limited with the result that only large area estimates of current biomass and the change in biomass can be reported reliably.

The United Nations collaborative program on Reducing Emissions from Deforestation and forest Degradation in developing countries (UN REDD) relies on establishment of efficient monitoring systems for purposes of estimating, reporting, and verifying changes in carbon stocks. In addition to carbon stocks, great interest has been expressed in monitoring the possible effects of climate change on forests such as changes in growth rates, accumulation of biomass in existing forest, and the invasion of trees on previously treeless areas. Pure field based inventory and monitoring systems such as NFIs can be and are used to estimate change over large areas such as regions and nations. However, because change phenomena of interest such as the invasion of forest into mountainous areas are often relatively rare, areas subject to specific changes and the consequent number of field plots are both small with the result that the precision of plot-based estimates is also small.

Airborne laser scanning (ALS) is a source of auxiliary data that can improve the efficiency of efforts to estimate change and can contribute to improving the precision of change estimates. Multiple approaches for using ALS data to estimate change may be considered. First, for areas of limited size, full area ALS coverage can be obtained, but for larger areas cost factors may constrain ALS data acquisition to strips. Strip sampling approaches, while less costly, require more complex statistical estimators and, of necessity, lead to less precise estimates because sampling uncertainty in the form of the strip-to-strip variability must be incorporated into variance estimates.

Second, methods for modeling and estimating change using multi-temporal field observations and corresponding multi-temporal ALS-data are often characterized as either direct or indirect (Sect. 15.3). With the direct method, change in the variable of interest is estimated directly from multi-temporal auxiliary data such as ALS height and density metrics. With the indirect method, change is estimated as the difference between estimates of the variable at two points in time.

Third, change can be estimated and relationships between change and auxiliary variables can be modeled at different spatial scales. For example, segment-based modeling and estimation focus on spatial units such as individual trees, whereas change modeling for grid cells focuses on regular, fixed area spatial units larger than the crowns of individual trees. Thus, inventory methods that use ALS data are frequently characterized as either single-tree or area-based (Næsset 2002, 2004). The success of single-tree methods depends largely on successful segmentation of crowns for individual trees (Hyyppä and Inkinen 1999; Popescu and Wynne 2004; Solberg et al. 2006a), a challenging task because the algorithms can both fail to detect some trees (omission errors) and produce false trees (commission errors) (Kaartinen et al. 2012; Vauhkonen et al. 2012). The accuracy of segmentation is influenced by both forest conditions (Falkowski et al. 2008) and the point density of the ALS data (Tesfamichael et al. 2009). Because omission and commission errors often vary between measurement occasions, they are potentially more serious for change estimation than for inventories of current conditions.

Area-based methods focus on constructing models of relationships between ground plot observations and corresponding ALS metrics that describe the vertical and horizontal distributions of the ALS pulse echoes. For areas with no ground observations, the models may be used to predict biomass or biomass change for grid cells that are ideally the same size or larger than the ground plot, but not necessarily the same shape. Biomass or biomass change for an entire area is then estimated as the sum of the model predictions for all grid cells. Area-based inventories conducted at two points in time facilitate estimation of change as the difference in estimates for both grid cells and coarser units such as forest stands or forests.

The remainder of the chapter is organized into three main sections. Section 15.2 consists of a review of the literature in three parts: single tree studies, canopy gap studies, and area-based studies. Section 15.3 focuses on modeling and mapping, primarily as preliminary and background information for Sect. 15.4 which addresses estimation. Although a map is often a useful end product, in a broader context it is simply an intermediate product on the path to an inference in the form of a confidence interval for a parameter of the population that the map only inaccurately depicts. Finally, although the focus is estimation of biomass or biomass change using ALS metrics, the approaches documented are generally appropriate for estimation using any remotely sensed data.

15.2   Literature Review

In characterizing the scientific method, Reichenbach (1938) emphasized the distinction between the context of *Discovery* and the context of *Justification*. Discovery is characterized by exploratory studies that estimate correlations between response and predictor variables and develop empirical models that account for variation in observations of the response variables. Whereas Discovery is a creative enterprise and is subject to few constraints, Justification has a formal logic that leads to scientific inferences characterized by statistical rigor. Thus, the chronological maturation of a discipline would be expected to be characterized by early exploratory and feasibility studies but shifting toward greater statistical rigor with the passage of time. As the following review indicates, such is the case for the use of ALS data to estimate forest change.

ALS-based change studies are of two general kinds, single-tree studies and area-based studies. Single-tree studies are characterized by individual crown segmentation algorithms, primary use of ALS maximum heights, and single-tree attributes such as tree height growth. Area-based studies are characterized by estimation for grid cells of sizes on the order of 100s of m^2, greater focus on multi-tree, below canopy attributes such as change in biomass per unit area, and use of distributions of heights of ALS pulse echoes.

15.2.1 Single Tree Studies

Because ALS data in their most basic form are height information, many early ALS change studies focused on exploring the relationships between ALS metrics and change in heights for single trees and aggregation of height changes for multiple trees. The first study using multi-temporal ALS data was conducted in Finland by Hyyppä et al. (2003) who estimated the height growth of single trees segmented from ALS data with pulse densities of 10 pulses/m^2. Single tree height growth was estimated as the difference between 1998 and 2000 maximum ALS heights within tree segments. Mean stand growth was estimated as the mean of single tree growth estimates within stand boundaries. Estimates of tree height growth based on differences in maximum ALS heights were within 15 cm of field observations of growth for approximately half the trees. At the stand-level, standard errors were less than 5 cm. Differences between field-based and ALS-based estimates of height growth were attributed to differences in the digital elevation models (DEM) between measurement occasions, failure to match trees between measurements, and suppressed trees that could not be identified. The important conclusion was that estimation of forest height growth is feasible using multi-temporal ALS data.

In a series of papers, Yu et al. (2003, 2004, 2005, 2006) used multi-temporal ALS data to estimate single-tree height change and related attributes for a study area in Finland. Yu et al. (2003, 2004) successfully detected 61 of 83 trees harvested between 1998 and 2000 using a tree matching algorithm, ALS data with pulse density of 10 pulses/m^2, and differences in maximum ALS heights. Trees that were not detected were generally below the dominant canopy. Means of single tree height growth estimates for plots and stands for the 2-year interval were estimated with accuracies of 5-cm and 10–15-cm, respectively. Yu et al. (2005) used an augmented dataset and an improved tree-to-tree matching algorithm to estimate height growth for 153 pine trees using differences in maximum ALS heights for

1998, 2000, and 2003. Statistically significant differences between field height growth and differences in maximum ALS heights were attributed to two causes. First, reception of an echo from the apex of a single coniferous tree is unlikely, even with dense ALS data, because the apex of a tree rarely has surface area sufficiently large to trigger an echo (Andersen et al. 2006). Second, differences can be partially attributed to the use of different sensors with different properties such as penetration rates at different times. Continuing rapid technological development diminishes the likelihood of using the same sensor at different times. Yu et al. (2006) estimated the correlation between observed height growth between 1998 and 2003 for segments corresponding to single trees and three ALS metrics: (i) differences between maximum ALS heights, (ii) mean and median of differences in digital surface models (DSM) corresponding to tree crowns, and (iii) differences in the 85th, 90th, and 95th ALS height percentiles. The single predictor variable that best explained the variation in observed growth was the difference between the maximum ALS heights. The two primary conclusions from this series of studies were that crown segmentation using ALS data is feasible, and that ALS metrics and changes in the metrics have potential for predicting individual tree and mean tree height growth.

15.2.2 Canopy Gap Studies

In multiple respects, canopy gap studies are intermediate between single-tree and area-based studies. Similar to single-tree studies, they typically focus on tree height and canopy-level response variables. Some canopy gap studies (e.g., Vastaranta et al. 2012) use distributions of ALS height metrics to segment individual crowns, whereas other studies (e.g., Kellner et al. 2009; Hirata et al. 2008) use ALS maximum heights and grid cells, albeit small grid cells ($0.25\text{--}25\ m^2$) that are closer in size to individual crowns than sizes used for more typical area-based studies.

St. Onge and Vepakomma (2004) used a region growing algorithm to assess differences in canopy height model predictions between 1998 and 2003 for 50-cm × 50-cm grid cells in Quebec, Canada. First echo ALS pulse densities were 0.3 for 1998 and 3.0 for 2003. Overall accuracy of the gap detection was 96 % based on comparisons with high resolution imagery. Vepakomma et al. (2008, 2010) used the same dataset for more detailed studies of gap and forest dynamics and concluded that gap sizes of $5\ m^2$ or greater could be reliably detected. Kellner et al. (2009) used 1997 and 2006 ALS data to characterize the dynamics and distribution of canopy gap sizes in Costa Rica. For each 5-m × 5-m grid cell, they calculated mean ALS height and estimated change in canopy height as differences in the means between 1997 and 2006. They then constructed a transition matrix for 1-m height classes for the 5-m × 5-m grid cells and reported three primary results: (i) ALS-based estimates of forest structure and dynamics were similar to ground-based estimates; (ii) most canopy gaps were produced by loss of vegetation below the canopy level; and (iii) the distribution of canopy heights was close to a steady-state equilibrium. Hirata et al. (2008) assessed canopy gap dynamics in Japan using ALS data acquired in 2001 and 2005 with densities in the range 32–50 pulses/m^2. Gaps were defined as areas for which canopy height was less than a specified threshold, and for each year gap areas were estimated. Based on changes in gap area estimates between 2001 and 2005, individual gaps were assigned to one of five change classes: appearance, enlargement, reduction, disappearance, and no change. For the reduction and disappearance classes, absolute values of estimates of area changes were positively correlated with initial gap sizes. The general conclusion from these canopy gap studies was that ALS data have considerable potential for quantitative evaluation of gaps and gap dynamics. More discussion on use of ALS data to assess canopy gaps can be found in Chap. 21.

15.2.3   Area-Based Studies

Whereas single tree studies focus on the use of ALS height data to segment individual trees, to predict attributes for single trees, and to predict means over single trees for larger areas, by contrast, area-based studies focus on estimation of plot and stand-level attributes using distributions of ALS heights without regard to individual trees.

Gobakken and Næsset (2004) compared height and density metrics obtained from 1999 to 2001 ALS first-pulse point clouds with densities of approximately 1 pulse/m^2 for 87 plots of size 300–400 m^2 in Norway. The plots were assigned to two categories, young forest and mature forest with poor site quality. Mean differences between 1999 and 2001 for all ALS metrics, except maximum ALS height in the mature stratum, were statistically significantly different from zero. Næsset and Gobakken (2005) used an extended dataset from the same study area and predicted change in mean stand tree height, stand basal area, and timber volume for selected stands in southeastern Norway using 1999 and 2001 ALS data with mean pulse densities of 0.87 and 1.18 pulses/m^2, respectively. For each year and each of three strata related to age class and site quality, regression models were constructed to represent relationships between the three response variables and ALS height percentiles. The models were used to predict the response variables for 350-m^2 grid cells that tessellated 56 stands, and change was estimated for each variable as the difference between the model predictions. Estimates of correlations between the response variables and nearly all ALS-based variables derived from the first-echo point cloud were statistically significantly greater than zero. However, estimates of correlations with last-echo variables were much smaller. Systematic deviations between growth predictions and field observations and large uncertainties were attributed to the short 2-year period between measurements. Nevertheless, the large correlations for both studies indicated that ALS metrics have the potential for predicting forest growth.

Solberg et al. (2006b) used multi-temporal ALS metrics to construct a map of change in leaf area index (LAI) as a measure of defoliation resulting from a pine-sawfly attack. ALS data with pulse densities in the range 3–10 pulses/m^2 were acquired in May, July, and September 2005 for a 21-km^2 area in southeastern Norway. For each date, a regression model was constructed to represent the relationship between field measured LAI and gap fraction calculated as the ratio of below canopy echoes and the total number of echoes. The models were then used to predict LAI for each 10-m × 10-m grid cell in the study area, and a defoliation map was constructed by geographically depicting differences in model LAI predictions for each grid cell. As expected, the maps indicated a decrease in LAI between May and July and a slight increase between July and September. The primary conclusion is that changes in gap fraction calculated from ALS data can be used to predict LAI change and to construct maps for identifying areas of defoliation.

Yu et al. (2008) used linear regression models to estimate relationships between plot-level mean canopy height and volume growth and three sets of ALS metrics. For ALS data acquired in 1998, 2000, and 2003 with pulse densities of 10 pulses/m^2, the metrics included differences between maximum ALS heights, mean and median of differences in DSMs corresponding to tree crowns, and differences in ALS height percentiles. Mean canopy height growth was best predicted by maximum ALS heights, and volume growth was best predicted by DSM differences. The primary conclusion was that multi-temporal ALS data can be used to estimate mean height and volume growth for spatial areas larger than individual trees.

Hopkinson et al. (2008) used ALS data to estimate plot-level height growth for a red pine plantation in Ontario, Canada. Their data consisted of field measurements of tree heights for 19 plots of 121-m^2 and ALS data with pulse densities in the range 2.5–3.6 pulses/m^2 acquired four times between 2000 and 2005. Differences in maximum ALS heights for 2000–2002, 2002–2004, and 2004–2005 were compared to corresponding observed mean plot-level height growth. The ALS-based height growth estimates correlated well with observed mean height growth for both methods, although changes in maximum ALS heights were, on average, smaller than observed changes. The reason is the same as for single tree studies, i.e., maximum ALS heights rarely capture canopy apices. The authors concluded that although uncertainties associated with annual estimates of plot-level change are unacceptably large, uncertainties on the order of 10 % for 3-year estimates of change are operationally acceptable.

Vastaranta et al. (2011) constructed a logistic regression model using ALS change data to predict the probability of snow damage for 85 grid cells of size 25-m^2 in Finland. Predictor variables were changes in height percentiles for ALS data acquired in 2007 with pulse density of 7 pulses/m^2 and in 2010 with pulse density of nearly 12 pulses/m^2. Overall accuracy with respect to predictions of the damage/no-damage class was 78.6 % when assessed using leave-one-out cross validation. Vastaranta et al. (2012) segmented damaged canopy areas using differences in canopy height model predictions between 2006/2007 and 2010. Detection of changes in all canopy layers due to causes such as storm damage or insect defoliation using full ALS height distributions and an area-based approach was judged to be superior to single-tree methods, particularly if the pulse density is small or if geometric alignment is a problem. These two studies demonstrated that abrupt structural changes to canopies can be detected using ALS-based height percentiles.

Nyström et al. (2013) used discriminant analysis with ALS metrics to classify field plots in the forest-tundra ecotone of northern Sweden with respect to three categories of harvest: (i) no harvest, (ii) harvest of 50 % of trees with heights greater than 1.5-m, (iii) harvest of all trees with height greater than 1.5-m. ALS data acquisitions were deliberately different to mimic the small likelihood of identical flight parameters over time. In 2008 ALS data were collected with a TopEye scanner with average pulse density of 12 pulses/m^2, and in 2010 data were acquired using an Optech scanner with average pulse density of 1.2 pulses/m^2. Histogram matching was used to calibrate the two sets of ALS variables. Classification accuracy was 0.88 using relative differences in ALS height percentiles and density metrics as predictor variables. The study demonstrated that histogram matching can alleviate difficulties associated with classification using ALS data acquired with different sensors with different flight parameters.

Bollandsås et al. (2013) evaluated three approaches for using ALS data to model and predict biomass change between 2005 and 2008 for a mountain forest in southeastern Norway. For each approach, the model predictor variables were either height percentiles or differences in height percentiles between 2005 and 2008 for ALS data with pulse densities of 3.4–4.7 pulses/m^2. The first approach was indirect and entailed constructing a model of the relationship between biomass observations and height percentiles aggregated for both years. Biomass change was then estimated as the difference between the model predictions for the 2 years. The second approach was direct and entailed constructing a model of the relationship between biomass change and differences in ALS height percentiles. The third approach consisted of constructing a model of the relationship between the average annual relative growth rate between 2005 and 2008 and corresponding average annual relative change in ALS height percentiles. Models based on the second and third approach that directly predicted growth variables were superior to the first approach that only indirectly predicted growth, and models based on the second approach that predicted actual growth were slightly superior to models based on the third approach that predicted relative growth. This study was among the very first to demonstrate that change in an important climate change variable, biomass, can be predicted using change in ALS metrics.

Næsset et al. (2013a) demonstrated that categories of biomass changes between 1999 and 2010 associated with different management activities in Norway could be distinguished, and that population estimates of biomass change could be calculated using observations for a sample of field plots and wall-to-wall ALS data with pulse densities of 1.2 and 7.3 pulses/m^2. The analyses consisted of four steps. First, non-forest areas and stands younger than 20 years were excluded, and aerial photography was then used to classify forest stands into four forest pre-strata based on stand age and species dominance in 1999. Second, based on changes between 1999 and 2010, three post-strata were defined: (i) recent clear cut representing deforestation, (ii) thinned representing forest degradation, and (iii) untouched representing natural growth with no harvest. Each of 176 plots was then assigned to a single post-stratum. A multinomial logistic regression model was constructed using the plot post-stratum assignments and corresponding ALS metrics from 1999 to 2010 and used to assign each 200-m^2 grid cell to a post-stratum. Third, for each post-stratum, three models were constructed: models of the relationship between biomass and ALS metrics for each of 1999 and 2010, and a model of the relationship between biomass change and change in ALS metrics. For each grid cell, biomass change was predicted both as the difference between the 2010 and 1999 biomass model predictions and from the biomass change model. Fourth, the design-based, model-assisted regression estimator was used to estimate the area of each post-stratum, biomass change for each post-stratum, and corresponding standard errors. The results were compared to corresponding estimates obtained using only field data. The standard errors for the post-strata area estimates obtained using the model-assisted estimator were 18–84 % smaller than standard errors obtained using only the field plot data. Similarly, standard errors for the post-strata mean biomass change estimates were 43–75 % smaller when obtained using the model-assisted estimator. The study demonstrated that deforested, thinned and degraded, and unchanged forest areas can be discriminated and that change in biomass for those areas can be precisely estimated using ALS metrics. Of more importance, the study was the first to use statistically rigorous methods to estimate and quantify bias, to estimate precision, and to quantify the utility of auxiliary ALS data for increasing the precision of estimates for areas consisting of multiple grid cells.

15.2.4 Summary

This brief review has illustrated general trends in the development of a discipline (Table 15.1). Initial studies focused on exploratory and feasibility investigations, estimation of correlations, and simple empirical models of relationships between response variables such as tree height and predictor variables such as ALS maximum heights. Greater maturity was reflected in construction of models for more complex response variables such as canopy gaps, snow damage, and classes of harvest. Enhancement and optimization of the models and their use for constructing maps represents a succeeding step. However, whereas map accuracy assessments focus on describing maps, mature disciplines focus on estimating parameters of the populations that maps only inaccurately depict. The latter studies emphasize estimating and adjusting for bias and estimating and maximizing precision.

15.3 Modeling and Mapping

Two categories of approaches for estimating forest attribute change from remotely sensed data are prominent. Trajectory methods use time series of three or more sets of remotely sensed data to assess the type, extent, and temporal patterns of change

Table 15.1 Literature review

Data years (Point density)	Reference	Approach	Response variable
1998, 2000 (10)	Hyyppä et al. (2003)	Indirect (object)	Growth of trees, mean mean height-growth stands
1998, 2000 (10)	Yu et al. (2003)	Indirect (object)	Harvested trees, growth of trees
1999, 2001 (1.1, 0.9)	Gobakken and Næsset (2004)	Indirect (grid cell)	Effect of growth on ALS-variables
1998, 2003 (0.3, 3.0)	St-Onge and Vepakomma (2004)	Indirect (grid cell, object)	Gap detection, growth of trees, mean growth on plots
1998, 2000 (10)	Yu et al. (2004)	Indirect (object)	Harvested trees, growth of trees
1998, 2000, 2003 (10)	Yu et al. (2005)	Indirect (object)	Growth of trees
1999, 2001 (1.1, 0.9)	Næsset and Gobakken (2005)	Indirect (grid cell)	Change of mean height, basal area, and volume
1998, 2003 (10)	Yu et al. (2006)	Indirect (object)	Growth of trees
2005 (4.4, 4.9, 4.6)	Solberg et al. (2006a, b)	Indirect (grid cell)	Leaf area index
2001, 2005 (32, 51)	Hirata et al. (2008)	Indirect (object)	Detection and classification of gaps
1998, 2000, 2003 (10)	Yu et al. (2008)	Direct (object, grid cell)	Mean height growth, volume growth
2000, 2005 (2.5, 3.6, 2.8, 3.2)	Hopkinson et al. (2008)	Direct (grid cell)	Mean height growth
1998, 2003 (0.3, 3.0)	Vepakomma et al. (2008, 2010)	Indirect (grid cell)	Gap dynamics
1997, 2006	Kellner et al. (2009)	Indirect (grid cell)	Detection of gaps
2007, 2010 (7.0, 11.9)	Vastaranta et al. (2011)	Direct (grid cell)	Classification of plots with snow-damage trees
2006/2007, 2010 (9.8/7.0, 11.9)	Vastaranta et al. (2012)	Direct (object)	Segmentation of damaged crown area
2008, 2010 (1.2, 12.0)	Nyström et al. (2013)	Direct (grid cell)	Classification of change type
2005, 2008 (3.4, 4.7)	Bollandsås et al. (2013)	Indirect, direct (grid cell)	Change of biomass
1999, 2010 (1.2, 7.3)	Næsset et al. (2013a)	Direct (grid cell)	Change of biomass

over time (Kennedy et al. 2007, 2010; Cohen et al. 2010; Zhu et al. 2012). Although trajectory methods have the potential to provide crucial details on the nature and exact time of change for individual grid cells, these methods are both data and computationally intensive. In addition, statistically rigorous methods for assessing uncertainty have not been developed. Bi-temporal methods are more common and entail analyses of separate sets of ground and remotely sensed data for two times and can further be separated into two subcategories, direct and indirect methods.

Direct bi-temporal methods use models to represent the relationship between change for a response variable and auxiliary data. Of necessity, models used with direct methods require training data consisting of two observations of the response and auxiliary variables for the same sample units. ALS auxiliary variables are often formulated as differences between corresponding metrics for the two time points; this approach is often characterized as *differencing*. However, using each metric separately is not prohibited.

For continuous response variables such as biomass change, models must produce both negative and positive predictions. Regardless of the mathematical forms of models and the ALS auxiliary variables, caution must be exercised when extrapolating beyond the range of the values of ALS independent variables in the training sample. Often, the ranges of values of the ALS independent variables may be greater in the population than in the training data used to construct the models. If so, extrapolating the models beyond the range of the sample data may produce both unrealistic negative and/or unrealistic positive predictions. Linear models have been popular, presumably because of their familiarity and their ease of use, but they are particularly susceptible to extrapolation errors. For example, Næsset et al. (2013a) found that linear model predictions of 11-year biomass change for some grid cells in the population were as large as 19,500 Mg/ha, whereas the maximum observed change for the field plot observations used as training data was only 462.3 Mg/ha. To compensate, they truncated the model predictions and used a design-based model-assisted estimator that includes a bias adjustment to obtain an estimate for the entire study area. Although construction of nonlinear models, particularly models with asymptotes, may be more difficult, judicious selection of such models may alleviate extrapolation errors.

For categorical response variables with classes such as deforestation, degradation, and no change, model procedures that produce continuous probabilities for each of the different possible classes are often used. Examples of such procedures include binomial or multinomial logistic regression models, discriminant analysis, and maximum likelihood methods. Nearest neighbor techniques can be used to produce either probabilities for classes or class predictions (Tomppo et al. 2009; McRoberts 2012).

Indirect bi-temporal methods use models to represent the relationship between the response and auxiliary variables for each of two separate times. Models for use with indirect methods can use training data consisting of either repeated observations of the same sample units at different times or observations of different sample units at different times. If the relationship between the response variable and the ALS metrics is stable over time, a single model may be constructed using data that has been aggregated for the two times. Change maps may be constructed by calculating the difference in model predictions for each grid cell.

For continuous response variables such as biomass, the same cautions regarding extrapolation errors must be exercised as for direct methods. McRoberts et al. (2013b) used a variation of a logistic model whose predictions cannot be negative and whose largest predictions are constrained by an asymptote to circumvent extrapolation errors. For categorical response variables, the same kinds of models used with direct methods may be used with indirect methods.

Change maps can be constructed using either direct or indirect methods. With direct methods, the model can be used to predict change in the response variable for each grid cell. With indirect methods, land cover class is predicted for each grid cell at each time, and the difference between the predictions constitutes the prediction of change. For many applications, the change map may be regarded as an end product such as when the location of change is as important as the overall estimate of change. In addition, a change map can be used as the basis for stratified sampling to produce precise estimates of attributes for each category of change. However, as per Sect. 15.4, the map may serve only as an intermediate product, albeit an important product, on the path toward an inference for a change parameter.

15.4   Inference and Estimation

15.4.1   Inference

The Oxford English Dictionary definition of infer is "to accept from evidence or premises" (Simpson and Weiner 1989). Because evidence in the form of complete enumerations of spatial populations is prohibitively expensive, if not logistically impossible, statistical procedures have been developed to infer values for population parameters from estimates based on observations from a sample of population units. In a sampling framework, inference requires expression of the relationship between a population parameter μ and its estimate $\hat{\mu}$ in probabilistic terms (Dawid 1983). For estimation problems, the probabilistic relationship is most often expressed as a confidence interval in the form,

$$\hat{\mu} \pm t \cdot \sqrt{\mathrm{Var}\,(\hat{\mu})},$$

where t reflects the probability that confidence intervals constructed using data for all possible samples will include μ. Thus, inference for estimation problems mostly reduces to the problem of using sample data to estimate both μ and $\mathrm{Var}\,(\hat{\mu})$.

15.4.2 *Estimation*

Both design-based and model-based inferential methods are appropriate for use with ALS data to estimate population parameters such as mean biomass per unit area or mean change in biomass per unit area. Both approaches require definition of the population of interest. For ALS studies, finite populations of interest are usually defined as consisting of N population units in the form of equal-size grid cells. The size of the cells is typically similar, if not the same, as the size of the ground plots. Similarity in sizes ensures consistency in ALS metrics based on scale-dependent height and density distributions calculated separately for plots and cells.

Design-based inference, also characterized as probability-based inference (Hansen et al. 1983; McRoberts 2011), is based on three assumptions: (1) units are selected for the sample using a randomization scheme; (2) the probability of inclusion into the sample for any unit is positive and known; and (3) the value of the response variable for each unit is a fixed value as opposed to a random variable. The validity of design-based estimators is based on random variation resulting from the probabilities of inclusion of units into the sample. Of importance for ALS studies, the sample units do not necessarily conform to the population units as previously defined. Whereas the plots that serve as sample units are typically circular to facilitate ground sampling, the grid cells that serve as population units are typically square to ensure mutual exclusivity. Further, because plot centers do not necessarily correspond to grid cell centers, plot area may partially extend into multiple grid cells. Nevertheless, if the relationship between the response variable and the ALS metrics is the same for both plots and grid cells, estimation and inference are not affected.

Design-based, model-assisted estimators use models to predict the response variable for each grid cell but rely on probability samples for validity. Multiple forms of these estimators have been reported for land cover applications using satellite data (McRoberts 2010, 2011, 2014; McRoberts and Walters 2012; Vibrans et al. 2013; Sannier et al. 2014) and for estimating volume and biomass using ALS and InSAR data (D'Oliveira et al. 2012; Gregoire et al. 2011; Gobakken et al. 2012; Næsset et al. 2011, 2013a, b; McRoberts et al. 2013a, b; Strunk et al. 2012).

The assumptions underlying model-based inference differ considerably from the assumptions underlying design-based inference. First, the observation for a population unit is a random variable whose value is considered a realization from a distribution of possible values, rather than a fixed value as is the case for design-based inference. Second, the basis for a model-based inference is the model, not the probabilistic nature of the sample as is the case for design-based inference. Third, randomization for model-based inference enters through the random realizations from the distributions for population units, whereas randomization for probability-based inference enters through the random selection of units into the sample. Several advantages accrue to model-based inference: (i) inference is possible for regions without probability samples; examples include tropical countries with no probability-based inventories but with data from management or other focused inventories; (ii) inference is possible for small areas for which sample sizes are too small to support design-based inference; and (iii) uncertainty can be estimated for each population unit which is not possible with design-based inference. The disadvantages are reliance on correct model specification, estimators with greater mathematical complexity, and greater computational intensity. Although model-based inference is not discussed further in this chapter, references include McRoberts (2006, 2010), and particularly Ståhl et al. (2011) and McRoberts et al. (2013a) who used ALS auxiliary data.

Regardless of whether a change is predicted and a map is constructed using direct or indirect methods, model prediction errors produce bias in estimators based on adding map unit predictions and uncertainty in estimates for areas consisting of multiple grid cells. Design-based methods for estimating and compensating for the bias and for estimating uncertainty require an estimation sample whose observations are of equal or greater quality with respect to factors such as resolution and accuracy than the training observations used to construct the models. This estimation sample may also be used as an accuracy assessment sample for estimating map accuracy measures, but when the emphasis is estimation rather than map assessment, the term estimation sample is more appropriate. If necessary, the sample used to acquire training data can also be used as the estimation sample (Næsset et al. 2011, 2013a), but the practice is discouraged because it may lead to underestimates of both bias and uncertainty. Finally, of importance for estimation purposes, the form of the design-based statistical estimators used to calculate estimates of bias and uncertainty must conform to the design of the estimation sample, not the design of the training sample.

The design-based estimators that follow are sufficient to produce the estimates of population parameters and their variances that are necessary for inference and are based on several assumptions. First, the ALS auxiliary data are assumed to represent wall-to-wall coverage of the population. Examples of estimators for ALS data acquired in strips are described in Ene et al. (2012). Second, estimation samples are assumed to have been acquired using simple random or systematic sampling designs. If simple random or systematic samples do not produce enough observations of change, then stratified sampling may be necessary. In the latter case, the estimators below may be used for within-strata estimation. Third, a continuous response variable such as biomass or biomass change is assumed as opposed to a categorical response variable such as snow damage or no snow damage (Vastaranta et al. 2011), harvest class (Nyström et al. 2013), or disturbance class (Næsset et al. 2013a). Fourth, change estimators are provided for two estimation sampling designs, a single sample with repeated observations and two independent samples. Finally, design-based estimators are provided only for full coverage ALS; Chap. 14 addresses estimation for partial coverage, such as acquisition of ALS data in strips.

15.4.3 Estimators for Single Estimation Sample with Repeated Observations

Generally, estimation is simplified if the estimation sample data can be acquired by observing the same population units at different times. The primary example is repeated observation of permanent NFI or other inventory plots whose locations are selected using a probability sampling design.

Denote a simple random sample of the population as $S = \{p_i : i = 1, \cdots, n\}$, where p denotes a population unit in the form of a grid cell, i indexes the sample, and n is the sample size. Further, let y_i^k denote the observation of biomass for p_i where $k = 1$ and $k = 2$ correspond to the two repeated observations. In addition, let $w_i = y_i^2 - y_i^1$ denote the observation of change in biomass for p_i with model prediction, $\widehat{w}_i$. An initial estimator of Δ, change in biomass per unit area, is,

$$\widehat{\Delta}_{\text{initial}} = \frac{1}{N} \sum_{i=1}^{N} \widehat{w}_i, \tag{15.1}$$

where N is the population size. However, this estimator may be biased as a result of systematic model prediction error. The bias of this estimator is estimated as,

$$\widehat{\text{Bias}}\left(\widehat{\Delta}_{\text{initial}}\right) = \frac{1}{n} \sum_{p_i \in S} (\widehat{w}_i - w_i). \tag{15.2}$$

One form of the model-assisted regression estimator (Section 6.5 in Särndal et al. 1992) for Δ is defined as the difference between the initial estimator and the expectation of its bias estimate which, under the assumptions that N is both large and much larger than n, is,

$$\widehat{\Delta} = \widehat{\Delta}_{\text{initial}} - \widehat{\text{Bias}}\left(\widehat{\Delta}_{\text{initial}}\right) = \frac{1}{N} \sum_{i=1}^{N} \widehat{w}_i - \frac{1}{n} \sum_{p_i \in S} (\widehat{w}_i - w_i) \tag{15.3}$$

with variance estimator,

$$\widehat{\text{Var}}\left(\widehat{\Delta}\right) = \frac{1}{n(n-1)} \sum_{p_i \in S} (\varepsilon_i - \bar{\varepsilon})^2, \tag{15.4}$$

where $\varepsilon_i = \widehat{w}_i - w_i$ is the model prediction error and $\bar{\varepsilon} = \frac{1}{n} \sum_{p_i \in S} \varepsilon_i$.

15.4.4 Estimators for Two Independent Estimation Samples

Two conditions may necessitate use of independent estimation samples. A first condition is when repeated observation of the same sample units is not possible. A first example is use of observations from temporary NFI plots; a second example is use of different sources for different dates. A second condition is when repeated observation of the same sample units fails to produce sufficient numbers of observations for some classes of a categorical response variable.

Denote the k-th estimation sample as $S^k = \{p_i^k : i = 1, \cdots, n\}$, where p denotes a population unit in the form of a grid cell, i indexes the sample, and n is the sample size. Further, let y_i^k denote the observation of biomass for p_i^k with model prediction, $\widehat{y}_i^k$. With independent estimation samples, mean biomass per unit area is estimated separately for each of $k = 1$ and $k = 2$, and mean biomass change per unit area is estimated as the difference between the two estimates. An initial estimator of mean biomass per unit area for the k-th time, denoted μ^k, is,

$$\widehat{\mu}_{initial}^k = \frac{1}{N} \sum_{i=1}^{N} \widehat{y}_i^k, \tag{15.5}$$

where N is the population size. However, systematic model prediction errors induce bias into this estimator which, for equal probability samples, can be estimated as,

$$\widehat{Bias}\left(\widehat{\mu}_{initial}^k\right) = \frac{1}{n} \sum_{p_i^k \in S^k} \left(\widehat{y}_i^k - y_i^k\right). \tag{15.6}$$

The same form of the model-assisted regression estimator may be used with independent samples to estimate μ^k, mean biomass per unit area, as was used with repeated observations of the same sample to estimate Δ, mean change in biomass per unit area (Section 6.5 in Särndal et al. 1992),

$$\widehat{\mu}^k = \widehat{\mu}_{initial}^k - \widehat{Bias}\left(\widehat{\mu}_{initial}^k\right).$$

$$= \frac{1}{N} \sum_{i=1}^{N} \widehat{y}_i^k - \frac{1}{n} \sum_{p_i^k \in S^k} \left(\widehat{y}_i^k - y_i^k\right). \tag{15.7}$$

Under the assumptions that N is both large and much larger than n, that the prediction errors for a particular year are independent, and that simple random sampling was used, the variance of $\widehat{\mu}^k$ can be estimated as,

$$\widehat{Var}\left(\widehat{\mu}^k\right) = \frac{1}{n(n-1)} \sum_{p_i^k \in S^k} \left(\varepsilon_i^k - \bar{\varepsilon}^k\right)^2, \tag{15.8}$$

where $\varepsilon_i^k = \widehat{y}_i^k - y_i^k$ is the model prediction error, and $\bar{\varepsilon}^k$ is the mean of the errors. When systematic sampling rather than simple random sampling is used, variances may be overestimated (Särndal et al. 1992, p. 83).

The estimator of mean change in biomass per unit area is,

$$\widehat{\Delta} = \widehat{\mu}^2 - \widehat{\mu}^1, \qquad (15.9)$$

and the estimator of $\mathrm{Var}\left(\widehat{\Delta}\right)$ is,

$$\widehat{\mathrm{Var}}\left(\widehat{\Delta}\right) = \widehat{\mathrm{Var}}\left(\widehat{\mu}^2 - \widehat{\mu}^1\right)$$
$$= \widehat{\mathrm{Var}}\left(\widehat{\mu}^2\right) - 2 \cdot \widehat{\mathrm{Cov}}\left(\widehat{\mu}^2, \widehat{\mu}^1\right) + \widehat{\mathrm{Var}}\left(\widehat{\mu}^1\right). \qquad (15.10)$$

Estimates $\widehat{\mu}^1$ and $\widehat{\mu}^2$ obtained using Eq. 15.7 are used with Eq. 15.9 to estimate mean biomass change per unit area, and estimates $\widehat{\mathrm{Var}}\left(\widehat{\mu}^1\right)$ and $\widehat{\mathrm{Var}}\left(\widehat{\mu}^2\right)$ obtained using Eq. 15.8 are used with Eq. 15.10 to estimate the variance of the forest change estimate. Of importance, Eq. 15.10 includes an estimate of $\mathrm{Cov}\left(\widehat{\mu}^2, \widehat{\mu}^1\right)$. For independent estimation samples, $\mathrm{Cov}\left(\widehat{\mu}^2, \widehat{\mu}^1\right) = 0$ may be assumed.

Although the estimators formulated as Eqs. 15.7, 15.8, 15.9, and 15.10 have been developed under the assumption of independent estimation samples, they may also be used with repeated observations for the same sample. For the latter case, the sample observations for k = 1 and k = 2 are considered separate samples. However, because the two sets of observations are for the same sample units, $\widehat{\mathrm{Cov}}\left(\widehat{\mu}^2, \widehat{\mu}^1\right) = 0$ cannot be assumed. Nevertheless, it can be estimated using the sample data as,

$$\widehat{\mathrm{Cov}}\left(\widehat{\mu}^2, \widehat{\mu}^1\right) = \frac{1}{n(n-1)} \sum_{p_i \in S} \left(\varepsilon_i^2 - \bar{\varepsilon}^2\right)\left(\varepsilon_i^1 - \bar{\varepsilon}^1\right). \qquad (15.11)$$

15.5 Conclusions

Although the number of published studies that report use of ALS data to estimate change in forest attributes is small, chronologically they represent a maturing process leading from early exploratory studies of correlations and simple relationships to construction of statistically rigorous inference. Of crucial importance, maturity requires distinguishing between modeling and mapping procedures and estimation and inferential procedures. Of importance, estimators that simply aggregate map values are inherently biased because of map errors resulting from prediction errors. When the objective is estimation and inference, this bias must be acknowledged, estimated, and incorporated into the final population estimates. In addition, the uncertainty that accrues as a result of map errors must be estimated. Design-based, model-assisted estimators for biomass change are intuitive and computationally easy and rely only on probability samples. Further, the samples may be in the form of independent samples for two different times or a single sample that is observed on two different occasions.

References

Andersen H-E, Reutebuch SE, McGaughey RJ (2006) A rigorous assessment of tree height measurements obtained using airborne lidar and conventional field methods. Can J Remote Sens 32:355–366

Bollandsås OM, Gregoire TG, Næsset E, Øyen B-H (2013) Detection of biomass change in a Norwegian mountain forest area using small footprint airborne laser scanner data. Stat Method Appl 22:113–129

Cohen WB, Yang Z, Kennedy R (2010) Detecting trends in forest disturbance and recovery using yearly Landsat time series: 2. TimeSync – tools for calibration and validation. Remote Sens Environ 114:2911–2924

Dawid AP (1983) Statistical inference I. In: Kotz S, Johnson NL (eds) Encyclopedia of statistical sciences, vol 4. Wiley, New York

D'Oliveira MVN, Reutebuch SE, McGaughey RJ, Andersen H-E (2012) Estimation of forest biomass and identifying low-intensity logging areas using airborne scanning lidar in Antimary State Forest, Acre State, Western Brazilian Amazon. Remote Sens Environ 124:479–491

Ene LT, Næsset E, Gobakken T, Gregoire TG, Ståhl G, Nelson R (2012) Assessing the accuracy of regional LiDAR-based biomass estimation using a simulation approach. Remote Sens Environ 123:579–592

Falkowski MJ, Smith AMS, Gessler PE, Hudak AT, Vierling LA, Evans JS (2008) The influence of conifer forest canopy cover on the accuracy of two individual tree measurement algorithms using lidar data. Can J Remote Sens 34:338–350

Gobakken T, Næsset E (2004) Effects of forest growth on laser derived canopy metrics. Int Arch Photogramm Remote Sens Spat Inf Sci 36–8(W2):224–227

Gobakken T, Næsset E, Nelson R, Bollandsås OM, Gregoire TG, Ståhl G, Holm S, Ørka HO, Astrup R (2012) Estimating biomass in Hedmark County, Norway using national forest inventory field plots and airborne laser scanning. Remote Sens Environ 123:443–456

Gregoire TG, Ståhl G, Næsset E, Gobakken T, Nelson R, Holm S (2011) Model-assisted estimation of biomass in a LiDAR sample survey in Hedmark County, Norway. Can J For Res 41:83–95

Hansen MH, Madow WG, Tepping BJ (1983) An evaluation of model-depending and probability-sampling inferences in sample surveys. J Am Stat Assoc 78:776–793

Hirata Y, Tanaka H, Furuya N (2008) Canopy and gap dynamics analysed using multi-temporal airborne laser scanner data in a temperate deciduous forest. In: Hill R, Rosette J, Suárez J (eds) Proceedings of SilviLaser 2008, 8th international conference on LiDAR applications in forest assessment and inventory, Heriot-Watt University, Edinburgh, 17–19 September 2008, pp 144–150

Höhne N, Wartmann S, Herold A, Freibauer A (2007) The rules for land use, land use change and forestry under the Kyoto Protocol – lessons learned for the future climate negotiations. Environ Sci Policy 10:353–369

Hopkinson C, Chasmer L, Hall RJ (2008) The uncertainty in conifer plantation growth prediction from multi-temporal lidar datasets. Remote Sens Environ 112:1168–1180

Hyyppä J, Inkinen M (1999) Detecting and estimating attributes for single trees using laser scanner. Photogramm J Finland 16:27–42

Hyyppä J, Yu X, Rönnholm P, Kaartinen H, Hyyppä H (2003) Factors affecting laser-derived object-oriented forest height growth estimation. Photogramm J Finland 18:16–31

Kaartinen H, Hyyppä J, Yu X, Vastaranta M, Hyyppä H, Kukko A, Holopainen M, Heipke C, Hirschmugl K, Morsdorf F, Næsset E, Pitkänen J, Popescu P, Solberg S, Wolf BM, Wu J (2012) An international comparison of individual tree detection and extraction using airborne laser scanning. Remote Sens 4:950–974

Kellner JR, Clark DB, Hubbell P (2009) Pervasive canopy dynamics produce short-term stability in a tropical rain forest. Ecol Lett 12:155–164

Kennedy RE, Cohen WB, Schroeder TA (2007) Trajectory-based change detection for automated characterization of forest disturbance dynamics. Remote Sens Environ 110:370–386

Kennedy RE, Yang Z, Cohen WB (2010) Detecting trends in forest disturbance and recovery using yearly Landsat time series: 1. LandTrendr – temporal segmentation algorithms. Remote Sens Environ 114:2897–2910

McRoberts RE (2006) A model-based approach to estimating forest area. Remote Sens Environ 103:56–66

McRoberts RE (2010) Probability- and model-based approaches to inference for proportion forest using satellite imagery as ancillary data. Remote Sens Environ 114:1017–1025

McRoberts RE (2011) Satellite image-based maps: scientific inference or pretty pictures? Remote Sens Environ 115(2):715–724

McRoberts RE (2012) Estimating forest attribute parameters for small areas using nearest neighbors techniques. For Ecol Manag 272:3–12

McRoberts RE (2014) Post-classification approaches to estimating change in forest area using remotely sensed auxiliary data. Remote Sens Environ. doi:10.1016/j.rse.2013.03.036

McRoberts RE, Walters BF (2012) Statistical inference for remote sensing-based estimates of net deforestation. Remote Sens Environ 124:394–401

McRoberts RE, Gobakken T, Næsset E (2013a) Inference for lidar-assisted estimation of forest growing stock volume. Remote Sens Environ 128:268–275

McRoberts RE, Næsset E, Gobakken T (2013b) Accuracy and precision for remote sensing applications of model-based inference. IEEE J Sel Topics Appl Earth Observ 6:27–34

Næsset E (2002) Predicting forest stand characteristics with airborne scanning laser using a practical two-stage procedure and field data. Remote Sens Environ 80:88–99

Næsset E (2004) Practical large-scale forest stand inventory using a small-footprint airborne scanning laser. Scand J For Res 19:164–179

Næsset E, Gobakken T (2005) Estimating forest growth using canopy metrics derived from airborne laser scanner data. Remote Sens Environ 96:453–465

Næsset E, Gobakken T, Solberg S, Gregoire TG, Nelson R, Ståhl G, Weydahl D (2011) Model-assisted regional forest biomass estimation using LiDAR and InSAR as auxiliary data: a case study from a boreal forest area. Remote Sens Environ 115:3599–3614

Næsset E, Bollandsås OM, Gobakken T, Gregoire TG, Ståhl G (2013a) Model-assisted estimation of change in forest biomass over an 11 year period in a sample survey supported by airborne LiDAR: a case study with post-stratification to provide "activity data". Remote Sens Environ 128:299–314

Næsset E, Gobakken T, Bollandsås OM, Gregoire TG, Nelson R, Ståhl G (2013b) Comparison of precision of biomass estimates in regional field sample surveys and airborne LiDAR-assisted surveys in Hedmark County, Norway. Remote Sens Environ 130:108–120

Nyström M, Holmgren J, Olsson H (2013) Change detection of mountain birch using multi-temporal ALS point clouds. Remote Sens Lett 4:190–199

Popescu SC, Wynne RH (2004) Seeing the trees in the forest: using lidar and multi-spectral data fusion with local filtering and variable windows size for estimating tree height. Photogramm Eng Remote Sens 70:589–604

Reichenbach H (1938) Experience and prediction: an analysis of the foundations and structure of knowledge. The Chicago University Press, Chicago

Rypdal K, Bloch VVH, Flugsrud K, Gobakken T, Hoem B, Tomter SM, Aalde H (2005) Emissions and removals of greenhouse gases from land use, land use-change, and forestry in Norway. NIJOS report 11/05, Ås, Norway

Sannier C, McRoberts RE, Fichet L-V, Makaga E (2014) Using the regression estimator with Landsat data to estimate proportion forest cover and net proportion deforestation in Gabon. Remote Sens Environ. doi:10.1016/j.rse.2013.09.015

Särndal C-E, Swensson B, Wretman J (1992) Model assisted survey sampling. Springer, New York

Simpson, JA, Weiner ESC (Preparers) (1989) The Oxford English Dictionary, 2nd edn. Clarendon Press, Oxford, pp 923–924

Solberg S, Næsset E, Bollandsås OM (2006a) Single tree segmentation using airborne laser scanner data in a structurally heterogeneous spruce forest. Photogramm Eng Remote Sens 12:1369–1378

Solberg S, Næsset E, Hanssen KH, Christiansen E (2006b) Mapping defoliation during a severe insect attack on Scots pine using airborne laser scanning. Remote Sens Environ 102:364–376

Ståhl G, Holm S, Gregoire TG, Gobakken T, Næsset E, Nelson R (2011) Model-based inference for biomass estimation in a LiDAR sample survey in Hedmark County, Norway. Can J For Res 41:96–107

St-Onge B, Vepakomma U (2004) Assessing forest gap dynamics and growth using multi-temporal laser scanner data. Int Arch Photogramm Remote Sens Spat Inf Sci 36:173–178

Strunk J, Reutebuch SE, Andersen H-E, Gould PJ, McGaughey RJ (2012) Model-assisted forest yield estimation with light detection and ranging. West J Appl For 27:53–59

Tesfamichael SG, Ahmed F, van Aardt JAN, Blakeway F (2009) A semivariogram approach for estimating stems per hectare in Eucalyptus grandis plantations using discrete-return lidar height data. For Ecol Manag 258:1188–1199

Tomppo EO, Gagliano C, De Natale F, Katila M, McRoberts RE (2009) Predicting categorical forest variables using an improved k-nearest neighbour estimator and Landsat imagery. Remote Sens Environ 113:500–517

Tomppo E, Gschwantner T, Lawrence M, McRoberts RE (2010) National Forest Inventories: pathways for harmonised reporting. Springer, Dordrecht

Vastaranta M, Korpela I, Uotila A, Hovi A, Holopainen M (2011) Area-based snow damage classification of forest canopies using bi-temporal lidar data. Int Arch Photogramm Remote Sens Spat Inf Sci 38:169–173

Vastaranta M, Korpela I, Uotila A, Hovi A, Holopainen M (2012) Mapping of snow-damaged trees based on bitemporal airborne LiDAR data. Eur J For Res 131:1217–1228

Vauhkonen J, Ene L, Gupta S, Heinzel J, Holmgren J, Pitkänen J, Solberg S, Wang Y, Weinacker H, Hauglin KM, Lien V, Packalén P, Gobakken T, Koch B, Næsset E, Tokola T, Maltamo M (2012) Comparative testing of single-tree detection algorithms under different types of forest. Forestry 85(1):27–40

Vepakomma U, St-Onge B, Kneeshaw D (2008) Spatially explicit characterization of borel forest gap dynamics using multi-temporal lidar data. Remote Sens Environ 112:2326–2340

Vepakomma U, Kneeshaw D, St-Onge B (2010) Interactions of multiple disturbances in shaping boreal forest dynamics: a spatially explicit analysis using multi-temporal lidar and high-resolution imagery. J Ecol 98:526–539

Vibrans AC, McRoberts RE, Moser P, Nicoletti AL (2013) Using satellite image-based maps and ground inventory data to estimate the area of the remaining Atlantic forest in the Brazilian state of Santa Catarina. Remote Sens Environ 130:87–95

Yu X, Hyyppä J, Rönnholm P, Kaartinen H, Maltamo M, Hyyppä H (2003) Detection of harvested trees and estimation of forest growth using laser scanning. In: Proceedings of the Scandlaser scientific workshop on airborne laser scanning of forests, Umeå, Sweden, 3–4 September, pp 115–124

Yu X, Hyyppä J, Kaartinen H, Maltamo M (2004) Automatic detection of harvested trees and determination of forest growth using airborne laser scanning. Remote Sens Environ 90:451–462

Yu X, Hyyppä J, Kaartinen H, Hyyppä H, Maltamo M, Rönnholm P (2005) Measuring the growth of individual trees using multi-temporal airborne laser scanning points clouds. In: Proceedings of the ISPRS WG III/3, III/4, V/3 workshop: laser scanning 2005, Enschede, Nederlands, pp 202–208

Yu X, Hyyppä J, Kukko A, Maltamo M, Kaartinen H (2006) Change detection techniques for canopy height growth measurements using airborne laser scanner data. Photogramm Eng Remote Sens 72:1339–1348

Yu X, Hyyppä J, Kaartinen H, Maltamo M, Hyyppä H (2008) Obtaining plotwise mean height and volume growth in boreal forests using multi-temporal laser surveys and various change detection techniques. Int J Remote Sens 29(5):1367–1386

Zhu A, Woodcock CE, Olofsson P (2012) Continuous monitoring of forest disturbance using all available landsat imagery. Remote Sens Environ 122:75–91

Chapter 16
Valuation of Airborne Laser Scanning Based Forest Information

Annika Kangas, Tron Eid, and Terje Gobakken

Abstract When an inventory is planned for, the decision makers typically strive for maximizing accuracy of the information with a given budget, or even maximize accuracy without considering any budget at all. Recent developments in airborne laser scanning and other remote sensing techniques facilitate the use of data obtained from such sources as an integrated part of the forest inventory process. However, a rational decision maker would not pay for information that is more expensive than the expected improvement in the value of the decisions based on the new information. The statistical accuracy that usually is provided in the remote sensing literature does not dictate the usefulness of the data in decision making. In this chapter we present methods to assess the value of information and go through the recent research related to this where we in different ways try to establish the links between the inventory effort level, decisions to be made and the value of information.

16.1 Introduction

Information has value to the decision maker if there is uncertainty concerning the correct decision. This value of information (VOI) can ex ante be calculated as the difference of expected value of a given decision with and without a source of new information (Lawrence 1999; Birchler and Bütler 2007). Some prior information is always assumed to be available, for instance in the form of old forest inventory data updated with growth models. Thus, VOI comes from the ability to make better

A. Kangas (✉)
Department of Forest Sciences, University of Helsinki, Helsinki, Finland
e-mail: annika.kangas@helsinki.fi

T. Eid • T. Gobakken
Department of Ecology and Natural Resource Management, Norwegian University
of Life Sciences, Ås, Norway

M. Maltamo et al. (eds.), *Forestry Applications of Airborne Laser Scanning: Concepts and Case Studies*, Managing Forest Ecosystems 27, DOI 10.1007/978-94-017-8663-8__16,
© Springer Science+Business Media Dordrecht 2014

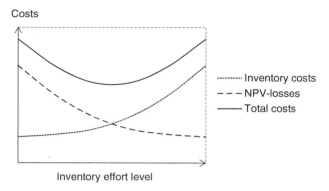

Fig. 16.1 Cost structure when considering the effort level of an inventory

decision with additional information, and a rational decision maker should not pay for information that is more expensive than the expected improvement in the value of the decision.

If the new information is without errors, i.e. perfect, then we can talk about the expected value of perfect information (EVPI). Usually the new information still contains some errors, however, and this means that we can talk about the expected value of imperfect information (EVII).

The problem can also be looked at from a different perspective by analyzing the losses due to sub-optimal decisions. Then, the value of information comes from the reduced expected losses when we have new information. This approach has been more commonly used in forestry, where different inventory methods have been compared using a so-called cost-plus-loss (CPL) analysis. In CPL, the total costs of an inventory consist of the inventory costs and the losses due to sub-optimal decisions, and the inventory method having the lowest total costs is defined as best (Hamilton 1978; Burkhart et al. 1978). Likewise, CPL could be used to define the optimal inventory intensity using a given method (Borders et al. 2008). The cost structure of a concept considering the effort level of an inventory is illustrated in Fig. 16.1. In general, the inventory costs increase and the losses in net present value (NPV) decrease when the effort level of the inventory increases. The ultimate goal in data acquisition planning should be to search for an effort level that minimises total costs.

In CPL, the benchmark is the perfect information, and the losses of a given inventory method are calculated related to that. In EVPI the benchmark is the given inventory, and the value of perfect information is calculated related to that. If we wish to consider the additional value of a new inventory method, it can be calculated if we know the expected losses (EL) or EVPI with prior information and with the new information as

$$EVII = EVPI_{prior} - EVPI_{new} = EL_{prior} - EL_{new}. \qquad (16.1)$$

An objective function is used to calculate the VOI, and the VOI is expressed in the same units as the objective function. Thus, if the objective is to maximize NPV from the harvests, the VOI is also expressed as monetary units. If the decision maker was maximizing a multi-criteria utility function, then also the VOI is expressed as utils (Kim et al. 2003; Kangas et al. 2010a). However, it is possible to calculate the respective value in monetary terms, if the utility function involves at least one criterion that can be expressed in monetary terms.

The VOI depends on the application of the information. For example, the use of forest stand inventory data spans from forest property valuation to management planning. Since a number of different decisions can be based on the same information, the information may have different value depending on type of decision. For instance, Duvemo et al. (2012) calculated the losses in a hierarchical planning case including both strategic and tactical level planning. They observed that the losses were much higher, when traditional inventory data was used in both levels (about 166 €/ha) than in the case where perfect data was utilized in tactical level and the traditional data in the strategic level (about 15 €/ha).

The VOI depends also on the constraints involved in the problem formulation, especially if the constraints have no tolerance. For instance, requiring that there is an exactly even flow of timber from the area requires better information than just maximizing the NPV. On the other hand, if the constraints have some tolerance for violations, their effect on VOI is lower.

When a new inventory is planned for, the decision makers typically strive for maximal accuracy with the given budget. However, the statistical accuracy does not dictate the usefulness of the data in decision making (Ketzenberg et al. 2007). For instance, the better the prior information available, the less is the value of additional information (e.g. Kangas et al. 2010b). Moreover, the more sensitive the decisions are to the information, the higher the value of information (e.g. Eid 2000; Holmström et al. 2003; Duvemo et al. 2012). For instance when maximizing NPV in stands that are well past the final harvest age, additional information may have little value as the decision is always to harvest immediately. Therefore, modeling the VOI as a function of forest characteristics may not be straightforward (Islam et al. 2010). Finally, the VOI is the higher the larger the number of decisions that are based on the information. In forestry this directly means that data is the more valuable, the longer it can be used for decision making.

The role of VOI in data acquisition planning is to ensure an optimal level of efforts. For instance in planning an optimal sample in a given stand to be harvested, the optimal number of sample plots is achieved when the marginal costs are equal to marginal benefits (e.g. Birchler and Bütler 2007), i.e. when the cost of the last sample plot is exactly the same as the additional value of information obtainable from that additional plot.

Most of the applications in forestry have dealt with defining the expected losses (or EVPI) from a given data acquisition method, such as sampling based forest inventory, traditional visual forest inventory, photo interpretation based forest inventory or airborne laser scanning (ALS) based forest inventory (e.g. Holmström et al. 2003; Eid et al. 2004; Mäkinen et al. 2010; Borders et al. 2008). It is, however, possible to calculate the value for any piece of information. For instance, it is possible to calculate the value for one or more specific variables (Eid 2000; Kangas et al. 2010b), which helps in determining to which variables the data acquisition efforts should be concentrated. It is also possible to evaluate the value of a single sample tree measurement or single sample plot, for instance in determining whether it is profitable to measure the past growth from the plots or not (Mäkinen et al. 2012). A given prediction model or a given predictor in a model can also be evaluated, which can help in determining what are the most profitable input variables for a model (Eid 2003). The totality of a data acquisition strategy can be evaluated as well. For instance in ALS inventory, it would be possible to define the optimal ALS acquisition settings, optimal number of field plots and measurements, and optimal calculation system for the results. Finally, it would be possible to evaluate decision support systems (DSS), e.g. a system of growth and yield models (Rasinmäki et al. 2013). While it is quite difficult to account for non-sampling errors such as measurement errors or tally tree prediction errors when defining the accuracy of a given sampling method, for instance, all these aspects can be easily accounted for in VOI analysis.

16.2 Methods for Analyzing Value of Information

16.2.1 VOI Estimates Based on Empirical Data or Simulation

The simplest way to calculate the VOI or expected losses is through simulation. If we assume that the decision maker is maximizing the NPV in his/her forest estate without constraints, the loss in each stand i can be calculated as the difference (NPV_{loss}) between the true NPV from the optimal decision (NPV_{opt}) and the true NPV from the sub-optimal decision (NPV_{err})

$$NPV_{loss\ i} = NPV_{opt\ i} - NPV_{err\ i}. \tag{16.2}$$

In order to calculate the true NPV from the sub-optimal decision, the optimal decision based on the erroneous data is first calculated, and then the NPV is calculated by applying these decisions on the correct data (Fig. 16.2). The NPV obtained from the erroneous data cannot be used for calculating the VOI: the difference between true NPV and the erroneous one reflects erroneous expectations rather than actual losses. In fact, it may be either lower or higher than the true optimum.

It is useful to restrict the decisions included into the analysis to the time period that the collected data will actually be used. For instance, if the data is to be used for 10 years, then only the losses occurring during that time should be accounted for. This can be done so that after the period in question, the decisions are assumed to be correct, given the decisions already made (Holmström et al. 2003). Another

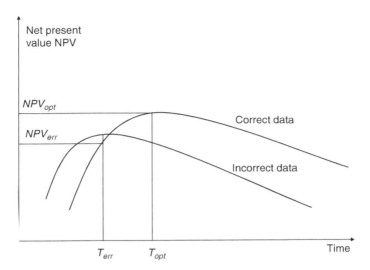

Fig. 16.2 The principle of cost-plus-loss analysis (see Ståhl 1994; Eid 2000). Time here means timing of a harvest, and the *curves* describe NPV obtainable with a given timing. *NPV* with the incorrect data is obtained by looking for the true *NPV* using erroneous timing

approach is to use an estimate of the value of the stand (i.e. net present value of expected future yields given the decisions so far) at the end of the period considered (Mäkinen et al. 2010; Pietilä et al. 2010).

In order to calculate the losses, the true and erroneous data both needs to be available for the areas in question. In some cases the test areas have been measured with several methods. Then, the most accurate method (typically a field sample) is considered as reference and ground truth, and the data obtained from e.g. remote sensing are assumed to be the erroneous data. The benefits of this approach are that the correlations between the different errors and their distribution, i.e. the error structure, will automatically be close to correct. The drawback of this approach is that perfect data for comparison can never be obtained. The field measurements also contain errors and the observed differences are a combination of errors from both methods. This means that if one of the methods is assumed to represent the truth, both sources of error are attributed to the other method. Another problem with this approach is that it is never possible to examine methods that do not yet exist, or even different variations of methods that do exist. It is impossible to determine, for instance, if it was profitable to measure some variables more accurately and some variables less accurately than with the currently used method.

The first study analyzing the VOI of forest inventory data obtained by means of ALS through CPL analysis was carried out by Eid et al. (2004). Based on actual data from practical management planning situations, they compared photo interpretation and ALS based inventory and used independent and intensive field sample plot inventories to represent the true values. The results were clear: while the inventory

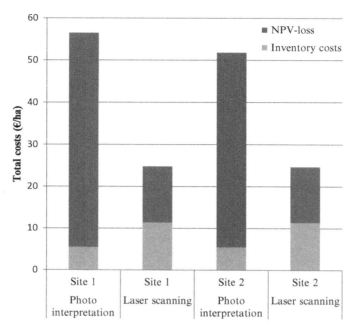

Fig. 16.3 The inventory costs and estimated NPV-loss of two inventory methods in two study areas (Eid et al. 2004)

costs of laser scanning were twice as expensive as that of the photo interpretation, the total costs were still less than those of photo interpretation (Fig. 16.3).

The above described example comparing photo interpretation and ALS is insufficient in the sense that it represents only one out many effort levels that could be applied for the methods. We do not know, for example, whether the effort levels of the ALS procedures carried for the two sites were near the optimum, or if they were too high or too low (see also Fig. 16.1). In general, the effort level and the VOI should be focused in any study related to innovative inventory methods. For ALS applications, both ALS acquisition settings (e.g. point density, footprint size, and scan angle) and efforts in field calibration (e.g. size of sample plots, number of sample plots, precision level required for georeferencing sample plots) could be considered in this context.

A more recent example of VOI analysis is the ALS study provided by Bergseng et al. (2013). The field data came from 23 large field plots in south-eastern Norway with georeferenced single trees collected. A single-tree growth and yield simulator was used in the analysis and NPV-losses when considering the timing of the final harvest were quantified. Losses were derived from four alternative inventory approaches, namely when using (1) mean values obtained by the area-based approach (ABA-MV), (2) diameter distributions obtained by the area-based approach (ABA-DD), (3) individual tree crown segmentation (ITC), and (4) an

Table 16.1 Maximum and mean loss in NPV (in percent) from the incorrect harvest timing decisions based on erroneous inventory data calculated by applying field reference data for the 23 plots

Inventory approach	NPV-loss (%)	
	Max	Mean
ABA-MV	23	5.4
ABA-DD	12	1.0
ITC	9	1.3
SITC	16	1.8

approach called semi-ITC, where field reference data within crown segments from the nearest neighboring segment are imputed. The results showed that using mean values from ABA (ABA-MV) may yield large NPV-losses (Table 16.1). When also inventory costs are taken into account, diameter distributions from ABA (ABA-DD) appear to provide a favorable combination of accuracy and costs.

Another option when applying CPL analysis would be to simulate the errors of different variables. Then, the field measurement is assumed to represent the true values, and errors are added to those values to obtain different erroneous datasets. Simulation allows for generating hundreds or even thousands of realizations or errors for each stand, and the estimate of stand-level loss is calculated as their mean. This enables a detailed analysis of the losses and their dependency on forest characteristics, and an analysis of the relative importance of different variables. It is thus possible to calculate the VOI for methods not yet available.

The first example on such a simulation approach was by Eid (2000). His results proved that for scheduling the final harvests for different forest stands, basal area and dominant height information had little value, while site index and age had much higher value. The low value of basal area, for instance, was due to the growth model used: if the growth models used do not utilize a variable as an input variable, then that variable does not have value. A similar approach was also used to evaluate and choose between different prediction models with different candidates regarding independent variables (Eid 2003). In addition to consider the usual fit statistics and independent data quantifying errors, the models were also evaluated by calculating NPV-losses due to errors related to field measurements of the candidate variables.

The simulation method has its pitfalls as well. In order to produce reliable results on the value of information, the structure and distribution of the errors need to be reliable. Assuming independent errors and normal distributions may produce misleading results. The shape of the marginal distributions and the dependencies between different errors can be accounted for using so-called copula approach (Mäkinen et al. 2010). The copula function directly describes the dependencies between the variables. In Gaussian copula, for instance, the dependencies are based on linear correlations as in the normal distribution, although the marginal distributions of each variable need not to be normal.

Mäkinen et al. (2010) tested the importance of different error structure approaches in a CPL analysis. They fitted a Gaussian copula with logit-logistic marginal distributions to observed errors of different forest variables based on ALS forest inventory, and used the estimated correlations and distributions to represent the true distributions in the simulation-based CPL analysis. They also

Table 16.2 The mean relative loss (%) estimated with different assumptions of the error distributions of ALS-based forest inventory for different forest operations in one test area inventory (Mäkinen et al. 2010)

Next operation	Assumption of the error structure			
	MLLTr	UN	ULLTr	MNTr
No operations	1.5	1.9	1.3	1.5
Thinning	6.4	5.5	6.6	6.7
Final harvest	9.8	5.0	9.7	9.6
Seed tree removal	34.2	26.4	33.2	34.8
All schedules	6.4	4.6	6.2	6.4

For the assumptions of error structure, see definitions in the text

accounted for the bias in the ALS assessments. The bias was observed to have a clear trend with large values generally underestimated and small values generally overestimated. They compared approaches where the errors were assumed to have a normal distribution, uncorrelated errors and no trend in the bias (UN), to have a multinormal distribution, correlated errors and trend in the bias (MNTr), to have logit-logistic distribution, uncorrelated errors and trend in the bias (ULLTr), or to have logit-logistic distribution, correlated errors and trend in bias (MLLTr).

It was clear that in the ALS data the estimated losses were markedly smaller when no trend in the bias was assumed (Table 16.2, column UN). The difference between multinormal or logit-logistic distribution assumptions was very small, indicating that the assumption of the shape of the distribution is not as important as the trend. Likewise, the difference between the correlated and uncorrelated logit-logistic distributions was small, so that the effect of correlations proved also quite small when compared to the effect of the trend. The effect of trend was so clear that for the case of final harvest the rank order between ALS and traditional visual forest inventory (results not shown here) changed when the trend was introduced (Mäkinen et al. 2010). It means that while with the UN assumption the best method was the ALS inventory, with the error structures actually estimated from observed errors (MLLTr), traditional forest inventory proved to be a better method due to the trend.

These results show that correct distributional assumptions are important, as those separate the different data acquisition methods from each other. If all the errors are assumed normally distributed, the only thing separating the different inventory methods would be the bias and standard error. In reality, the error structure of an ALS-based inventory is, however, clearly different from sample plot based field inventory, for example.

16.2.2 Bayesian VOI Analysis

In general, the ex-ante VOI can be calculated with (Lawrence 1999, p. 17, see also Eq. 9 in Kangas 2010)

$$VOI(S) = E_y \max_a E_{x|y} \pi(x, a) - \max_a E_x \pi(x, a) \qquad (16.3)$$

where S denotes the information source (e.g. inventory method), x denotes the possible states of nature concerning the uncertain variable (the prior information), y denotes the possible messages from an information source (message denoting one piece of information), $\max_a E_x\pi(x,a)$ the expected payoff of the optimal decision a without the additional information, and $\max_a E_{x|y}\pi(x,a)$ is the expected payoff of the optimal decision (e.g. max NPV) with a given message y and state x, which is further averaged over all possible messages y. This is called preposterior analysis (Lawrence 1999 p. 65). In fact what is needed in such analysis is the prior distribution of the posterior mean of the uncertain x (Ades et al. 2004). If the information source S is perfect (i.e. message precisely describe the state of nature), the formula can be presented simply as

$$VOI(S) = E_x \max_a \pi(x,a) - \max_a E_x \pi(x,a). \tag{16.4}$$

Then, the posterior distribution of x given y is not needed, and Eq. 16.4 provides the EVPI, the upper bound for the acquisition cost of information y (e.g. Ades et al. 2004). This formula is closely related to calculating the mean loss from the simulated stand-level losses, only here the expectation is taken on the optimal decisions with true values and the decisions with incorrect or prior values rather than the differences of them, i.e. the losses. It is essentially the same as Eq. 16.1.

The Bayesian analysis method has not been used in forestry for many cases. The first application of Bayesian analysis in forest inventory was published by Ståhl et al. (1994). A more recent example is provided by Kangas et al. (2010b) where the value of wood quality information (information concerning the height of the living and dead crown) in a bidding situation was analyzed. The value proved to depend highly on the prior information available: if a model for the crown height was utilized as prior information, additional field information was only valuable in the potentially most valuable stands. However, when the prior information was poor, i.e. only the range of possible variation in wood quality was assumed known, the quality information obtained from crown height measurements could be valuable.

16.2.3 Value of Information Under Constrained Optimization

In most studies, VOI has been analyzed in a setting where the utility of decision maker only consists of maximizing NPV without any constraints. In practical decision-making, constraints are, however, common. Then, if there is uncertainty about the coefficients of the decision making problem, the feasibility of the optimal solution in all possible states of nature cannot be guaranteed. This complicates the VOI analysis: it is not enough to analyze the possible losses in the objective function value, but the costs of violating the constraints need to be accounted for as well. The infeasible solutions may well have a higher objective function value (i.e. higher NPV) than the true optimal decision. If the costs of violating of constraints are not accounted, VOI may then be negative.

In forestry decision-making situation with constraints, Islam et al. (2009) studied the losses, when the decision maker required a given level of growing stock at the end of the planning period and/or the incomes to be even between the planning periods. In order to solve the problem due to violated constraints, Islam et al. (2009) calculated the NPVs under uncertain information by using modified constraints. The constraints were modified as much as required to find a feasible solution as close as possible to the original problem setting. Even then, the costs of violating the original constraints were not included in the analysis.

The costs of violating the constraint depend on the nature of the constraint. The first aspect of importance is whether the constraint is a real constraint (external constraint) or part of the decision maker's utility (internal constraint). For instance, requiring even flow of incomes may be part of the forest owner's utility rather than a real constraint, i.e. the decision maker is giving up some of the objective function value to get an even flow of incomes. Then, the costs due to violating the constraint depend on the relative importance of the incomes from different periods. They can be defined from the shadow prices or the weights of the incomes from different periods in a utility function. An example of this approach is the study of Gilabert and McDill (2010). They maximized the NPV of the harvests plus the value of the forest at the end of the planning period, with the flow constraints for the harvests and minimum average ending age of the stands, and penalized the violations of constraints using shadow prices.

If the constraints are real, the costs of violating the constraints should also be real. For instance, in the case of budget constraint it might be possible to borrow more money, and thus the cost of violation of the constraint is the interest of that loan (e.g. Kangas et al. 2012). Then, optimal decision may be to violate the constraints to some extent and pay the penalty rather than requiring the constraints to be strictly fulfilled.

Whether the constraints are internal (set by the DM) or external (i.e. real), one possible avenue to solve these problems and to calculate the VOI is stochastic programming. In stochastic programming the decision is made to maximize the expected value of the decision over all possible scenarios of the state of nature, while requiring that the constraints will be met in every possible scenario. Thus, it means that if the constraints are strict, decisions are made in order to be prepared for the worst possible scenario. In such a case information can have very high value. If the violation of the constraints is possible but penalized, the problem could be solved by stochastic goal programming (or stochastic programming with simple recourse), i.e. by not requiring the constraints to be met in every possible state of nature. Then, very poor but highly improbable scenarios do not dictate the decision.

If the uncertainty can be resolved at some later point of time, the decision-making problem can be solved using, for instance, two-stage stochastic programming (e.g. Boychuck and Martell 1996). It means that the first stage decisions are carried out to maximize the expected value of the decision as above, but when data becomes available, the second stage decisions can be done to adjust to the observed scenario (see Eriksson 2006). Another possibility would be to use chance constrained programming (e.g. Bevers 2007) or robust programming (Palma and

Nelson 2009). In these approaches, the feasible solution is sought for by requiring a safety margin in the constraint. It means that the decision maker gives up some of the objective function value to be sure that the constraints are met under (nearly) all possible states of nature. In this case, the VOI stems from the possibility of reducing the safety margin or removing it altogether, which allows the decision maker to improve the objective function value.

An example of using stochastic programming to calculate the VOI is given in the following. The problem is to maximize the NPV from the first planning period, with the requirement that the NPV from the subsequent periods $k = 1, \ldots, K$ is at least as high as in the first. It is assumed that the constraints are not strict, but they form part of the decision maker's utility, and the deviations from the even flow are penalized using shadow prices of the constraints as penalties. This problem can be presented as a stochastic goal programming formulation. Assuming there are I different possible realizations of the uncertain coefficients, the stochastic goal programming formulation of the above problem is

$$\max z = \sum_{j=1}^{n} E\left(c_{j1}\right)x_{j1} - \frac{1}{I}\sum_{i=1}^{I}\sum_{k=1}^{K-1}\left(p_k^+ d_{ik}^+ + p_k^- d_{ik}^-\right)$$

subject to

$$\sum_{j=1}^{n} c_{jik+1}x_{jk+1} - \sum_{j=1}^{n} c_{jik}x_{jk} + d_{ik}^+ - d_{ik}^- = 0, \forall i = 1, \ldots, I, k = 1, \ldots, K-1$$

$$d_{ik}^-, d_{ik}^+ \geq 0 \forall \ i = 1, \ldots, I, k = 1, \ldots, K - 1 \tag{16.5}$$

where n is the number of stands, $E(c_j)$ is the expected net income (€/ha) from the stand j during the first planning period, c_{jik} is the net income in stand j realization i and period k and d_{ik}^+ is the negative deviation in realization i and period k and d_k^- is the respective positive deviation. The differences of this formulation and the ordinary goal programming formulation are that there are even flow constraints for all the I scenarios, meaning that there are I times as many constraints in the stochastic problem compared to the ordinary problem. Moreover, the penalties of not achieving the goals (p_k) are averaged over all the scenarios, i.e. the expected value of the penalties is minimized.

In case the true values of the uncertain coefficients are known, the optimal solution for the problem for realization i can be calculated as

$$\max z = \sum_{j=1}^{n} c_{ji1}x_{ji1} - \sum_{k=1}^{K-1}\left(p_k^+ d_{ik}^+ + p_k^- d_{ik}^-\right)$$

subject to

$$\sum_{j=1}^{n} c_{jik+1}x_{jik+1} - \sum_{j=1}^{n} c_{jik}x_{jik} + d_{i1}^+ - d_{i1}^- = 0$$

$$d_{ik}^-, d_{ik}^+ \geq 0 \forall k = 1, \ldots, K - 1$$

$$\tag{16.6}$$

For each possible realization i ($i = 1, \ldots, I$), the optimal solution for problem, z_i, is calculated where the optimal values of the decision variables are x_{ji1}. This is the optimal solution with full information. The value of stochastic solution can be calculated with the same objective function, but using the stochastic solution denoted with $\tilde{x}_{ji1}$, and the respective deviations $\tilde{d}_{ik}^-$ and $\tilde{d}_{ik}^+$ as

$$\tilde{z}_i = \sum_{j=1}^{n} c_{ij1} \tilde{x}_{ij1} - \sum_{k=1}^{K-1} \left(p_k^- \tilde{d}_{ik}^- + p_k^+ \tilde{d}_{ik}^+ \right) \tag{16.7}$$

The loss due to incorrect information in realization i is then $loss_i = z_i - \tilde{z}_i$, and the value of perfect information EVPI is the mean loss over all the scenarios. Thus, essentially the VOI is calculated similarly to as in the unconstrained case, but in addition to the difference between the net incomes from the solutions with and without information, also the effect of penalties due to deviations from the constraints is accounted for. However, in stochastic programming the solution without information is made for all the scenarios at the same time, to account for the constraints.

The effect of constraints was calculated for a forest estate of 29 stands and 67.29 ha. It was assumed that stand volume has an uncertain initial value, and the development of a stand was predicted based on that information with a growth model for volume. In the case where the error estimate of the initial volume of the stands was 25 %, the optimal value of the objective function with true values was 80,608 €, and with the stochastic solution 79,926 €, meaning that the loss was 815 €, i.e. 12 € per hectare. If the solution were calculated using normal goal programming, the violations of the constraints would be clearly higher as the solution would not be selected so as to adjust to the different scenarios. In that case, the observed loss would have been 360 €/ha. In the case that the NPV of the stands was maximized without constraints, the respective value of information would have been 22 €/ha. Thus, the potential losses from the penalties may have a substantial effect on the value of information. In addition, just using stochastic programming might reduce the losses and thus reduce the value of information. It means that a stochastic solution may have value as such. In Kangas (2013) this case was extended to a case where the accuracy of basal area and height measurements were considered.

16.3 How Long Can the Data Be Used?

16.3.1 Direct Effect of Time

In forestry, time has many direct effects on the VOI, as the decisions considered in a plan typically cover a longer period, for instance 10 years or more. The first direct effect is that VOI is positively related to the number of decisions that can be made.

Table 16.3 Sub-optimality losses (€/ha) from four different inventory methods in two different test areas and with two different interest rates (2 and 4 %) (Holmström et al. 2003)

	Site 1		Site 2	
Method	2 %	4 %	2 %	4 %
k nearest neighbor imputation, stand record information	49	48	82	161
k nearest neighbor imputation, aerial photograph interpretation	44	39	63	138
Field inventory, 5 plots per stand	13	18	21	17
Field inventory, 10 plots per stand	8	11	14	9

It means that VOI will be higher if also thinning decisions are considered instead of just final harvest decisions. It also means that VOI will be higher for an area where there are many decisions, compared to an area with a lot of young stands where harvest decisions are not relevant. Finally, it means that the longer the same data can be used for decision making, the higher the VOI. This needs to be taken into account when comparing different studies on VOI. It could therefore be useful to calculate VOI per year in addition to per hectare, to emphasize the effect of time.

Another direct effect of time is the consequence of discounting: the further away the decisions are, and the higher the interest rate, the less are the costs of making sub-optimal decisions for the decision maker. However, the higher interest rate shortens the optimal rotation times, i.e. moves the final harvest decisions to earlier periods. The observed losses will be influenced by both these factors. Effects of interest rate were evaluated by Holmström et al. (2003) when they compared the sub-optimality losses from thinning and final harvest decisions (Table 16.3).

16.3.2   Growth Prediction Errors

Predictions concerning the future development of a stand are made with growth models. The errors of initial data and growth predictions propagate in time, meaning that the longer the prediction period, the lower the quality of the data at the end of the period (Gertner 1987; Mowrer 1991; Kangas 1997). As the errors propagate through the system, the expected losses will therefore increase and at some time the expected losses increase to a level where collecting new data is more profitable than using the old data. This defines the optimal inventory interval, i.e. the life span of the initial data (Pietilä et al. 2010; Mäkinen et al. 2012). There may also be interactions between the prediction errors and the initial accuracy, meaning that an initial data set with given accuracy is more valuable when the growth models are more accurate (c.f. Ståhl and Holm. 1994).

Pietilä et al. (2010) used a stand-level growth and yield simulator (SIMO) and error models that were based on observed errors in predictions. The development predicted with the stand-level models for basal area and dominant height were assumed true, and errors were simulated around that prediction to analyze the value of growth information. The erroneous growth predicted was generated by the error model as

$$\widehat{I}_t = I_t + \varepsilon_t, \tag{16.8}$$

where $\widehat{I}_t$ is predicted erroneous growth, I_t is true growth and ε_t is the growth prediction error. The growth prediction error was assumed to consist of stand effect (inter-stand variation) and period effect (intra-stand, or between growth period variation). Thus the growth prediction error in each stand was described as

$$\varepsilon_t = u_t + e_t, \tag{16.9}$$

where ε_t is total growth prediction error, u_t is stand effect and e_t is period effect within the stand. Intra-stand error, e_t, was assumed to follow the first-order autoregressive process (AR(1)). This model was used to predict the variation around the development of basal area and dominant height in each stand predicted with SIMO, resulting in a 6.5 % RMSE for basal area and 10 % RMSE for dominant height after 5 years. With an inventory interval varying from 5 to 60 years, the losses varied from 230 €/ha/60 year up to 860 €/ha/60 year, i.e. from 3.8 to 14.3 €/ha/year.

16.3.3 Life Span of Forest Information

Mäkinen et al. (2012) used a similar error model as Pietilä et al. (2010) except that the between-period variation was not assumed autocorrelated, and the autocorrelation was solely due to the stand effect (inter-stand variation). They introduced inventory errors between 0 and 25 % to the dominant height and basal area in addition to the errors due to growth model. The costs of the inventory were described with a hypothetical model to show the effect of the inventory cost structure on the life-span of inventory data with different parameters.

Quite obviously, the losses were smallest with shortest inventory interval (meaning smallest possible growth prediction errors) and perfectly accurate inventory with relative RMSE 0 % (Fig. 16.4). The difference between the 0 % RMSE and 5 % RMSE was, however, quite small. It is notable that when the losses with the 5-year interval increase from 188 to 591 €/ha (3.14-fold) as the inventory accuracy decreases, they increase only from 420 to 695 €/ha (1.65-fold) when the inventory interval is 30 years. In other words, with the least accurate inventories the losses do not much increase as inventory interval increases, but with the most accurate inventories the trend is very clear.

When the inventory costs were assumed to be 8 €/ha for the least accurate inventory and perfect inventory was assumed to cost 324 €/ha, the optimal life-span of the data proved to be 15 years with 10 % RMSE of basic forest information (Table 16.4). For a less accurate data the optimal life-span shortens, and for more

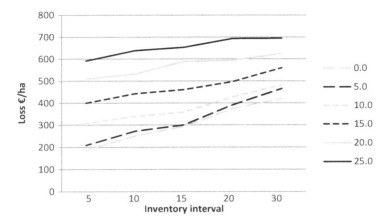

Fig. 16.4 The dependency of the losses on the inventory interval and the accuracy of the initial inventory data (Mäkinen et al. 2012)

Table 16.4 The average absolute total costs plus losses, in €/ha with different inventory interval (years) and accuracy levels (σ)

Accuracy level	Inventory interval (years)				
(σ)	5	10	15	20	30
0	1575.7	995.3	825.9	878.0	744.2
5	1017.7	706.3	610.4	685.4	655.1
10	718.6	560.1	**516.5**	577.2	579.1
15	570.9	534.7	527.3	559.6	599.7
20	567.6	563.0	611.0	615.4	638.9
25	625.5	656.9	666.8	706.7	703.3

The optimal accuracy and inventory period is given in bold (Mäkinen et al. 2012)

accurate data the life-span lengthens. The approach enables analysing the optimal life span with a fixed accuracy data or to optimize both the inventory interval and the data collection accuracy optimally at the same time.

16.4   Conclusions

In this chapter, we have dealt with different aspects regarding valuating forest information. Most of the theory presented applies to all kinds of data. However, actual VOI depends heavily on the data source used and the approach used to calculate the inventory results. ALS data may be offered as area based information and single tree information, or through diameter distributions as an approach somewhere in between. These approaches will differ in the error structure and therefore also in VOI. The differences between the approaches can be quite large, even if the data collected were the same. For each approach, the specifications used for collecting the ALS data, the field data used for interpreting the ALS data and all the models and methods used in different phases of the calculations have their effect on VOI. The importance of these specifications on the VOI has not yet been considered, but in the future the sensitivity of the VOI to them needs to be analyzed in order to maximize the benefits of the collected information.

The VOI also depends on its application, i.e. types of decisions to be made, and how long the data are used, i.e. we should search for a solution where the expected losses have increased to a level where collecting new data is more profitable than using the old data. Although quantifying VOI is not always is an easy task, the links between inventory effort level, decisions to be made and the VOI should be considered in any study related to innovative forest inventory methods including ALS applications. The accuracy measures usually presented in the remote sensing literature to describe a method are not providing a complete picture regarding the usefulness of the data.

References

Ades AE, Lu G, Claxton K (2004) Expected value of sample information calculations in medical decision modeling. Med Decis Mak 24:207–227

Bergseng E, Ørka HO, Næsset E, Gobakken T (2013) Valuation of information obtained from different forest inventory approaches and remote sensing data sources (Unpublished manuscript)

Bevers M (2007) A chance constrained estimation approach to optimizing resource management under uncertainty. Can J For Res 37:2270–2280

Birchler U, Bütler M (2007) Information economics. Routledge advanced texts in economics and finance. Routledge, London, 462 p

Borders BE, Harrison WM, Clutter ML, Shiver BD, Souter RA (2008) The value of timber inventory information for management planning. Can J For Res 38:2287–2294

Boychuck D, Martell DL (1996) A multistage stochastic programming for sustainable forest-level timber supply under risk of fire. For Sci 42:10–26

Burkhart HE, Stuck RD, Leuschner WA, Reynolds MA (1978) Allocating inventory resources for multiple-use planning. Can J For Res 8:100–110

Duvemo K, Lämås T, Eriksson L-O, Wikström P (2012) Introducing cost-plus-loss analysis into a hierarchical forestry planning environment. Ann Oper Res. doi:10.1007/s10479-012-1139-9

Eid T (2000) Use of uncertain inventory data in forestry scenario models and consequential incorrect harvest decisions. Silva Fenn 34:89–100

Eid T (2003) Model validation by means of cost-plus-loss analyses. In: Amaro A, Reed D, Soares P (eds) Modelling forest systems. CABI Publishing, Wallingford, UK, pp 295–305

Eid T, Gobakken T, Næsset E (2004) Comparing stand inventories for large areas based on photo-interpretation and laser scanning by means of cost-plus-loss analyses. Scand J For Res 19: 512–523

Eriksson L-O (2006) Planning under uncertainty at the forest level: a systems approach. Scand J For Res 21:111–117

Gertner G (1987) Approximating precision in simulation projections: an efficient alternative to Monte Carlo methods. For Sci 33:230–239

Gilabert H, McDill M (2010) Optimizing inventory and yield data collection for forest management planning. For Sci 56:578–591

Hamilton DA (1978) Specifying precision in natural resource inventories. In: Integrated inventories of renewable resources: proceedings of the workshop. USDA Forest Service, General technical report RM-55, pp 276–281

Holmström H, Kallur H, Ståhl G (2003) Cost-plus-loss analyses of forest inventory strategies based on kNN-assigned reference sample plot data. Silva Fenn 37:381–398

Islam N, Kurttila M, Mehtätalo L, Haara A (2009) Analyzing the effects of inventory errors on holding-level forest plans: the case of measurement error in the basal area of the dominated tree species. Silva Fenn 43:71–85

Islam N, Kurttila M, Mehtätalo L, Pukkala T (2010) Inoptimality losses in forest management decisions caused by errors in an inventory based on airborne laser scanning and aerial photographs. Can J For Res 40:2427–2438

Kangas A (1997) On the prediction bias and variance of long-term growth predictions. For Ecol Manag 96:207–216

Kangas A (2010) Value of forest information. Eur J For Res 129:863–874

Kangas A (2013) Effect of sustainability constraints on the value of forest information (in review)

Kangas A, Horne P, Leskinen P (2010a) Measuring the value of information in multi-criteria decision making. For Sci 56:558–566

Kangas A, Lyhykäinen H, Mäkinen H (2010b) Value of quality information in timber bidding. Can J For Res 40:1781–1790

Kangas A, Lyhykäinen H, Mäkinen H, Lappi J (2012) Value of quality information in selecting stands to be purchased. Can J For Res 42:1347–1358

Ketzenberg ME, Rosenzweig ED, Marucheck AE, Metters RD (2007) A framework for the value of information in inventory replenishment. Eur J Oper Res 182:1230–1250

Kim JB, Hobbs BF, Koonce JF (2003) Multicriteria Bayesian analysis of lower trophic level uncertainties and value of research in Lake Erie. Hum Ecol Risk Assess 9:023–1057

Lawrence DB (1999) The economic value of information. Springer, New York, 393 p

Mäkinen A, Kangas A, Mehtätalo L (2010) Correlations, distributions, and trends in forest inventory errors and their effects on forest planning. Can J For Res 40:1386–1396

Mäkinen A. Kangas A, Nurmi M (2012) Using cost-plus-loss analysis to define optimal forest inventory interval and forest inventory accuracy. Silva Fenn 46:211–226

Mowrer HT (1991) Estimating components of propagated variance in growth simulation model projections. Can J For Res 21:379–386

Palma CD, Nelson JD (2009) A robust optimization approach protected harvest scheduling decisions against uncertainty. Can J For Res 39:342–355

Pietilä I, Kangas A, Mäkinen A, Mehtätalo L (2010) Influence of growth prediction errors on the expected losses from forest decisions. Silva Fenn 44:829–843

Rasinmäki J, Kangas A, Mäkinen A, Kalliovirta J (2013) Value of information in DSS: models & data. In: Tuček J, Smreček R, Majlingová A, Garcia-Gonzalo J (eds) Implementation of DSS tools into the forestry practice, Reviewed conference proceedings. Published by Technical University in Zvolen, Slovakia, 167p. ISBN 978-80-228-2510-8

Ståhl G (1994) Optimizing the utility of forest inventory activities. Swedish University of Agricultural Sciences, Department of Biometry and Forest Management, Umeå. Report 27

Ståhl G, Holm S (1994) The combined effect of inventory errors and growth prediction errors on estimations of future forestry states. Manuscript. In: Ståhl G (1994) Optimizing the utility of forest inventory activities. Swedish University of Agricultural Sciences, Department of Biometry and Forest Management, Umeå. Report 27

Ståhl G, Carlson D, Bondesson L (1994) A method to determine optimal stand data acquisition policies. For Sci 40:630–649

Part III
Ecological Applications

Chapter 17
Assessing Habitats and Organism-Habitat Relationships by Airborne Laser Scanning

Ross A. Hill, Shelley A. Hinsley, and Richard K. Broughton

Abstract Three-dimensional structure is a fundamental physical element of habitat. Because of the well-recognised link between vegetation structure and organism-habitat associations, many published studies that make use of airborne LiDAR for forest applications have results of potential relevance for habitat assessment. This chapter reviews those published studies that have made direct use of airborne LiDAR data for habitat assessment of individual species or groups of species in a woodland or forest context. This is followed by a case study of the authors' own work at a study site in eastern England, Monks Wood National Nature Reserve. We show that airborne LiDAR has a proven ability for supplying a range of forest structural measures that are key elements of an organism's habitat at the meso-scale. Examined in combination with detailed field ecology data on species distributions, abundances or biological activity, airborne LiDAR data can be used as an exploratory tool to advance ecological understanding by quantifying how forest structure impacts habitat use and thereby influences habitat quality.

17.1 Introduction

The term habitat is typically taken to mean the physical environment in which an organism or community of organisms live, and includes both biological and geo-physical characteristics. A habitat can be specific to an individual or a population and can be assessed for a single species or a species assemblage. Any physical

R.A. Hill (✉)
School of Applied Sciences, Bournemouth University, Talbot Campus, Poole,
Dorset BH12 5BB, UK
e-mail: rhill@bournemouth.ac.uk

S.A. Hinsley • R.K. Broughton
Centre for Ecology and Hydrology, Maclean Building, Benson Lane, Crowmarsh Gifford,
Wallingford, Oxfordshire OX10 8BB, UK

M. Maltamo et al. (eds.), *Forestry Applications of Airborne Laser Scanning: Concepts and Case Studies*, Managing Forest Ecosystems 27, DOI 10.1007/978-94-017-8663-8__17,
© Springer Science+Business Media Dordrecht 2014

space has the potential to be a habitat, and so the term only takes on true ecological meaning if considered in relation to its inhabitants. There is thus a long history in ecology of seeking organism-habitat relationships or associations, and this lies at the heart of much ecological conservation and management (Johnson 2007).

Three-dimensional structure is a fundamental physical element of habitat and has long been identified as a key determinant of biological diversity, particularly in forests (MacArthur and MacArthur 1961). Because airborne LiDAR can supply detailed information about the vertical structure of a forest and its spatial variability, its use in habitat assessment was quickly realised once methods of deriving forest structural variables from airborne LiDAR became commonplace. Chapter 18 provides an overview of the ecological rationale of various LiDAR-derived forest structural variables, and their applicability extends to the assessment of habitat as much as to biodiversity. Vierling et al. (2008, pp. 95–96) helpfully identified the more relevant LiDAR-derived forest structure variables as including: "canopy height, roughness, volume, stand density and age, number of snags and downed trees, number of large trees, understory/middlestory height and density, ground surface texture, patch and edge characteristics, and the matrix in which a habitat occurs". These can be obtained by combining LiDAR data with ancillary data, typically acquired by field collection. Recent reviews by Bergen et al. (2009) and Shugart et al. (2010) have considered vegetation structural measures, focussing on: how these can be derived from LiDAR as well as radar and optical data; their significance for investigating forest biodiversity, habitat and function; and the implications of this for the design of future satellite missions.

Because of the well-recognised link between vegetation structure and organism-habitat associations, many published studies that make use of airborne LiDAR for forest characterisation have expressed habitat implications of their results (e.g. Hyde et al. 2005, 2006; Newnham et al. 2012). However, without expanding this to specify relevance at an individual species or species assemblage level, the use of the term habitat in many such studies can remain somewhat abstract. Examples include the modelling of mature hardwood forest distribution across four US states (Weber 2011), the assessment of forest structural complexity in both the Pacific Northwest of the USA (Kane et al. 2010) and the Cradle Coast region of Tasmania, Australia (Miura and Jones 2010), the discrimination of vegetation layers in a Mediterranean forest (Morsdorf et al. 2010), the appraisal of scrub encroachment in forest mires in central Switzerland (Waser et al. 2008), the categorization of canopy gaps in old-growth boreal forests (Vehmas et al. 2011), the mapping of forest maturity classes in Maryland, USA (Weber and Boss 2009) and forest successional stages in Idaho, USA (Falkowski et al. 2009), the classification of mature herb-rich forest stands in the Koli National Park, Finland (Vehmas et al. 2009), and the evaluation of large woody debris recruitment to the Narraguagas River from riparian forest in coastal Maine, USA (Kasprak et al. 2012) and of pine beetle infestation in lodgepole pine (*Pinus contorta*) stands in British Columbia, Canada (Coops et al. 2009). All of these are identified as potential habitat descriptors or determinants and thus as key information for the management and conservation of the habitat and its wildlife populations. As a final example here, Bässler et al. (2011) used airborne LiDAR data to map four Natura 2000 forest habitat classes based on subtle differences in structure between the habitat types. They showed that airborne LiDAR-derived structure variables can predict forest habitat types with a similar accuracy as ground data on soil, vegetation composition and climate. In this and many of the examples above, each identified habitat type would have its own associated species assemblage.

This chapter reviews those published studies that have made use of airborne LiDAR data for habitat assessment of individual species or groups of species in a woodland or forest context. The review follows an order of firstly the prediction of species distribution based on known habitat requirements and airborne LiDAR modelling of where those conditions occur; followed by the assessment of habitat characteristics for individual species and groups of species from airborne LiDAR data extracted from known occurrence locations; and finally the quantification of habitat quality from airborne LiDAR based on field data relating to biological activity levels, such as feeding or reproduction. This represents a progression towards gaining increased ecological understanding of organism-habitat relationships and how structure can influence habitat use. The review is followed by a case study of the authors' own work at a study site in eastern England, some of which is covered in earlier reviews by Bradbury et al. (2005) and Newton et al. (2009). The purpose of the case study is to further exemplify how airborne LiDAR data can be utilised to improve our knowledge of what drives organism-habitat relationships at a single site that has been studied in great detail. An overall consideration of what airborne LiDAR data have been able to contribute to habitat assessment (and using what derived metrics) is provided at the end of the chapter.

17.2 Predicting Species Distribution Based on Known Habitat Requirements

A first step in assessing habitat with airborne LiDAR data is to derive an independently validated geospatial model of one or more elements of forest vertical or horizontal structure. A frequent second step is then to use such models to describe or predict habitat suitability for a specific organism based on known habitat requirements. An early example of this was by Nelson et al. (2005) who identified 53 potential habitat sites for the Delmarva fox squirrel (*Sciurus niger cinereus*) from 1,304 km of airborne laser profiling data over Delaware, USA. Potential habitat was identified based on the criteria of having 120 m of contiguous forest (within each flight-line of data), with an average canopy height of >20 m and average canopy closure of >85 %, and with no impervious surfaces or open water bodies present. Site visits were made to two-thirds of the identified potential habitat sites and field data recorded, which in three-quarters of cases supported the predicted habitat suitability. It should be noted that Nelson et al. (2005) were clear to specify that they had modelled potential, and not actual, habitat. Similarly, potential habitat for over-wintering mule deer (*Odocoileus hemionus*) was assessed in British Columbia, Canada using airborne LiDAR data by Coops et al. (2010). They made use of field ecological information which had characterised good winter feeding habitat for mule deer as being on moderately steep slopes with a southerly aspect and with moderate to high canopy cover of old-growth Douglas fir (*Pseudotsuga menziesii*). Their aim was to compare the capabilities of discrete return LiDAR data with the common approach to mapping winter mule deer habitat (as applied by the BC Ministry of Forests and Range) which involved the use of forest cover and inventory maps interpreted manually from aerial photographs. Both topographic and forest canopy structure information from the LiDAR data (totalling 16 variables) were used as input in decision-tree models, which were able to explain between 50 and 75 % of variance in the inventory-derived descriptors of mule deer habitat. Coops et al. (2010) thus concluded that airborne LiDAR is an effective tool that can complement standard aerial photography for forest inventory mapping and the assessment of winter mule deer habitat, helping to reduce the subjectivity and potential inconsistency in manual interpretation approaches.

For mixed conifer forest in northern Idaho, USA, Martinuzzi et al. (2009) firstly investigated the use of airborne LiDAR data to map understorey and snag diameter classes, and then used this information to derive habitat associations for four bird species for which the presence of understorey and snags are known to be key habitat features: the dusky flycatcher (*Empidonax oberholseri*), hairy woodpecker (*Picoides villosus*), Lewis's woodpecker (*Melanerpes lewis*), and downy woodpecker (*Picoides pubescens*). Using discrete return LiDAR data, forest inventory plot data and a random forest algorithm, they were able to identify which LiDAR structure metrics out of 19 extracted either directly indicated the presence of shrubs or snags or which influenced their likely distribution (e.g. factors relating to canopy gaps/cover and forest succession). They achieved classification accuracies of between 83 and 88 % for mapping understorey shrubs and snag classes and then used this additional detail to improve habitat suitability maps for the four bird species.

In the Cedar River Municipal Watershed, Washington State, USA, the known habitat preferences of two species of anadromous fish, Coho salmon (*Oncorhynchus kisutch*) and Steelhead (*Oncorhynchus mykiss*), along with information on forest cover type and river channel geomorphology (derived from airborne spectral and LiDAR data respectively) were used to identify riparian conifer stands where habitat restoration would have the greatest benefit for salmonid population recovery (Mollot and Bilby 2008). The most favourable sites for anadromous fish spawning and rearing were identified as low gradient and unconfined streams with mature conifer riparian vegetation which can provide large woody debris to the stream channel. Suitable sites for restoration were those areas capable of supporting conifer trees long term. In this example, salmonid habitat suitability, or the potential restoration thereof, was mapped at a watershed scale and with a greater level of detail, accuracy and objectivity than had been achievable previously.

17.3 Characterising the Habitat of Single Species Based on Field Distribution Data

An alternative to the use of airborne LiDAR data products to predict habitat suitability and species distribution based on known habitat requirements, is to utilise species distribution and abundance data to quantify habitat requirements directly from occupied areas. Associated with this is the subsequent potential to make predictions over a wider geographical area for which LiDAR data may be available but species occurrence data may not. As outlined by Vierling et al. (2008) this makes use of LiDAR data as an explanatory tool to increase understanding (and quantification) of resource selection by species of known distribution. Such studies are most frequently focussed on a single bird species, particularly in North America and Western Europe where strong ornithological traditions exist. Field data typically take the form of presence and absence (generally pseudo-absence), with presence recorded along transects or at bird count locations or identified via the presence of nest sites. A spatial sampling frame is then used to extract various forest structural metrics from the LiDAR data and establish bird-habitat relationships, which can then be mapped across the spatial extent of the LiDAR data. There are numerous examples of this approach for individual bird species habitat assessment. For example, Bellamy et al. (2009) described habitat selection of a rapidly-declining bird species, the willow warbler (*Phylloscopus trochilus*), at three woods with contrasting management in eastern and southern England. Height profiles derived from discrete return LiDAR were compared for areas of willow warbler presence and for randomly selected parts of the woods to gain an understanding of both the breadth of habitat occupation and optimum habitat. Vegetation height was quantified as a key habitat parameter (<6 m canopy height being preferred) but only where this occurs in patches >0.5 ha in size and not along external woodland edges. The three different management regimes (recent low intervention, active coppicing, and high forest with clear fells) impacted the structure of the woods but the derived habitat suitability models, generated using logistic regression, did not differ between the three woodland sites as habitat preference was consistently for early successional or open canopy woodland. Goetz et al. (2010) focussed on a New World warbler, the black-throated blue warbler (*Dendroica caerulescens*), in the northern hardwood forests of Hubbard Brook, New Hampshire, USA. They used 4 years of field data from 371 point-count stations and canopy vertical structure metrics derived from full waveform LiDAR data to model habitat quality based on prevalence, taking a multiple decision tree clustering approach (i.e. random forest). They found that canopy height, elevation and canopy complexity were key characteristics in the frequently occupied habitat, explaining 47 % of variation in multi-year prevalence.

Graf et al. (2009) investigated the habitat of an endangered grouse, the caper-caillie (*Tetrao urogallus*), in a forest reserve in the Swiss Pre-Alps. They used generalised linear models (GLM) to establish relationships between field data on presence-absence and seven LiDAR-derived metrics capturing the forest horizontal and vertical structure. They demonstrated that horizontal structure (i.e. relative tree canopy cover) was the key habitat variable determining the presence of capercaillie, with intermediate values representing the most suitable habitat. From this they were able to make conservation-oriented management recommendations at a local scale that were informed by a landscape level assessment of an entire population. Taking a different approach, Garcia-Feced et al. (2011) used nest location data to characterise the breeding habitat of the Californian spotted owl (*Strix occidentalis occidentallis*) in the Sierra Nevada Mountains, California, USA. They identified the location of four nest trees from field survey and used airborne LiDAR data to assess the canopy height profile in a 200 m radius around each nest site. From this they were able to quantify the habitat structure in the core nesting area of these four pairs of Californian spotted owls at this particular site, providing indicative information for conservation management.

The studies of Wilsey et al. (2012) and Smart et al. (2012) are notable for their integrated use of several environmental datasets in addition to vegetation structure derived from airborne LiDAR in their modelling of bird species distributions and habitat associations. Wilsey et al. (2012) used field data on presence-absence of the black-capped vireo (*Vireo atricapilla*) in the Fort Hood Military Reservation, Texas, USA, along with information on soil, vegetation type (mapped from aerial imagery) and vegetation structure (derived from airborne LiDAR). They used a non-parametric machine learning algorithm (called *cforest*) to build predictive habitat suitability models and found that vegetation type and soil were more important predictive variables than vegetation structure (i.e. mean height, canopy cover, and edge density). Nonetheless, vegetation structure was shown to be a useful predictive variable along with soil in the absence of vegetation type information. Smart et al. (2012) examined field data on occupancy in habitat use zones (i.e. nesting and foraging sites) for the red-cockaded woodpecker (*Picoides borealis*, Vieillot) in a forested catchment contained within US military land in North Carolina, USA. They used geospatial data on elevation, landcover, and hydrography together with airborne LiDAR-derived measures of forest vertical structure (e.g. mean and standard deviation of canopy height) and horizontal patterns of vertical structure (assessed by both semivariograms and lacunarity analysis) in a maximum entropy approach to species distribution modelling. They found that the addition of the LiDAR structure variables to the Maxent model improved the predictive power by just 8 %. Nonetheless, the LiDAR data were found to be highly useful in describing the forest structural characteristics of the two habitat use zones. Thus, as is often inferred in studies that focus only on LiDAR data for characterising organism-habitat relationships, these latter two studies showed that whilst vegetation structure is a key component of habitat, it is not the only characteristic of the physical or biological environment that influences bird-habitat associations.

To-date, studies that take a single species approach to habitat characterisation (based on field data of known occurrence) for organisms other than birds are relatively few. However, published studies exist that explore organism-habitat relationships for individual species of both monkey and weasel. Palminteri et al. (2012) used full waveform airborne LiDAR data to investigate the forest structural characteristics of areas habituated by bald-faced saki monkeys (*Pithecia irrorata*) in south-eastern Peruvian Amazonia. They observed the home ranges of five groups of saki monkeys over a 3 year period, building up an extensive geospatial record of occurrence frequency. Binary logistic regression was used to test if threshold levels of either the mean or standard deviation of canopy height (in 30 × 30 m cells) could be identified which would explain saki occupancy. They found a strong preference for areas containing higher stature and more homogeneous canopies, with saki monkeys avoiding areas <15 m and preferring areas in the 25–35 m height range. Very tall forest canopy areas were avoided. This occurrence frequency pattern was related to ease of movement through the canopy and availability of food source trees, with a preferred canopy structure more consistent with floodplain than terra firme forest. Zhao et al. (2012) studied the Pacific fisher (*Martes pennanti*), a small carnivorous mammal, in the Sierra Nevada Mountains, California, USA. They used Classification and Regression Trees (CART) to explore relationships with LiDAR-derived forest structure and topographic variables (extracted over different sized spatial sampling frames) for trees identified in the field as having resting dens for the Pacific fisher and randomly selected trees which did not. Interestingly, they found that tree height and slope were important variables within a 20 m radius of a denning tree, but forest structure and complexity became more important explanatory variables between 20 and 50 m. Zhao et al. (2012) concluded that airborne LiDAR has great utility for characterising forest structure and thereby forest mammal habitat and so recommend that this technology is applied to the study of other mammal species.

17.4 Characterising the Habitat of Multiple Species Based on Field Distribution Data

As with the study of single-species organism-habitat relationships quantified from airborne LiDAR and field data, there is an overwhelming bias towards birds in those studies which have taken a multi-species focus. An excellent example is the study of Müller et al. (2009) which investigated 23 forest passerine bird species in a mixed montane forest in the Bavarian Forest National Park, southeast Germany. They evaluated the use of airborne LiDAR data (in combination with aerial photography) to model abundance of each species and the composition of species assemblages. Both were based on point-count bird data from 223 plots and the subsequent use of canonical correlation analysis to identify relationships. They were able to explain up to 40 % of variance in individual species and also ca. 20 % of variance in species composition, with key LiDAR variables being mean and standard deviation of canopy height and penetration rate between 1 and 5 m and between 2 and 10 m. Trends in bird species distributions were related to gradients in open to dense forest stands and homogeneous to highly structured forest stands. Seavy et al. (2009) also used bird point-count data for forest passerine species, but in their study examined a riparian forest habitat in the Cosumnes River Preserve in central

California, USA. They used logistic regression to identify habitat associations for 16 bird species based on LiDAR-derived forest structure (i.e. mean, standard deviation, and coefficient of variation of canopy height) at a range of spatial scales (from 0.2 to 50 ha). The best logistic regression models were considered to perform well for ten of the 16 passerine species. Of these ten models, it was noted that six were based on vegetation structure measurements at the 0.2–3 ha scale and four at the 50 ha scale. Hence, as with the findings of Zhao et al. (2012) for the Pacific fisher, the results of Seavy et al. (2009) demonstrate the need to take a multi-scale approach to fully understand how individual organisms use a habitat.

Clawges et al. (2008) used small footprint, discrete return airborne LiDAR data to investigate relationships between vegetation structural characteristics and bird species diversity, density and occurrence in the Black Hills Experimental Forest in South Dakota, USA. As with the above studies this work was based on bird point-count data, here from 51 plots recorded during 2004–2005. They found significant positive relationships between foliage height diversity and bird species diversity (for 43 species) and between vegetation volume and bird count (i.e. density). Furthermore, they found that bird species counts were highest where the vegetation structure contained a higher proportion of foliage in lower strata (i.e. <5 m tall). They went on to assess the habitat preferences of two bird species at the site, the dark-eyed junco (*Junco hyemalis*) and warbling vireo (*Vireo gilvus*), which are known to forage and nest in lower vegetation, and in both cases found a significant association with the 0.5–2.0 m high vegetation layer. This study thus showed the benefits of habitat heterogeneity for increasing bird species richness and further demonstrated the ability to quantify organism-habitat associations (in terms of habitat structure) using airborne LiDAR data.

A study by Hinsley et al. (2009a) compared the canopy structure characteristics for 23 bird species across two woods in southern England, Bradfield Wood in Suffolk and Sheephouse Wood in Buckinghamshire. These two woods had contrasting management and structure (active rotational coppicing versus mature oak woodland) and thus were shown to have very different canopy height profiles as modelled from airborne LiDAR data. Nonetheless, based on territory mapping, the 23 bird species common to both woods showed consistency in their rank positions across the two canopy height gradients. The species with the most habitat-limited distribution were those associated with early pre-canopy closure stages of forest growth. It was concluded that airborne LiDAR data can be used to predict bird composition and distribution in lowland British woodland, thereby providing a means of assessing habitat availability for a range of bird species.

The earlier single-species work of Goetz et al. (2010) was extended by Swatantran et al. (2012) by comparing satellite multispectral imagery (Landsat ETM+), airborne LiDAR data (both small footprint discrete return, and large footprint full waveform), and airborne radar data (L-band polarimetric) in predictive models of prevalence for eight bird species in the Hubbard Brook Experimental Forest, New Hampshire, USA. The modelling made use of random forest regression models based on 10 years of bird observation data at 371 field plots. The models with the greatest predictive power were based on multi-sensor input variables, explaining between 54 and 75 % of variation in prevalence across the eight bird species. The two most commonly used predictive variables in the various models were stem density from the discrete return LiDAR data and phenology from the time-series multispectral data. The derived spatial maps of prevalence for each species were found to be consistent with previously described habitat preferences, and were considered a benefit for forest management by providing fine-scale information at a landscape level.

Currently only one study has looked at multi-species habitat associations for a taxon other than birds. A detailed study of carabid beetles and the influence of surface topography in determining the prevalence of females (taken as indicators of preferred reproductive habitat) under boreal forest in northwest Quebec was conducted by Work et al. (2011). Carabid beetles of eight species types were collected in pitfall traps in forest areas representing a gradient of natural disturbance and succession. Topographic variables (e.g. slope, azimuth, curvature and flow accumulation) were extracted from discrete return airborne LiDAR data at four different spatial scales (grid cell size ranging from 4 to 32 m), thus capturing potential micro-scale to hillside-scale topographic influences on beetle distributions. Relationships were assessed using logistic regression. Work et al. (2011) found that north facing slopes with high flow accumulation were the most important predictor variables for the proportion of females, and that larger scale topography gave better predictions than smaller scale micro-topography. Although they stress that imprecision of their models limits their predictive applications, they conclude that topography may influence the distribution of females by maintaining more mesic habitats and attenuating wildfire.

17.5 Quantifying Habitat Quality Based on Field Ecological Activity Data

The most ecologically sophisticated use of airborne LiDAR data for the assessment of habitats is its use along with field recorded data on one or more aspects of biological activity, such as foraging, hunting, reproduction, etc. This enables habitat quality to be quantified at the species or organism level based on how vegetation structure impacts upon that particular biological activity. This establishes a link between ecological process and function and its relationship with habitat structure. Studies that combine airborne LiDAR data with biological activity data thus generate a more detailed understanding of how a habitat is being used, and what features within the habitat either impede or facilitate its use. Such studies therefore not only address the question of what is good (or bad) habitat for a certain species or group of species, but implicitly embedded in the analysis is the issue of why.

Two recent studies have investigated forest dwelling bats and the impacts of forest structure on their foraging activities. Müller et al. (2012) studied the foraging activities of 12 species of insectivorous bat in the Bavarian Forest National Park, southeast Germany. They assessed vegetation density using terrestrial laser scanning (TLS), prey density using light and pitfall traps, and bat density using bat call recorders. The bat species were examined in terms of their foraging guilds; open habitat, edge habitat and closed habitat foragers. The results demonstrated that as forest vegetation density increased (from a minimum of 2 % canopy coverage of a 20 m radius plot to a maximum of 79 %) so prey abundance increased and the occurrence of open habitat foragers decreased. The occurrence of edge habitat and closed habitat foragers was not affected by vegetation density, with bat species in these guilds being able to forage across the spectrum of habitat density assessed. Thus, the open habitat foragers were excluded from the habitat with highest overall prey abundance (i.e. the densest forest) and instead showed an aggregative response to prey abundance within the more open habitat areas. In a similar study, Jung et al. (2012) explored the occurrence and activity levels of insectivorous bat species in the Schorfheide-Chorin biosphere reserve in Brandenburg, northeast Germany. They carried out acoustic monitoring in 50 plots and assessed forest structural characteristics from full waveform airborne LiDAR data. A total of 85 structural parameters were extracted from the LiDAR data for 1 ha plots centred on the monitoring locations. Random forest regression, generalised linear candidate models and canonical correspondence analysis were carried out to identify a set of 20 uncorrelated structural parameters with explanatory power for the whole bat assemblage and to explore relationships with forest structure for multiple and single species occurrence and activity levels. They found that bat occurrence and activity in general and the occurrence of multiple species of bats increased with canopy height, canopy surface roughness and edge fraction; indicating the significance of forest structural heterogeneity, in particular for older forest stands surrounded by managed stands. At the species level, Jung et al. (2012) showed that open space foraging species (which have high flight speed and low maneuverability) were most strongly related to LiDAR metrics relating to ground area and gaps, whilst edge space foraging species (which have higher maneuverability) were associated with LiDAR variables related to closed canopy with high levels of understorey vegetation. They were thus able to determine species-specific relationships between bat occurrence and activity and forest structural parameters relating to differently managed production forest types. This had clear implications for the use of airborne LiDAR data to help inform forest management practices to maintain bat diversity and ecosystem functioning.

17.6   Case Study: The Use of Airborne LiDAR for Habitat Assessment in Monks Wood, UK

Monks Wood (52°24′N, 0°14′W) is a 157 ha deciduous woodland in Cambridgeshire, eastern England. It was established as a National Nature Reserve (NNR) in 1953, and is also a Site of Special Scientific Interest (SSSI) (Massey and

Fig. 17.1 Three-dimensional view of Monks Wood NNR, looking northwards, showing hyperspectral imagery acquired with a HyMap sensor draped over a LiDAR-derived surface model

Welch 1993). The site is managed by Natural England with a view of maintaining and enhancing its biodiversity, in particular its butterfly, beetle and bird populations and also rare plants such as the greater butterfly orchid (*Platanthera chlorantha*), violet helleborine (*Epipactis purpurata*) and crested cow-wheat (*Melampyrum cristatum*). Monks Wood is an ash-oak woodland; with ash (*Fraxinus excelsior*) being the most common and widespread tree species, and oak (*Quercus robur*) occurring less frequently because of intense felling in the early part of the Twentieth Century. Field maple (*Acer campestre*) and birch (*Betula pendula*) are found scattered throughout the woods, whilst aspen (*Populus tremula*) and small-leaved elm (*Ulmus minor*) form occasional clusters on the wetter soils. The dominant shrub species making up the understorey and woodland fringes are hawthorn (*Crataegus* spp.), blackthorn (*Prunus spinosa*), hazel (*Corylus avellana*), plus scattered dogwood (*Cornus sanguinea*) and wild privet (*Ligustrum vulgare*).

Multiple airborne LiDAR and hyperspectral datasets have been acquired for Monks Wood (Fig. 17.1). Initial habitat assessment for Monks Wood NNR used discrete return airborne LiDAR data from leaf-on conditions to derive a canopy height model for the overstorey (Fig. 17.2a) (Gaveau and Hill 2003). Subsequent data acquisitions under both leaf-on and leaf-off conditions enabled the extraction of an understorey height model (Fig. 17.2b) (Hill and Broughton 2009). The integration of airborne LiDAR and hyperspectral data was used for the patch-based mapping of species composition and structure relevant to the National Vegetation Classification scheme of Great Britain and reflecting woodland successional and management processes (Hill and Thomson 2005). As a separate study, a time series of airborne multispectral data were used to produce a map of overstorey tree species distributions across Monks Wood (Hill et al. 2010).

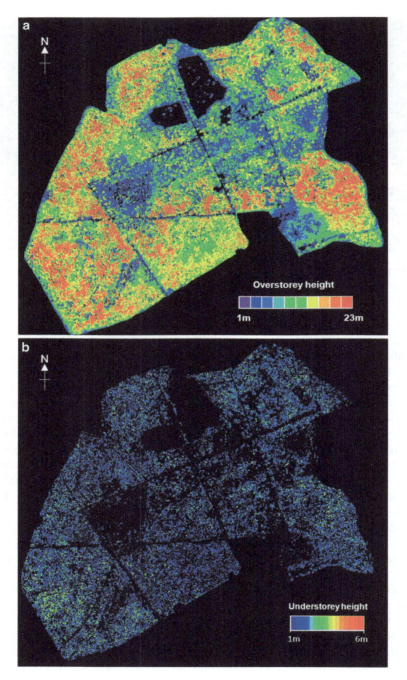

Fig. 17.2 Overstorey canopy height model (*top*, **a**) and understorey canopy height model (*bottom*, **b**) for Monks Wood NNR. Areas of height <1 m or no data areas are shown in *black* (The data area shown covers 1.7 × 1.6 km)

An immediate possibility with the habitat-relevant information derived from airborne LiDAR and spectral data, was the production of habitat suitability models based on the known habitat requirements of species of interest. As with many other studies, we have focussed exclusively on bird species because of their importance in woodland ecosystems and to utilise their presence, abundance, distribution and biological activity as indicators of habitat productivity at lower trophic levels. In Fig. 17.3, we applied ornithological knowledge to the map of forest structure and species composition, considering patch size metrics and surrounding context to derive a series of habitat suitability maps for a range of bird species commonly found within Monks Wood NNR. The maps shown in Fig. 17.3 are thus the spatial manifestation of combined ecological knowledge, enabled through the acquisition of remotely sensed data. Such maps may be useful for management purposes but do not particularly expand ecological knowledge. This can be achieved, however, by combining the remote sensing derived habitat-relevant data layers with field data on bird species distribution to quantify the characteristics of bird territories (or territory cores). As an example, we studied territory characteristics of a bird species of conservation concern, the marsh tit (*Poecile palustris*), initially for a single breeding season (Broughton et al. 2006), but subsequently by building up data on territory occupation from 5 years of field-based territory mapping (Broughton et al. 2012a). We found that habitat occupation by marsh tits responded to the vegetation characteristics through the full vertical profile of woodland structure (Fig. 17.4), which unified previously conflicting results from studies involving ground-based field plot data of vegetation structure. The most frequently occupied habitat had a mature canopy, with tall trees and high canopy closure but with a well-developed shrub layer beneath. Marsh tits were found to avoid areas dominated by young and immature trees or woodland edges, but were apparently indifferent to the tree species composition of the upper canopy. In an extension to this work, the question of habitat versus social cues in nest placement within bird territories was addressed for marsh tits by relating nest locations to habitat structure and composition and to conspecific activities (Broughton et al. 2012b). It was found that there was no evidence for conspecific factors determining nest placement, which could instead be explained by habitat differences within the territory, with nest sites situated in areas of localised taller overstorey with more ash and field maple and less understorey than other parts of the territory. Studies such as this help to extend ecological knowledge at the individual species level by quantifying habitat preferences in greater detail and over larger areas than previously possible.

Greater expansion of ecological knowledge can be achieved by combining the airborne LiDAR-derived habitat-relevant data layers with field data on bird biological activity. An example of this is the investigation carried out at Monks Wood NNR on relationships between habitat characteristics and reproductive success for pairs of blue tits (*Cyanistes caeruleus*) and great tits (*Parus major*) breeding in nest boxes. Examined initially for both species in a single breeding season, we found a positive relationship between the mean weight of blue tit nestlings and

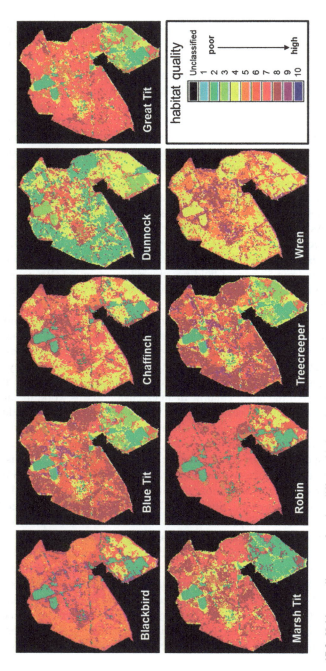

Fig. 17.3 Habitat quality maps for nine different bird species that are commonly found in Monks Wood NNR. Habitat quality is scored over a range of 0 (uninhabitable) to 10 (best quality) based on a consideration of the structure and composition of each patch together with patch size and context

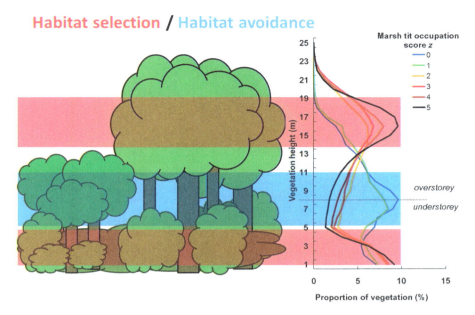

Fig. 17.4 Distribution of vegetation height surfaces in relation to areas defined by marsh tit occupation score (denoted as Z). Z has a range of 0 (never occupied) to 5 (occupied in all 5 years of field survey). Vegetation surfaces are represented as a percentage of pooled values of total understorey and overstorey height, and characterised in a schematic diagram

mean canopy height for the core territory area surrounding the nest box, but in contrast, a negative relationship for great tit nestlings (Hinsley et al. 2002). The weight of nestlings reflects food abundance and availability within the territory and the ability of the parent birds to find it and deliver it to the nestlings, thus the measure of nestling body mass provides a direct, ecologically-determined means for quantifying foraging habitat quality. This can be expressed geospatially as habitat quality maps (Fig. 17.5) by applying the derived relationship to each grid cell in the LiDAR-derived canopy height model (Hill et al. 2004). Interestingly, no other single structure variable or tree species composition variable was found to have a significant relationship with nestling body mass. By examining these relationships for great tits in more detail over a series of breeding seasons, we demonstrated that the structure conferring best habitat quality differed between years depending on the prevailing weather conditions. Thus the weight of nestlings declined with mean canopy height in the core territory area in cold, late springs, but increased with canopy height in warm, early springs (Hinsley et al. 2006). The nature of this nestling body mass and canopy height relationship was found to relate linearly to the North Atlantic Oscillation (NAO) Winter Index with a 1 year time lag, and so we were able to elucidate the regional-scale drivers of habitat quality determination in this case.

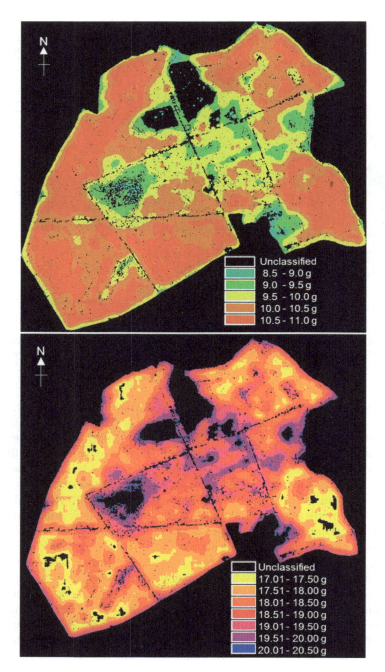

Fig. 17.5 Predictive map of blue tit (*top*) and great tit (*bottom*) nestling body mass for spring 2001 based on a relationship between nestling body mass and canopy height. Note that areas of Monks Wood NNR with a mean canopy height beyond the range encountered in the sample areas for nestboxes occupied in 2001 were unclassified (shown in *black*)

An alternative example of combining remote sensing derived, habitat-relevant data layers with field data on bird biological activity, is the examination of daily energy expenditure of parent birds feeding young in the nest (Hinsley et al. 2008). This again focussed primarily on great tits, but provided a comparative analysis of woodland habitat (Monks Wood NNR) with open parkland (Bute Park, Cardiff). We found that great tits in the park reared fewer young and that mean body mass was also low compared with woodland, probably due to a relative scarcity of food resources in the park. Thus, parent birds of the same species were shown to work (via their daily energy expenditure) 64 % harder per chick reared in open parkland than in woodland. A key tipping point in the parkland was related to a canopy cover of 65 %, below which daily energy expenditure increased disproportionately above average demand, even within the parkland habitat taken in isolation. Interestingly, whilst none of the territories in Monks Wood had a canopy cover anywhere near as low as 65 %, a similar effect of higher than average daily energy expenditure was detected for parent birds with core territory areas with less than ca. 30 % canopy cover of oak. The ecological significance of these findings in terms of habitat use and the impacts of landscape modification for generalist and specialist bird species are considered in Hinsley et al. (2009b).

This case study from a single field site demonstrates considerably increased avian ecological knowledge and highlights the requirement of detailed field ecological data to achieve this. Also evident from this body of work is the benefits of integrating airborne LiDAR and spectral data to maximise information available on forest structure and composition.

17.7 Discussion and Conclusions

Airborne LiDAR has a proven ability to supply highly detailed, extensive and accurate vegetation structure data, which has long been identified as a key element of organisms' habitats. This approach to habitat assessment is most apposite in woodland where the three-dimensional complexity of the habitat is greatest and where its characterisation by field data collection can be most difficult. The term habitat is frequently used in remote sensing studies, but only takes on direct ecological meaning when considered in relation to an individual organism, species or species assemblage. To-date, the attention of studies that have made use of airborne LiDAR data to assess the habitat characteristics of named species has fallen heavily, but not exclusively, on bird species. The potential exists to expand this to a range of other forest-dwelling taxa, and this is increasingly being realised and fulfilled.

In addition to habitat assessment, airborne LiDAR data have also been used to explore relationships between woodland structure and species richness (i.e. diversity). This is the theme of Chap. 18 and so has not been reviewed in detail here. It is worth emphasising, however, that a similar pattern is evident with considerable focus on bird species richness (e.g. Goetz et al. 2007; Clawges et al.

2008; Müller et al. 2010; Flaspohler et al. 2010; Lesak et al. 2011) and less emphasis on other taxa (e.g. spiders: Vierling et al. 2011; plant species: Simonson et al. 2012; Wolf et al. 2012). Amongst these studies a distinction can be drawn between those that seek to predict species diversity (e.g. Goetz et al. 2007; Lesak et al. 2011) and those that seek to predict species composition (e.g. Müller et al. 2010; Vierling et al. 2011). From a habitat assessment point of view, perhaps the most interesting of the above listed species diversity studies is that of Flaspohler et al. (2010). They examined the underpinning concepts of island biogeography and habitat theory by studying a series of forest fragments isolated by volcanic activity in Hawaii. They investigated species richness of both native and exotic bird species in the forest fragments in relation to information on fragment size and vegetation structure. They found the frequently reported relationship between species richness and fragment size, but also discovered that the native and exotic species were disproportionately located in fragments sized either smaller than or greater than 3 ha respectively. Furthermore, fragment size was also strongly correlated with vegetation volume, maximum tree height and canopy height heterogeneity, which were therefore also strong predictors of bird species richness. Thus although this study explored bird species richness, its findings have tremendous significance in terms of understanding habitat quality in a highly fragmented landscape and how this could be managed in respect of native or exotic species.

The nature of both airborne LiDAR systems and subsequent data processing has become increasingly sophisticated over recent time. Standard metrics that have been extracted from LiDAR data for forest structure and habitat assessment have included the mean, maximum, standard deviation, and coefficient of variation of canopy height in regular grid cells or sample areas relating to field plots, count station or pitfall trap locations, territories, etc. Other frequently extracted canopy structure metrics from airborne LiDAR have included measures of skewness and kurtosis, height percentiles, and the percentage of returns (or return energy) within specified height bands. More developed metrics include canopy cover (calculated as vegetation returns as a proportion of all returns), canopy closure (calculated as the percentage of returns above a specified gap height threshold), canopy permeability (calculated as the proportion of laser pulses for which there are multiple returns), canopy penetration ratio (calculated between specified height bands), foliage height diversity (calculated as the proportion of returns in specified height strata), and vertical distribution ratio (calculated as a normalised ratio between maximum and median height). Use of more complex canopy structure metrics has become increasingly prevalent as studies have progressed from using LiDAR data in the form of rasterised canopy height models to working directly with terrain-normalised point clouds. Additional metrics can be extracted from full waveform LiDAR data, for example height of median energy return (HOME) as used in the study of Goetz et al. (2007). The intensity data available in both discrete return and full waveform LiDAR data has seldom been studied in those published papers with a focus on habitat assessment. Terrain and surface geomorphology variables extracted from airborne LiDAR data, such as elevation, slope, aspect, curvature, flow accumulation and river channel dimensions, have been shown to be highly relevant habitat determinants, in some cases of equal or greater importance than vegetation structure variables.

The spatial scale of a habitat can be at the micro-, meso- or macro- level depending on the organism of interest. These may be considered as organisms with territory areas that can best be measured in units of (i) square centi- or deci-metres, (ii) square metres or hectares, and (iii) square kilometres respectively. Airborne LiDAR thus has greatest relevance at the meso-scale, with terrestrial LiDAR providing capability at the micro-scale and any future satellite-based LiDAR providing potential for macro-scale habitat assessment.

As with any environmentally-focussed study, remote sensing techniques such as airborne LiDAR offer just one component of the data requirements necessary to expand our understanding. In general, the more detailed the ecological field data (from presence-absence to prevalence and biological activity data), the greater the advancement in ecological understanding that can be attained through the use of airborne LiDAR data as an exploratory tool. A cautionary note was signalled in the conclusions of Work et al. (2011, p. 636) that *"While it is initially seductive to mine high resolution remote sensing data for correlations with* [species] *assemblages, this seems more of a preliminary exercise in calibration of the technique than an end goal. The utility of LIDAR will be borne out when the force of this data is realized through mechanistic hypotheses related to habitat requirements of plants and animals."* Moving beyond presence/absence or prevalence data and taking an ecological hypothesis driven approach, we can investigate how the vegetation structural components of a habitat influence its quality by determining the way in which organisms use and prosper within that habitat.

References

Bässler C, Stadler J, Müller J, Forster B, Gottlein A, Brandl R (2011) LiDAR as a rapid tool to predict forest habitat types in Natura 2000 networks. Biodivers Conserv 20:465–481

Bellamy PE, Hill RA, Rothery P, Hinsley SA, Fuller RJ, Broughton RK (2009) Willow Warbler (*Phylloscopus trochilus*) habitat in woods with different structure and management in southern England. Bird Study 56:338–348

Bergen KM, Goetz SJ, Dubayah RO, Henebry GM, Hunsaker CT, Imhoff ML, Nelson RF, Parker GG, Radeloff VC (2009) Remote sensing of vegetation 3-D structure for biodiversity and habitat: review and implications for lidar and radar spaceborne missions. J Geophys Res Biogeosci 114:G00E06

Bradbury RB, Hill RA, Mason DC, Hinsley SA, Wilson JD, Balzter H, Anderson GQA, Whittingham MJ, Davenport IJ, Bellamy PE (2005) Modelling relationships between birds and vegetation structure using airborne LiDAR data: a review with case studies from agricultural and woodland environments. Ibis 147:443–452

Broughton RK, Hinsley SA, Bellamy PE, Hill RA, Rothery P (2006) Marsh Tit (*Poecile palustris*) territories in a British broad-leaved wood. Ibis 148:744–752

Broughton RK, Hill RA, Freeman SN, Hinsley SA (2012a) Describing habitat occupation by wood land birds with territory mapping and remotely sensed data: an example using the Marsh Tit (*Poecile palustris*). Condor 114:812–822

Broughton RK, Hill RA, Henderson LJ, Bellamy PE, Hinsley SA (2012b) Patterns of nest placement in a population of Marsh Tits (*Poecile palustris*). J Ornithol 153:735–746

Clawges R, Vierling K, Vierling L, Rowell E (2008) The use of airborne lidar to assess avian species diversity, density, and occurrence in a pine/aspen forest. Remote Sens Environ 112:2064–2073

Coops NC, Varhola A, Bater CW, Teti P, Boon S, Goodwin N, Weiler M (2009) Assessing differences in tree and stand structure following beetle infestation using lidar data. Can J Remote Sens 35:497–508

Coops NC, Duffe J, Koot C (2010) Assessing the utility of lidar remote sensing technology to identify mule deer winter habitat. Can J Remote Sens 36:81–88

Falkowski MJ, Evans JS, Martinuzzi S, Gessler PE, Hudak AT (2009) Characterizing forest succession with lidar data: an evaluation for the Inland Northwest, USA. Remote Sens Environ 113:946–956

Flaspohler DJ, Giardina CP, Asner GP, Hart P, Price J, Lyons CK, Castaneda X (2010) Long-term effects of fragmentation and fragment properties on bird species richness in Hawaiian forests. Biol Conserv 143:280–288

Garcia-Feced C, Tempel DJ, Kelly M (2011) LiDAR as a tool to characterize wildlife habitat: California spotted owl nesting habitat as an example. J For 109:436–443

Gaveau DLA, Hill RA (2003) Quantifying canopy height underestimation by laser pulse penetration in small-footprint airborne laser scanning data. Can J Remote Sens 29:650–657

Goetz S, Steinberg D, Dubayah R, Blair JB (2007) Laser remote sensing of canopy habitat heterogeneity as a predictor of bird species richness in an eastern temperate forest, USA. Remote Sens Environ 108:254–263

Goetz SJ, Steinberg D, Betts MG, Holmes RT, Doran PJ, Dubayah R, Hofton M (2010) Lidar remote sensing variables predict breeding habitat of a neotropical migrant bird. Ecology 91:1569–1576

Graf RF, Mathys L, Bollmann K (2009) Habitat assessment for forest dwelling species using LiDAR remote sensing: capercaillie in the Alps. For Ecol Manag 257:160–167

Hill RA, Broughton RK (2009) Mapping the understorey of deciduous woodland from leaf-on and leaf-off airborne LiDAR data: a case study in lowland Britain. ISPRS J Photogramm Remote Sens 64:223–233

Hill RA, Thomson AG (2005) Mapping woodland species composition and structure using airborne spectral and LiDAR data. Int J Remote Sens 26:3763–3779

Hill RA, Hinsley SA, Gaveau DLA, Bellamy PE (2004) Predicting habitat quality for Great Tits (*Parus major*) with airborne laser scanning data. Int J Remote Sens 25:4851–4855

Hill RA, Wilson AK, George M, Hinsley SA (2010) Mapping tree species in temperate deciduous woodland using time-series multi-spectral data. Appl Veg Sci 13:86–99

Hinsley SA, Hill RA, Gaveau DLA, Bellamy PE (2002) Quantifying woodland structure and habitat quality for birds using airborne laser scanning. Funct Ecol 16:851–857

Hinsley SA, Hill RA, Bellamy PE, Balzter H (2006) The application of lidar in woodland bird ecology: climate, canopy structure, and habitat quality. Photogramm Eng Remote Sens 72:1399–1406

Hinsley SA, Hill RA, Bellamy PE, Harrison NM, Speakman JR, Wilson AK, Ferns PN (2008) Effects of structural and functional habitat gaps on breeding woodland birds: working harder for less. Landsc Ecol 23:615–626

Hinsley SA, Hill RA, Fuller RJ, Bellamy PE, Rothery P (2009a) Bird species distributions across woodland canopy structure gradients. Commun Ecol 10:99–110

Hinsley SA, Hill RA, Bellamy P, Broughton RK, Harrison NM, Mackenzie JA, Speakman JR, Ferns PN (2009b) Do highly modified landscapes favour generalists at the expense of specialists? An example using woodland birds. Landsc Res 34:509–526

Hyde P, Dubayah R, Peterson B, Blair JB, Hofton M, Hunsaker C, Knox R, Walker W (2005) Mapping forest structure for wildlife habitat analysis using waveform lidar: validation of montane ecosystems. Remote Sens Environ 96:427–437

Hyde P, Dubayah R, Walker W, Blair JB, Hofton M, Hunsaker C (2006) Mapping forest structure for wildlife habitat analysis using multi-sensor (LiDAR, SAR/InSAR, ETM plus, Quickbird) synergy. Remote Sens Environ 102:63–73

Johnson MD (2007) Measuring habitat quality: a review. Condor 109:489–504

Jung K, Kaiser S, Bohm S, Nieschulze J, Kalko EKV (2012) Moving in three dimensions: effects of structural complexity on occurrence and activity of insectivorous bats in managed forest stands. J Appl Ecol 49:523–531

Kane VR, Bakker JD, McGaughey RJ, Lutz JA, Gersonde RF, Franklin JF (2010) Examining conifer canopy structural complexity across forest ages and elevations with LiDAR data. Can J For Res 40:774–787

Kasprak A, Magilligan FJ, Nislow KH, Snyder NP (2012) A LiDAR-derived evaluation of watershed-scale large woody debris sources and recruitment mechanisms: coastal Maine, USA. River Res Appl 28:1462–1476

Lesak AA, Radeloff VC, Hawbaker TJ, Pidgeon AM, Gobakken T, Contrucci K (2011) Modeling forest songbird species richness using LiDAR-derived measures of forest structure. Remote Sens Environ 115:2823–2835

MacArthur RH, MacArthur J (1961) On bird species diversity. Ecology 42:594–598

Martinuzzi S, Vierling LA, Gould WA, Falkowski MJ, Evans JS, Hudak AT, Vierling KT (2009) Mapping snags and understory shrubs for a LiDAR-based assessment of wildlife habitat suitability. Remote Sens Environ 113:2533–2546

Massey ME. Welch RC (eds) (1993) Monks Wood National Nature Reserve. The experience of 40 years 1953–1993. English Nature, Peterborough

Miura N, Jones SD (2010) Characterizing forest ecological structure using pulse types and heights of airborne laser scanning. Remote Sens Environ 114:1069–1076

Mollot LA, Bilby RE (2008) The use of geographic information systems, remote sensing, and suitability modeling to identify conifer restoration sites with high biological potential for anadromous fish at the Cedar River Municipal Watershed in western Washington, USA. Restor Ecol 16:336–347

Morsdorf F, Marell A, Koetz B, Cassagne N, Pimont F, Rigolot E, Allgower B (2010) Discrimination of vegetation strata in a multi-layered Mediterranean forest ecosystem using height and intensity information derived from airborne laser scanning. Remote Sens Environ 114:1403–1415

Müller J, Moning C, Bässler C, Heurich M, Brandl R (2009) Using airborne laser scanning to model potential abundance and assemblages of forest passerines. Basic Appl Ecol 10:671–681

Müller J, Stadler J, Brandl R (2010) Composition versus physiognomy of vegetation as predictors of bird assemblages: the role of lidar. Remote Sens Environ 114:490–495

Müller J, Mehr M, Bässler C, Fenton MB, Hothorn T, Pretzsch H, Klemmt HJ, Brandl R (2012) Aggregative response in bats: prey abundance versus habitat. Oecologia 169:673–684

Nelson R, Keller C, Ratnaswamy M (2005) Locating and estimating the extent of Delmarva fox squirrel habitat using an airborne LiDAR profiler. Remote Sens Environ 96:292–301

Newnham GJ, Siggins AS, Blanchi RM, Culvenor DS, Leonard JE, Mashford JS (2012) Exploiting three dimensional vegetation structure to map wildland extent. Remote Sens Environ 123:155–162

Newton AC, Hill RA, Echeverria C, Golicher D, Benayas JMR, Cayuela L, Hinsley SA (2009) Remote sensing and the future of landscape ecology. Prog Phys Geogr 33:528–546

Palminteri S, Powell GVN, Asner GP, Peres CC (2012) LiDAR measurements of canopy structure predict spatial distribution of a tropical mature forest primate. Remote Sens Environ 127:98–105

Seavy NE, Viers JH, Wood JK (2009) Riparian bird response to vegetation structure: a multiscale analysis using LiDAR measurements of canopy height. Ecol Appl 19:1848–1857

Shugart HH, Saatchi S, Hall FG (2010) Importance of structure and its measurement in quantifying function of forest ecosystems. J Geophys Res Biogeosci 115:G00E13

Simonson WD, Allen HD, Coomes DA (2012) Use of an airborne lidar system to model plant species composition and diversity of Mediterranean oak forests. Conserv Biol 26:840–850

Smart LS, Swenson JJ, Christensen NL, Sexton JO (2012) Three-dimensional characterization of pine forest type and red-cockaded woodpecker habitat by small-footprint, discrete-return lidar. For Ecol Manag 281:100–110

Swatantran A, Dubayah R, Goetz S, Hofton M, Betts MG, Sun M, Simard M, Holmes R (2012) Mapping migratory bird prevalence using remote sensing data fusion. PLoS ONE 7:e28922

Vehmas M, Eerikainen K, Peuhkurinen J, Packalen P, Maltamo M (2009) Identification of boreal forest stands with high herbaceous plant diversity using airborne laser scanning. For Ecol Manag 257:46–53

Vehmas M, Packalen P, Maltamo M, Eerikainen K (2011) Using airborne laser scanning data for detecting canopy gaps and their understory type in mature boreal forest. Ann For Sci 68: 825–835

Vierling KT, Vierling LA, Gould WA, Martinuzzi S, Clawges RM (2008) Lidar: shedding new light on habitat characterization and modelling. Front Ecol Environ 6:90–98

Vierling KT, Bässler C, Brandl R, Vierling LA, Weiss I, Müller J (2011) Spinning a laser web: predicting spider distributions using LiDAR. Ecol Appl 21:577–588

Waser LT, Baltsavias E, Ecker K, Eisenbeiss H, Feldmeyer-Christe E, Ginzler C, Kuchler M, Zhang L (2008) Assessing changes of forest area and shrub encroachment in a mire ecosystem using digital surface models and CIR aerial images. Remote Sens Environ 112:1956–1968

Weber TC (2011) Maximum entropy modeling of mature hardwood forest distribution in four U.S. States. For Ecol Manag 261:779–788

Weber TC, Boss DE (2009) Use of LiDAR and supplemental data to estimate forest maturity in Charles County, MD, USA. For Ecol Manag 258:2068–2075

Wilsey CB, Lawler JJ, Cimprich DA (2012) Performance of habitat suitability models for the endangered black-capped vireo built with remotely-sensed data. Remote Sens Environ 119: 35–42

Wolf JA, Fricker GA, Meyer V, Hubbell SP, Gillespie TW, Saatchi SS (2012) Plant species diversity is associated with canopy height and topography in a neotropical forest. Remote Sens 4: 4010–4021

Work TT, St Onge B, Jacobs JM (2011) Response of female beetles to LIDAR derived topographic variables in Eastern boreal mixedwood forests (Coleoptera, Carabidae). ZooKeys 147:623–639

Zhao F, Sweitzer RA, Guo Q, Kelly M (2012) Characterizing habitats associated with fisher den structures in the Southern Sierra Nevada, California using discrete return lidar. For Ecol Manag 280:112–119

Chapter 18
Assessing Biodiversity by Airborne Laser Scanning

Jörg Müller and Kerri Vierling

Abstract Estimating biodiversity in complex habitats, particularly in forests, is still a major challenge for ecologists and conservationists. In ground-breaking work, Robert MacArthur and his colleagues quantified relationships between bird and vertical vegetation diversity, and found that the diversity of vegetation structure strongly influenced bird species diversity. However, they were limited in spatial extent when describing vertical vegetation structure due to the labor-intensive nature of data collection. Current remote sensing techniques, such as LiDAR, can describe ecologically relevant measurements of forest structure across broad extents, and thus, there are increasing efforts to examine relationships between LiDAR-derived data and patterns of animal biodiversity. LiDAR-based data have been utilized for silvicultural assessments for over a decade, but LiDAR use in biodiversity studies is more recent. LiDAR data can assist in the assessment of local animal diversity across taxa, and might assist in larger scale biodiversity assessments in remote and rugged environments. In the following chapter, we first briefly discuss the role of vegetation structure in biodiversity studies, followed by a description of the variables that are most commonly used in biodiversity studies. We then give an overview of biodiversity studies that have utilized LiDAR in forests to date. Although there is a growing body of literature relating LiDAR-derived variables to single species distributions and habitat quality, we focus this chapter solely on studies that address animal species diversity in forested landscapes. We conclude with a discussion of future directions concerning biodiversity assessments in forested systems that might benefit from the use of LiDAR-based data.

J. Müller (✉)
Conservation and Research Department, Bavarian Forest National Park, Freyunger Str. 2, 94481 Grafenau, Germany
e-mail: joerg.mueller@npv-bw.bayern.de

K. Vierling
Department of Fish and Wildlife Resources, University of Idaho, Moscow, ID, USA

M. Maltamo et al. (eds.), *Forestry Applications of Airborne Laser Scanning: Concepts and Case Studies*, Managing Forest Ecosystems 27, DOI 10.1007/978-94-017-8663-8_18, © Springer Science+Business Media Dordrecht 2014

18.1   Introduction

The biodiversity of a system can be classified in multiple ways. Biodiversity may include the diversity of species, genes and habitats (Wilson 1987). Scientists interested in species diversity have further utilized the number of species and their abundances to calculate various measures of species diversity within (α-diversity) and between (β-diversity) habitats (Whittaker 1960). More recently species diversity measures have been expanded to a focus on species traits (functional diversity) and relatedness (phylogenetic diversity) (Tilman et al. 1997).

Patterns of species diversity at local scales are often examined as a function of habitat characteristics. Vegetation structure and composition are characteristics that are often utilized in species diversity studies (e.g. Lassau et al. 2005a, b; Lassau and Hochuli 2008; Müller et al. 2010). Many taxa, and particularly mobile animals, are sensitive to vertical structure, and LiDAR has shown to be a useful tool for describing the vertical vegetation characteristics that might influence animal distributions and diversity (see Vierling et al. 2008 for a review). The influence of vertical structure on animal diversity might be increasingly important in forested habitats. Increasing structural diversity in forests is correlated with increases in animal diversity for multiple taxa, based in part on the fact that there are more habitat niches in structurally diverse forests. MacArthur and MacArthur (1961) were the first to examine this question quantitatively for bird species, and labor-intensive efforts were required to measure the fine scale distribution of leaves across the vertical strata of a forest stand. Their quantification of foliage height diversity and its influence on patterns of bird species diversity was a critical step in our attempt to understand the factors that influence local scale biodiversity. However, the difficulty associated with quantifying fine scale foliage height diversity across broad spatial extents has limited our ability to extrapolate fine scale animal-habitat relationships to broader landscapes.

Although foliage height diversity has been noted to be an important determinant of local animal diversity, other factors can influence local scale forest animal diversity as well. For some animals, plant species composition may be an important factor influencing local diversity. For instance, plant species composition is a better predictor of arthropod communities than structural variables (Schaffers et al. 2008; Ter Braak and Schaffers 2004). Within bird communities, the relative importance of foliage height diversity may be mediated by plant species composition and/or landscape context. For instance, Rosenzweig (1995) noted that foliage height diversity might be a less important factor in explaining bird species diversity in the tropics, presumably because of different evolutionary histories. Müller et al. (2010) noted that vertical structure was a more important factor influencing bird diversity in mixed montane forests in Germany, and thus, the relative importance of vertical diversity and plant species composition on bird diversity may depend on the habitat types (Fleishman and Mac Nally 2006).

The overall objective of this chapter is to provide an overview of LiDAR variables commonly utilized in diversity studies, summarize findings of recent studies that utilized LiDAR to examine local diversity, and discuss challenges and opportunities related to the use of LiDAR in assessing diversity. Because the majority of LiDAR-based diversity studies in forests thus far have focused upon animals, we will emphasize these studies in the review. However, there exist a small handful of studies that address plant diversity and we provide a brief overview of those studies as well.

18.2   The Ecological Rationale of LiDAR Variables

The great amount of data available from LiDAR acquisitions can be a challenge for scientists who are considering the use of LiDAR data for biodiversity assessments. Bergen et al. (2009) have previously published an excellent review of the lidar and radar variables that are important for biodiversity and single-species assessments. Below, we provide a summary of the most common lidar-variables that have been included in biodiversity studies within forested landscapes, many of which were previously described by Bergen et al. (2009). Some of these variables are direct measurements of environmental variables of ecological interest (e.g. tree or shrub height), while others variables such as penetration rate may be correlated with environmental conditions of interest (canopy gaps).

18.2.1   Canopy Openness

One of the major LiDAR-derived variables used in many diversity assessments is penetration rate, which describes how many of the laser returns come from a specific stratum. Penetration rate has been assumed to be a measure of canopy openness and thereby of sun intensity near the ground within a forest stand (Müller and Brandl 2009). Penetration rate might be particularly important for arthropod communities because in forested habitats, increasing sun intensity near the ground will influence plant diversity and biomass production, and arthropods are noted to respond to these factors (Blakely and Didham 2010).

As with multiple variables available through LiDAR, it is important to compare LiDAR data with ground measurements to assess their accuracy. Multiple studies exist which assess the relationship between LiDAR-derived vegetation structure and field measurements (e.g. Clawges et al. 2008; Müller et al. 2009; Vierling et al. 2011). Wünsche (2012) compared the penetration rate captured by airborne LiDAR at a height of 2 m above ground with the relative solar radiation (diffuse site factor) measured with fisheye photographs (Kodak, DCSPro14n with a fisheye-objective Nikkor1:2.8). The penetration rate and diffuse site factor were compared within 0.1 ha circular plots (n = 75) under cloudy sky conditions (Wünsche 2012), and were highly correlated (Fig. 18.1).

Fig. 18.1 Comparison of penetration rate measured at 2 m above ground with airborne LiDAR and solar radiation measured with hemispherical image captured by fish-eye photography at 75 plots

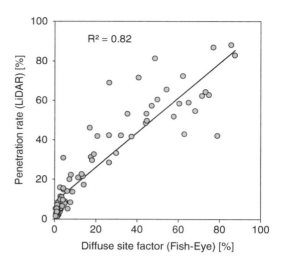

The identification of canopy gaps with LiDAR-derived penetration rate data can be used to identify locations that will likely experience more direct sun and/or will be warmer. These microclimatic differences are particularly important for many insects and particularly for those who are sensitive to warm/dry conditions compared to cool/moist conditions. For instance, open forest gaps are likely to favor arthropod communities that prefer dry conditions (Grimbacher and Stork 2009), and these same sites will likely be avoided by species adapted to moist habitats such as fungi (Bässler et al. 2010). Because LiDAR-derived variables are able to identify these sites at potentially sub-meter resolutions, penetration rate may be applicable for examining issues that affect arthropod diversity and the diversity of other taxa that respond to environmental variables at a small spatial scale. For instance, penetration rate has been correlated with the number of syrphid species in a mixed montane forest in Germany (Fig. 18.2), and has been utilized in other studies examining arthropod communities (e.g. Vierling et al. 2011; Müller and Brandl 2009).

18.2.2 Foliage Height Diversity and Biomass Distribution

Foliage height diversity is perhaps the most commonly used variable in biodiversity studies, and multiple authors have noted the importance of lidar-derived measurements of foliage height diversity to biodiversity studies (e.g. Vierling et al. 2008; Bergen et al. 2009). Foliage height diversity was initially described by MacArthur and MacArthur (1961) as

$$-\sum_i p_i \log_e p_i \tag{18.1}$$

Fig. 18.2 Number of syrphid species sampled at 35 plots versus penetration rate in a mixed montane forest in Germany

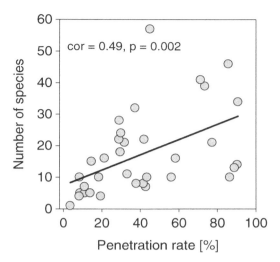

(with p_i is the proportion of the total foliage which lies in the ith of the chosen horizontal layers) and subsequent studies have incorporated foliage height in different forms. Several studies of diversity within forests have used the standard deviation of foliage heights as an important variable (Müller et al. 2009). Alternatively, some researchers have used a vertical distribution ratio (VDR), which provides an index of the vertical distribution of biomass (Goetz et al. 2007). Jung et al. (2012) more recently utilized measures of the diversity of the outer canopy surface in their assessment of bat communities, which emphasizes that the diversity of specific layers in the canopy might be important for different diversity assessments.

In addition to measures of foliage height diversity and canopy complexity, LiDAR data can be binned to represent the distribution of vegetation biomass within different strata. The strata of ecological importance can be defined by the researcher, as was done originally by MacArthur and MacArthur (1961). For instance, Clawges et al. (2008) examined bird diversity as a function of different heights of the understory strata (0–2 m, 2–5 m, etc.), and found that bird species diversity was highest with more foliage layers within 5 m of the forest floor in a relatively species-poor forest. Jung et al. (2012) utilized a height class of 0–5 m to represent the understory layer in their study of bat activity, and Vogeler et al. (2014) utilized vegetation biomass between 1 and 2.5 m above ground to represent the understory layer that might influence bird diversity for their study sites. In contrast, Lesak et al. (2011) utilized a proportional approach in defining ecologically relevant understory layers as opposed to using absolute height thresholds. The height of the understory layer that is ecologically significant for diversity might therefore differ with the forest type, the taxa being studied, the succession stage of the forest, and the questions being posed by researchers.

18.2.3 Individual Tree Characteristics: Height, Size, and Snags

Tree height is another variable from LiDAR data that is ecologically significant, in part because tree height is often correlated with tree size. Large trees are particularly important when they become decadent, because the cavities that form in snags become habitat for a variety of animals. For instance, species diversity of lichens, hollow tree beetles and mites has been noted to be high in snags, and multiple taxa living near slow decaying trees will colonize these habitats (Ohlson et al. 1997; Lindo and Winchester 2008). Vertebrate species are also likely to respond positively to the presence of large snags in forested systems. In forests where natural cavity formation is limited, the activity of woodpeckers is particularly important because these woodpeckers function as ecosystem engineers (e.g., Daily et al. 1993). The resulting nest webs (interactions between trees, primary cavity excavators and those that use cavities) are fundamentally based upon the availability of suitable nest trees which are typically large snags (Gentry and Vierling 2008; Martin et al. 2004; Blanc and Walters 2008; Cooke and Hannon 2011).

The identification of snag size and distribution within a landscape might therefore be of interest in biodiversity assessments (see Chap. 19 for more details). The identification of individual trees (and snags) is difficult (Ferster et al. 2009, 2011), and most studies that have included snags have been related to carbon stock assessments (Ferster et al. 2009; Kim et al. 2009). However, Martinuzzi et al. (2009) developed snag distribution models for different size classes of snags in a northern Idaho coniferous forest, and related those snag distributions to wildlife habitat distributions. The inclusion of snag data and distributions in biodiversity assessments is likely to improve the predictive capability of models for specific guilds. For instance, models that addressed cavity excavator diversity in northern Idaho forests performed poorly, and this may be due to the lack of snag data in the models (Vogeler et al. 2014).

18.2.4 Tree Species

Many forests in the temperate and boreal zone are dominated by only a few tree species, characterized by their typical crown structure. Recently, new methods have enabled scientists to identify single coniferous and broadleaf trees species using high resolution (20 pulses/m^2) LiDAR data (Wei et al. 2012). Furthermore the classification in tree species opens new possibilities also for biodiversity assessment, because many arthropods utilize only a few tree species or genera (Brändle and Brandl 2001). Thus biodiversity assessment of these specialists may be assisted with the inclusion of single species tree data. However, these methods may be more favorable in forest ecosystems where the number of tree species is relatively limited, such as in certain montane or boreal forests. For example, the amount

Fig. 18.3 Composition of saproxylic beetle species in 1-ha grids with a growing stock of more than 280 m³/ha in a montane forest separated by the classification as broadleaf, or conifer specialists (Müller and Brandl 2009) (Note the significant decrease of conifer and increase of broadleaf specialist's proportion along an increasing amount of broadleaf tree volume estimated by LiDAR single tree species detection)

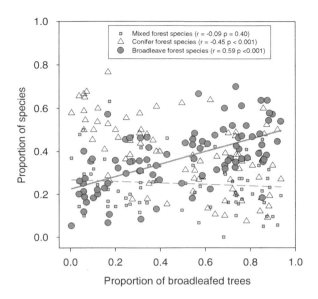

and composition of broadleaf trees in a forest has been noted to influence the diversity of saproxylic beetles (Fig. 18.3) and birds (Siitonen and Martikainen 1994; Moning and Müller 2008). Additionally, the opening of the canopy is one of the major determinants of plant species in the herb layer and can be used for diversity assessment of plants (Lucas et al. 2010). The differentiation of plant species via LiDAR data combined with other remotely sensed products may therefore advance our understanding of biodiversity patterns, particularly as they are related to arthropods or other specialists.

18.2.5   Elevation

Elevation is an important LiDAR-derived variable, and is particularly important in montane environments where altitude is regularly correlated with important ecological variables such as temperature or precipitation (Körner 2007). Temperature and precipitation are correlated with biodiversity, particularly with regard to arthropods. For insects, the influence of altitudinal gradients has been reviewed by Hodkinson (2005), and the use of LiDAR-derived altitudes has been applied in several studies (Müller and Brandl 2009; Vierling et al. 2011). Furthermore elevation strongly determines the composition and richness of plants, by changing temperature, precipitation and productivity.

18.2.6   Topography

LiDAR-derived landscape variables that are associated with topography may assist with biodiversity modelling, because these variables can be associated with the wetness of a patch or levels of solar radiation. Programs such as SAGA GIS (System for Automated Geoscientific Analyses) can be used to derive variables (Böhner et al. 2006). The local effect of exposure to microclimates and microhabitats has long been recognized; for instance, exposure was used as a correction factor for species occurrence on south- or north-facing slopes for plants in mountain forests (Sendtner 1860). The importance of such microhabitats has recently resurfaced in the framework of extinction risk assessment by global warming (Hof et al. 2011b; Scherrer et al. 2011).

18.3   Current Biodiversity Assessments in Forests Using LiDAR

Although LiDAR has been utilized increasingly to address single-species distributions, only a handful of biodiversity-focused studies have been conducted using LiDAR in conjunction with other remotely sensed data. Birds have been the primary taxa examined (7 studies to date; see Table 18.1), and there have been only three other studies thus far relating LiDAR-derived variables to diversity of other taxa. These include a single study on bats (Jung et al. 2012), spiders (Vierling et al. 2011), and beetles (Müller and Brandl 2009). The studies to date have all noted the importance of LiDAR-derived data in assessing biodiversity patterns across landscapes, and variables that influence biodiversity across the studies have generally supported the hypothesis that foliage height diversity is important. For example, Müller et al. (2010) found that vegetation physiognomy was a more important predictor of bird species richness than plant species composition.

Additionally, the majority of studies that incorporate LiDAR-derived variables and biodiversity assessments have been conducted in temperate regions. The studies are distributed across coniferous forests in Germany (2 studies), Idaho, and South Dakota (Table 18.1). Additionally, several of these studies have been conducted in temperate deciduous-dominated forested systems, but only one has been thus far conducted within tropical areas (Flashpoler et al. 2010).

Other environmental variables have been shown to influence species richness. For instance, Vogeler et al. (2014) found that NDVI and terrain effects (slope and elevation) contributed strongly to patterns of bird diversity at one of their study sites, and that canopy complexity was a less important factor. Flashpoler et al. (2010) found that the influence of vertical vegetation structure diversity was mediated with forest fragment area. As more studies are conducted with LiDAR data, it will become increasingly important to identify the factors that repeatedly influence biodiversity and the spatial scales and habitats in which they operate.

Table 18.1 Applications of LiDAR for biodiversity assessment in forests

Taxon	Location	Major habitat type	Major finding	Reference
Plants				
Species richness	Horn Island, Mississippi, U.S.A.	Marsh, meadow and woodland habitats in a coastal barrier island	Vegetation structure described with LiDAR explained species richness best within habitats, not among all habitats	Lucas et al. (2010)
Species richness, Species composition	Germany	Mixed montane forests	Laser data and hyperspectral data were used to model plant species richness, and models based on laser data were superior to models based on hyperspectral data	Leutner et al. (2012)
Diversity of forest species	Portugal	Mediterranean oak forests	Diversity of forest species was significantly associated with vegetation height	Simonson et al. (2012)
Beetles				
Individuals, Richness, Diversity, Composition	Germany	Mixed montane forests dominated by spruce (*Picea abies*), beech (*Fagus sylvatica*) and fir (*Abies alba*), with considerable parts of bark beetle disturbance	The diversity of ground and near-ground flying beetles was better explained using LiDAR-derived measurements than ground-based measurements. In particular, mean body size and species composition were well-predicted using LiDAR-derived variables	Müller and Brandl (2009)
Spiders				
Diversity, richness, composition and body size	Germany	Mixed montane forests (see above box)	LiDAR was effective at describing diversity parameters of spider assemblages, Predictability of single species depends on habitat niche position	Vierling et al. (2011)
Birds				
Species richness	Idaho, U.S.A.	Coniferous forests dominated by Douglas fir (*Pseudotsuga menziesii*), Grand fir (*Abies grandis*), pine (*Pinus* spp.), and spruce (*Picea* spp.)	LiDAR-derived vegetation and terrain metrics varied in their importance at the two study sites. Specific ecological predictor variables also differed among avian nesting guilds and across the two study sites	Vogeler et al. (2014)

(continued)

Table 18.1 (continued)

Taxon	Location	Major habitat type	Major finding	Reference
Species richness	South Dakota, U.S.A.	Conifer forests dominated by Ponderosa pine (*Pinus ponderosa*) with small stands of aspen (*Populus tremuloides*)	Shrub and sapling heights from 0 to 2 m agl were strongest predictors of bird diversity	Clawges et al. (2008)
Species richness	Wisconsin, U.S.A.	Deciduous forests dominated by oak (*Quercus* spp.), maple (*Acer* spp.), hickory (*Carya* spp.) and ash (*Fraxinus* spp.)	Songbird species richness was most correlated with canopy and midstory height LiDAR variables	Lesak et al. (2011)
Species richness	Maryland, U.S.A.	Deciduous forests	Models based LiDAR explained more variation than did models based solely on optical variables (NDVI)	Goetz et al. (2007)
Species richness	Hawaii, U.S.A.	Mature ohia (*Metrosideros polymorpha*)-dominated wet forest	The area of forest fragments influenced the relationship between LiDAR-derived variables and species richness. As area increased, bird species richness increased linearly with vegetation volume, whereas the influence of maximum tree height and the height variability did not continue to influence species richness in areas > ~3 ha	Flashpoler et al. (2010)
Passerine composition	Germany	Mixed montane forests (see above box)	LiDAR derived variables better explained bird diversity than plant composition in montane temperate forests	Müller et al. (2010)
Passerine composition	Germany	Mixed montane forests (see above box)	LiDAR derived variables better explained bird diversity and composition than aerial photo and field measurements in montane temperate forests	Müller et al. (2009)
Mammals				
Bat occurrence and activity	Germany	Coniferous forests comprised of mixtures of young pine (*Pinus sylvestris* L.), beech (*Fagus sylvatica* L.), and oak (*Quercus rubor* L.)	LiDAR data well described activity of bats and could be related to wing morphology and foraging strategy in temperate forests in Germany	Jung et al. (2012)

18.4   Future Challenges and Opportunities

18.4.1   *Taxonomic and Geographic Biases*

There currently exists a taxonomic and geographic bias in the biodiversity studies that have been conducted thus far. Studies relating bird diversity with LiDAR-derived variables dominate the studies to date, and the number of studies addressing biodiversity patterns for other taxa is limited. For example, there are no studies to date that have examined small mammal diversity in forested systems using LiDAR data, nor have there been any studies that incorporate amphibian or reptilian diversity measures. The assessment of factors that influence biodiversity patterns for different taxonomic groups is necessary to investigate the general hypothesis that vertical structural diversity is an important predictor of biodiversity for all taxa.

Almost all of the current studies have been conducted in temperate regions, and we encourage more studies to be conducted in the tropics for multiple reasons. First, the tropics represent the most species rich regions of the globe (Gaston 2000), and factors that influence local biodiversity patterns in the tropics are likely going to differ from those that influence local biodiversity patterns in temperate regions (see Rosenzweig 1995 for a summary). For example, in tropical regions, tree canopies provide important habitat for many arthropods, and the upper canopy can harbour a high diversity of taxa (Didham and Fagan 2004). In such environments, LiDAR data will be increasingly important to assess the overall diversity of tropical forests as well as the patterns of biodiversity within different strata of these forests because high beetle diversity in tree crowns has been noted to drive overall diversity patterns at some sites (Erwin 1982).

Additionally, we can begin to utilize LiDAR to examine issues about biodiversity within more of a theoretical framework. For instance, Rosenzweig (1995) summarized multiple studies relating to patterns of species diversity and foliage height diversity, and patterns in bird diversity in tropical areas sometimes differ from patterns in temperate areas. MacArthur and MacArthur (1961) arbitrarily classified three vegetation layers that they felt represented the major structural diversity that birds might respond to, but different taxa and different landscape and evolutionary histories are likely to influence our basic assumptions concerning the influence of foliage height diversity on local scales of biodiversity. The factors and fundamental theoretical questions that may be explored in a temperate vs. tropical framework using LiDAR may include questions that address the role of evolutionary history in shaping diversity patterns and the relative role of physiognomy versus tree species composition. Several different forest types are represented in the reviewed biodiversity studies, but boreal forests and other forest types are not well represented in these studies. Additional studies which include a diversity of forest types will contribute towards a better understanding of the roles of physiognomy and tree species composition.

18.4.2 Management-Relevant Data

Managers of forested landscapes require biodiversity assessments for planning purposes, and a critical need for these managers is a way to identify areas of high diversity. Although there exist multiple frameworks that address biodiversity patterns across broad spatial extents (e.g. Hof et al. 2011a; Schuldt and Assmann 2010), these assessments have not incorporated vertical structure, and other studies (e.g. Goetz et al. 2007; Müller et al. 2009) have noted that the inclusion of lidar data improves our ability to describe biodiversity patterns over optical sensor data alone. Vogeler et al. (2014) developed a map of bird diversity for 80,000 ha of managed forest, which highlights for managers where biodiversity hotspots based upon LiDAR data occur, and these maps are critical tools for management planning. Although the Vogeler et al. (2014) study focused upon birds, imputation methods (e.g. Hudak et al. 2008) can assist in the mapping of any type of diversity across forested landscapes based upon LiDAR data, and these maps can then be used by managers who need spatially-explicit information concerning patterns of biodiversity.

LiDAR is a powerful tool with which to examine biodiversity patterns due to its ability to capture ecologically important variables for organisms whose body mass and habitat use differ by orders of magnitude (e.g. spiders to birds). A potential avenue of future study involves the examination of spatially-explicit patterns of biodiversity across different taxonomic groups. Finch et al. (2012) recently examined the role of spatial scale in maps of carbon biomass, tree species diversity, and bird diversity in managed coniferous forests. Maps that simultaneously identify biodiversity hotspots of different taxa are critical tools for multiple management and conservation assessments, and studies that incorporate multiple diversity assessments in the same landscape based upon LiDAR data are limited at the present time (Finch et al. 2012). There is a developing interest in mapping the spatial distribution of ecosystem services across landscapes (e.g. Turner et al. 2007), and biodiversity, although not explicitly noted to be an ecosystem service is considered the umbrella under which ecosystem services are provisioned (MEA 2005).

Although LiDAR has been shown to be effective at evaluating and identifying biodiversity hotspots at relatively various spatial scales (up to thousands of hectares), forests are a dynamic environment in which harvest, disease outbreaks and/or fire can all influence forest structure and function. Flashpoler et al. (2010) found that the size of forest fragments in Hawaii mediated the influence of vertical vegetation structure, and that the influence of vertical vegetation structure was most important in patches <3 ha large. Therefore, assessments of biodiversity with LiDAR-derived data need to be put into a larger landscape context to better understand the relative importance of vertical structural diversity relative to other important environmental factors (Bergen et al. 2009).

18.4.3 Community Composition

Changes in forest structure and composition due to new and ever-shifting combinations of climate regimes, human management, and other disturbances (e.g. McDowell et al. 2008; Allen et al. 2010) can profoundly affect 'ecosystem engineers', species that disproportionately influence local biodiversity by altering the environment (Jones et al. 1994, 1997, 2010). This review has focused primarily on biodiversity, and it is important to acknowledge that LiDAR will likely be able to assist in our understanding of how ecosystem engineer distributions are related to overall species diversity. For example, ecosystem engineers can create keystone structures (*sensu* Tews et al. 2004) such as tree cavities that provide resources for myriad other species. Woodpeckers, flickers, nuthatches, and sapsuckers all serve keystone functions in conifer forests because their tree excavation activities produce cavities that are used by a wide variety of avian and non-avian species for purposes of roosting, breeding, and predator escape (Martin et al. 2004; Daily et al. 1993; Blanc and Walters 2008; Gentry and Vierling 2008). Woodpeckers generally excavate cavities in snags, and the role of woodpeckers is especially important in areas where natural decay processes and/or human management affects cavity availability (Remm and Lohmus 2012). Because cavity users are among the most threatened groups of forest biota worldwide (Imbeau et al. 2001), it is increasingly important to identify (1) the patterns of biodiversity supported by snags and the tree cavities contained within, (2) the processes that drive these patterns, and (3) the interactions of these patterns and processes across multiple spatial scales.

We encourage more in depth assessments of community composition in future studies. Several of the reviewed studies incorporated guild-level assessments (Goetz et al. 2007; Vogeler et al. 2014) and others included assessments of native versus non-native species (Flashpoler et al. 2010). Additional assessments on community composition patterns and niche-specific assessments with LiDAR data will help to advance our understanding of functional relationships within ecological communities.

18.4.4 Different Sources of LiDAR Data

Our discussion thus far has focused upon airborne sources of LiDAR data, but recent advances in LiDAR technologies, particularly with terrestrial LiDAR systems, are important to discuss. Terrestrial LiDAR systems are typically mounted on tripods and can collect data that includes, but is not limited to, needle/leaf orientations and branch orientation (e.g. Clawges et al. 2007; Eitel et al. 2010). Combinations of airborne LiDAR with terrestrial LiDAR data are possible (Parker et al. 2004), and the resulting quantification of canopy structure can be related to diversity of insects and their predators (Müller et al. 2012). With these new sources of data, a deeper understanding of the influence of three dimensional structures in forests on food webs, functional diversity and biodiversity is likely.

Airborne LiDAR data have provided multiple opportunities for biodiversity assessments within forested systems, and these opportunities will likely be expanded with future satellite-based LiDAR products. To date, most applications of LiDAR data are restricted to research projects and applications by site managers are still somewhat limited (Seavy et al. 2009; Müller and Brandl 2009), mainly due to the high cost of acquisition and data processing. Satellite-based LiDAR will be important for detection of multiple ecological patterns and processes across the globe (Bergen et al. 2009), and current investigations are occurring to determine the relative utility of these types of data. For instance, vertical structure characterizations of vegetation are possible using data from the Geoscience Laser Altimeter System (GLAS) lidar instrument aboard the ICESat satellite, which was in operation between 2003 and 2009 (Lefsky et al. 2005; Lefsky 2010; Harding and Carabajal 2005). Vierling et al. (2013) attempted to map distributions of Red-naped sapsuckers (*Sphyrapicus nuchalis*) with these data, and efforts such as these should help us to expand our understanding of how vegetation structure detected from space, airborne, and terrestrial platforms influence species distributions and diversity in forested systems.

18.4.5 Beyond Animals and Species Diversity

LiDAR applications in species diversity studies have been biased toward animals, and our review focused upon those studies that have been published that related diversity to LiDAR derived variables. Although few studies have addressed plant diversity as a function of LiDAR derived variables, it is likely that LiDAR may be a tool that can advance our understanding of plant diversity and distributions as well. Most ground plants are strongly determined by the light regime, which can be measured precisely by lidar (Fig. 18.1). Therefore, the ability of LiDAR to help understand patterns of plant richness is not surprising (e.g. Lucas et al. 2010; Leutner et al. 2012). The inclusion of LiDAR data to examine diversity and distributions of other taxa growing in the canopy or at the forest floor such as fungi, bryophytes or lichens may also be possible, since canopy structure may a strong determinant of diversity (Unterseher et al. 2005).

In addition to the potential expansion of LiDAR variables to help better understand taxonomic-specific patterns of diversity and distribution, it seems fruitful to expand LiDAR-diversity studies to address functional and phylogenetic diversity (Pausas and Verdu 2010). Considering that natural and anthropogenic disturbance regimes (fires, windthrows, insects, logging) in forests strongly affects the canopy structure, LiDAR offers new ways to get a deeper understanding for ecological processes and for derivation of early warning signals. For the latter, a higher sensitivity has been demonstrated for community approaches beyond species diversity neglecting their identity (Gossner et al. 2013).

18.5 Conclusion

Airborne LiDAR data have become increasingly available over the last decade, but the applications of vertical measures in biodiversity studies are still very limited. The studies we reviewed have shown that LiDAR provides useful information to model diversity across many animal taxa from hyperdiverse arthropods to birds. To date, these studies are still limited in spatial extent, and more research is needed to understand the fundamental relationships and drivers of diversity in forested systems. Expansion of diversity studies across taxa, across scales, and across different forest types (temperate vs. tropical) will allow for ecologists to advance our understanding on the relative role of forest structure as a driver of diversity. Furthermore, expanding species diversity studies to include functional diversity may a useful way to deepen our understanding on the complex relationship of species composition and forest structure.

References

Allen CD, Macalady AK, Chenchouni H (2010) A global overview of drought and heat-induced tree mortality reveals emerging climate change risks for forests. For Ecol Manage 259:660–684

Bässler C, Müller J, Dziock F, Brandl R (2010) Microclimate and especially resource availability are more important than macroclimate for assemblages of wood-inhabiting fungi. J Ecol 98:822–832

Bergen KM, Goetz SJ, Dubayah RO, Henebry GM, Hunsaker CT, Imhoff ML, Nelson RF, Parker GG, Radeloff VC (2009) Remote sensing of vegetation 3-D structure for biodiversity and habitat: review and implications for lidar and radar spaceborne missions. J Geophys Res 114:G00E06

Blakely TJ, Didham RK (2010) Disentangling the mechanistic drivers of ecosystem-size effects on species diversity. J Anim Ecol 79:1204–1214

Blanc LA, Walters JR (2008) Cavity-nest webs in a longleaf pine ecosystem. Condor 110:80–92

Böhner J, McCloy KR, Strobl J (2006) SAGA-analysis and modelling applications. Göttinger Geographische Abhandlungen 115:1–130

Brändle M, Brandl R (2001) Species richness of insects and mites on trees: expanding Southwood. J Anim Ecol 70:491–504

Clawges R, Vierling L, Calhoon M, Toomey M (2007) Use of a ground-based scanning lidar for estimation of biophysical properties of western larch (Larix occidentalis). J Remote Sens 28:4331–4344

Clawges R, Vierling K, Vierling L, Rowell E (2008) The use of airborne lidar to assess avian species diversity, density, and occurrence in a pine/aspen forest. Remote Sens Environ 112:2064–2073

Cooke HA, Hannon SJ (2011) Do aggregated harvests with structural retention conserve the cavity web of old upland forest in the boreal plains. For Ecol Manage 261:662–674

Daily GC, Ehrlich PR, Haddad NM (1993) Double keystone bird in a keystone species complex. Proc Natl Acad Sci 90:592–594

Didham RK, Fagan LL (2004) Forest canopies. In: Burley J, Evans J, Youngquist J (eds) Encyclopaedia of forest sciences. Academic Press/Elsevier Science, London, pp 68–80

Eitel JUH, Vierling LA, Long DS (2010) Simultaneous measurements of plant structure and chlorophyll content in broadleaf saplings with a terrestrial laser scanner. Remote Sens Environ 114:2229–2237

Erwin TL (1982) Tropical forests: their richness in Coleoptera and other arthropod species. Coleopt Bull 36:74–75

Ferster CJ, Coops NC, Trofymow JA (2009) Aboveground large tree mass estimation in a coastal forest in British Columbia using plot-level metrics and individual tree detection from LiDAR. Can J Remote Sens 35:270–275

Ferster CJ, Trofymow JA, Coops NC, Chen BZ, Black TA, Gougeon FA (2011) Determination of ecosystem carbon-stock distributions in the flux footprint of an eddy-covariance tower in a coastal forest in British Columbia. Can J For Res 41:138–1393

Finch S, Vierling LA, Vierling KT, Hudak AT (2012) A case study using field surveys and LiDAR to quantify aboveground carbon, bird diversity, and tree species richness to prioritize conservation based on multiple ecosystem services. In: Proceedings of SilviLaser 2012 conference, Vancouver, BC, Canada, pp 16–19

Flashpoler DJ, Giardina CP, Asner GP, Hart P, Price JT, Lyons CK, Castaneda X (2010) Long-term effects of fragmentation and fragment properties on bird species richness in Hawaiian forests. Biol Conserv 143:280–288

Fleishman E, Mac Nally R (2006) Patterns of spatial autocorrelation of assemblages of birds, floristics, physiognomy, and primary productivity in the central Great Basin, USA. Divers Distrib 12:236–243

Gaston KJ (2000) Global patterns in biodiversity. Nature 405(6783):220–227

Gentry DJ, Vierling KT (2008) Reuse of cavities during the breeding and nonbreeding season in old burns in the Black Hills, South Dakota. Am Midl Nat 160:413–429

Goetz S, Steinberg D, Dubayah R, Blair B (2007) Laser remote sensing of canopy habitat heterogeneity as a predictor of bird species richness in an eastern temperate forest, USA. Remote Sens Environ 108:254–263

Gossner MM, Lachat T, Brunet J, Isacsson G, Bouget C, Brustel H, Brandl R, Weisser WW, Müller J (2013) Current "near-to-nature" forest management effects on functional trait composition of saproxylic beetles in beech forests. Conserv Biol 27:605–614. doi:10.1111/cobi.12023

Grimbacher PS, Stork NE (2009) How do beetle assemblages respond to cyclonic disturbance of fragmented tropical rainforest landscape? Oecologia 161:591–599

Harding D, Carabajal C (2005) ICESat waveform measurements of within-footprint topographic relief and vegetation vertical structure. Geophys Res Lett 32:L21S10

Hodkinson ID (2005) Terrestrial insects along elevation gradients: species and community response to altitude. Biol Rev 80:489–513

Hof C, Araujo MB, Jetz W, Rahbek C (2011a) Additive threats from pathogens, climate and land-use change for global amphibian diversity. Nature 480:516–519

Hof C, Levinsky I, Araujo MB, Rahbek C (2011b) Rethinking species' ability to cope with climate change. Glob Change Biol 17:2987–2990

Hudak AT, Crookston NL, Evans JS, Hall DE, Falkowski MJ (2008) Nearest neighbor imputation of species-level, plot-scale forest structure attributes from LiDAR data. Remote Sens Environ 112:2232–2245

Imbeau L, Monkkonen M, Desrochers A (2001) Long-term effects of forestry on birds of the eastern Canadian boreal forests: a comparison with Fennoscandia. Conserv Biol 15:1151–1162

Jones CG, Lawton JH, Shachak M (1994) Organisms as ecosystem engineers. Oikos 69:373–386

Jones CG, Lawton JH, Shachak M (1997) Positive and negative effects of organisms as physical ecosystem engineers. Ecology 78:1946–1957

Jones CG, Gutierrez JL, Byers JE, Crooks JA, Lambrinos JG, Talley TS (2010) A framework for understanding physical ecosystem engineering by organisms. Oikos 119:1862–1869

Jung K, Kaiser S, Böhm S, Nieschulze J, Kalko EKV (2012) Moving in three dimensions: effects of structural complexity on occurrence and activity of insectivorous bats in managed forest stands. J Appl Ecol 49:523–531

Kim Y, Yang Z, Cohen WB, Pflugmacher D, Lauver CL, Vankat JL (2009) Distinguishing between live and dead standing tree biomass on the North Rim of Grand Canyon National Park, USA using small-footprint lidar data. Remote Sens Environ 113:2499–2510

Körner C (2007) The use of 'altitude' in ecological research. Trends Ecol Evol 22:569–574

Lassau SA, Hochuli DF (2008) Testing predictions of beetle community patterns derived empirically using remote sensing. Divers Distrib 14:138–147
Lassau SA, Cassis G, Flemons PKJ, Wilkie L, Hochuli DF (2005a) Using high-resolution multi-
  spectral imagery to estimate habitat complexity in open-canopy forests: can we predict ant
  community patterns? Ecography 28:495–504
Lassau SA, Hochuli DF, Cassis G, Reid CAM (2005b) Effects of habitat complexity on forest
  beetle diversity: do functional groups respond consistently? Divers Distrib 11:73–82
Lefsky M (2010) A global forest canopy height map from the Moderate Resolution Imaging
  Spectroradiometer and the Geoscience Laser Altimeter System. Geophys Res Lett 37:L15401
Lefsky M, Harding D, Keller M, Cohen W, Carabajal C (2005) Estimates of forest canopy height
  and aboveground biomass using ICESat. Geophys Res Lett 32:L22S02
Lesak AA, Radeloff VC, Hawbaker TJ, Pidgeon AM, Gobakken T, Contrucci K (2011) Modeling
  forest song bird species richness using LiDAR-derived measures of forest structure. Remote
  Sens Environ 115:2823–2835
Leutner BF, Reineking B, Müller J, Bachmann M, Beierkuhnlein C, Dech S, Wegmann M (2012)
  Modelling forest α-diversity and floristic composition – on the added value of LiDAR plus
  hyperspectral remote sensing. Remote Sens 4:2818–2845
Lindo Z, Winchester NN (2008) Scale dependent diversity patterns in arboreal and terrestrial
  oribatid mite (Acari: Oribatida) communities. Ecography 31:53–60
Lucas KL, Raber GT, Carter GA (2010) Estimating vascular plant species richness of Horn Island,
  Mississippi using small-footprint airborne LiDAR. J Appl Remote Sens 4:033545
MacArthur RH, MacArthur J (1961) On bird species diversity. Ecology 42:594–598
Martin K, Aitken KEH, Wiebe KL (2004) Nest sites and nest webs for cavity-nesting communities
  in interior British Columbia, Canada: nest characteristics and niche partitioning. Condor
  106:5–19
Martinuzzi S, Vierling LA, Gould WA, Falkowski MJ, Evans JS, Hudak AT, Vierling KT (2009)
  Mapping snags and understory shrubs for a LiDAR-based assessment of wildlife habitat
  suitability. Remote Sens Environ 113:2533–2546
McDowell N, Pockman WT, Allen CD (2008) Mechanisms of plant survival and mortality
  during drought: why do some plants survive while others succumb to drought? New Phytol
  178:719–739
MEA (2005) Ecosystems and human well-being: synthesis. Island Press, Washington, DC
Moning C, Müller J (2008) Environmental key factors and their thresholds for the avifauna of
  temperate montane forests. For Ecol Manag 256:1198–1208
Müller J, Brandl R (2009) Assessing biodiversity by remote sensing and ground survey in
  mountainous terrain: the potential of LiDAR to predict forest beetle assemblages. J Appl Ecol
  46:897–905
Müller J, Moning C, Bässler C, Heurich M, Brandl R (2009) Using airborne laser scanning to
  model potential abundance and assemblages of forest passerines. Basic Appl Ecol 10:671–681
Müller J, Stadler J, Brandl R (2010) Composition versus physiognomy of vegetation as predictors
  of bird assemblages: the role of lidar. Remote Sens Environ 114:490–495
Müller J, Mehr M, Bässler C, Fenton MB, Hothorn T, Pretzsch H, Klemmt H-J, Brandl R (2012)
  Aggregative response in bats: prey abundance versus habitat. Oecologia 169:673–684
Ohlson M, Söderström L, Hörnberg G, Zackrisson O, Hermansson J (1997) Habitat qualities versus
  long-term continuity as determinants of biodiversity in boreal old-growth swamp forests. Biol
  Conserv 81:221–231
Parker GG, Harding DJ, Berger ML (2004) A portable LIDAR system for rapid determination of
  forest canopy structure. J Appl Ecol 41:755–767
Pausas JG, Verdu M (2010) The jungle of methods for evaluating phenotypic and phylogenetic
  structure of communities. Bioscience 60(8):614–625
Remm J, Lohmus A (2012) Tree cavities in forests: the broad distribution of a keystone structure
  for biodiversity. For Ecol Manag 262:579–585
Rosenzweig ML (1995) Species diversity in space and time. Cambridge University Press,
  Cambridge/New York

Schaffers AP, Raemakers IP, Sýkora KV, ter Braak CJF (2008) Arthropod assemblages are best predicted by plant species composition. Ecology 89:782–794

Scherrer D, Schmid S, Körner C (2011) Elevational species shifts in a warmer climate are overestimated when based on weather station data. Int J Biometeorol 55:645–654

Schuldt A, Assmann T (2010) Invertebrate diversity and national responsibility for species conservation across Europe – a multi-taxon approach. Biol Conserv 143:2747–2756

Seavy NE, Viers JH, Wood JK (2009) Riparian bird response to vegetation structure: a multiscale analysis using LiDAR measurements of canopy height. Ecol Appl 19:1848–1857

Sendtner O (1860) Die Vegetations-Verhältnisse des Bayerischen Waldes nach den Grundsätzen der Pflanzengeographie. Literarisch-artistische Anstalt, München, p 505

Siitonen J, Martikainen P (1994) Occurrence of rare and threatened insects living on decaying *Populus tremula*: a comparison between Finnish and Russian Karelia. Scand J For Res 9:89–95

Simonson WD, Allen HD, Coomes DA (2012) Use of an airborne Lidar system to model plant species composition and diversity of Mediterranean oak forests. Conserv Biol 26:840–850

Ter Braak CJF, Schaffers AP (2004) Co-correspondence analysis: a new ordination method to relate two community compositions. Ecology 85(3):834–846

Tews J, Brose U, Grimm V, Tielbörger K, Wichmann MC, Schwager M, Jeltsch F (2004) Animal species diversity driven by habitat heterogeneity/diversity: the importance of keystone structures. J Biogeogr 31:79–92

Tilman D, Knops J, Weldin D, Reich P, Ritchie M, Sieman E (1997) The influence of functional diversity and composition on ecosystem processes. Science 277:1300–1302

Turner WR, Brandon K, Brooks TM, Costanza R, da Fonseca GAB, Portela R (2007) Global conservation of biodiversity and ecosystem services. BioScience 57:868–873

Unterseher M, Otto P, Morawetz W (2005) Species richness and substrate specificity of lignicolous fungi in the canopy of a temperate, mixed deciduous forest. Mycol Prog 42:117–132

Vierling KT, Vierling LA, Gould WA, Martinuzzi S, Clawges RM (2008) Lidar: shedding new light on habitat characterization and modeling. Front Ecol Environ 6:90–98

Vierling KT, Bässler C, Brandl B, Vierling LA, Weiß M, Müller I (2011) Spinning a laser web: predicting spider distributions using lidar. Ecol Appl 21:577–588

Vierling LA, Vierling KT, Adam P, Hudak AT (2013) Using satellite and airborne LiDAR to model woodpecker habitat occupancy at the landscape scale. PLoS ONE 8(12):e80988

Vogeler JC, Hudak AT, Vierling LA, Evans J, Green P, Vierling KT (2014) Terrain and vegetation structural influences on local avian species richness in two mixed-conifer forests. Remote Sens Environ (in press)

Wei Y, Krzystek P, Heurich M (2012) Tree species classification and estimation of stem volume and DBH based on single tree extraction by exploiting airborne full-waveform LiDAR data. Remote Sens Environ 123:368–380

Whittaker RH (1960) Vegetation of the Siskiyou Mountains, Oregon and California. Ecol Monogr 30:279–338

Wilson EO (1987) An urgent need to map biodiversity. Scientist 1:11

Wünsche A (2012) Erfassung der lichtökologischen Situation im Nationalpark Bayerischer Wald. Technical University Dresden, Dresden

Chapter 19
Assessing Dead Wood by Airborne Laser Scanning

Matti Maltamo, Eveliina Kallio, Ole Martin Bollandsås, Erik Næsset, Terje Gobakken, and Annukka Pesonen

Abstract This chapter reviews the existing research concerning coarse woody debris (CWD) characterization by means of airborne laser scanning (ALS) data. The research so far has concentrated on modelling and mapping CWD on different scales, from single tree to stands, and to the use of ALS as auxiliary data in sampling-based inventories. In general the accuracy of ALS-based CWD models has varied considerably, from accurate predictions in different nature conservation areas to hardly statistically significant models in managed forests with a very low amount of CWD. A large-scale case study carried out in Norway showed that the classification of the presence of notable CWD quantities is more reliable than characterization of CWD volumes.

19.1 Introduction

Decaying wood is a key factor for biodiversity and many rare and specialized species are dependent on it (e.g. Siitonen 2001; Lonsdale et al. 2008; Bradshaw et al. 2009). Dead wood is also important for forest fuels, carbon sequestration and nutrient cycling; it is a long-term source of organic matter and nutrients; it forms the bases for cycling of photosynthetic energy, and creates nesting and feeding habitats

M. Maltamo (✉) • E. Kallio
School of Forest Sciences, University of Eastern Finland, Joensuu, Finland
e-mail: Matti.maltamo@uef.fi

O.M. Bollandsås • E. Næsset • T. Gobakken
Department of Ecology and Natural Resource Management, Norwegian University
of Life Sciences, Ås, Norway
e-mail: Erik.naesset@umb.no

A. Pesonen
Blom Kartta Oy, Helsinki, Finland

M. Maltamo et al. (eds.), *Forestry Applications of Airborne Laser Scanning: Concepts and Case Studies*, Managing Forest Ecosystems 27, DOI 10.1007/978-94-017-8663-8_19,
© Springer Science+Business Media Dordrecht 2014

for birds (Harmon et al. 1986; Esseen et al. 1997; Siitonen 2001; Karjalainen and Kuuluvainen 2002; Enrong et al. 2006). Moreover, dead wood has a role in stand dynamics by arboreal regeneration; functions such as erosion control, and it provides substrates and habitats for fungi, insects, lichens, mosses and invertebrates (e.g. Hyvönen and Ågren 2001; Jonsson et al. 2005; Hottola and Siitonen 2008; Lonsdale et al. 2008; Aakala 2010). A comprehensive recent review of biodiversity in dead wood can be found in the textbook by Stokland et al. (2011).

The term 'coarse woody debris' (CWD) refers to a variety of sizes and types of woody material, including downed dead trees, logs, large branches and roots, standing dead trees, snags and stumps (Newton 2007). Some ecologists make a distinction between fine and coarse woody debris, depending on the diameter of the dead wood pieces (e.g. Eaton and Lawrence 2006), but in some studies the term CWD is used for all dead wood irrespective of size (e.g. Rouvinen 2002). However, no ecological study has provided a clear threshold for distinguishing between fine and coarse woody debris.

Intensive forest management, the loss and fragmentation of forests, the prevention of forest fires, the use of firewood and the removal of damaged trees to prevent insect infestation have reduced the numbers of dying and dead trees in managed forests to a small fraction of those in natural, unmanaged forests (e.g. Vaillancourt et al. 2008). More specifically, CWD volumes in Fennoscandian managed forests have decreased to less than 10 % of those in old-growth forests (Siitonen 2001). The main factors that determine the suitability of dead wood for saproxylics, i.e. species that are dependent on CWD, are tree species, decay stage, the fungal species decaying the wood, size, dead wood material (e.g. snag, log, stump) and environmental conditions (Siitonen 2001; Lonsdale et al. 2008). Large diameters of dead wood pieces are usually preferable for many species (Samuelsson et al. 1994). Many large-sized species prefer large dead wood, but also some small species are restricted to it (Brin et al. 2013; Gossner et al. 2013).

The quantity of CWD is a commonly used measure of the 'ecological goodness' of an area (Ståhl et al. 2001). Thresholds for species occurrence or assemblage compositions have been shown (Müller and Bütler 2010). Usually, distinctions are made between standing dead trees (whole or snags) and downed dead trees (whole downed trees, logs and branches). The characteristics reported for dead trees include size measures (diameter and length or height, volume, biomass), species, decay stage and amount of bark remaining (Samuelsson et al. 1994). The volumes of single dead trees and CWD pieces are calculated using taper curves or volume functions developed for living trees, or using a sectioning method, i.e. partitioning a tree into several pieces.

Nowadays CWD can usually be considered as a rare forest phenomenon in managed forests. Therefore, accurate characterization of CWD quantities of a forest area is usually difficult and costly. The general usefulness of airborne laser scanning (ALS) for CWD characterization is based on the fact that the height distribution of ALS data is related to the vertical structure of the vegetation (Magnussen and Boudewyn 1998). Furthermore, ALS systems enable data collection from beneath closed tree canopies and allow measurement of the roughness of the forest floor. Usually ALS data have wall-to-wall coverage of the whole inventory area, allowing large-scale mapping of the CWD.

In the following, we will review the background of CWD assessments and different ALS-based CWD applications. The main emphasis is on the quantitative inventory measures, whereas in Chap. 18 a more ecological biodiversity-related overview on CWD is given. We also carried out a large-scale case study in Norway, which we report on in Sect. 19.3. No distinction between fine and coarse woody debris will be made here, and all sizes of dead wood will be referred to hereafter as CWD, unless we refer to studies where only some component of CWD is considered.

19.2 Inventory of CWD

19.2.1 General Background to CWD Inventories

The amount and quality of CWD can be characterized by detailed and time-consuming field measurements. These measurements are usually restricted to a limited number of small-sized plots. Since accurate description of CWD is expensive this has led to different remote sensing assisted experiments.

In general the role of remote sensing in CWD characterization can be related to the modelling and mapping of CWD quantities from tree level to larger areas. This can be done by discriminating living and dead trees or by predicting CWD volumes or biomasses. In sampling-based inventories remote sensing can be used in a design phase in pre- or post- stratification, or to guide field plot locations. Correspondingly, in an estimation phase remote sensing can also be applied, e.g., in model-assisted estimation.

Some optical remote sensing studies propose that aerial photographs can be successfully used for assessing standing dead trees (e.g. Uuttera and Hyppänen 1998; Haara and Nevalainen 2002; Pasher and King 2009). The same is also true for mapping and quantification of large snags, even though canopy closure has been observed to significantly decrease the detection ability (Bütler and Schlaepfer 2004). Recently, Eskelson et al. (2012) utilized satellite data for the estimation of snag density and snag quality attributes. The accuracy was not very good, but in that study the main emphasis was to compare parametric and imputation methods in the prediction. Poorer results have also been reported for the prediction of downed dead wood volumes using aerial photographs (Pasher and King 2009). One of the main problems affecting the use of optical remote sensing data is the fact that weather conditions may have notable effects on image quality, and the CWD characteristics visible in images from the same area differ over time. Thus, the applicability of optical remote sensing data is usually restricted to the mapping of standing dead trees or snags.

Studies based on field information report that CWD volume is correlated with living tree characteristics (Sippola et al. 1998; Siitonen 2001). Consequently, models for predicting CWD volume from living tree characteristics have been developed. For example, Ranius et al. (2004) estimated the amount of dead wood from the growth of living trees, tree mortality rate and CWD decomposition rate. Correspondingly, Pesonen et al. (2008) applied basal area of living trees to predict CWD. Nevertheless, the accuracy of the models for predicting CWD volume using characteristics of living trees as predictors has been rather poor.

Many countries have recently also included dead wood measurements in their national forest inventories (NFI). Currently, estimates of the quantity and quality of CWD are calculated from NFI data in approximately 13 % of countries around the world, accounting for a total of 41 % of the area of the world's forests (Woodall et al. 2009). The sampling of rare and clustered forest characteristics such as CWD is not cost-efficient if inventory methods originally designed for living trees are applied (Kangas 2006).

During the last 15 years, numerous more efficient field-based sampling methods for assessing the quantity and quality of CWD have been applied and developed (Ståhl et al. 2001, 2010; Ducey et al. 2013; Gove et al. 2013). Strip sampling was a common inventory method for collecting information about living trees in the past, but nowadays it is mainly used for rare forest characteristics (Stehman and Salzer 2000), while fixed-sized sample plots, which are widely used in forest inventories, are also common in CWD inventories (Ståhl et al. 2001; Gove and van Deusen 2011). Other methods for assessing CWD include, for example, line-intersect sampling (Warren and Olsen 1964), transect relascope sampling (Ståhl 1998), point relascope sampling (Gove et al. 1999), adaptive cluster sampling (Thompson 1990), perpendicular distance sampling (Williams and Gove 2003), critical point relascope sampling (Gove et al. 2005), critical length sampling (Ståhl et al. 2010), distance-limited perpendicular distance sampling (Ducey et al. 2013) and guided transect sampling (Ståhl et al. 2000). Some of the above-mentioned methods are specified to the sampling of downed CWD. The variable of interest (e.g. volume, length or number of CWD pieces) and forest conditions, like visibility, have been observed to have an effect on the preferred sampling method for CWD (Ringvall and Ståhl 1999; Jordan et al. 2004; Woldendorp et al. 2004).

19.2.2   CWD Modeling and Mapping by ALS

Research related to ALS-based CWD characterization is still rather rare. The first studies to characterize the relationship between ALS and CWD were based on visual assessments of ALS data or the use of descriptive statistics. In this context Seielstad and Queen (2003) observed that ALS data can be used for assessing the roughness of the forest floor. Their study was related to forest fuels and fire behaviour and they characterized lower canopy layers in forest areas with large amounts of CWD. One of their conclusions was that it might be possible to also map volumes of CWD with

ALS. Correspondingly, Sherrill et al. (2008) found a correlation between ALS mean and maximum heights and standing dead trees biomass.

In general, the characterization of CWD by ALS can be considered according to basic techniques to utilize ALS data in the forest inventory context, i.e. a area-based approach or single-tree detection. The area-based approach can be applied to model amounts of CWD or used to classify plots according to some CWD existence-based rules. Single-tree detection is a natural choice for the mapping of standing dead trees and snags, while characterization of downed CWD is problematic using that specific method. The approaches to quantify CWD by ALS can also be separated according to the study area because most of the studies have been conducted in different nature conservation areas while assessments in managed forest areas can be highly challenging.

Using the area-based method, Pesonen et al. (2008) modelled volumes separately for standing and downed dead trees in a national park in Finland. The approach was based on the relationship between forest structure and CWD. The existence of canopy gaps was expected to be an indicator of the presence of CWD and, thus, standard deviation of ALS height values was used as the main predictor in the fitted models. The accuracy of modelling downed CWD volume was moderate, while for standing CWD the accuracy was poorer. However, ALS metrics had more explaining power of CWD than the characteristics of living trees.

The same approach was also examined in managed forest areas on a larger scale in Finland by Pesonen et al. (2009). The modelling of the CWD volume was found to be almost impossible for managed forests, where canopy gaps are more likely to be a consequence of forest-thinning operations rather than indicators of CWD. This also resulted in the large amount of zero values in the study data.

In Canada, Bater et al. (2009) modelled the cumulative wildlife tree (decay) class distribution in a multi-use area where both harvested and mature forests occurred. These classes, nine altogether, describe the growth and decay class, starting from healthy trees and ending up to rotting stumps. The coefficient of variation of the ALS height data was found to be the best predictor.

Correspondingly, Kim et al. (2009) assessed standing dead trees in Arizona, USA. They used ALS-derived intensities to model the standing CWD biomass in a national park area. The density distributions of the intensity were found to be bimodal in plots that included both living and dead standing trees. They hypothesized that lower and higher intensity peaks in ALS returns were related to dead and live materials, respectively, and found that the low intensity distribution peak frequency was a good predictor of standing CWD biomass. Finally, Martinuzzi et al. (2009) used ALS height and topographic metrics to assess the presence of snags in different diameter classes. The classification accuracies were rather good but varied between diameter classes.

In the case of single-tree detection the critical point is to find metrics which discriminate living and dead standing trees. Korpela et al. (2010) considered dead standing trees in their study on tree species recognition, utilizing intensity values. They stated that the intensity values from dead trees were approximately 40–60 % lower than those from living trees. The average intensities from trees that were deteriorated were 10–20 % lower than for 'normal' trees. The reasoning for the use of intensity is based on the fact that for dead trees the values are based on reflections from stem and branches, whereas they originate from the tree crown in the case of living trees (Reitberger et al. 2009).

Mücke et al. (2012) conducted point cloud analysis for standing dead trees using leaf-on and leaf-off full-waveform ALS data. They found differences in the echo distributions and the echo amplitudes between living and dead trees.

The first attempt to detect dead single trees from high-density full-waveform ALS data was made by Yao et al. (2012). They ended up with classification accuracies of 71 and 73 % and kappa values of 0.27 and 0.45 in classifying living and dead trees for leaf-off and leaf-on data, respectively. The discrimination between living and dead trees was based on seven metrics: describing the tree crown's outer geometry; intensity; pulse width; the average number of pulses between first and last reflection; the ratio between below canopy pulses and total pulses; the ratio between first and intermediate pulses divided by the last and single pulses; and the maximum diameter of the tree crown. The most important metrics were the echo ratios and the outer geometry of the tree crown.

Correspondingly, Wing et al. (2012) developed semi-automated methods for snag detection. They utilize 2D and 3D local-area ALS point filters focused on spatial location and intensity information to identify individual ALS points associated with snags and to eliminate points associated with living trees.

Detection of individual downed dead trees from ALS data is a rather novel approach. Blanchard et al. (2011) applied object-based image analysis for that. In the study by Mücke et al. (2012) dense (about 20 points per m^2) full-waveform ALS data were used to characterize downed dead trees in the Natura 2000 area in Germany. The proposed workflow begins with the selection of the ALS echoes with a height of less than 2 m. Waveform echo-widths smaller than 4.5 ns are then used to measure surface roughness, i.e. to separate rough shrubs from smoother CWD. The digital height model is then constructed for remaining areas and the digital terrain model is subtracted from it. The remaining image contains the elongated area representing the downed wood and spot-like features of standing trees which were still removed. As a result of this procedure, the final vector map represents the outlines of the identified downed stems. Using this algorithm, 37.3 % of downed stems of the study area were fully detected and 33.2 % were partly detected.

To sum up, the studies based on the area-based approach have so far concentrated on modelling CWD quantities such as volumes, biomasses and number of trees. An exception is the approach by Bater et al. (2009) which also allows the characterization of standing CWD decay class at plot level. The most powerful ALS metrics have been those related to intensity values and standard deviation of ALS heights. In the case of single-tree detection the metrics applied to discriminate living and dead trees have somewhat differed between the conducted studies. However, intensity, echo ratios and variables describing the outer geometry of the tree crown have most often applied. The choice of the metrics is also related to the proportion of dead trees, the decay stage of dead trees and the forest type considered. It also seems that full-waveform data is suitable for CWD characterization also allowing the mapping of downed dead trees.

19.2.3   The Use of ALS in Sampling-Based CWD Inventory

Although numerous studies and different sampling methods have been presented and applied for field-based CWD inventory in recent years, only the work by Pesonen (2011) has so far linked the use of ALS to CWD sampling. She applied different methods using ALS data to guide CWD inventories for commercially managed forests in Finland in order to increase the efficiency of sample-based field inventory methods for assessing CWD at plot level. This means that prior to a field-based inventory, ALS metrics were used to guide the location of field plots to be measured in the most promising places by means of simulation. The sampling methods, including simple random sampling, cluster sampling, strip sampling, adaptive cluster sampling, line-intersect sampling, point relascope sampling and transect relascope sampling, were compared in terms of the accuracy of the estimated mean CWD volume with a fixed input effort specified in fieldwork hours (Pesonen et al. 2009). How much the use of auxiliary information derived from ALS data or other sources could improve the sampling efficiency, i.e. reduce the standard error of the mean given the same budget as in an inventory based solely on field information, was also studied. ALS data were used to form probability layers which were based on the relationship between CWD values and ALS metrics, either in the form of a logistic model or as a correlation between the CWD and ALS variables. The auxiliary information was used either in the design phase, for probability proportional to size (PPS) sampling, or in the estimation phase, for model-assisted ratio or regression estimation (Pesonen et al. 2010b).

The sample-based inventory methods especially developed for assessing CWD or other rare characteristics were observed to be the most efficient field inventory methods, and in particular the relascope-based sampling methods were highly efficient. The use of PPS sampling notably improved the efficiency of the CWD inventory compared to simple random sampling, but efficiency was modest when auxiliary information from ALS was used in the estimation phase. The improvement in efficiency varied considerably between different inventory methods and CWD components. Although the efficiency of other inventory methods could be improved more by introducing PPS sampling, relascope-based sampling methods remained the most efficient methods for assessing CWD. It was also observed that the sampling efficiency was not markedly better if ALS data were combined with either aerial photographs or stand-register data, and it was usually preferable to use ALS data as the sole source of the auxiliary information (Pesonen et al. 2010a).

19.3   Case Study: CWD Classification and Modeling in Hedmark County, Norway

19.3.1   Background

Most of the earlier attempts to characterize CWD by ALS have been based on geographically rather small study areas and a limited number of field sample plots. Study areas have usually been nature conservation areas where CWD cannot be considered as a rare forest phenomenon. There are also some studies where CWD prediction has been applied in managed forests (Pesonen et al. 2009). While in conservation areas highly accurate results have been obtained, it has been difficult to even formulate statistically significant models in managed forests. A study area in Hedmark County (HC), Norway, offers a unique opportunity to study CWD characteristics in a large-scale commercial forest area.

In Norway and Finland CWD assessments were included in the NFIs in 1994 and 1996, respectively. According to latest published results from the Finnish NFI, the average CWD volume for the whole country was estimated to be 5.4 m^3ha^{-1}. Corresponding estimates for southern and northern Finland were 3.2 m^3ha^{-1} and 7.6 m^3ha^{-1}, respectively (Ihalainen and Mäkelä 2009). CWD volumes varied notably between areas managed for commercial timber production (4.7 m^3ha^{-1}) and in conservation areas (12.7 m^3ha^{-1}). In Norway, the corresponding average CWD volume for the whole country is about 8.7 m^3ha^{-1} (Larsson and Hylen 2007). The presence and quantity of CWD in HC appears to be somewhat higher than in corresponding managed areas in, for example, Finland. The reason for that lies mainly in the silvicultural regimes not favouring thinning operations in Norway.

In the following, we characterize CWD in two different ways. The modelling approach included (1) classifying the sample plots into two classes according to the presence/absence of the CWD material (downed and standing CWD material together) and (2) estimating the volume of the CWD. For classifying the plots into 'present' and 'absent' classes, two different limits for the volume of the CWD were explored: (a) 0.00 m^3 and (b) 5 % of the volume of the total stock, i.e. growing stock and CWD together. If the volume of the CWD on a plot exceeded these limits the plot was classified into the class 'CWD present', otherwise they were classified into the class 'CWD absent'. For estimating the volume of the CWD, both linear regression modelling and k-NN (k-nearest neighbour) imputation were employed. In addition to the ALS metrics we also utilized different additional information available from stand level and large-scale NFI inventories. This information included, for example, the stand development class, damages and tree species proportions. The information concerning the stand development class was used to stratify the data to construct separate models for each stratum.

19.3.2 Study Area, Materials and Methods

The study area comprised HC which is located in southeastern Norway. The total area of the county is 27 390 km^2. However, ALS data does not cover the whole area. Circular 250 m^2 field plots systematically distributed on a 3 km × 3 km grid from the NFI survey of Norway were used. On each plot, all the trees with DBH >5 cm were callipered and on average ten sample trees per plot were selected for height measurements. The Norwegian NFI measured all dead wood on their permanent sample plots from 1994 to1998 as a part of the seventh cycle of the NFI. These measurements constitute a baseline for subsequent measurement and estimation of dead wood. In the subsequent following inventories carried out every 5 years the dead wood assessment was of those trees that had died after the previous inventory. We used data from 481 sample plots inventoried in 2005–2007. The basic statistics of the volumes of the growing stock and total, downed and standing CWD on the 481 sample plots, are presented in Table 19.1. Also the mean, maximum and the standard deviation of the proportion of the CWD volume of the sum of the growing stock and CWD volumes are presented in the table. The location of the centre of each of the sample plots was georeferenced with differential Global Positioning System and Global Navigation Satellite System measurements according to the procedure suggested by Næsset (2001). The estimated precision of the planimetric plot coordinates ranged from 0 to 2 m, with an average of 0.05 m (Gobakken et al. 2012). The plots were divided into four development classes according to maturity, i.e. (1) regenerated areas and young forests, (2) young thinning stands, (3) advanced thinning stands and (4) mature forests. All other NFI plots were removed from the study data.

The independent variables for the classifications, the regression models and the k-NN imputation models were selected from among a set of variables calculated from ALS echo data and variables defined by the NFI. Further details regarding the ALS acquisition can be found in Gobakken et al. (2012). The ALS metrics used in this study included laser echo height (percentiles; $h5$, $h10$, $h20$, ..., $h90$, $h95$) and vertical density metrics (canopy density at different vertical levels; $p5$, $p10$, $p20$, ..., $p90$, $p95$), which were created separately for the first and the last echoes reflected from the tree canopy. The range calibrated intensity distribution included the corresponding percentiles ($int5$, $int10$, $int20$, ..., $int90$, $int95$). Also the standard deviation (sd), the coefficient of variation ($coeffvar$), the maximum (max), the mean ($mean$) and the proportion of the echoes on the vegetation (veg) were calculated for the intensity and for the heights of the first and the last echoes. The height limit for the variable veg was 2 m. Values of the intensity distribution were further divided correspondingly as in Kim et al. (2009) for lower and higher values to separate dead and living trees. Means of these lower (int_{low}) and higher (int_{high}) values were then calculated. Logarithms and the inverse of all the ALS metrics were also included.

Table 19.1 Basic statistics of the sample plots

Attribute	Development class				
	All	2	3	4	5
N_{obs}	481	147	95	106	133
Mean					
Total CWD	7.81	5.08	4.40	9.28	12.07
Standing	5.09	4.00	2.60	5.32	7.91
Downed	2.71	1.09	1.80	3.96	4.16
CWD Vol%	9.95	16.44	4.38	7.31	8.85
Vol	110.21	29.62	124.0	154.9	153.79
Maximum					
Total CWD	117.52	73.68	30.32	100.9	117.52
Standing	99.64	73.68	27.52	99.64	99.20
Downed	67.16	14.56	26.64	60.72	67.16
CWD Vol%	96.56	96.56	52.95	77.88	70.12
Vol	646.28	217.00	352.8	646.3	594.32
Minimum					
Gstock	0.24	0.24	14.88	3.88	11.48
Standard deviation					
Total CWD	13.88	10.37	6.81	14.46	18.58
Standing	11.33	9.84	4.86	12.42	14.36
Downed	6.57	2.49	4.98	7.69	8.86
CWD Vol%	16.48	23.65	7.89	11.80	11.73
Vol	108.59	30.97	75.96	121.6	122.69

N_{obs} = observed number of plots, Total CWD = volume (m^3ha^{-1}) of the standing and downed CWD, Standing = volume (m^3ha^{-1}) of the standing CWD, Downed = volume (m^3ha^{-1}) of the downed CWD, CWD Vol% = proportion of the total CWD volume relative to the total CWD and growing stock volume, Vol = volume (m^3ha^{-1}) of the growing stock. The minimum values for total, downed and standing CWD are 0.00 m^3 in all the development classes

In addition, variables, such as productivity (*prod*), reservation (*res*), treatment (*treat*), damage (*dge*), site index (*boni*), municipality (*muni*), the age of the growing stock (*age*), height above sea level (*hasl*), slope (*sl*), slope azimuth (*slAzi*), volume of the growing stock (*vol*), number of stems (*stems*) and the proportion of the different tree species (*pine, spruce, deciduous, aspen*) were defined for each plot from the NFI data. The first six of the NFI variables mentioned above were categorical variables. 'Productivity' describes whether a plot is located in productive forest or in other types of areas. Correspondingly, 'reservation' describes whether a plot belongs to a nature reservation area or is used for forestry without constraints. Most of the plots are in productive forests which are used for forestry. The variable treatment describes whether any silvicultural operations have been performed in the last 5 years or not. Furthermore, the variable damage describes the presence of stand damage in general. In the case of species proportions, deciduous tree species are combined but additionally the proportion of aspen is calculated since the presence of aspen trees is usually seen as a biodiversity indicator.

Linear discriminant analysis (LDA) was employed to classify the sample plots into the 'CWD present' and 'CWD absent' classes. The aim of the LDA is to find an optimal linear combination of the predictors whose variance is as small as possible within groups and as large as possible between groups (e.g. Lebart et al. 1984; Venables and Ripley 2002). The LDA was carried out with equal prior probabilities of the class memberships using the LDA function in the MASS package (Venables and Ripley 2002) in the R environment (www.r-project.org). The predictors for the classifications were selected using correctness rate, kappa-value analysis, and stepwise variable selection based on the correctness rate (Garczarek 2002). With kappa-value it is possible to compare the estimated classification with a random classification (Rosenfield and Fitzpatrick-Lins 1986). Stepwise variable selection is an estimator for the correctness of a classification rule when adding individual variables to the discriminator vector. The stepwise selection was implemented with the *stepclass* function of the klaR package in the R environment that uses tenfold cross validation. LDA was employed separately for all plots and separately for different development classes and classification rules, resulting in a total of ten models.

In the parametric modelling approach the relationship between ALS-metrics and NFI-variables and the square root of the observed CWD volumes was characterized by means of linear regression modelling (LM). Fitting the regression line to the square roots of the volumes ensures non-negative estimates of the volume. The fitting of the LM was carried out in the R environment using the STATS package. The independent variables were selected by means of the stepwise function step implemented in the STATS package. The function uses the AIC statistics for selecting between the models which were formulated by allowing both inclusion and exclusion of the independent variables in the candidate models. The number of the independent variables in the LM was controlled by a k parameter, and was set between 2 and 6. The LM was formulated for the whole study data and for each development class separately, resulting in five individual models.

In k-NN the dependent variable for a target plot is estimated by means of finding its k nearest neighbour plots in the training data and assigning the value of the variable to be the weighted average of the values of the neighbours. Nearness of the plots is measured with the independent variables and is defined in terms of the weighted Euclidean distance. In multivariable k-NN several dependent variables are estimated simultaneously. Both downed and standing CWD volumes were dependent variables. In the imputation approach the estimate of the total CWD volume was the sum of the downed and standing CWD estimates. The imputation was carried out with the yaImpute package (Crookston and Finley 2008) in the R environment. The selection of the independent variables was carried out by first simulating an initial model with the LM and step function and then manually correcting the model until the accuracy improved any longer. The k ranged between 2 and 5.

Table 19.2 Accuracies and the independent variables (*var*) of the discriminant analysis

Cl	$0\ m^3$			$5\ \%$		
	CR	κ	var	CR	κ	var
All	0.784	0.205	$\log(fh_{std})$, fh_{20}^{-1}, dge, $hasl$, $\log(intmax)$, fp_{80}^{-1}	0.665	0.297	vol, $hasl$, $\log(fh_{30})$, int_{70}^{-1}, lh_5
2	0.721	0.339	$hasl$, $treat$, $\log(lh_{40})$, $\log(fp_{90})$, fh_{30}^{-1}	0.694	0.386	$lhmean$, lh_{70}^{-1}, $\log(fh_{70})$, $aspen$, $\log(lh_{80})$, fh_{70}^{-1}
3	0.779	0.447	fp_{40}, fp_{20}^{-1}, fh_{60}^{-1}, $\log(fh_{mean})$, fh_{80}	0.842	0.525	dge, $\log(fh_{95})$, lh_{20}^{-1}, fp_{90}, lp_5^{-1}, lp_{90}
4	0.877	0.258	fh_{mean}, f_{veg}, fh_5^{-1}, fh_{60}, $muni$, $inthigh$	0.726	0.406	$\log(lh_{30})$, int_{20}, int_{60}^{-1}, $lh_{coeffvar}$, int_{10}^{-1}, h_{80}
5	0.850	0.190	fh_{mean}, fh_{std}, fh_{20}, fh_5, fh_{60}, $\log(int_5)$	0.662	0.312	lh_{70}, lh_{10}, lp_5^{-1}, fp_{80}, fp_{20}^{-1}

Cl development class, *CR* correctness rate, κ kappa value. The two different classification rules employed in the LDA were: $0\ m^3$ = total CWD volume is over $0.00\ m^3$ and $5\ \%$ = the proportion of the CWD volume of the sum of the growing stock and CWD volume is over $5\ \%$

Since no separate validation data were available, the accuracy assessment of the classification and the volume estimations were carried out by means of leave-one-out cross validation (loocv). In the loocv, each of the observations in turn was used as validation data and the rest of the data as training data. The final accuracy measurement was the mean of the accuracies of the estimations of the individual observations. The LDA function in the R environment calculates the loocv results automatically. When estimating the volume of the CWD the loocv has to be computed within a programming loop.

The accuracy of the classification results was assessed by means of the correctness rate and the kappa value. The accuracy of the estimation of the CWD volumes was assessed with RMSE and RMSE%.

19.3.3 Results

The correctness rates (CR) and kappa values for the LDA classifications are presented in Table 19.2. The CRs ranged between 0.72 and 0.88 and between 0.66 and 0.84 when using the $0\ m^3$ and $5\ \%$ classification rules, respectively. The kappa values ranged between 0.19 and 0.45 and between 0.30 and 0.53 using the $0\ m^3$ and $5\ \%$ classification rules, respectively. The number of the independent variables selected to the classification models was between 5 and 6.

The numbers of plots in each development class with and without CWD are presented in Table 19.3. The classification accuracy varied between the 'CWD absent' and 'CWD present' classes. On average, 33 % and 81 % of the plots in the 'CWD absent' class were classified correctly in the different development classes when using the CWD volume limits of $0\ m^3$ and $5\ \%$, respectively. Correspondingly,

Table 19.3 Number of the observed (N_{obs}) and estimated (N_{est}) sample plots using two different classification rules: total CWD volume is over 0 m^3 and the proportion of the total CWD volume of the sum of the growing stock and CWD volume is over 5 %

Cl	0 m^3		5 %	
	N_{obs}	N_{est}	N_{obs}	N_{est}
All	364	454	204	165
2	96	111	75	80
3	64	75	25	14
4	93	100	42	33
5	111	129	62	51

Cl development class

Table 19.4 Absolute and relative root mean square errors (RMSE, RMSE%) and the independent variables of the linear regression models for CWD volume

Cl	RMSE	RMSE%	Independent variables
All	13.48	172.68	$vol, dge, pine, sl, hasl, int_5{}^{-1}, fh_5$
All[a]	12.89	165.15	
2	10.33	203.32	$pine, hasl, sl_{Azi}, \log(int_{std})$
3	5.64	128.14	$dge, \log(lh_{20}), lp_{90}{}^{-1}, fp_5{}^{-1}$
4	13.19	142.14	$int_{std}{}^{-1}, lh_{10}{}^{-1}, lp_{60}, \log(lh_{30}), int_{60}{}^{-1}, int_{80}, lh_{70}$
5	17.93	148.54	$vol, sl, fh_{90}{}^{-1}, fh_{95}{}^{-1}$

Cl development class
[a]Combined with development class specific models

on average 93 % and 58 % of the plots in the 'CWD present' class were classified correctly when using the CWD volume limits of 0 m^3 and 5 %, respectively. As it can be seen, almost all of the 481 plots were classified to the 'CWD present' class in the case of the classification rule 0 m^3. The situation was more realistic in the case of rule 5 % where the classification was closer to the reference values.

The accuracies and the independent variables of the estimation of the CWD volume by means of regression modelling and k-NN imputation are presented in Tables 19.4 and 19.5 respectively. The number of the variables using LM ranged between 3 and 7 and between 8 and 24 when using k-NN. For more details of the independent variables, see Sect. 19.3.2.

In general, all obtained RMSE% values were rather high (>100 %). The average RMSEs (RMSE% in parenthesis) of the total CWD volume estimates over the development classes when using LM or multivariate k-NN were 12.89 m^{3}ha^{-1} (165.15 %) and 13.46 m^{3}ha^{-1} (172.4 %), respectively. The corresponding RMSEs of the estimation of the total CWD volume when development classes are ignored were 13.48 m^{3}ha^{-1} (172.68 %) and 14.7 m^{3}ha^{-1} (188.3 %) respectively. So there do not appear notable differences between modelling approaches or whether CWD is estimated separately for different development classes or for the whole data.

Table 19.5 Absolute and relative root mean square errors (RMSE, RMSE%) of the multivariate k-NN imputation of the downed and standing CWD volumes[a]. Total CWD volume estimates were calculated as the sum of the standing and downed CWD volumes

Cl	k	Total RMSE	Total RMSE%	Downed RMSE	Downed RMSE%	Standing RMSE	Standing RMSE%
All	5	14.70	188.3	6.34	233.8	12.47	244.8
All[b]		13.46	172.4	6.52	240.5	10.83	212.7
2	4	10.27	202.0	2.51	231.0	9.57	239.3
3	2	6.80	154.5	3.61	200.8	5.62	215.8
4	5	13.72	147.9	8.23	207.8	11.63	218.8
5	4	18.92	156.7	9.11	218.9	13.99	176.9

Cl development class

[a]Independent variables used in the imputations

[b]Combined with development class specific models

All: *vol, age, dge, prod, sl, hasl, pine, deci, fh_std, veg*, $\log(int_5)$, int_{40}, lh_{mean}, $\log(fh_5), fh_{10}, \log(lh_{10}), fh_{20}, \log(lh_{50}), lh_{60}, \log(lh_{70}), \log(fh_{80}), \log(fh_{90}), fp_{70}, lp_{70}^{-1}$

2: *boni, spruce, pine, dge, sl_Azi, stems*, $\log(f_{veg})$, int_5, int_{80}^{-1}, fh_{10}^{-1}, int_{70}, fh_{60}^{-1}, int_{mean}

3: *sl, dge, vol, age, muni, stems*, int_{max}, $int_{coeffvar}^{-1}$, $\log(int_{veg})$, int_{high}^{-1}, int_{20}^{-1}, $int_5, fh_5^{-1}, lh_{40}^{-1}, lh_{95}^{-1}$

4: *res, vol, std*, fp_{30}-1, fp_{70}^{-1}, fp_{80}, $\log(lh_{40}), \log(lh_{20})$, int_{max}^{-1}

5: *pine, treat, aspen, sl, dge*, $int_{coeffvar}^{-1}$, int_5^{-1}, $fh_{10}, fh_{20}, lh_{40}, fh_{90}^{-1}$, $fh_{95}^{-1}, \log(fp_{70}), lp_{70}^{-1}$

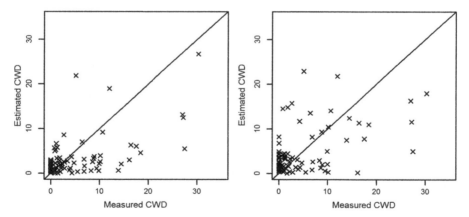

Fig. 19.1 Measured total CWD volume (m^3ha^{-1}) vs. values estimated with the linear regression modelling (*left*) and k-NN imputation (*right*)

As an example, the fitting of the employed models in development class 3 is presented in Fig. 19.1. The number of the plots in the data sets is 95. Although both approaches were also able to predict larger CWD volumes, the variation in the estimates is still very large. The large number of zero values is also notable.

19.4 Discussion and Conclusions

19.4.1 Findings of the Case Study

In the case study we characterized CWD presence and volume by using ALS metrics and stand register information. Firstly, we classified CWD presence (present-absent) either by 0.00 m³ or 5 % of the volume of the total stock (growing stock and CWD volumes together) limit by means of linear discriminant analysis. Secondly, we modelled the CWD volume directly by applying linear regression modelling and k-NN imputation. While the results were satisfactory for classification of CWD present and CWD absent classes, the predictions of CWD volumes were rather poor and the RMSE% values were above 100 %.

In the case of CWD classification both the use of the 5 % volume proportion limit and the separate estimation for different stand development classes increased the classification accuracy considerably compared to the analysis of the whole data. This was especially true for the obtained kappa values which were always higher when utilizing the above-mentioned rules. At its highest, the kappa value was slightly over 0.5 for the development class young mature growing stock. For remote sensing based CWD classification this level of accuracy can be considered as rather good.

The applied CWD existence limit of 5 % of the total tree stock is, of course, subjective. However, this limit can be easily modified for different applications. The proportion of CWD also describes the role of CWD in relation to the total tree stock. For example, in young stands a smaller amount of CWD forms a larger proportion of the total tree stock. Due to the ecological importance it is also reasonable to leave the smallest values of CWD out of the CWD present class.

In the constructed models we also applied information which is not directly based on ALS: for example, stand development classes used for stratification of the study area. Correspondingly, volume, number of stems, age of the growing stock, species information, damage and treatments of the stand represent such variables. The idea of using this information is that prior to the CWD inventory there would be an operational ALS-based forest inventory (Næsset 2002). The above-mentioned variables could be products from such an inventory for subsequent use in a CWD assessment. However, these estimates would then include some errors which would also affect the CWD estimates. Additionally, we also applied some of the topography variables from the NFI, but they can easily be estimated from ALS data. However, in all of the fitted models most of the independent variables were ALS metrics including height, density and intensity metrics. It seems that in using stratification classification in particular was improved. Thus, it was more effective to analyse the relationship between the presence of CWD and ALS metrics when the variation in the growing stock was reduced.

In all the models different area-based ALS metrics, including height, density and intensity variables, were applied. Additionally, we calculated corresponding intensity variables for lower and higher values which Martinuzzi et al. (2009)

successfully applied to separate dead and living standing trees. Although these variables were statistically significant in the current study, in some of the models they did not actually improve the accuracy. The reason might be that in most of the plots the proportion of standing CWD is minor.

In this study, no major differences in the accuracy between regression models and k-NN imputation were found. In imputations on the stand development class level the low number of reference plots may have affected the results. An alternative for the prediction of CWD volumes would have been a two-stage approach where first the presence of CWD is predicted and then in the second stage CWD volume is predicted for the plots where CWD is predicted to be present. However, if no new information is available in the second stage, no accuracy improvements can be expected. One possibility would also have been to estimate volumes of the living trees and CWD simultaneously.

In our case study we selected an approach where we first classified the existence of CWD and then modelled the quantity of CWD. This allows the mapping of CWD. An alternative would have been a sampling-based approach. The sampling design of our field data (Norwegian NFI data) would have allowed model-assisted estimation. The use of the regression estimator would also have enabled the wall-to-wall mapping of CWD quantities at the inventory area. However, in the study by Pesonen et al. (2010b) the regression estimation improved the sampling efficiency only by 6 % compared to random sampling, so it would have been probable that the effect of model-assisted estimation would also have been minor in this case.

Although the classification of CWD was partly successful there were still high RMSE values for the estimation of CWD volume. The absolute errors were not very high but since the level of magnitude of this characteristic was on average very low, the resulting RMSEs were over 100 %. In this study, we obtained considerably lower RMSE% values than in the previous studies (e.g. Pesonen et al. 2009 and Temesgen et al. 2008). Still the applicability of these kinds of estimates is questionable. So, it seems that the characterization of CWD volumes in managed forest areas is highly challenging, regardless of the applied silvicultural regimes.

19.4.2   Recommendations for ALS-Based CWD Inventories

According to the studies that have been conducted, the accuracy of CWD characterizations seems to vary considerably a lot from accurate predictions in different nature conservation areas to hardly statistically significant models in managed forests with a very low amount of CWD. The classification of the presence of CWD is more reliable than the estimation of CWD quantities. Most of the studies consider boreal conditions but some studies conducted in North America and Central Europe suggest corresponding applicability in broadleaved forests. In tropical conditions the studies have so far been concentrated on general biomass or carbon storage mapping (e.g. Asner et al. 2010).

The two basic methods for utilizing e ALS data, the area-based approach and single-tree detection, have both been used in a CWD context. While the area-based approach has been applied in modelling CWD quantities or for plot level classification, single-tree detection enables discrimination of single living and dead trees. Related to this, conventional height distribution metrics have been used, but correspondingly, as in tree species recognition, the role of the intensity has been found to be very important in separating CWD and living trees, especially in the case of standing CWD. Some studies still suggest that intensity may not be usable in all conditions. For example, Yao et al. (2012) suggest that in mixed stands neighbouring trees of different tree species may oversmooth intensity differences between the tree crowns and the stem. Yet another possible metric for CWD characterization is tree crown outer geometry whose usability is also dependent on the stage of tree decay. The results of Yao et al. (2012) also show that leaf-on data provides better possibilities for discrimination than leaf-off data. Data fusion of ALS with aerial images has already been examined but in future studies the role of hyperspectral data may also be evaluated in this context. Finally, most of the studies have so far been based on discrete-return sensor data but high-point density full-waveform data may even allow recognition of downed CWD.

In a sampling-based CWD inventory ALS can be successfully used as auxiliary data, either in the design phase to guide the selection of plot locations or for estimation and inference using model-assisted estimators. The use of low-cost auxiliary information can reduce the inventory costs and/or increase the accuracy of the estimates considerably whenever CWD is to be assessed. The studies included in the doctoral thesis by Pesonen (2011) introduced the use of ALS data in a CWD sampling context. The recent development of applying ALS as auxiliary data in sampling to estimate forest resources has been rapid (e.g. Andersen et al. 2011; Gregoire et al. 2011; Ståhl et al. 2011; McRoberts et al. 2012, 2013; Wulder et al. 2012; Ene et al. 2013; Næsset et al. 2013 – see further details in Chap. 14). Some of these studies have taken a design-based approach while model-based applications have also been demonstrated. Even for sampling and estimation of CWD a model-based approach is worthwhile considering and may be subject to future studies. Correspondingly, auxiliary ALS information could be used in the design phase and also in the pre-stratifying the inventory area (Maltamo et al. 2011, Chap. 14), but if no data for stratification are available prior to field inventory, such data can also be used afterwards for post-stratification (McRoberts et al. 2012, 2013; Næsset et al. 2013). Also, suitable new methods developed for field-based CWD sampling should be applied with ALS. One obvious topic is also the use of ALS in CWD surveys within inventories such as an NFI without wall-to-wall cover. However, ALS applications in a NFI context are, in general, still in the development phase (e.g. McRoberts et al. 2010). In future surveys, the inventory planning for CWD and the living tree stock should be coordinated and conducted simultaneously.

To sum up, there are already some studies where ALS has been used to characterize CWD either by means of modelling CWD quantities, classifying at tree or plot level or in a sampling context. There is a strong relationship between ALS metrics and many commonly used vegetation properties but the characterization of

CWD by ALS is considerably more challenging. The advantage of ALS data over field observations is usually the full area coverage but field measurements are still always needed as training data.

References

Aakala T (2010) Tree mortality and deadwood dynamics in late-successional boreal forests. Dissertationes Forestales 100. 41 p

Andersen H-E, Strunk J, Temesgen H, Atwood D, Winterberger K (2011) Using multilevel remote sensing and ground data to estimate forest biomass resources in remote regions: a case study in the boreal forests of interior Alaska. Can J Remote Sens 37:596–611

Asner GP, Powell GVN, Mascaro J, Knapp DE, Clark JK, Jacobsen J, Kennedy-Bowdoin T, Balaji A, Paez-Acosta G, Victoria E, Secada L, Valqui M, Hughes RF (2010) High-resolution forest carbon stocks and emissions in the Amazon. Proc Natl Acad Sci 107:16738–16742

Bater CW, Coops NC, Gergel SE, LeMay V, Collins D (2009) Estimation of standing dead tree class distributions in northwest coastal forests using lidar remote sensing. Can J For Res 39:1080–1091

Blanchard SD, Jakubowski MK, Kelly M (2011) Object-based image analysis of downed logs in disturbed forested landscapes using lidar. Remote Sens 2:2420–2439

Bradshaw CJA, Warkentin IG, Sodhi NS (2009) Urgent preservation of boreal carbon stocks and biodiversity. Trends Ecol Evol 24:541–554

Brin A, Bouget C, Valladares L, Brustel H (2013) Are stumps important for the conservation of saproxylic beetles in managed forests? – insights from a comparison of assemblages on logs and stumps in oak-dominated forests and pine plantations. Insect Conserv Divers 6:255–264

Bütler R, Schlaepfer R (2004) Spruce snag quantification by coupling colour infrared aerial photos and a GIS. For Ecol Manag 195:325–339

Crookston NL, Finley A (2008) yaImpute: an R package for kNN imputation. J Stat Softw 23:1–16

Ducey MJ, Williams MS, Gove JH, Roberge S, Kenning RS (2013) Distance-limited perpendicular distance sampling for coarse woody debris: theory and field results. Forestry 86:119–126

Eaton JM, Lawrence D (2006) Woody debris stocks and fluxes during succession in a dry tropical forest. For Ecol Manage 232:46–55

Ene LT, Næsset E, Gobakken T, Gregoire TG, Ståhl G, Holm S (2013) A simulation approach for accuracy assessment of two-phase post-stratified estimation in large-area LiDAR biomass surveys. Remote Sens Environ 133:210–224

Enrong Y, Xihua W, Jianjun H (2006) Concept and classification of coarse woody debris in forest ecosystems. Front Biol China 1:76–84

Eskelson BNI, Hagar JC, Temesgen H (2012) Estimation of snag density and snag quality attributes in western Washington and Oregon. For Ecol Manag 272:26–34

Esseen P-A, Ehnström B, Ericson L, Sjöberg K (1997) Boreal forests. Ecol Bull 46:16–47

Garczarek UM (2002) Classification rules in standardized partition spaces. Doctoral thesis, University of Dortmund, Germany. Available from http://hdl.handle.net/2003/2789

Gobakken T, Næsset E, Nelson R, Bollandsås OM, Gregoire TG, Ståhl G, Holm S, Ørka HO, Astrup R (2012) Estimating biomass in Hedmark County, Norway using national forest inventory field plots and airborne laser scanning. Remote Sens Environ 123:443–456

Gossner MM, Thibault L, Brunet J, Isacsson G, Bouget C, Brustel H, Brandl R, Weisser WW, Müller J (2013) Current near-to-nature forest management effects on functional trait composition of saproxylic beetles in beech forests. Conserv Biol 27:605–614

Gove JH, van Deusen PC (2011) On fixed-area sampling for downed coarse woody debris. Forestry 82:109–117

Gove JH, Ringvall A, Ståhl G, Ducey MJ (1999) Point relascope sampling of downed coarse woody debris. Can J For Res 29:1718–1726

Gove JH, Williams MS, Ståhl G, Ducey MJ (2005) Critical point relascope sampling for unbiased volume estimation of downed coarse woody debris. Forestry 78:417–431

Gove JH, Ducey MJ, Valentine HT, Williams MS (2013) A comprehensive comparison of perpendicular distance sampling methods for sampling downed coarse woody debris. Forestry 86:129–143

Gregoire TG, Ståhl G, Næsset E, Gobakken T, Nelson R, Holm S (2011) Model-assisted estimation of biomass in a lidar sample survey in Hedmark county, Norway. Can J For Res 41:83–95

Haara A, Nevalainen S (2002) Detection of dead or defoliated spruces using digital aerial data. For Ecol Manage 160:97–107

Harmon ME, Franklin JF, Swanson FJ, Sollins P, Gregory SV, Lattin JD, Andersson NH, Cline SP, Aumen NG, Sedell JR, Lienkaemper GW, Cromack K Jr, Cummins KW (1986) Ecology of coarse woody debris in temperate ecosystems. Adv Ecol Res 15:133–302

Hottola J, Siitonen J (2008) Significance of woodland key habitats for polypore diversity and red-listed species in boreal forests. Biodivers Conserv 17:2559–2577

Hyvönen R, Ågren GI (2001) Decomposer invasion rate, decomposer growth rate, and substrate chemical quality: how they influence soil organic matter turnover. Can J For Res 31:1594–1601

Ihalainen A, Mäkelä H (2009) Kuolleen puuston määrä Etelä- ja Pohjois-Suomessa 2004–2007. Metsätieteen aikakauskirja 1:35–56 (In Finnish)

Jonsson BG, Kruys N, Ranius T (2005) Ecology of species living on dead wood: lessons for dead wood management. Silva Fenn Monogr 39:289–309

Jordan GJ, Ducey MJ, Gove JH (2004) Comparing line-intersect, fixed-area and point relascope sampling for dead and downed coarse woody material in a managed northern hardwood forest. Can J For Res 34:1766–1775

Kangas A (2006) Sampling rare populations. In: Kangas A, Maltamo M (eds) Forest inventory. Methodology and applications, vol 10, Managing forest ecosystems. Springer, Dordrecht

Karjalainen L, Kuuluvainen T (2002) Amount and diversity of coarse woody debris within a boreal forest landscape dominated by Pinus sylvestris in Vienansalo wilderness, eastern Fennoscandia. Silva Fenn 36:147–167

Kim Y, Yang Z, Cohen WB, Pflugmacher D, Lauver CL, Vankat JL (2009) Distinguishing between live and dead standing tree biomass on the North Rim of Grand Canyon National Park, USA using small-footprint lidar data. Remote Sens Environ 113:2499–2510

Korpela I, Ørka H-O, Maltamo M, Tokola T, Hyyppä J (2010) Tree species classification in airborne LiDAR data: influence of stand and tree factors, intensity normalization and sensor type. Silva Fenn 44:319–339

Larsson JY, Hylen G (2007) Statistics of forest conditions and forest resources in Norway. Viten fra Skog og landskap 1/07. 91 p

Lebart L, Morineau A, Warwick KM (1984) Multivariate descriptive statistical analysis: corresponding analysis and related techniques for large matrices. Wiley, New York

Lonsdale D, Pautasso M, Holdenrieder O (2008) Wood-decaying fungi in the forest: conservation needs and management options. Eur J For Res 127:1–22

Magnussen S, Boudewyn P (1998) Derivation of stand heights from airborne laser scanner data with canopy-based quantile estimators. Can J For Res 28:1016–1031

Maltamo M, Bollandsås OM, Næsset E, Gobakken T, Packalén P (2011) Different sampling strategies for field training plots in ALS-assisted forest inventory. Forestry 84:23–31

Martinuzzi S, Vierling LA, Gould WA, Falkowski MJ, Evans JS, Hudak AT, Vierling KT (2009) Mapping snags and understory shrubs for a LiDAR-based assessment of wildlife habitat suitability. Remote Sens Environ 113:2533–2546

McRoberts RE, Tomppo E, Næsset E (2010) Advances and emerging issues in national forest inventories. Scand J For Res 25:368–381

McRoberts RE, Gobakken T, Næsset E (2012) Post-stratified estimation of forest area and growing stock volume using lidar-based stratifications. Remote Sens Environ 125:157–166

McRoberts RE, Næsset E, Gobakken T (2013) Inference for lidar-assisted estimation of forest growing. Remote Sens Environ 128:268–275

Mücke W, Hollaus M, Pfeifer N (2012) Identification of dead trees using small footprint full-waveform airborne laser scanning data. In: Proceedings of Silvilaser 2012, Vancouver

Müller J, Bütler R (2010) A review of habitat thresholds for dead wood: a baseline for management recommendations in European forests. Eur J For Res 129:981–992

Næsset E (2001) Effects of differential single- and dual-frequency GPS and GLONASS observations on point accuracy under forest canopies. Photogramm Eng Remote Sens 67:1021–1026

Næsset E (2002) Predicting forest stand characteristics with airborne scanning laser using a practical two-stage procedure and field data. Remote Sens Environ 80:88–99

Næsset E, Bollandsås OM, Gobakken T, Gregoire TG, Ståhl G (2013) Model-assisted estimation of change in forest biomass over an 11 year period in a sample survey supported by airborne LiDAR: a case study with post-stratification to provide "activity data". Remote Sens Environ 128:299–314

Newton AC (2007) Forest ecology and conservation: a handbook of techniques. Oxford University Press, Oxford

Pasher J, King DJ (2009) Mapping dead wood distribution in a temperate hardwood forest using high resolution airborne imagery. For Ecol Manage 258:1536–1548

Pesonen A (2011) Comparison of field inventory methods and use of airborne laser scanning for assessing coarse woody debris. Dissertationes Forestales 113. 56 p

Pesonen A, Maltamo M, Eerikäinen K, Packalén P (2008) Airborne laser scanning-based prediction of coarse woody debris volumes in a conservation area. For Ecol Manage 255:3288–3296

Pesonen A, Leino O, Maltamo M, Kangas A (2009) The comparison of field sampling methods and the use of airborne laser scanning as auxiliary information for assessing coarse woody debris. For Ecol Manage 257:1532–1541

Pesonen A, Kangas A, Maltamo M, Packalén P (2010a) The effects of auxiliary data source and inventory unit size on the efficiency of sample-based coarse woody debris inventory. For Ecol Manage 259:1890–1899

Pesonen A, Maltamo M, Kangas A (2010b) The comparison of airborne laser scanning-based probability layers as auxiliary information for assessing coarse woody debris. Int J Remote Sens 31:1245–1259

Ranius T, Jonsson BG, Kruys N (2004) Modeling dead wood in Fennoscandian old-growth forests dominated by Norway spruce. Can J For Res 34:1025–1034

Reitberger J, Schnörr Cl, Krzystek P, Stilla U (2009) 3D segmentation of single tree exploiting full waveform LIDAR data. ISPRS J Photogramm Remote Sens 64:561–574

Ringvall A, Ståhl G (1999) On the field performance of transect relascope sampling for assessing downed coarse woody debris. Scand J For Res 14:552–557

Rosenfield GH, Fitzpatrick-Lins K (1986) A coefficient of agreement as a measure of thematic classification accuracy. Photogramm Eng Remote Sens 52:223–227

Rouvinen S (2002) Amount, diversity and spatio-temporal availability of dead wood in old forests in boreal Fennoscandia. PhD dissertation, University of Joensuu, Faculty of Forestry

Samuelsson J, Gustafsson L, Ingelog T (1994) Dying and dead trees: a review of their importance for biodiversity. Swedish Threatened Species Unit, Uppsala

Seielstad CA, Queen LP (2003) Using airborne laser altimetry to determine fuel models for estimating fire behaviour. J For 101:10–15

Sherrill KR, Lefsky MA, Bradford JB, Ryan MG (2008) Forest structure estimation and pattern exploration from discrete-return lidar in subalpine forests of the central Rockies. Can J For Res 38:2081–2096

Siitonen J (2001) Forest management, coarse woody debris and saproxylic organisms: Fennoscandian boreal forests as an example. Ecol Bull 49:11–41

Sippola A-L, Siitonen J, Kallio R (1998) Amount and quality of coarse woody debris in natural and managed coniferous forests near the timberline in Finnish Lapland. Scand J For Res 13:204–214

Ståhl G (1998) Transect relascope sampling – a method for the quantification of coarse woody debris. For Sci 44:58–63

Ståhl G, Ringvall A, Lämås T (2000) Guided transect sampling for assessing sparse populations. For Sci 46:108–115

Ståhl G, Ringvall A, Fridman J (2001) Assessment of coarse woody debris – a methodological overview. Ecol Bull 49:57–70

Ståhl G, Gove JH, Williams MS, Ducey MJ (2010) Critical length sampling: a method to estimate the volume of downed coarse woody debris. Can J For Res 129:993–1000

Ståhl G, Holm S, Gregoire TG, Gobakken T, Næsset E, Nelson R (2011) Model-based inference for biomass estimation in a lidar sample survey in Hedmark County, Norway. Can J For Res 129:96–107

Stehman SV, Salzer D (2000) Estimating density from vegetation surveys employing unequal-area belt transects. Wetlands 20:512–519

Stokland JN, Siitonen J, Jonsson BG (2011) Biodiversity in dead wood. Ecology, biodiversity and conservation. Cambridge University Press, Cambridge

Temesgen H, Barrett T, Latta G (2008) Estimating cavity tree abundance using nearest neighbor imputation methods for western Oregon and Washington forests. Silva Fenn 42:337–354

Thompson SK (1990) Adaptive cluster sampling. J Am Stat Assoc 85:1050–1059

Uuttera J, Hyppänen H (1998) Determination of potential key-biotope areas in managed forests of Finland using existing inventory data and digital aerial photographs. For Landsc Res 1:415–429

Vaillancourt M, Drapeau P, Gauthier S, Robert M (2008) Availability of standing trees for large cavity-nesting birds in the eastern boreal forest of Québec, Canada. For Ecol Manage 255:2272–2285

Venables WN, Ripley BD (2002) Modern applied statistics with S, Statistics and computing. Springer, New York

Warren WG, Olsen PF (1964) A line intersect technique for assessing logging waste. For Sci 10:267–276

Williams MS, Gove JH (2003) Perpendicular distance sampling: an alternative method for sampling downed coarse woody debris. Can J For Res 33:1564–1579

Wing BM, Ritchie MW, Boston K, Cohen WB, Olsen MJ (2012) Individual snag detection using airborne lidar data and 3D local-area point-based intensity filtration. In: Proceedings of Silvilaser 2012, Vancouver

Woldendorp G, Keenan RJ, Barry S, Spencer RD (2004) Analysis of sampling methods for coarse woody debris. For Ecol Manage 198:133–148

Woodall CW, Rondeux J, Verkerk PJ, Ståhl G (2009) Estimating dead wood during national forest inventories: a review of inventory methodologies and suggestions for harmonization. Environ Manage 44:624–631

Wulder MA, White JC, Nelson R, Næsset R, Ørka HO, Coops NC, Hilker T, Bater CV, Gobakken T (2012) Lidar sampling for large-area forest characterization: a review. Remote Sens Environ 121:196–209

Yao W, Krzystek P, Heurich M (2012) Identifying standing dead trees in forest area based on 3D single tree detection from full waveform lidar data. ISPRS Ann Photogramm Remote Sens Spat Inf Sci I–7:359–364

Chapter 20
Estimation of Canopy Cover, Gap Fraction and Leaf Area Index with Airborne Laser Scanning

Lauri Korhonen and Felix Morsdorf

Abstract Forest canopy cover and gap fraction are commonly used metrics in forest ecology. Airborne laser scanning is capable of measuring both very accurately, but slightly different estimation methods should be used as these metrics are defined differently. In canopy cover estimation the proportion of vertical gaps between the crowns is needed for a specific area. Canopy gap fraction includes all gaps observed from a single point with some angular view range. Canopy cover can be estimated with high accuracy as the fraction of first echoes above a specified height threshold, because only the large gaps are considered. In gap fraction estimation also last echoes should be used so that the effect of the smaller gaps within the crowns is considered. Leaf area index can be estimated from the gap fraction using a logarithmic model with a single coefficient representing leaf orientation. However, sensor effects have a strong influence on the estimates, and therefore validation with high-quality field data is recommended.

20.1 Introduction

The main focus of forest inventories has traditionally been the monitoring of timber resources, but also the ecological aspects of forests have become increasingly important. The canopy acts as the functional interface between the forest and the atmosphere, and it also has an important role in the maintenance of biodiversity (Ozanne et al. 2003). Nowadays forest inventories also need to provide information on the canopy structure, but accurate measurements of the canopy from the ground

L. Korhonen (✉)
School of Forest Sciences, University of Eastern Finland, Joensuu, Finland
e-mail: lauri.korhonen@uef.fi

F. Morsdorf
Department of Geography, University of Zürich, Zürich, Switzerland

M. Maltamo et al. (eds.), *Forestry Applications of Airborne Laser Scanning: Concepts and Case Studies*, Managing Forest Ecosystems 27, DOI 10.1007/978-94-017-8663-8__20, © Springer Science+Business Media Dordrecht 2014

are laborious and expensive. The crowns are easier to observe from above, and therefore remote sensing is commonly used to measure canopy structure. Airborne laser scanners (ALS) that directly measure the distances to the tree crowns creating a 3D point cloud are especially useful for this purpose. Throughout this chapter we assume that the reader has a basic knowledge of ALS sensors and their use in forested environments (see Chap. 1).

Vertical canopy cover, canopy gap fraction and leaf area index are some of the most commonly used forest canopy characteristics. They are all related to the amount of gaps in the canopy and are highly correlated with each other. The nomenclature related to these metrics is not yet fully established (Korhonen 2011), so in this chapter we will first briefly go through the most commonly used definitions. We also discuss the field measurements that are needed to validate the remotely sensed estimates. The rest of the chapter deals with the estimation of vertical canopy cover, gap fraction and leaf area index using commercially available small footprint airborne laser scanners. Experimental or spaceborne sensors such as GLAS and LVIS are not discussed here.

Vertical canopy cover (VCC) is defined as the "the proportion of the forest floor covered by the vertical projection of the tree crowns" (Jennings et al. 1999). This definition was further extended by Gschwantner et al. (2009) in an article aimed at establishing common tree and forest definitions for European national forest inventories. They defined canopy cover through the crown projection area: Canopy cover is "the aggregation of the crown projection areas of individual trees (without double-counting of overlapping crown projection areas) divided by the stand area". The crown projection is defined as "outermost perimeter of the crown on the horizontal plane". This means that small gaps inside of the crown perimeter are considered to be a part of the crown. If the within-crown gaps are observed, the term "effective canopy cover" can be used (Rautiainen et al. 2005). The crown perimeter is fractal and hence there are no ways to unambiguously distinguish within-crown and between-crown gaps. However, a reasonably meaningful separation of the two types can be done in forests where the trees are straight, vertical and have a regular crown outline. In practical measurements the crown perimeter can only be delineated with an accuracy of about 10 cm. If unbiased estimates of VCC are needed, only vertical observations are allowed. Note that VCC is related to some specific area, which could be a sample plot, a previously delineated forest stand or a grid cell. In forest area mapping, the size of the reference area may affect the results; see Eysn et al. (2012) for a discussion of these effects.

Jennings et al. (1999) also defined another concept, the canopy closure, which is different from the canopy cover. It is defined as "the proportion of sky hemisphere obscured by vegetation when viewed from a single point". Thus canopy closure is related to a point (or a set of points) from which a larger area of the canopy is observed simultaneously, whereas canopy cover is related to a predefined area (Fig. 20.1). For instance, skyward-looking canopy photographs provide an estimate of the angular canopy closure (ACC). All gaps that are visible are usually observed in ACC measurements. The angle of view can be anything from 0° to 180° and therefore the observation angles should always be provided together with the

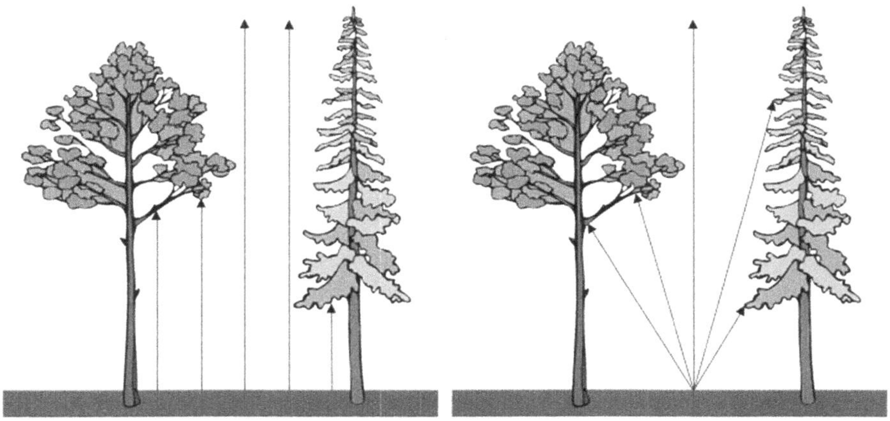

Fig. 20.1 Canopy cover (*left*) is measured in vertical direction and is defined for a specified area. Canopy closure (*right*) is measured in perspective projection and is unique to the measured point and view angle (Figure reprinted from Korhonen et al. 2006)

estimated ACC. Canopy gap fraction (CGF) is the antonym of canopy closure. Often it is easier to use CGF instead of ACC, as it is more commonly used in studies of canopy structure and easier to understand correctly.

VCC and ACC are highly correlated and can therefore be used to quantify the same properties. For instance, modeling of plant and animal habitats or forest regeneration are well known applications (e.g. Anderson et al. 1969; Stancioiu and O'Hara 2006). As the definitions indicate, ACC should have a better correlation with illumination and microclimate under the forest canopy, because it takes into account the radiation incoming from all directions (Alexander et al. 2013). Canopy cover is also used as a basis for the international definition of a forest (FAO 2004); therefore it is needed for instance in national forest inventories and REDD+ applications that focus on the monitoring of forest area (GOFC-GOLD 2009).

Leaf area index (LAI) is commonly defined as half of the total intercepting leaf area per unit ground surface (m^2/m^2) (Chen and Black 1992). For flat leaves this means the one-sided area, and for a conifer needles 50 % of the total surface. LAI is widely used to describe the canopy structure, and also one of the essential climate variables defined by the United Nations (UNFCCC 2003). LAI is commonly used as an explanatory variable in different biosphere-atmosphere interaction models and is strongly related to gross primary production (Running and Coughlan 1988; Turner et al. 2004). It is also used in quantification and monitoring of defoliation and insect damage (Solberg et al. 2006).

LAI can only be measured directly by using destructive sampling or litter traps, so in practice theoretical models are often used instead. If just basic forest inventory data are available, it is possible to use allometric equations to model leaf biomass and convert it to leaf area (Turner et al. 2000), but such models may be inaccurate and cannot be used to monitor canopy phenology or disturbance (Majasalmi et al.

2013). Thus, LAI is commonly obtained from optical measurements of the canopy gap fraction using the well-known Beer-Lambert law of transmission. When the canopy is assumed to consist of a homogenous layer of randomly positioned leaves, the probability that a ray of light penetrates it (i.e., canopy gap fraction) is exponentially dependent on the leaf area index (Eq. 20.1)

$$T_\theta = \exp\left(-G\left(\theta\right) \; LAI / \sin\left(\theta\right)\right), \tag{20.1}$$

where T is the transmission or canopy gap fraction, $G(\theta)$ is a function describing the leaf angle distribution, and θ is the view angle. The G-function is the unknown parameter, but it is simplified out of the equation if the gap fraction of the whole hemisphere is observed (Miller 1967) (Eq. 20.2)

$$LAI = 2 \int_0^{\pi/2} - \ln\left(T\left(\theta\right)\right) \cos\left(\theta\right) \sin\left(\theta\right) d\theta. \tag{20.2}$$

In practice the integral is approximated with the sum

$$LAI = 2 \sum_{i=1}^{n} - \ln\left(\overline{T_i}\right) \cos\left(\theta_i\right) w_i, \tag{20.3}$$

where the hemisphere is divided into n concentric rings (often $15°$), from which the ring-wise gap fractions $\overline{T_i}$ are derived, and w_i is the weight of the ring i:

$$w_i = \frac{\sin \theta_i}{\displaystyle\sum_{j=1}^{n} \sin \theta_j}. \tag{20.4}$$

In forests the underlying assumption of a homogeneous medium is not fulfilled, because the leaves are clumped into branches, branches into trees, and trees into stands. Furthermore, in conifer forests the needles are clumped into shoots in a very regular manner so that they shade each other more than if the orientation was random (Stenberg 1996). The woody components of the trees are also intercepting light and hence classified as if they were leaves. Thus, the LAI derived from optical measurements with Eq. 20.3 differs from the true LAI, and is more correctly called the effective LAI (LAI_e) or plant area index (PAI). In many applications the LAI_e can be used instead of the true LAI but if necessary the true LAI can be estimated from the LAI_e by using different correction methods, which have been developed both at shoot level for conifers (Chen 1996; Stenberg 1996; Majasalmi et al. 2013) and tree level (Pisek et al. 2011). If true LAI is estimated using these corrections, the gap fractions from individual measurements should be averaged before the LAI_e is calculated to avoid overcorrection (Ryu et al. 2010). The LAI that can be derived from ALS corresponds to LAI_e.

20.2 Validation of the Remotely-Sensed Canopy Metrics

20.2.1 Canopy Cover

Airborne laser scanning provides very detailed, wall-to-wall maps of the canopy within the area of interest. Obtaining a similar level of detail in the field is very laborious, but validating the accurate ALS measurements with field measurements that have a large uncertainty should be avoided. Hence special attention must be paid to the quality of the validation or modeling data. For instance, ocular estimates should not be used as validation, because their reliability is difficult to confirm.

Field-based estimation of VCC is primarily a sampling problem. An accurate estimate of VCC requires that the area of interest must be covered representatively with vertical measurements. Vertical sighting tubes have been commonly used for this kind of mapping. In the dot count method, a regular grid of sampling points is established within a sample plot with the help of a tape measure (Sarvas 1953). The sighting tube is used to check if there is a crown above the point or not, and the estimate is obtained as the average of the repeated 0/1 measurements. Estimating the accuracy of estimates from systematic samples can be problematic, but measuring 200 points per plot should provide very accurate results (Sarvas 1953; Jennings et al. 1999; Rautiainen et al. 2005). Alternatively, line intersect sampling can be used with the sighting tube. In this approach the length of line segments below the canopy is recorded (Korhonen et al. 2006; Gregoire and Valentine 2007).

It is also possible to measure the locations of the trees and map the dimensions of the crowns with the sighting tube (Lang and Kurvits 2007; Korhonen et al. 2011; Gatziolis 2012). The canopy cover can then be estimated from the map. A potential problem is that in reality the crowns are asymmetric and irregular, so a detailed map would require many more crown radius measurements than is practical, and assuming e.g. an elliptic crown shape between the measurement directions may simplify too much (Lang and Kurvits 2007; Korhonen et al. 2011). Dot count and line intersect sampling methods are free of such assumptions, so if just a canopy cover estimate is required they are easier to acquire and still provide a sufficient precision.

20.2.2 Canopy Closure, Gap Fraction and Leaf Area Index

Angular canopy closure and gap fraction are commonly estimated from canopy photography (e.g. Anderson 1964; Jonckheere et al. 2005) or with specific light sensors such as the LAI-2000/2200 plant canopy analyzer (Li-Cor Inc 1992). LAI-2000 is a quantum sensor that records the intensity of the incoming radiation at the blue region of the spectrum (320–490 nm) within five concentric annular rings (0°–75°). One sensor is used for the measurements in the forest while another is placed above the canopy or into a nearby open place. Afterwards the data from the two sensors are combined, and the gap fraction is obtained as the radiation difference between the two sensors. The underlying assumption is that the canopy elements are optically black, which is closest to reality in the blue part of the spectrum, as the leaves absorb blue light efficiently and the level of scattering is small.

The LAI values obtained with the LAI-2000 sensor have been perceived to have a good agreement with the true LAI (Chen et al. 2006), and they have also provided stronger correlations with ALS-based estimates than hemispherical images (Solberg et al. 2006, 2009). However, the instrument requires uniform sky conditions, i.e. overcast cloud cover or twilight. Rain or rapidly varying cloud cover make the measurements impossible, as well as direct sunlight, which can to some extent be avoided by using view restrictors. It is recommended that at least ten measurements per plot should be taken (Weiss et al. 2004; Majasalmi et al. 2012). When a large angle of view is used the sampling does not need to be as intensive as with the sighting tube, because the variance between the individual measurements is smaller.

Digital cameras and modern image processing methods have made photography a popular method for measuring the canopy structure. SLR (single lens reflex) cameras with hemispherical lenses can provide plenty of additional information of the surrounding canopy, such as gap size distribution (Pisek et al. 2011). Obtaining high quality gap fraction data from the images requires uniform sky conditions, as with the LAI-2000. Variable cloud cover can be corrected for, however, this makes the image analysis more complicated. Special attention must be paid to the exposure settings of the camera: in dense canopies the automatic settings will expose the image too much, which leads to overestimation of gap fraction (Zhang et al. 2005; Pueschel et al. 2012). Often the case is that near the zenith the image is overexposed and near the horizon underexposed, which complicates the image processing and may lead to errors in the gap fraction data. Some cameras can shoot several exposures and merge them into one evenly exposed image (auto HDR mode), which could reduce or remove this problem.

The first step in the analysis of a canopy photograph is usually thresholding, i.e. classifying the pixels as either canopy or sky. Gap fraction is obtained as the proportion of sky pixels within the desired zenith angle range. The blue RGB band is usually used for the same reasons as with the LAI-2000. However, the blue band is also the most sensitive to saturation and overexposure, so in such cases the red band may provide better results (Pueschel et al. 2012). The selection of the thresholding method is not straightforward. Earlier manual thresholding was commonly applied, but nowadays automatic algorithms are preferred as they provide more consistent results. Several papers have studied different thresholding algorithms (e.g. Jonckheere et al. 2005; Pueschel et al. 2012) and the results show that the exposure and the thresholding method have a strong effect on which algorithm provides results closest to LAI-2000 (Korhonen et al. 2011; Pueschel et al. 2012).

The only way of confirming the quality of the image-based estimates is a comparison with LAI-2000 or destructive measurements. Standardization of image acquisition and analysis so that the result is comparable to LAI-2000 is an ongoing research effort. Pueschel et al. (2012) obtained a good agreement by using an Isodata thresholding algorithm with a stack of differently exposed images. Lang et al. (2010) estimated the sky brightness distribution from the canopy gaps, after which the analysis could be made based on the brightness differences as with LAI-2000. This way the problem of finding the best threshold could be avoided. These methods are promising but further testing is needed.

Finally, if cameras with hemispherical lenses are not available, standard "point-and-shoot" cameras can also be used. They have a limited angle of view, but can be used to map the near-vertical ACC quickly as a tripod is not necessarily needed. They can also be used to estimate VCC, if the angle of view is kept smaller than 30°, within-crown gaps are excluded and small biases in the estimates are acceptable (Korhonen and Heikkinen 2009). If the point-and-shoot camera is placed at an angle of 57° from the zenith, an unbiased estimate of LAI can also be obtained (Lang 1987). Analysis methods similar to hemispherical images can be used.

20.2.3   Terrestrial Laser Scanning

Terrestrial laser scanners (TLS) are a relatively new class of instruments that are especially well suited for detailed canopy structure measurements and validation of remotely-sensed metrics. A TLS instrument is typically mounted on a tripod, and it creates a 3D point cloud of the surrounding vegetation (see Chap. 2). There are two types of TLSs: continuous wave and time-of-flight systems. Time-of-flight TLS is principally similar to the typical ALS sensors, whereas a continuous wave TLS scans the surroundings with a single beam and measures the distances based on the phase of the returning signal. In general the continuous wave sensors are less expensive and faster to use, but the return signal is noisy and therefore it is more difficult to measure the canopy gaps (Newnham et al. 2012). Thus, time-of-flight sensors are usually preferred in canopy measurements. The most recent systems developed for canopy measurements can digitize full waveforms and/or use two wavelengths (Zhao et al. 2011; Danson et al. 2012).

Compared with passive optical sensors, the TLS has several advantages. The information contained in the point cloud enables, for instance, separation of wood and foliage echoes and better estimation of clumping effects (Zhao et al. 2011; Danson et al. 2012). Also a TLS is insensitive to sky and weather conditions as long as rain and fog are avoided. The problems with the TLS are mainly technical ones – the data files can be huge and their processing requires powerful computers and algorithms. Also the price of the sensors is still relatively high.

Validation of TLS-based forest characteristics is currently an active area of research. Zhao et al. (2011) compared estimates from an Echidna TLS with several optically based LAI estimation techniques and found no statistically significant differences, but the dispersion between the LAI-2000 and the two TLS-based estimates was relatively large ($R^2 = 0.39$ and 0.42). Danson et al. (2012) found that the TLS-based LAI had a strong correlation with image-based LAI, but the estimates were too high when both woody and foliage elements were taken into account. Korhonen et al. (2010) used a continuous wave TLS to estimate VCC. The results showed that the TLS-based estimates were considerably smaller than the VCCs estimated with the sighting tube. The reason was that although 4–8 individual scans per plot were merged into one point cloud, large parts of the canopy were missing from the canopy map because of occlusion.

TLS applications are becoming more and more popular in the field of canopy research, but more work is still needed to quantify the effects of different sensor settings on the estimated canopy metrics. Although the information content in the TLS data is much larger than can be obtained with traditional methods, the traditional methods can still provide adequate information regarding the simple canopy metrics such as canopy cover and gap fraction. If more complex information, such as woody area index, clumping-corrected LAI or vertical distribution of foliage is also needed, TLS sensors are capable of providing that.

20.3 Estimation Using Airborne Laser Scanning

20.3.1 Canopy Cover

As was mentioned earlier, in VCC estimation only the vertical between-crown gaps should be observed, and the whole reference area should be sampled representatively. The point cloud derived from ALS is very well suited for this kind of mapping: the plot is covered wall-to-wall with nearly vertically emitted pulses that reliably record the coordinates of the first echo. Thus, it is easy to provide an estimate of canopy cover that is similar to the field-based dot count method: all pulses that provide an echo above a predefined height are classified as canopy observations, and the rest as ground. If the discrete return ALS data is classified into single, first-of-many, intermediate, and last-of-many echoes, this index can be written as

$$\mathrm{FCI} = \frac{\sum \mathrm{Single}_{\mathrm{canopy}} + \sum \mathrm{First}_{\mathrm{canopy}}}{\sum \mathrm{Single}_{\mathrm{All}} + \sum \mathrm{First}_{\mathrm{All}}}. \tag{20.5}$$

The FCI here means "first echo cover index"; also terms such as "vegetation index" or simply "canopy cover" can be used. Note that only the surface of the canopy needs to be observed, as the small gaps within the crowns are considered to be part of the canopy. Thus, it is sufficient to use only the first echoes (Korhonen et al. 2011). The height limit of canopy echoes should be such that shrubs and other understory plants do not usually exceed it. Thresholds between 1 and 2 m are commonly used, as the field data is commonly obtained from a height level corresponding to breast height (Smith et al. 2009; McLane et al. 2009).

There are just a few studies where FCI has been compared with reliable field-based estimates of VCC. Holmgren et al. (2008) found that the FCIs were significantly larger than the VCCs derived from field measurements of crown radii. However, a regression equation VCC = 0.77 FCI yielded a RMSE of 4.9 %. Korhonen et al. (2011) noted that the FCI estimates were very similar to the estimates obtained with the sighting tube, but usually the VCC was overestimated by 3–5 %. The scale of the overestimation had a strong correlation with the scan zenith angle: the oblique pulses have a larger probability of hitting a crown than the vertical ones. This result was in line with the simulation study by Holmgren et al. (2003). Also the ALS-derived maximum tree height affected the results: longer crowns have a larger likelihood of intercepting oblique pulses. Another issue likely to affect first echo based VCC estimates is the beam divergence and subsequent footprint size of the ALS instrument. Even if within crown gaps are considered to be part of the canopy, the size of the crowns might be overestimated, as edges of the crowns can trigger first echoes even if the footprint is not fully overlapping with the crown. This effect will obviously be more pronounced for larger footprint diameters ($\sim$1 m) and might not be noticeable when using smaller diameters (e.g. <0.5 m).

If the scan zenith angle is retained in the ALS data, it can be used to normalize the bias caused by the non-zero scan zenith angle. Korhonen et al. (2011) proposed the following correction models:

$$VCC = FCI - 0.6233 \times \theta_{scan}, \tag{20.6}$$

$$VCC = FCI - 0.0253 \times \theta_{scan} \times F_{max}, \tag{20.7}$$

where θ_{scan} is the mean scan angle (degrees) and F_{max} is the height of the highest echo above the digital terrain model (m). Equation 20.6 indicates that for instance $\theta_{scan} = 10°$ equals 6.2 % overestimation. Note that if the plot has echoes from several ALS strips, only the closest one should be used. According to the binomial theorem, the reliability of the FCI-based VCC estimate can be evaluated based on the number of samples (echoes). Thus, even low density ALS data give a reliable estimate as long as the number of echoes per plot is sufficient.

Different raster maps of canopy cover have also been used in canopy cover estimation. Canopy mapping requires higher density laser data (>1 pulse/m^2) than the simple FCI calculations. Canopy height models (CHMs) are commonly created by assigning the value of the highest echo within a cell to the cell (Hyyppä et al. 2001). In the VCC estimation the height threshold is applied to the CHM similar to the FCI index methods. Assuming that the crowns are evenly covered from all sides and that the XY coordinates of the echoes are accurate, the CHM-based VCC estimates are free from the bias caused by the oblique scanning. However, the selection of pixel size is critical, as a single canopy echo located at the edge will classify the entire pixel as canopy. With the commonly used 0.5–1.0 m pixel sizes this effect results in a significant overestimation of the VCC (Lee and Lucas 2007; Korhonen et al. 2011).

Fig. 20.2 A canopy map derived from ALS data using morphological image analysis (Image reprinted from Korhonen 2011)

Nevertheless, Gatziolis (2012) reported that no errors larger than 3 % were obtained in a comparison of VCCs derived from field-based canopy map and ALS-based canopy map. He used dense (8 echoes per m^2, all echoes considered) ALS data while the resolution was set equal to the mean distance between the echoes, and cells without echoes were ignored in the calculation. Korhonen et al. (2011) used morphological opening and closing operations to create canopy maps with 0.1 m resolution (Fig. 20.2). Four different high density data sets were examined, and in three of them the RMSE was better than with the simple FCI. However, even better results were obtained by data decimation: using a 1–1.5 m grid size, one (first or single) echo was selected randomly from each cell, and the FCI was then calculated based on the selected echoes. This way the concentration of echoes to the crowns could be reduced, and for the cells located at the edges of the crowns the probability of selecting a canopy echo was related to cover percent within the cell.

As a summary, VCC estimation can be done fairly reliably using low density ALS data. The ALS sensors can capture both the canopy surface and the ground elevation (if the foliage is not extremely dense) with reasonably good precision, after which the canopy cover can be estimated. The simple fraction of pulses that produced canopy echoes slightly overestimates the VCC, but the scan angle information or grid-based approaches can be used to decrease the bias. Considering how laborious accurate field measurements of VCC are, they should only be collected for validation of the ALS-based estimates. In some applications the precision of ALS-based VCC estimates may be sufficient even without field validation.

20.3.2   Angular Canopy Closure and Gap Fraction

Canopy gap fraction and ACC are more difficult to map with ALS than the VCC. First, the difference in view geometry is much larger. In ALS applications the maximum scan zenith angle is usually around 15°, while the commonly used LAI-2000 sensor maps the canopy gap fraction at zenith angles of 0°–75°. Secondly, all gaps within the angular range should be observed. Hence just mapping the outer surface of the canopy will result in overestimation of the ACC (Lovell et al. 2003; Morsdorf et al. 2006). Thus, the information in the pulses that partially penetrated the canopy must also be utilized.

The ALS sensors can in theory be used to map the near-vertical CGF only if the observed angular range is similar to the scan zenith angle, which usually corresponds to the first detector ring of the LAI-2000 sensor (0°–15°). The discrete-return ALS sensors commonly used in forestry have a footprint diameter of 10–50 cm at the ground level, and can usually record 1–4 echoes per pulse. If the pulse hits dense foliage, e.g. at the top of the tree crown, usually only a single echo is recorded, i.e. there is no penetration. If the foliage is sparse, more echoes may be recorded from the lower parts of the canopy or the ground. It has been suggested that the ALS footprint size and peak pulse power affect the pulse penetration (Hopkinson 2007), but the exact mechanisms how the scanners digitize echoes from the returning waveform are not known to the users. Thus, it is difficult to assess which cover value should be assigned to pulses that produced both canopy and ground echoes, although there have been some attempts to study it (Korpela et al. 2013). Some information is included in the intensity, but its utilization is not straightforward (see below).

Near-vertical gap fraction can again be estimated based on the proportion of pulses that totally or partially penetrated the canopy, i.e. produced echoes under the predefined height threshold. The selection of the threshold is somewhat arbitrary, but as the reference measurements (LAI-2000, hemispherical photography) are often made at the breast height, 1.3 or 2.0 m height thresholds are commonly applied. Solberg et al. (2009) estimated near-vertical canopy closure using the following index, labeled here as SCI ("Solberg's cover index") (Eq. 20.8):

$$SCI = \frac{\sum Single_{canopy} + 0.5\left(\sum First_{canopy} + \sum Last_{canopy}\right)}{\sum Single_{All} + 0.5\left(\sum First_{All} + \sum Last_{All}\right)}. \tag{20.8}$$

This cover index has an intuitive interpretation. Single echoes have either 100 % or 0 % cover. Pulses that produced a first echo in the canopy are assigned 50 % cover if the last echo came from the ground, or 100 % cover if the last echo also came from the canopy. The constant 0.5 for the partially penetrating echoes is an arbitrary approximation and is strongly dependent on the scanner (Solberg et al. 2009). In their original paper Solberg et al. (2009) found that the gap fraction was overestimated, i.e. partially penetrating pulses should have been assigned higher cover values. Korhonen et al. (2011) used the same index and found that 0.5 was close to correct; the best absolute RMSE for the near-vertical gap fraction was 8.1 % and bias 1.0 %.

An index based on calculating all echo types above the height threshold (ACI) was used by Morsdorf et al. (2006) and Richardson et al. (2009):

$$ACI = \frac{\sum All_{canopy}}{\sum All}$$

$$= \frac{\sum Single_{canopy} + First_{canopy} + Intermediate_{canopy} + Last_{canopy}}{\sum All}. \quad (20.9)$$

ACI is similar to the FCI except that all echoes are observed. Thus, the estimated cover is smaller because the fraction of ground echoes increases when the last echoes are also included. In practice ACI and SCI are close to each other; when both the SCI and ACI were calculated from the same data, the absolute RMSD between them was 2–3 % (unpublished data from the authors).

A different approach to gap fraction estimation was tested by Hopkinson and Chasmer (2009). They used the intensity information to determine the gap fraction, obtaining an $R^2 = 0.75$ while the model slope was close to unity and intercept close to zero. The intensity cover index (ICI) that they used takes into account the transmission losses in the canopy:

$$ICI = 1 - \frac{\left(\frac{\sum I_{ground\ single}}{\sum I_{total}}\right) + \sqrt{\frac{\sum I_{ground\ last}}{\sum I_{total}}}}{\left(\frac{\sum I_{first} + \sum I_{single}}{\sum I_{total}}\right) + \sqrt{\frac{\sum I_{intermediate} + \sum I_{last}}{\sum I_{total}}}}, \quad (20.10)$$

where I is the intensity of the echo. However, the use of the intensity information in practical applications is not easy because of the varying sensor effects. The intensity values are affected e.g. by the distance from the sensor to the target and automatic gain control, which is applied in some topographic discrete-return ALS sensors to increase the number of ground echoes (Korpela et al. 2010). Sensor effects are discussed in detail in Sect. 20.3.4.

Finally, the indices discussed above have a linear relationship with the near-vertical gap fraction or canopy closure (Hopkinson and Chasmer 2009; Korhonen et al. 2011). However, the canopy closure or gap fraction is commonly estimated using 150° angle of view (zenith angles 0°–75°). The increasing angular range means that usually also the closure increases, as a smaller proportion of the large between-crown gaps is visible. The effect is large in forests where the near-vertical ACC is close to 50 %, but not so pronounced if the closure is close to 0 % or 100 %. Thus, the relationship between the near-vertical ALS-based indices and the large-angle canopy closure becomes curvilinear, and e.g. exponential models are needed for the modeling (Korhonen et al. 2011). Also the optimal radius for intersecting the ALS echoes is in this case more difficult to determine, as it depends on the angle of view and tree height. Usually several radii should be compared to find the best alternative (Lovell et al. 2003; Morsdorf et al. 2006; Solberg et al. 2006).

20.3.3 Leaf Area Index

Leaf area index can be estimated from the ALS data using either univariate or multivariate regression. As the optical LAI_e is directly related to the gap fraction, the various cover indices described above are useful as predictors in the LAI models (e.g. Morsdorf et al. 2006; Solberg et al. 2006; Richardson et al. 2009). The basic model shape proposed by Solberg et al. (2006) is based on the exponential relationship between the CGF and LAI_e:

$$LAI_e = \beta \left(-\ln \left(CGF \right) \right), \qquad (20.11)$$

where CGF is the near-vertical gap fraction and β is the coefficient that is estimated in the regression. If written $\beta = 1/k$, the k is commonly called the extinction coefficient (Richardson et al. 2009). The CGF does not need to be unbiased because the regression will ensure that the estimated LAI_e is unbiased. Thus, any of the cover indices described above can be used as a proxy for CGF. However, if the CGF is close to the field-measured gap fraction, the β has a physically based interpretation: it is related to the leaf orientation within the scattering medium, i.e. the mean projection of unit foliage area. If the leaf orientation is random, the value of the β is 2. If $\beta > 2$ the leaf orientation is vertical, i.e. there are more gaps near the zenith than near the horizon, and similarly $\beta < 2$ means horizontal leaf orientation. Richardson et al. (2009) compared four different methods for estimating the LAI, and this approach provided the best results.

The values of β have been studied with ALS most intensively in boreal forests (Solberg et al. 2009; Solberg 2010; Korhonen et al. 2011). In these studies the estimates of β have ranged from 2.33 to 2.71, indicating erectophile foliage angle distribution. This result is typical for boreal forests that commonly have large vertical gaps between the long and relatively narrow crowns. Richardson et al.

(2009) used the ACI index in a heterogeneous urban forest, and found that β was close to 2, so in their study area the assumption of a random foliage angle distribution was valid.

If the CGF could be estimated reliably without field data, also the LAI could be obtained assuming that the β is known for structurally similar forests. For instance, Korhonen et al. (2011) proposed that in boreal forests $\beta \approx 2.6$. Note that the β can also be estimated e.g. from the hemispherical images without linking them with ALS data, and for some areas β may be available in the literature. Nevertheless, the safest way is still to estimate β directly from the ALS data using well measured and positioned field data.

As was discussed in 20.3.2, the SCI and ACI indices are more sensitive to scanner effects than the FCI. Therefore the FCI-based LAI estimates could be more reliable if no field data are available for calibration. Based on the results from Norway and Finland, Korhonen et al. (2011) proposed that in boreal forests $\beta \approx 1.6$ could be used in the LAI_e estimation with the FCI index.

Morsdorf et al. (2006) used a proxy for LAI_e that was based on the ratio of particular echo types. The idea for this proxy was based on the observation that the amount of single and last echoes within the canopy was higher for areas with higher (field measured) LAI. Since only first and last echo data were available at the time of the study, they used the ratio of single and last echoes to all echoes within the canopy. Canopy echoes were discriminated from ground returns using a height threshold as in the VCC computations. A correlation of this proxy with field estimates of LAI showed moderately strong agreement, with an R^2 of 0.69. In theory, the computation of this LAI_e proxy should be independent of vegetation cover, as its computation is only based on vegetation echoes, making it a canopy (tree) based estimate. In order to obtain plot level estimates of LAI_e, this proxy needs to be multiplied with the VCC of the particular stand (Morsdorf et al. 2006).

The LAI can also be estimated using the area based method just like any other variable of interest (Riaño et al. 2004; Jensen et al. 2008; Peduzzi et al. 2012). The use of multiple predictors may increase the accuracy of the estimates, but the model interpretation becomes more difficult. For instance, Peduzzi et al. (2012) obtained an R^2 of 0.69 using four predictor variables in a temperate mixed forest. Multiple predictor approach also allows inclusion of data from several sensors, which can sometimes improve the models. Peduzzi et al. (2012) found that adding GeoSAR X-band metrics improved the multivariate LAI models, especially in hardwood forests. However, significant improvements have not been observed by including optical data into the models (Jensen et al. 2008; Zhao and Popescu 2009).

20.3.4 Sensor Effects and Multitemporal Data Acquisition

The cover indices may be significantly influenced by the sensor specific differences. If sensor settings and survey configurations do not change, plot level derived vegetation metrics (such as height and cover) are stable between repeated acquisitions,

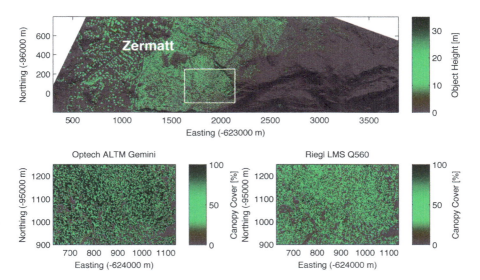

Fig. 20.3 ACI vegetation cover (using all echoes) for a coniferous forest in the area of Zermatt, southern Switzerland (46° 1′ N, 7°45′E). The axes are in Swiss national coordinates. The two acquisitions using the both sensors (Optech ALTM GEMINI and RIEGL LMS Q560) were 2 weeks apart, with likely no actual change of vegetation cover in between. However, the average ACI for the Optech (*left image*) is 64 %, while it is 50 % for the RIEGL instrument (*right image*)

unless the number of echoes per plot is very small (Bater et al. 2011). However, if different sensors are used or if the survey configuration is changed, the ALS metrics might be impacted significantly and it is not always clear which setting differences actually cause changes. Earlier studies from Norway have shown that the canopy density variables derived from the last echo data are typically more sensitive to changes in survey settings than those obtained by using only first echoes (Næsset 2004, 2009; Solberg et al. 2010). Here we report the results from two additional tests in Switzerland and Finland, where the same area was scanned twice within a small time interval.

Figure 20.3 shows two maps of ACI (Eq. 20.9), a vegetation cover estimate based on all ALS echoes. The first survey of the coniferous forest was made with an Optech ALTM GEMINI, and the second one with a RIEGL LMS Q 560. All surveying configurations were the same (e.g. flying altitude, point density, max. scan angle), but still the mean estimates of ACI are different by 14 % (64 % for the Optech compared to 50 % for the Riegl). This difference may have been caused by differences in the systems such as: (i) wavelength (1,064 nm with Optech; 1,550 nm with Riegl), (ii) echo detection method (the Optech was used in discrete return mode, the Riegl is a full-waveform scanner) or (iii) differences in beam divergence/footprint size. The observable differences might also vary between different forest ecosystems.

Table 20.1 Differences
between scans made in 2007
and 2006 for three cover
indices

Cover index	RMS D.	Average D.	Min D.	Max D.
FCI, %	3.9	1.7	−5.5	9.4
SCI, %	4.2	3.2	−3.2	10.0
ACI, %	3.7	2.9	−2.6	8.3

The data are from 60 plots in Hyytiälä, Finland

On the other hand, in Finland the differences between two discrete-return scanners were relatively small. Table 20.1 displays the differences in the three cover indices between two scans with 1 year interval. The first scan was made 25 July 2006 using an Optech ALTM 3100, having a 1,064 nm wavelength, max. 14° scan zenith angle, 25 cm footprint and 6.4 pulses m^{-2}. The second scan was made with a 1,064 nm Leica ALS 50-II on 4 July 2007, with scan parameters being 15°, 12 cm, and 10.3 m^{-2}, respectively. A total of 62 circular 20-m radius plots were available, but two of them were removed as outliers: the first was treated between the scans and the second was a rapidly growing seedling stand. The absolute root mean squared differences between the two scans were only 3.7–4.2 % depending on the cover index, with maximum errors no larger than 10 %. On average all indices indicated larger coverage in 2007, which is logical because of the growth and a slightly larger scan angle. However, the average difference was larger for the SCI and ACI than for the FCI, which could mean that the ALTM 3100 registered slightly more last echoes from the ground than the ALS50-II.

The utilization of intensity and waveform data is even more dependent on the scanner settings. The largest problem in exploiting ALS based intensities is the ambiguity of intercepted canopy area (elements) and the optical properties of these canopy elements. Even full-waveform systems can measure only the so-called cross-section, which is a product of intersected (leaf) area and canopy reflectance (Wagner 2010). This problem is not solvable, it can only be mitigated by the assumption that the reflectance of canopy elements does not change (e.g. vertically or within species) and thus the observed intensity differences are not changed by leaf reflectance, but only by intercepted area. Then the intensity could be exploited to derive gap fraction estimates.

Another major issue with this approach is the inability of ALS systems to resolve spatial information within a footprint. In vegetation, the return signal is represented in the data by one or several echoes, with the intensity of each echo depending on the intersected area and the reflectance of canopy elements (see Chap. 2). Thus, the energy distribution among echoes of a single pulse is more or less randomly based on the amount of intersected canopy material for each shot. In addition, the available energy for reflection is reduced by each previous interaction of the pulse with the canopy, so that less energy is available for a second echo than for a first one. Korpela et al. (2012) tried to correct multiple-echo ALS based intensity for transmission losses using detailed canopy reconstructions and ray-tracing, but ultimately failed to do so. According to them, the main issues are transmission losses, which were not recorded by the discrete-return instrument. It may well be that there are canopy-pulse interactions, which provide too little energy back to the sensor, so that the energy level is below the detection threshold of the system. Thus, so far mainly single echo intensity data is used in ALS based vegetation studies.

It is expected that full-waveform systems could potentially allow for the correction of transmission losses of multiple echo intensity data, since they record a better representation of the backscattered signal than discrete return systems. However, further studies are needed to provide validation of this hypothesis. In addition, it is very likely that the beam-divergence, small-scale canopy structure (e.g. leaf orientation) and the choice of wavelength impact the correction of transmission losses. For instance, the amount of multiple scattering is different at 1,560 nm compared to 1,064 nm.

These sensor specific effects pose a large problem to multi-temporal studies, as they might affect the comparability of results. As the forward propagation of sensor effects into biophysical products is challenging, currently the most practical way to obtain transferable results is to calibrate the measurements with independent field data for each of the acquisitions.

20.4   Conclusion

ALS data are very useful for estimation of the canopy structural variables. Several methods have been developed for the estimation of canopy cover, canopy gap fraction and leaf area index, and high correlations are consistently achieved. For these variables, ALS will usually produce better estimates than optical imagery or radars, but ALS data are more expensive and difficult to obtain in many parts of the world. Some studies indicate that even if the ALS data were collected for other purposes and no validation data for canopy structure data are available, at least canopy cover can still be estimated fairly reliably.

The canopy metrics derived from ALS are strongly affected by the sensor and operational characteristics, such as the scan zenith angle and altitude, beam divergence, wavelength, automatic gain control and echo detection method. Therefore it is recommended to always obtain high-quality field measurements for calibrating the ALS-based estimates. Data from ALS sensors are capable of estimating the canopy structure wall-to-wall very accurately, and therefore also the validation measurements must be made with similar level of precision using comprehensive sampling. The field work is laborious and mensuration errors do happen, so one aim for future research is to reduce the amount of field work. It might be possible to develop new calibration methods that are based on a combination of field trials (e.g. different sensor systems on the same site), laboratory calibrations and physical modeling. However, the requirement is that system providers and surveying companies provide the instrument specifications and survey settings, and that the handling of these is documented throughout the processing chain.

Acknowledgements This study was supported by the strategic funding of the University of Eastern Finland. Ilkka Korpela provided the data for Table 20.1. We are thankful to Valerie Thomas, Svein Solberg, and Erik Næsset for their useful comments.

References

Alexander C, Moeslund JE, Bøcher PK, Arge L, Svenning JC (2013) Airborne laser scanner (LiDAR) proxies for understory light conditions. Remote Sens Environ 134:152–161

Anderson MC (1964) Studies of the woodland light climate: I. The photographic computation of light conditions. J Ecol 52:27–41

Anderson RC, Loucks OL, Swain AM (1969) Herbaceous response to canopy cover, light intensity, and throughfall precipitation in coniferous forests. Ecology 50:255–263

Bater C, Wulder M, Coops N, Nelson R, Hilker T, Næsset E (2011) Stability of sample-based scanning-LiDAR-derived vegetation metrics for forest monitoring. IEEE Trans Geosci Remote Sens 49:2385–2392

Chen JM (1996) Optically-based methods for measuring seasonal variation of leaf area index in boreal conifer stands. Agric For Meteorol 80:135–163

Chen JM, Black TA (1992) Defining leaf area index for non-flat leaves. Plant Cell Environ 15:421–429

Chen JM, Govind A, Sonnentag O, Zhang Y, Barr A, Amiro B (2006) Leaf area index measurements at Fluxnet-Canada forest sites. Agric For Meteorol 140:257–268

Danson FM, Armitage RP, Gaulton R, Gunawan OT, Ramirez FA (2012) Field trials of a full-waveform terrestrial laser scanner to measure forest canopy dynamics. Abstract. In: Proceedings of Silvilaser 2012, Vancouver, Canada

Eysn L, Hollaus M, Schadauer K, Pfeifer N (2012) Forest delineation based on airborne LIDAR data. Remote Sens 4:762–783

FAO (2004) Global forest resources assessment update 2005. Terms and definitions (Final version). Forest resources assessment programme working paper 83/E, Rome, Italy

Gatziolis D (2012) Comparison of lidar- and photointerpretation-based estimates of canopy cover. In: McWilliams W, Roesch FA (eds) Monitoring across borders: 2010 joint meeting of the Forest Inventory and Analysis (FIA) symposium and the southern mensurationists. e-Gen. Technical report SRS-157. U.S. Department of Agriculture, Forest Service, Southern Research Station, Asheville, NC, pp 231–235

GOFC-GOLD (2009) Reducing greenhouse gas emissions from deforestation and degradation in developing countries: a sourcebook of methods and procedures for monitoring, measuring and reporting, GOFC-GOLD Report version COP14-2, GOFC-GOLD Project Office, Natural Resources Canada, Alberta, Canada

Gregoire T, Valentine H (2007) Sampling strategies for natural resources and the environment, vol 3, Applied environmental statistics. Chapman & Hall/CRC, Boca Raton, 474 p

Gschwantner T, Schadauer K, Vidal C, Lanz A, Tomppo E, di Cosmo L, Robert N, Duursma DE, Lawrence M (2009) Common tree definitions for national forest inventories in Europe. Silva Fenn 43:303–321

Holmgren J, Nilsson M, Olsson H (2003) Simulating the effects of lidar scanning angle for estimation of mean tree height and canopy closure. Can J Remote Sens 29:623–632

Holmgren J, Johansson F, Olofsson K, Olsson H, Glimskär A (2008) Estimation of crown coverage using airborne laser scanner. In: Hill RA, Rosette J, Suárez J (eds) Proceedings of SilviLaser 2008: 8th international conference on LiDAR applications in forest assessment and inventory, 17–19 September. Heriot-Watt University, Edinburgh, pp 50–57

Hopkinson C (2007) The influence of flying altitude, beam divergence, and pulse repetition frequency on laser pulse return intensity and canopy frequency distribution. Can J Remote Sens 33:312–324

Hopkinson C, Chasmer L (2009) Testing LiDAR models of fractional cover across multiple forest ecotones. Remote Sens Environ 113:275–288

Hyyppä J, Kelle O, Lehikoinen M, Inkinen M (2001) A segmentation-based method to retrieve stem volume estimates from 3-dimensional tree height models produced by laser scanner. IEEE Trans Geosci Remote Sens 39:969–975

Jennings SB, Brown ND, Sheil D (1999) Assessing forest canopies and understorey illumination: canopy closure, canopy cover and other measures. Forestry 72:59–74

Jensen JJ, Humes KS, Vierling LA, Hudak AT (2008) Discrete return lidar-based prediction of leaf area index in two conifer forests. Remote Sens Environ 112:3947–3957

Jonckheere I, Nackaerts K, Muys B, Coppin P (2005) Assessment of automatic gap fraction estimation of forests from digital hemispherical photography. Agric For Meteorol 132:96–114

Korhonen L (2011) Estimation of boreal forest canopy cover with ground measurements, statistical models and remote sensing. Dissertationes Forestales, 56 p

Korhonen L, Heikkinen J (2009) Automated analysis of in situ canopy images for the estimation of forest canopy cover. For Sci 55:323–334

Korhonen L, Korhonen KT, Rautiainen M, Stenberg P (2006) Estimation of forest canopy cover: a comparison of field measurement techniques. Silva Fenn 40:577–588

Korhonen L, Kaartinen H, Kukko A, Solberg S, Astrup R (2010) Estimating vertical canopy cover with terrestrial and airborne laser scanning. In: Proceedings of Silvilaser 2010, 14–17 September, Freiburg, Germany

Korhonen L, Korpela I, Heiskanen J, Maltamo M (2011) Airborne discrete-return LiDAR data in the estimation of vertical canopy cover, angular canopy closure and leaf area index. Remote Sens Environ 115:1065–1080

Korpela I, Ørka HO, Heikkinen V, Tokola T, Hyyppä J (2010) Range- and AGC normalization of LIDAR intensity data for vegetation classification. ISPRS J Photogramm Remote Sens 65:369–379

Korpela I, Hovi A, Morsdorf F (2012) Understory trees in airborne LiDAR data – selective mapping due to transmission losses and echo-triggering mechanisms. Remote Sens Environ 119:92–104

Korpela I, Hovi A, Korhonen L (2013) Backscattering of individual LiDAR pulses from forest canopies explained by photogrammetrically derived vegetation structure. ISPRS J Photogramm Remote Sens 83:81–93

Lang ARG (1987) Simplified estimate of leaf area index from transmittance of the sun's beam. Agric For Meteorol 41:179–186

Lang M, Kurvits V (2007) Restoration of tree crown shape for canopy cover estimation. For Stud 46:23–34

Lang M, Kuusk A, Mõttus M, Rautiainen M, Nilson T (2010) Canopy gap fraction estimation from digital hemispherical images using sky radiance models and a linear conversion method. Agric For Meteorol 150:20–29

Lee AC, Lucas RM (2007) A LiDAR-derived canopy density model for tree stem and crown mapping in Australian forests. Remote Sens Environ 111:493–518

LI-COR Inc (1992) LAI-2000 plant canopy analyzer instruction/operating manual

Lovell JJ, Jupp DLP, Culvenor DS, Coops NC (2003) Using airborne and ground-based ranging lidar to measure canopy structure in Australian forests. Can J Remote Sens 29:607–622

Majasalmi T, Rautiainen M, Stenberg P, Rita H (2012) Optimizing the sampling scheme for LAI-2000 measurements in a boreal forest. Agric For Meteorol 154–155:38–43

Majasalmi T, Rautiainen M, Stenberg P, Lukes P (2013) An assessment of ground reference methods for estimating LAI of boreal forests. For Ecol Manag 292:10–18

McLane AJ, McDermid GJ, Wulder MA (2009) Processing discrete-return profiling lidar data to estimate canopy closure for large-area forest mapping and management. Can J Remote Sens 35:217–229

Miller JB (1967) A formula for average foliage density. Aust J Bot 15:141–144

Morsdorf F, Kötz B, Meier E, Itten KI, Allgöwer B (2006) Estimation of LAI and fractional cover from small footprint airborne laser scanning data based on gap fraction. Remote Sens Environ 104:50–61

Næsset E (2004) Effects of different flying altitudes on biophysical stand properties estimated from canopy height and density measured with a small-footprint airborne scanning laser. Remote Sens Environ 91:243–255

Næsset E (2009) Effects of different sensors, flying altitudes, and pulse repetition frequencies on forest canopy metrics and biophysical stand properties derived from small-footprint airborne laser data. Remote Sens Environ 113:148–159

Newnham G, Goodwin N, Armston J, Muir J, Culvenor D (2012) Comparing time-of-flight and phase-shift terrestrial laser scanners for characterising topography and vegetation density in a forest environment. In: Proceedings of Silvilaser 2012, Vancouver, Canada. 6 p

Ozanne CMP, Anhuf D, Boulter SL, Keller M, Kitching RL, Körner C, Meinzer FC, Mitchell AW, Nakashizuka T, Silva Dias PL, Stork NE, Wright SJ, Yoshimura M (2003) Biodiversity meets the atmosphere: a global view of forest canopies. Science 301:183–186

Peduzzi A, Wynne RH, Thomas VA, Nelson RF, Reis JJ, Sanford M (2012) Combined use of airborne lidar and DBInSAR data to estimate LAI in temperate mixed forests. Remote Sens 4:1758–1780

Pisek J, Lang M, Nilson T, Korhonen L, Karu H (2011) Comparison of methods for measuring gap size distribution and canopy nonrandomness at Järvselja RAMI (RAdiation transfer Model Intercomparison) test sites. Agric For Meteorol 151:365–377

Pueschel P, Buddenbaum H, Hill J (2012) An efficient approach to standardizing the processing of hemispherical images for the estimation of forest structural attributes. Agric For Meteorol 160:1–13

Rautiainen M, Stenberg P, Nilson T (2005) Estimating canopy cover in Scots pine stands. Silva Fenn 39:137–142

Riaño D, Valladares F, Condés S, Chuvieco E (2004) Estimation of leaf area index and covered ground from airborne laser scanner (Lidar) in two contrasting forests. Agric For Meteorol 124:269–275

Richardson J, Moskal LM, Kim S (2009) Modeling approaches to estimate effective leaf area index from aerial discrete-return LiDAR. Agric For Meteorol 149:1152–1160

Running S, Coughlan JC (1988) A general model of forest ecosystem processes for regional applications: 1. Hydrologic balance, canopy gas exchange and primary production processes. Ecol Model 42:125–154

Ryu Y, Nilson T, Kobayashi H, Sonnentag O, Law BE, Baldocchi DD (2010) On the correct estimation of effective leaf area index: does it reveal information on clumping effects? Agric For Meteorol 150:463–472

Sarvas R (1953) Measurement of the crown closure of the stand. Commun Inst For Fenn 41(6): 1–13

Smith AMS, Falkowski MJ, Hudak AT, Evans JS, Robinson AP, Steele CM (2009) A cross-comparison of field, spectral, and lidar estimates of forest canopy cover. Can J Remote Sens 35:447–459

Solberg S (2010) Mapping gap fraction, LAI and defoliation using various ALS penetration metrics. Int J Remote Sens 31:1227–1244

Solberg S, Næsset E, Hanssen KH, Christiansen E (2006) Mapping defoliation during a severe insect attack on Scots pine using airborne laser scanning. Remote Sens Environ 102:364–376

Solberg S, Brunner A, Hanssen KH, Lange H, Næsset E, Rautiainen M, Stenberg P (2009) Mapping LAI in a Norway spruce forest using laser scanning. Remote Sens Environ 113:2317–2327

Solberg S, Næsset E, Lange H (2010) Comparing canopy penetration of repeated ALS acquisitions. In: Proceedings of Silvilaser 2010, Freiburg, Germany

Stancioiu PT, O'Hara KL (2006) Regeneration growth in different light environments of mixed species, multiaged, mountainous forests of Romania. Eur J For Res 125:151–162

Stenberg P (1996) Correcting LAI-2000 estimates for the clumping of needles in shoots of conifers. Agric For Meteorol 79:1–8

Turner DP, Acker SA, Means JE, Garman SL (2000) Assessing alternative allometric algorithms for estimating leaf area of Douglas-fir trees and stands. For Ecol Manag 126:61–76

Turner DP, Ollinger SV, Kimball JS (2004) Integrating remote sensing and ecosystem process models for landscape- to regional-scale analysis of the carbon cycle. Bioscience 54:573–584

UNFCCC (2003) The second report on the adequacy of the global observing systems for climate in support of the UNFCCC, GCOS-82, WMO/TD No. 1143, Geneva, Switzerland.

Wagner W (2010) Radiometric calibration of small-footprint full-waveform airborne laser scanner measurements: basic physical concepts. ISPRS J Photogramm Remote Sens 65:505–513

Weiss M, Baret F, Smith GJ, Jonckheere I, Coppin P (2004) Review of methods for in situ leaf area index (LAI) determination. Part II. Estimation of LAI, errors and sampling. Agric For Meteorol 121:37–53

Zhang Y, Chen JM, Miller JR (2005) Determining digital hemispherical photograph exposure for leaf area index estimation. Agric For Meteorol 133:166–181

Zhao K, Popescu S (2009) Lidar-based mapping of leaf area index and its use for validating GLOBCARBON satellite LAI product in a temperate forest of the southern USA. Remote Sens Environ 113:1628–1645

Zhao F, Yang X, Schull MA, Román-Colón MO, Yaoa T, Wang Z, Zhang Q, Jupp DLB, Lovell JL, Culvenor DS, Newnham GJ, Richardson AD, Ni-Meister W, Schaaf CL, Woodcock CE, Strahler AH (2011) Measuring effective leaf area index, foliage profile, and stand height in New England forest stands using a full-waveform ground-based lidar. Remote Sens Environ 115:2954–2964

Chapter 21
Canopy Gap Detection and Analysis with Airborne Laser Scanning

Benoît St-Onge, Udayalakshmi Vepakomma, Jean-François Sénécal, Daniel Kneeshaw, and Frédérik Doyon

Abstract The opening and closure of "gaps" in forest canopies plays an important role in the structure, turnover, and overall ecological processes of natural forests. Gap characterization was until recently mostly based on field studies and relied on sampling approaches. ALS (Airborne Laser Scanning) has now revolutionized this field of scientific enquiry by giving researchers the capacity to detect and measure gaps rapidly over large areas. We first provide a brief scientific background on gaps and then succinctly review field and other conventional remote sensing methods to characterize them. We then turn our attention to the principles of ALS-based gap detection and review different methods of automated gap delineation and measurement. We explain how gap types can be automatically classified, and how multitemporal ALS can be used to not only monitor gap dynamics, but also to reveal the complex role of gaps in influencing tree growth within and around them.

21.1 Introduction

While the main focus of forest-related airborne laser scanning (ALS) has been logically directed towards the description of tree, stand, and canopy attributes, the study of the local absence of canopy trees within a continuous canopy has

B. St-Onge (✉)
Department of Geography, University of Québec at Montréal, Montréal, Canada
e-mail: st-onge.benoit@uqam.ca

U. Vepakomma
FP Innovations, Pointe-Claire, Canada

J.-F. Sénécal • D. Kneeshaw
Department of Biological Sciences, University of Québec at Montréal, Montréal, Canada

F. Doyon
Temperate Forest Sciences Institute (ISFORT), Ripon, Canada

M. Maltamo et al. (eds.), *Forestry Applications of Airborne Laser Scanning: Concepts and Case Studies*, Managing Forest Ecosystems 27, DOI 10.1007/978-94-017-8663-8_21,
© Springer Science+Business Media Dordrecht 2014

also received attention. Such openings in the canopy, formally termed "gaps," are indicators of forest characteristics, such as young/old growth status, and play a critical role in the overall dynamics of forest ecosystems (Runkle 1985). The advent of ALS has strongly increased our capacity to study gap attributes and dynamics, enabling precise wall-to-wall mapping over large landscapes. Such abilities strongly contrast with past studies that used only sparse yet painstaking field sampling. In this chapter, we first define forest gaps and briefly explain their importance and function in forested environments. We then present how they can be detected and measured using ALS, distinguishing monotemporal and multitemporal cases, the latter of which enable the direct observation of gap dynamics. We also show how ALS can be used to study regeneration within canopy gaps, together with the growth response of trees in the immediate vicinity of these disturbed areas and beyond. Although gaps are also important in the dynamics of tropical forests, this chapter places greater emphasis on studies pertaining to boreal and temperate canopies.

21.2   Forest Gaps

21.2.1   Gap Definition

A forest gap can be defined as a small opening within a continuous and relatively mature canopy, in which trees are absent or markedly smaller than their immediate neighbours. Although normal fluctuations of canopy height (designated by "A" in Fig. 21.1) or subcanopy gaps ("B") can be considered as empty spaces, we are essentially interested in gaps that create a sharp breach in the canopy surface ("C1, C2"). Inherent to the definition of a gap is the fact that they are localized, and not part of an open-ended system, such as a river or a large burnt area. Although there

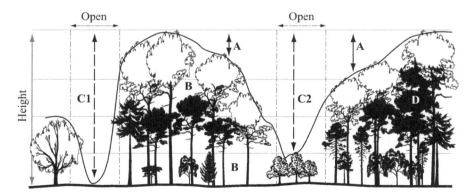

Fig. 21.1 Cross-section of a forest canopy showing crowns visible from the sky (*white polygons*) and shaded crowns (*solid black*). The surface line is a schematic representation of the ALS digital surface model. (A) indicates overstory gaps; (B): a subcanopy gap; (C): gaps extending through all strata (C1 – Edaphic gap, C2 – Developmental gap); and (D): absence of a gap

has been some consideration of imitating gaps in silvicultural systems (Coates and Burton 1997; Seymour et al. 2002) shapes and sizes are more regular and patterns of creation are pre-determined. This review thus focuses on the more diverse (temporally and spatially) naturally created gaps which have been the focus of much ecological literature. However, we do provide a few examples in the following sections to show the potential use for evaluating partial cutting regimes.

Different upper size limits for considering an opening as being a gap have been suggested, ranging from 200–250 m^2 (McCarthy 2001) to 1,000 m^2 (Yamamoto 1992). Additionally, small natural spaces between trees, which are quite common in coniferous canopies, are excluded from the definition. Of particular importance in the context of ALS-based analysis of gaps is the distinction between a "canopy gap" and an "expanded gap" (Runkle 1992). The canopy outline of a gap corresponds to the edges of tree foliage on its periphery, while the expanded outline refers to the area extending to the bases of the surrounding trees. ALS can directly sense the canopy outline, but it remains to be demonstrated whether it could be further used to infer the expanded outline.

The existence of gaps can be attributed to two broad causes: developmental events and edaphic conditions. In the former case, gaps are ephemeral and formed by the fall of a part of a crown, of one or more overstory trees due to disease, decay, insects, herbivores, windthrow, or snow weight, or a combination of these. While most studies focus on gap dynamics and thus on this type of gap, a significant proportion of gaps may be due to localized edaphic conditions (such as adverse drainage, rock outcrops, etc.) that prevent or limit the establishment of tree regeneration (Lertzman et al. 1996). To simplify the text, the word "gap" hereafter designates "developmental gaps," unless otherwise specified.

21.2.2   Gap Dynamics and Its Ecological Role

Gaps are a characteristic of forest dynamics in older forests in which gap openings that are caused by tree fall are subsequently closed by tree recruitment and growth. Closure can be vertical when advance regeneration within the gap grows to a height equivalent to that of the surrounding canopy. Lateral closure designates the closing of a gap by inward crown expansion of the peripheral trees. A single gap may result from multiple tree mortality events that have occurred over a period of time. Complex situations may arise in which, for example, part of a gap is closing while opening continues on the other side, leading to spatial displacement of a gap (Vepakomma et al. 2012).

Gaps are important in driving the dynamics of old-growth forests (Pickett and White 1985). Traditionally, they have been considered more common in temperate forests than in boreal forests due to the importance of fires and because the area of old-growth stands have been underestimated in the latter systems. However, this myth was dispelled by recent studies that have shown many regions of the boreal forest have long fire cycles and large expanses of old-growth forest (Kneeshaw and Gauthier 2003; Bergeron and Fenton 2012). In boreal forests, gap dynamics are the mechanism that permits the maintenance of forest conditions when fire cycles are long and species longevity is short (Kneeshaw and Bergeron 1998; Pham et al. 2004; St-Denis et al. 2010; Kneeshaw et al. 2011). In both temperate and boreal forest types larger openings may lead to the recruitment of early successional species, thus retarding succession, or conversely, small openings can release shade-tolerant advance regeneration, thereby accelerating succession (Chen and Popadiouk 2002). Small gaps may also be closed through lateral expansion of peripheral trees such that multiple gap events are required for the recruitment of understory stems to the overstory. Due to low sun angles in high latitude forests, the displacement of gaps over multiple periods has been implicated in the maintenance of early successional species in old-growth forests (Vepakomma et al. 2012).

The array of gap sizes, shapes and configurations creates heterogeneous canopy conditions, changes biomass accumulation, and also modifies conditions for tree growth (Messier et al. 1999; Paré and Bergeron 1995). Gap dynamics are considered to be a key process in carbon dynamics and plays a role in wildlife habitat quality.

21.2.3 Field Measurement of Gaps

Since the early days of canopy gap research in the 1970s, most studies have relied upon field measurements of gaps. Gaps are typically inventoried along intersect or strip transects that are established on foot (Schliemann and Bockheim 2011). The presence/absence of gaps is generally recorded, together with their sizes, i.e., the linear extent of the gap that is traversed by the transect or the length of the minor and major axes of the gap, which is modelled as an ellipse. Since gaps are most often irregularly shaped, precise delineation of the gap to measure its area would be preferred to an elliptical approximation (Gagnon et al. 2004). However, full delineation of gap boundaries from a ground perspective by vertically projecting the outline of peripheral trees is not a trivial task. An alternative that was based on hemispherical photographs had been proposed by Hu and Zhu (2009) to quantify the three-dimensional shapes of gaps. Regardless of what approach is used, ground-based methods are costly and, for this reason, were often limited to single surveys of relatively small study areas (Seymour et al. 2002). Before the advent of accurate 3D mapping methods such as ALS, our knowledge of gap disturbances and their influences on forest dynamics was based on limited spatial and temporal scales.

21.3 Remote Sensing of Forest Gaps

21.3.1 Overview

Remote sensing offers a very efficient alternative to field-based tree height or gap identification, by providing a means of scaling measurements across two or more spatial scales of observation (e.g., tree or plot to landscape or region) over multiple time intervals. The primary benefits include synoptic (i.e., spatially complete) coverage, repeat measurements, high cost-effectiveness, and coverage of inaccessible areas. Although simple and readily available, airphoto interpretation is unreliable in detecting smaller (<100 m^2) and deeper gaps (Nakashizuka et al. 1995; Miller et al. 2000; Fujita et al. 2003; Betts et al. 2005). It was also observed that the degree of error is associated with canopy height (or gap depth) and increases with higher topographic relief (Tanaka and Nakashizuka 1997). Multispectral image classification is not a viable alternative because of the heterogeneity of the vegetation found inside gaps and the effects of shadows, among other problems, that preclude the use of well-defined spectral signatures. InSAR (Interferometric Synthetic Aperture Radar) techniques have the capacity to generate high resolution digital surface models (DSMs), but are hindered by their side-looking acquisition geometry, which creates shadows and occlusion effects that complicate the detection and delineation of gaps.

The potential of airborne laser remote sensing for gap detection became evident in the early 1980s. Gaps are conspicuous in lidar profiles in early studies published by Arp and Tranarg (1982), Nelson et al. (1984), and Krabill et al. (1984), for example. The advent of scanning airborne lasers later permitted the creation of wall-to-wall coverage, although the first attempts were characterized by a rather low return density. When densities passed a certain threshold, however, interest in detecting individual forest objects, initially trees and then gaps, grew rapidly. In a study comparing InSAR, photogrammetry and ALS, the last technique was identified as superior for characterizing detailed canopy features such as gaps (Andersen et al. 2003). Thus, it is not surprising that recent remote sensing studies focusing on forest gaps have relied on ALS. In the following section, we first discuss the principles of ALS acquisition and data preparation relevant for gap analysis. We then review gap detection and measurement methods, as well as the classification of gap types.

21.3.2   Gap Detection and Measurement with ALS

21.3.2.1   Principles of ALS Data Acquisition and Preparation for Gap Analysis

Due to the characteristics of modern ALS sensors and assuming typical parameters for data acquisition, ALS can in principle provide a very reliable census of gaps. Routine data acquisition uses a laser beam cross-section at half maximum energy of around 20 cm, a resolution of a few returns/m^2, and X, Y, and Z accuracy on the order of a few decimetres. The gap to be detected is itself a highly contrasted object, i.e., an opening several metres deep with a size significantly greater than the ALS point cloud resolution. Moreover, its detection is usually straightforward. However, blind areas within gaps remain possible in the case of non-nadir pulses.

The occluded margin within the gap near its edge is of $tan\theta * D$ (where θ is the off-nadir angle and D is the gap depth), which, for example amounts to 2.6 m in the case of a 15 m deep gap and a 10-degree view angle. This occluded margin essentially occurs in the cross-track direction (parallel to the scanning axis), although small deviations from zero aircraft pitch may occasionally induce similar, albeit very small, effects in the along-track direction. It should be noted that carrying out an ALS survey with at least 50 % overlap between adjacent swaths should result in the absence of blind areas, except in the case of very deep and narrow gaps that are located some distance away from the flight lines. In the case where swath overlap is kept to a minimum, blind areas within gaps will be present to some extent, resulting in a slight underestimation of gap area, or increasing artifacts in the multitemporal comparison of gap outlines. This bias in gap size will remain much lower compared to what can be achieved through stereomatching of aerial images or InSAR.

Gap detection is normally performed on a raster canopy height model (CHM), which is derived from the ALS first returns (surface model) and ground hits (terrain model). Rasterization raises questions of optimal pixel size, together with the interpolation method appropriate for subsequent processing steps and its parameter values. Few authors have formally addressed this question despite the fact that edges, such as the outlines of gaps, are the elements that are most sensitive to interpolation errors. The choice of pixel size is conditioned by a trade-off between having several ALS points in each grid cell, which can lead to information loss, and choosing a relatively small pixel size, which results in a greater proportion of void cells that need to be interpolated. Vepakomma et al. (2008) proposed a cross-validation approach for choosing the optimal interpolation method (e.g., Goovaerts 1997), and a selection strategy for the best pixel size that was based on minimizing both spurious gaps and loss of original lidar points in the DSM. Seven different interpolation methods and six different pixel sizes (from 0.1 to 1 m on a side) were evaluated. They found that local polynomial and Inverse Distance Weighting (IDW) methods had the lowest magnitude of error, and that rasterization errors increased significantly when pixel size exceeded 0.25 m for an ALS point density ranging from 0.1 to 4.2 hits/m^2. In earlier studies of ALS-based gap detection, researchers used various interpolation methods, and pixels size ranging from 0.5 to 1.5 m (for point densities ranging from 0.3 to 11.4 returns/m^2; see Table 21.1), but most did not provide an explicit rationale for choosing their interpolation methods or parameters.

21.3.2.2 Gap Detection and Delineation

The first analysis step for characterizing gaps in a given forested landscape, and frequently the only one, has been the detection and delineation of gaps (see Fig. 21.2, for an example). The gap detection methods themselves logically follow from the operational definition of gaps. The various definitions that have been proposed in previous studies all use canopy height differences as their main

Table 21.1 Literature summary of previous gap detection methods

Authors	Forest type	ALS system	Returns/m² (max. scan angle, degrees)	Method
St-Onge and Vepakomma (2004)	Boreal mixed wood – natural old-growth	ALTM 1020 ALTM 2050	0.3 (10°); 3 (15°)	Absolute height plus additional constraints/single and multiple dates. Height below 10 m, minimum of 5 m² and 3 hits per gap
Koukoulas and Blackburn (2004)	Semi-natural broadleaved deciduous	ALTM 1020	0.5 (20°)	Absolute height plus additional constraints/single date. Height below 4 m and slope over 60°
Yu et al. (2004)	NA	Toposys-1	4.5 (10°)	Threshold on height difference/multiple dates
Vepakomma et al. (2008)	Boreal mixed wood – natural old-growth	ALTM 1020 ALTM 2050	0.3 (10); 3 (15°)	Absolute height plus additional constraints/single and multiple dates. Height below 5 m, minimum of 5 m² and 3 hits per gap object
Zhang (2008)	Mangrove	ALTM 1233	Return spacing: 0.6 m × 2 m	Fixed height and morphological method
Kellner and Asner (2009)	Neotropical	NA	NA	Absolute height. Cumulated gaps at height <H, H varying
Latif and Blackburn (2010)	Semi-natural deciduous	ALTM 3033	NA	Absolute height. Heights equal to or below 4 m
Gaulton and Malthus (2010)	Sitka spruce plantations, U.K.	ALTM 3033 ALTM 3100	1.2 (20°) and 11.4 (12°) resp.	Relative height/single date. (1) All pixels below 66 % of the highest point in a window; (2) point cluster of crowns followed by convex hull extraction for gap outlines. Both raster CHMs and point clouds were used
Kane et al. (2011)	Conifer-dominated across Pacific Northwest	DATIS II	1	Absolute height/single date
Sénécal (2011)	Temperate deciduous	ALTM 3100	>2 (20°)	Relative threshold/single date. Heights below two relative thresholds determined by height values in a 0.25 ha neighbourhood

(continued)

B. St-Onge et al.

Table 21.1 (continued)

Authors	Forest type	ALS system	Returns/m² (max. scan angle, degrees)	Method
Vehmas et al. (2011)	Boreal mature – semi natural and managed	ALTM 3100	3.9 (11°)	Absolute height/single date. Gap class evaluation using points and landscape metrics
Vepakomma et al. (2012)	Boreal mixed wood – natural old-growth	ALTM 1020 ALTM 2050 ALTM 3120	0.3 (10°), 0.5(20°), 3.9 (11°)	Absolute height plus additional constraints/single and multiple date. Height below 5 m, minimum of 5 m² and 3 hits per gap object;
Asner et al. (2013)	Tropical amazonian	CAO-Alpha	Return spacing: 1.1 m (19°)	Absolute height thresholds (≤1 m, ≤20 m)
Jung et al. (2013)	Tropical rain forest	ALTM 3100EA	3.3 (10°)	Analysis of vertical profiles generated by summing ALS return intensities in 5 m × 5 m × 1 m voxels (1 m = bin height)

NA Not Available

Fig. 21.2 Illustration of a
delineated forest canopy gap,
based on a high resolution
CHM

criterion. In the case of single-date studies, this height differential is between the
gap surface and the height of the surrounding forest (e.g., Gaulton and Malthus
2010). In the case of bitemporal studies, the criterion for identifying gap openings
is the amount of canopy height decrease between dates (e.g., Vepakomma et al.
2008). In all cases, large openings in the canopy that are not considered gaps, e.g.,
water bodies, recently harvested forests, and other features, are masked out prior
to detection of the gaps. Gap detection methods can be classified as using either
an absolute or a relative height threshold. Table 21.1 summarizes the gap detection
methods that were adopted in several published studies.

In the case of absolute height (H), thresholds typically vary from $H \leq 3$ m (Kane
et al. 2011) to $H \leq 5$ m (Vepakomma et al. 2008). In one tropical study, however,
two classes of gaps were detected using thresholds of ≤ 1 m and ≤ 20 m, the latter
serving to detect gaps that were caused by crown failures in the upper canopy
(Asner et al. 2013). Researchers usually choose cut-off values empirically based
on field observations, with consideration of the known characteristics of the forests
under study. Additional constraints are often used to exclude non-gap objects. For
example, Koukoulas and Blackburn (2004) retained gaps when $H \leq 3$ m and the
slope of the gap edges was over 60°, while Vepakomma et al. (2008) required that
gap objects included at least three lidar returns, and had a minimum size of 5 m^2
and a minimum width of 1.5 m (additional details are in Vepakomma et al. 2012).
Thresholds on object sizes help in filtering out small spaces that naturally occur
between trees.

Because a gap is naturally recognized as a part of the forest canopy that is
significantly lower than its surroundings, some authors have preferred to use relative
height differences. For example, Gaulton and Malthus (2010) considered a CHM
pixel as being part of a gap if its height was below 66 % of the highest pixel in

a given window. Sénécal (2011) used a combination of the two following relative thresholds:

$$H_i < \left[H_{\frac{1}{4}\text{ha}} - \left(1.5^* SD_{H\frac{1}{4}\text{ha}} \right) \right] \quad \text{AND} \quad H_i < \left(0.2^* H_{\max \frac{1}{4}\text{ha}} \right)$$

where H_i is the CHM height of the pixel being evaluated, $H_{\text{¼ha}}$ is the average CHM value in a circular neighbourhood of 0.25 ha, and $SD_{H\text{¼ha}}$ and $H_{\max\text{¼ha}}$ are respectively the standard deviation and the maximum of the CHM heights in the same neighbourhood.

Almost all gap studies using ALS have focused on locally depressed canopy surfaces visible from the sky (types C1 and C2 in Fig. 21.1). In at least one instance, researchers tried to identify subcanopy gaps (type B in Fig. 21.1). Jung et al. (2013) performed such a task by summing the ALS return intensities in voxels having an XYZ resolution of 5 m × 5 m × 1 m in a tropical forest area of Costa Rica to create vertical intensity profiles. These were then transformed into two metrics, (1) the ratio between the sum of the vertical length of subcanopy voids in a given 5 × 5 m cell and the maximum canopy height for that cell, and (2) the number of distinct subcanopy voids in a cell. Using these metrics, structural differences between selectively logged, secondary growth and old forests were identified.

Validation of gap detection results has been quite rare, possibly reflecting the confidence researchers have that ALS is an accurate, high-resolution remote sensing technique. Gap outlines are not always clear, or clearly visible from a ground perspective. The canopy surface of a regenerating gap is often above the field observer and effective delimitation may be either obscured by dense understory vegetation or difficult to measure due to its irregular shape. For validation purposes, Vepakomma et al. (2008) mapped gaps along four field transects and found from 83 to 100 % correspondence in the number of gaps between field observations and those detected by ALS. Gaulton and Malthus (2010) used a pixel-based comparison between detected gaps and those that were outlined in field plots, resulting in accuracies between 71 and 85 % in the case of the raster CHM. Accuracy improved by 2 % when gap detection was performed directly on the ALS points rather than upon the raster CHM. It should be noted that precise geopositioning of field observations under a dense canopy remains an issue, which can cause spurious false positives or false negatives during the gap detection validation process, thereby possibly causing an overestimation of the gap detection error.

21.3.2.3 Gap Measurement

Gap shapes and size distributions help to determine the extent of the disturbance and the availability of resources (Runkle 1985; Denslow and Spies 1990). Once gaps have been delineated, a number of gap measurements can be performed using standard GIS procedures, either for each individual gap or for global gap characteristics in a given landscape. At the individual gap level, gap area is commonly measured, but other geometric characteristics are also studied, such as tree height-to-gap diameter ratios, gap elongation, and orientation. Some researchers have also attempted to characterize individual gap characteristics by analyzing canopy height distributions within the gaps. For example, Vehmas et al. (2009) were able to detect the presence of dead wood using this approach. Sénécal et al. (2010) attempted to detect multiple events within single gaps by determining height distributions within the gaps.

For gap regimes at the landscape scale, metrics such as gap fraction[1] (Vepakomma et al. 2008) and size distributions are used. Gap fraction is measured to determine the proportion of forest area under gaps (Runkle 1985):

$$GF = \sum_{k=1}^{M} AG_k \Big/ (aT)$$

where AG_k is the area (m^2) of the kth gap object, a is the pixel size and T is the total number of pixels in the study area that do not belong to an open-ended system. Size distributions of gaps are quantified to determine the relative proportions of different-sized openings, due to their influence on resource availability and species recruitment. They are usually presented using histograms or statistics, such as the standard deviation of gap size. Size distributions can be employed to provide a portrait of the overall influence of gaps or to reveal the effect of terrain conditions on the gap regime. Asner et al. (2013), for example, fitted a Riemann Zeta probability density function to the size frequencies of gaps in a Peruvian forest and compared the resulting λ values for different types of soil substrates. They showed a high similarity in gap-size frequency distributions indicating structural responses to canopy failure independent of regional variation in soil and other conditions. Finally, to our knowledge, no geostatistical analysis has been attempted based on ALS gap identification despite the potential of this approach in ecological studies for highlighting spatial processes.

21.3.2.4   Gap Classification

Since the main motive for studying forest gaps is to better understand the evolution of forest structure and composition in older gap driven forests attention has been mainly directed towards openings that are created by tree falls. However, longer-term or permanent gaps exist that are caused by various factors such as locally poor drainage, rock outcrops, and the like. In studies concerning forest dynamics, it is important to distinguish short-term "developmental gaps" that will regenerate after the opening from "non-regenerating" gaps that are either caused by adverse local edaphic or drainage conditions, or by outcompeting recalcitrant understory vegetation that prevents trees from establishment (sensu Royo and Carson 2006).

---

[1] Not be confused with the smaller scale foliage "gap fraction" measured from hemispherical photographs or terrestrial laser scanning (see Chap. 20).

Sénécal (2011) developed a method that was based on ALS alone for classifying gaps as being either developmental or non-regenerating in a deciduous forest in southeastern Canada. Gaps were first detected using an adaptive height threshold algorithm. Intra-gap vegetation height distributions were then characterized from the CHM, and a topographical index that was based on local concavity was computed from the ALS DTM, for each detected gap. Furthermore, stream locations were predicted from the DTM and the distance to the nearest stream was calculated for each gap. A subset of detected gaps were field-inspected to determine their true class and separated into training and validation data. A mixed-effects logistic regression model classification was employed to discriminate between the two gap types, by using the three features described above. The classification error was 17 % for the validation data and 26 % for the training dataset, at a 0.5 cut-off value.

This approach using data from a single flight, although clearly successful, has been tested in only one type of forested landscape and, thus, should be validated in other landscapes. Multi-temporal ALS data can, in contrast, unambiguously discriminate between highly dynamic developmental gaps and longer-lived non-regenerating gaps.

21.3.3 Gap Dynamics Characterization with Multi-date ALS

Using multi-date ALS considerably decreases the difficulty of distinguishing between developmental and edaphic gaps, and further allows for several types of measurements that cannot be done with single-date ALS, such as determining the rate at which gaps open and the rate of height regrowth within gaps.

21.3.3.1 Gap Opening and Closure

The first multi-temporal ALS studies that were applied to gaps date from the early 2000s. Yu et al. (2004, 2005), as well as St-Onge and Vepakomma (2004), showed that gap openings can be detected, even in the case of a single tree fall or partial harvest. In these studies, harvested trees were identified by comparing canopy heights at two dates and assigning change to removal, or they were detected directly on the CHM difference layer. Yu et al. (2004, 2005) verified gap openings using field data. St-Onge and Vepakomma (2004) only verified gap detection accuracy at the second date, but they used concomitant high-resolution aerial and satellite images to confirm the gap opening results (Fig. 21.3). "New" gaps were identified using this approach, given that those openings did not exist in the reference year and that persistent "old" gaps present in both years of analysis could thus be discounted (Fig. 21.4, and Vepakomma et al. 2010). More recently, Vastaranta et al. (2012) successfully used height differences in ALS datasets acquired in 2006–2007, and 2010, to quantify the level of crown damage caused by snow.

Fig. 21.3 Gap openings between 1998 and 2003 in a southern boreal forest. *Top row*: high-resolution images (1997, 2003) showing the disappearance of large crowns (*circles*) and of a single tree (*arrow*). *Bottom row*: corresponding ALS canopy height model (1998, 2003; brightness proportional to height) showing the canopy height changes caused by gap openings and by the fall of a single tree

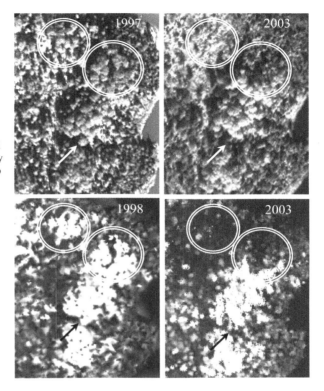

Gap closure is defined as the stage at which the gap vegetation, due to its vertical or lateral growth, ceases to be distinct from the surrounding forest canopy (higher than an arbitrary low threshold). Therefore, a gap that is detected at time t_1, and which is no longer identified as a gap at time t_2, is said to have "closed." Closure due to vertical growth occurs when the difference in canopy height over time is positive. When the rate of height growth is clearly above growth reported in a neighbourhood of high surrounding canopy, closure is attributed to lateral growth. ALS studies conclusively demonstrate that closure, at least in mixed boreal forests, results from both vertical and lateral growth (Vepakomma et al. 2011).

21.3.3.2   Spatial Contiguity and Continuity of Canopy Gaps

The formation of canopy gaps is a continuous process. They appear, persist, expand, shrink, displace, or eventually disappear, all of which can occur simultaneously at various levels. Increased stress at gap edges can lead to further mortality and, thus, to expansion and maintenance of openings for longer periods. Large gaps may shrink in part due to closure, while persisting or expanding in other directions, thereby displacing the location of gap area (Vepakomma et al. 2012). While bi-temporal

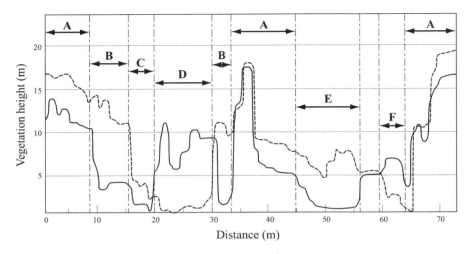

Fig. 21.4 Illustration of different gap-related events. Vertical profile along a random transect from multi-temporal lidar CHMs showing the changes between 1998 (*solid line*) and 2003 (*dotted line*). (A) represents a high canopy where canopy height (H) is over 5 m in both years; (B) a region where a gap present in 1998 is laterally closed by the adjacent high canopy by 2003; (C) an old gap that is still open in 2003; (D) a gap expansion from an earlier gap; (E) a gap closure from below due to regeneration; (F) a random new gap

ALS can efficiently characterize new openings and closures, understanding of the continuous pervasiveness of gaps requires observations over two moments in time. Vepakomma et al. (2012) use the concepts of random set theory to characterize dynamic changes, after delineating canopy gaps at three time steps (*viz.*, 1998, 2003, and 2007). A simple guide for assessing gap dynamics using ALS data when three dates are available is:

t_1–t_2: Detect old and new gap openings and closures.

t_2–t_3: Detect persistence, pervasiveness, expansion and displacement.

21.3.3.3   Tree Height Growth Response due to Gap Openings

The use of ALS time series is not only shown to be important in characterizing gap regimes and their dynamics across landscapes, but also to have great potential in understanding their impacts on vegetation. By identifying individual trees on the t_1 (1998) and t_2 (2003) ALS surfaces, Vepakomma et al. (2011) noted that increased resource availability that was brought about by canopy openings not only enhances the height growth of matured trees in the immediate gap periphery but also deep into the forest for up to a distance of 30 m. The gradient of growth rates, however, decreased with distance from the opening. An independent study in the same area revealed that neighbouring trees may interconnect with one another through grafted root systems (Tarroux and Desrochers 2011). This communal root system improves the use of resources by redistributing them among trees, which leads to improve growth. As ALS acquisition becomes more widespread and even systematically applied in some jurisdictions, datasets with at least three dates will become more commonplace, increasing our opportunities to advance knowledge regarding these complex interactions.

21.3.3.4 Technical Considerations for Multitemporal Analysis

Multi-temporal studies require that individual datasets be geometrically co-registered in 3D. As technology improves, the absolute geolocation of ALS returns is calculated with greater and greater accuracy. However, failure to verify that all datasets are registered to the same vertical datum could lead to spurious results. Subtle changes over the local definition of the geoid, for example, have to be verified. Moreover, pre-2000 lidar acquisition may have been performed with sub-optimal GPS constellations, leading to uncertainty in X, Y and Z measurements. The problem is compounded by the fact that shifts in X and Y, however small, may result in artifactual discrepancies in Z. For this reason, it is preferable to first assess X-Y coregistration before Z-adjustments are verified. Cross-correlation for different levels of experimentally introduced X-Y shifts between DTMs that were acquired on two different dates should be the highest when the DTMs are properly co-registered (provided that there is no physical change in topography). Vertical shifts can then be verified at selected locations such as hard and exposed surfaces with low gentle slopes (e.g., rock outcrops). Using such techniques, Vepakomma et al. (2008) found no discrepancy in X and Y, but a 22 cm shift in Z between ALS data that were acquired in 1998 and 2003. Such maladjustments should be less common in the future, but because the objects of analysis are small changes in height, especially when growth in or around gaps is measured, such verifications should always be performed in multi-date studies.

As development of technologically advanced sensors continues, notably in increased pulse rates to acquire dense point clouds or operational considerations leading to reduced sampling density to cover larger areas, time series of ALS datasets are very likely to be disparate and inconsistent density-wise. Accurate delineation of spatial objects like gaps and measurement of meaningful changes in them are influenced by the resultant sampling density. Vepakomma et al. (2011) compared the frequency of detected gaps using ALS datasets having respective densities of 7, 2 and 0.4 hits/m^2. They found, as could be expected theoretically, that the largest differences between densities occurred for small gaps (5–10 m^2). For this size class, the difference was much greater between 2 and 0.4 hits/m^2 than between 7 and 2 hits/m^2. No significant differences in results were observed for gaps larger than 50 m^2 for these three levels of density. Moreover, in testing on a progressively decreasing sampling density (7–0.5 hits/m^2), Vepakomma and Fortin (in review) were able to show that 3 hits/m^2 is optimal for accurate delineation of gap objects and gap density. Combinations of datasets with 3 or more hits per m^2 minimized artifacts incurred in dynamic changes such as gap displacement, expansion, and shrinkage.

It is probable that other differences in the characteristics of heterogeneous ALS systems used in multitemporal studies may cause artifactual effects in gap detection. For example, it can be hypothesized that differences in the number of returns recorded by pulse, the length of the blind pulse trajectory segments caused by the detector's downtime, or the intensity of the incident pulse (as a function of pulse power and range) could cause some noise in the multitemporal data. We however surmise that, because canopy gaps are large objects both horizontally and vertically compared to ALS resolution in XYZ, the effects of these other system parameters should be almost non significant relative to those of density.

21.4   Concluding Remarks

The characterization of gap regimes at the landscape scale is a difficult task when gaps are measured, either in the field or through the interpretation of high-resolution images. The advent of ALS has revolutionized gap studies by providing a means for creating wall-to-wall 3D representations of gaps at fine scales and with high accuracy. We can now reliably and automatically delineate and characterize gaps over entire landscapes, and monitor their dynamics, openings, expansions, and closures over time. It is also possible to classify gaps and to measure the effects of gap openings on the rates of height growth of surrounding trees. This has helped the in understanding the role of gaps in ecological processes such as forest succession, which led to refuting pre-conceived hypotheses of directional succession in the boreal forest.

Due to the young age of high-resolution ALS images, multi-temporal datasets are still rare, despite being excellent tools for rapidly acquiring information on canopy dynamics. Many regions throughout the world are currently surveying vast areas with ALS. There is a good chance that some or many of these areas will be re-surveyed in the near future, thereby providing opportunities for more multi-temporal studies. ALS coverage in tropical regions is in its infancy, but it will eventually lead to new insights, as gap dynamics in these areas lead to rapid changes in forest composition and structure. We are only starting to uncover complex processes in forest canopies using ALS and future research will certainly bring greater insights. Increased spatial and temporal coverage, more advanced technology, and enhanced analysis methods, focusing for example on the effects of small-scale fragmentation, should bring about not only new scientific knowledge but also better informed forest management.

References

Andersen H-E, McGaughey RJ, Carson WW, Reutebuch SE, Mercer B, Allan J (2003) A comparison of forest canopy models derived from LIDAR and IFSAR data in a Pacific Northwest conifer forest. In: Proceedings, ISPRS working group III/3 workshop "3-D reconstruction from airborne laser scanner and InSAR data", Dresden, Germany, 8–10 October 2003, pp 211–217

Arp H, Tranarg C (1982) Mapping in tropical forests: a new approach using the laser APR [Airborne Profile Recorder]. Photogramm Eng Remote Sens 48:91–100

Asner GP, Kellner JR, Kennedy-Bowdoin T, Knapp DE, Anderson C, Martin RE (2013) Forest canopy gap distributions in the southern Peruvian Amazon. PLoS ONE 8(4):e60875. doi:10.1371/journal.pone.0060875

Bergeron Y, Fenton N (2012) Boreal forests of eastern Canada revisited: old growth, nonfire disturbances, forest succession, and biodiversity. Botany 90:509–523

Betts HD, Brown LJ, Stewart GH (2005) Forest canopy gap detection and characterisation by the use of high-resolution Digital Elevation Models. N Z J Ecol 29:95–103

Chen HY, Popadiouk RV (2002) Dynamics of North American boreal mixedwoods. Environ Rev 10:137–166

Coates DK, Burton PJ (1997) A gap-based approach for development of silvicultural systems to address ecosystem management objectives. For Ecol Manage 99:337–354

Denslow JS, Spies T (1990) Canopy gaps in forest ecosystems: an introduction. Can J For Res 20:619

Fujita T, Itaya A, Miura M, Manabe T, Yamamoto SI (2003) Long-term canopy dynamics analysed by aerial photographs in a temperate old-growth evergreen broad-leaved forest. J Ecol 91:686–693

Gagnon JL, Jokela EJ, Moser WK, Huber DA (2004) Characteristics of gaps and natural regeneration in mature longleaf pine flatwoods ecosystems. For Ecol Manage 187:373–380

Gaulton R, Malthus TJ (2010) LiDAR mapping of canopy gaps in continuous cover forests: a comparison of canopy height model and point cloud based techniques. Int J Remote Sens 31:1193–1211

Goovaerts P (1997) Geostatistics for natural resources evaluation, Applied geostatistics. Oxford University Press, New York

Hu L, Zhu J (2009) Determination of the tridimensional shape of canopy gaps using two hemispherical photographs. Agric For Meteorol 149:862–872

Jung J, Pekin BK, Pijanowski BC (2013) Mapping open space in an old-growth, secondary-growth, and selectively logged tropical rainforest using discrete return LIDAR. IEEE J Sel Top Appl Earth Obs Remote Sens 6:2453–2461. doi:10.1109/jstars.2013.2253306

Kane VR, Gersonde RF, Lutz JA, McGaughey RJ, Bakker JD, Franklin JF (2011) Patch dynamics and the development of structural and spatial heterogeneity in Pacific Northwest forests. Can J For Res 41:2276–2291

Kellner JR, Asner GP (2009) Convergent structural responses of tropical forests to diverse disturbance regimes. Ecol Lett 12:887–897

Kneeshaw DD, Bergeron Y (1998) Canopy gap characteristics and tree replacement in the Southeastern Boreal Forest. Ecology 79:783–794

Kneeshaw DD, Gauthier S (2003) Old-growth in the boreal forest at stand and landscape levels. Environ Rev 11:s99–s114

Kneeshaw D, Bergeron Y, Kuuluvainen T (2011) Forest ecosystem structure and disturbance dynamics across the circimboreal forest. In: Millington A, Blumler M, Schickhoff U (eds) The SAGE handbook of biogeography. Sage, London

Koukoulas S, Blackburn GA (2004) Quantifying the spatial properties of forest canopy gaps using LiDAR imagery and GIS. Int J Remote Sens 25:3049–3072

Krabill W, Collins J, Link L, Swift R, Butler M (1984) Airborne laser topographic mapping results. Photogramm Eng Remote Sens 50:685–694

Latif ZA, Blackburn GA (2010) Extraction of gap and canopy properties using LiDAR and multispectral data for forest microclimate modelling. In: 2010 6th international Colloquium on Signal Processing & Its Applications (CPSA), Melaka, Malaysia, 21–23 May 2010. IEEE Malaysia Section, pp 1–5

Lertzman KP, Sutherland GD, Inselberg A, Saunders SC (1996) Canopy gaps and the landscape mosaic in a coastal temperate rain forest. Ecology 77:1254–1270

McCarthy J (2001) Gap dynamics of forest trees: a review with particular attention to boreal forests. Environ Rev 9:1–59

Messier C, Doucet R, Ruel JC, Claveau Y, Kelly C, Lechowicz MJ (1999) Functional ecology of advance regeneration in relation to light in boreal forests. Can J For Res 29:812–823

Miller DR, Quine CP, Hadley W (2000) An investigation of the potential of digital photogrammetry to provide measurements of forest characteristics and abiotic damage. For Ecol Manage 135:279–288

Nakashizuka T, Katsuki T, Tanaka H (1995) Forest canopy structure analyzed by using aerial photographs. Ecol Res 10:13–18

Nelson R, Krabill W, MacLean G (1984) Determining forest canopy characteristics using airborne laser data. Remote Sens Environ 15:201–212

Paré D, Bergeron Y (1995) Above-ground biomass accumulation along a 230-year chronosequence in the southern portion of the Canadian boreal forest. J Ecol 83:1001–1007

Pham AT, Grandpré LD, Gauthier S, Bergeron Y (2004) Gap dynamics and replacement patterns in gaps of the northeastern boreal forest of Quebec. Can J For Res 34:353–364

Pickett STA, White PS (1985) The ecology of natural disturbance and patch dynamics. Academic, San Diego

Royo AA, Carson WP (2006) The formation of dense understory layers in forests worldwide: consequences and implications for forest dynamics, biodiversity, and succession. Can J For Res 36:1345–1362

Runkle JR (1985) Disturbance regimes in temperate forests. In: Pickett STA, White PS (eds) The ecology of natural disturbance and patch dynamics. Academic, San Diego

Runkle JR (1992) Guidelines and sampling protocol for sampling forest gaps. USDA Forest Service Pacific Northwest Research Station general technical report PNW-GTR-283, Portland, Oregon, USA

Schliemann SA, Bockheim JG (2011) Methods for studying treefall gaps: a review. For Ecol Manage 261:1143–1151

Sénécal J-F (2011) Dynamique spatio-temporelle des trouées en forêt feuillue tempérée par télédétection Lidar. Dissertation, Université du Québec à Montréal, Montréal, QC, Canada

Sénécal J-F, Doyon F, St-Onge B (2010) Is it possible to predict gap age in North American temperate deciduous forest using LiDAR data? In: Proceedings of SilviLaser 2010: the 10th international conference on LiDAR applications for assessing forest ecosystems, Freiburg, Germany, 14–17 September 2010, pp 99–108

Seymour RS, White AS, deMaynadier PG (2002) Natural disturbance regimes in northeastern North America – evaluating silvicultural systems using natural scales and frequencies. For Ecol Manage 155:357–367

St-Denis A, Kneeshaw D, Bergeron Y (2010) The role of gaps and tree regeneration in the transition from dense to open black spruce stands. For Ecol Manage 259:469–476

St-Onge B, Vepakomma U (2004) Assessing forest gap dynamics and growth using multitemporal laser scanner data. In: Proceedings of the laser-scanners for forest and landscape assessment – instruments, processing methods and applications international conference, Frieburg im Breisgau, 3–6 October 2004, pp 173–178

Tanaka H, Nakashizuka T (1997) Fifteen years of canopy dynamics analyzed by aerial photographs in a temperate deciduous forest, Japan. Ecology 78:612–620

Tarroux E, DesRochers A (2011) Effect of natural root grafting on growth response of jack pine (*Pinus banksiana*; Pinaceae). Am J Bot 98:967–974

Vastaranta M, Korpela I, Uotila A, Hovi A, Holopainen M (2012) Mapping of snow-damaged trees based on bitemporal airborne LiDAR data. Eur J For Res 131:1217–1228

Vehmas M, Packalén P, Maltamo M (2009) Assessing deadwood existence in canopy gaps by using ALS data. In: Proceedings of Silvilaser 2009, College Station, Texas, USA, 7 p

Vehmas M, Packalén P, Maltamo M, Eerikäinen K (2011) Using airborne laser scanning data for detecting canopy gaps and their understory type in mature boreal forest. Ann For Sci 68:825–835

Vepakomma U, Fortin MJ (in review) Optimising lidar return density to understand forest dynamics – a study using an event-based approach

Vepakomma U, St-Onge B, Kneeshaw D (2008) Spatially explicit characterization of boreal forest gap dynamics using multi-temporal lidar data. Remote Sens Environ 112:2326–2340

Vepakomma U, St-Onge B, Kneeshaw DD (2010) Interactions of multiple disturbances in shaping boreal forest dynamics – a spatially explicit analysis using multi-temporal lidar data and high resolution imagery. J Ecol 98:526–539

Vepakomma U, St-Onge B, Kneeshaw DD (2011) Boreal forest height growth response to canopy gap openings – an assessment with multi-temporal lidar data. Ecol Appl 21:99–121

Vepakomma U, Kneeshaw D, Fortin MJ (2012) Spatial contiguity and continuity of canopy gaps in mixed wood boreal forests: persistence, expansion, shrinkage and displacement. J Ecol 100:1257–1268

Yamamoto SI (1992) The gap theory in forest dynamics. Bot Mag Tokyo 105:375–383

Yu X, Hyyppä J, Kaartinen H, Maltamo M (2004) Automatic detection of harvested trees and determination of forest growth using airborne laser scanning. Remote Sens Environ 90:451–462

Yu X, Hyyppä J, Kaartinen H, Hyyppä H, Maltamo M, Rönnholm P (2005) Measuring the growth of individual trees using multi-temporal airborne laser scanning point clouds. In: Proceedings of the ISPRS workshop "Laser scanning 2005", Enschede, The Netherlands, 12–14 September 2005, pp 12–14

Zhang K (2008) Identification of gaps in mangrove forests with airborne LIDAR. Remote Sens Environ 112:2309–2325

Chapter 22
Applications of Airborne Laser Scanning in Forest Fuel Assessment and Fire Prevention

John Gajardo, Mariano García, and David Riaño

Abstract Forest fire management requires accurate, spatially explicit and up-to date information on forest fuels and their vertical structure. Airborne laser scanning (ALS) provides 3-D vegetation models to map accurate fuel properties critical for modelling fire behaviour. Laser point cloud data stratified into height intervals

John Gajardo, Mariano García and David Riaño contributed equally to this chapter.

J. Gajardo
Center for Spatial Technologies and Remote Sensing (CSTARS), University of California, Davis, One Shields Avenue, 139 Veihmeyer Hall, Davis, CA 95616, USA

Department of Geography and Geology, University of Alcalá, Alcalá de Henares, 28801 Madrid, Spain

Associated Research Unit GEOLAB, University of Alcalá-Spanish National Research Council (CSIC), Alcalá de Henares, 28801 Madrid, Spain/Albasanz 26-28 28037 Madrid, Spain

Faculty of Forest Sciences, University of Talca, 2 Norte # 685 Talca, Chile

M. García
Center for Spatial Technologies and Remote Sensing (CSTARS), University of California, Davis, One Shields Avenue, 139 Veihmeyer Hall, Davis, CA 95616, USA

Remote Sensing Area, Complutig-University of Alcalá, Alcalá de Henares, 28801 Madrid, Spain

Department of Geography and Geology, University of Alcalá, Alcalá de Henares, 28801 Madrid, Spain

Associated Research Unit GEOLAB, University of Alcalá-Spanish National Research Council (CSIC), Alcalá de Henares, 28801 Madrid, Spain/Albasanz 26-28 28037 Madrid, Spain

D. Riaño (✉)
Center for Spatial Technologies and Remote Sensing (CSTARS), University of California, Davis, One Shields Avenue, 139 Veihmeyer Hall, Davis, CA 95616, USA

Associated Research Unit GEOLAB, University of Alcalá-Spanish National Research Council (CSIC), Alcalá de Henares, 28801 Madrid, Spain/Albasanz 26-28 28037 Madrid, Spain
e-mail: driano@ucdavis.edu

M. Maltamo et al. (eds.), *Forestry Applications of Airborne Laser Scanning: Concepts and Case Studies*, Managing Forest Ecosystems 27, DOI 10.1007/978-94-017-8663-8__22, © Springer Science+Business Media Dordrecht 2014

coupled with spectral information can provide accurate fuel type maps, especially if non-parametric classifiers are used. Canopy bulk density (CBD) depends on ALS metrics related to canopy volume and biomass to yield regression models ranging between 0.77 and 0.94 in R^2. ALS estimates canopy base height (CBH) after the identification of the gap in the canopy that describes the beginning of the tree crown. Due to laser point density among other factors, CBH for individual trees is usually less accurate than for plots. Penetration through the upper canopy and the low height of the surface fuels combined with a low laser pulse density constrain the estimation of surface canopy height (SCH). ALS together with optical sensors map the above mentioned fuels properties more accurately than with any of these sensors alone due to the synergy of the structural and spectral information collected by each sensor. Despite the fact that most attempts at using ALS for fire management have been focused on characterization of fuels at the pre-fire stage or during the fire, multi-temporal ALS data also have high potential at the post-fire stage to estimate burn severity and vegetation regeneration after fire.

22.1 Introduction

Wild-land fires are an integral and major forest disturbance factor, playing an important role in the vegetation succession cycle as well as in the ecosystem's structure and its functionality (Koutsias and Karteris 2003; Gitas et al. 2012). Fuels, oxygen and a heat source form the fire fundamentals triangle and they coexist to drive fire occurrence. Once the fire is initiated, it behaves according to the physical factors triangle: weather, topography and fuel availability (Pyne et al. 1996). Fuels are any material that can burn during a fire. Their spatial arrangement and the live and dead biomass loads that constitute them impact fire spread, intensity and severity (Burgan and Rothermel 1984). Therefore, a good understanding of these components is key when considering all aspects of wild-land fire management, including fuel accumulation reduction; fire suppression; smoke control; wild-land/urban interface; forest health and ecosystem management; and ultimately global climate change (Pyne et al. 1996).

Satellite earth observation has proved to be a suitable tool for mapping fuel properties (Chuvieco et al. 2003). Remote sensing is a sound approach for mapping fuel types at different spatial scales, since it provides a spatially explicit representation of vegetation characteristics. Despite this potential shown, there are still some limitations which can be summarized as: (1) incapacity to detect surface fuels due to canopy occlusion and (2) limited capability to discriminate fuel types that differ only in height, since there is no direct relation between height and reflectance (Salas and Chuvieco 1995; Keane et al. 2006). As ALS data can provide 3-D information on vegetation structure it is useful to overcome these limitations.

This chapter reviews how airborne laser scanning (ALS) technology contributes to fuel characterization at the different pre, during and post-fire management phases. The following characteristics are considered: fuel type mapping, canopy bulk density (CBD), canopy base height (CBH), surface canopy height (SCH); fire severity and vegetation regeneration. In addition, the potential of ALS technology is reviewed either alone or in combination with multispectral/hyperspectral data, taking advantage of the structural information provided by ALS and the spectral information provided by passive optical sensors.

22.2 Fuel Type Mapping

A fuel type is "an identifiable association of fuel elements of distinctive species, form, size, arrangement, and continuity that will exhibit characteristic fire behaviour under defined burning conditions" (Merrill and Alexander 1987). Richard Rother-mel was the first to model fire behaviour, quantifying the fire spread rate and fire intensity in surface fires (Rothermel 1972). He also developed equations to model crown fires (Rothermel 1991). His formulations, still in use, consider the following fuel parameters to drive fire behaviour: fuel loading of fine, medium and large dead fuels and herbaceous and woody live fuels; surface to volume ratio of fine dead fuels and herbaceous and woody live fuels; fuel bed depth; moisture of extinction and heat content of live and dead fuels. Measuring these fuel parameters requires labour-intensive destructive sampling, so researchers at the Northern Forest Fire Laboratory (NFFL) in Missoula (Montana, USA) identified 13 distinctive fuel parameter combinations (fuel types) out of the infinite possibilities, depending on the type of fuel that carried the fire (Albini 1976; Anderson 1982). Fuel types 1st–3rd were herbaceous fuels, 4th–6th shrubs, 7th was considered non-burnable, 8th–10th were dead leaves under tree canopy and the 11th–13th were slash residues and basal accumulation material (Fig. 22.1). The authors sampled the fuels and

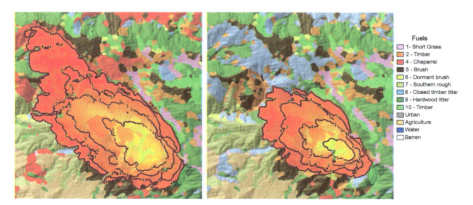

Fig. 22.1 A theoretical example of fire perimeter growths simulated with Wildfire Analyst (wildfireanalyst.com) using NFFL fuel type classes. The same meteorological conditions, fire ignition point and fuel type map were applied, except fuel type 4 (*left*) was changed for fuel type 8 (*right*)

established a photo guide to be able to identify each fuel type in the field based on its vertical structure and fire carrier without having to destructively collect fuels. Other fuel type classifications and photo guides including a wider range of fuel types and accommodated to other ecosystems across the globe have been developed since then (Chuvieco et al. 2003).

Field surveys provide very detailed data on the fuel types, but generating comprehensive fuel type maps with enough spatial and temporal coverage is time consuming and human-resource intensive (Chuvieco et al. 2003). Optical remote sensing has been used to cover large areas (Chuvieco et al. 2003, 2009), but these sensors are not able to map the vertical fuel structure (Riaño et al. 2002a). Riaño et al. (2002a) found significant confusion between three shrub fuel types that had a different fire behaviour according to shrub height, less than 0.6 m, 0.6–2 m or more than 2 m. This distinction in shrub heights served to map fuel types according to the classification scheme developed under the frame-work of the European project Prometheus (Prometheus 2000). In this study difficulties were found in distinguishing a tree canopy fuel type with understory shrub canopy versus another fuel type without one. These limitations can be overcome with ALS. Seielstad and Queen (2003) successfully identified two NFFL fuel types of the same tree canopy species, only differing due to the presence or absence of understory. The authors revisited their field plots to show that field surveyors had wrongly classified their fuel type in some cases, whereas ALS identified them correctly. Even though field surveys based on photo guides for determining fuel types might be considered subjective, calibration and validation of methodologies to generate fuel type maps from ALS still rely on field identification of fuel types. A more objective method, but one that would require further processing could be Terrestrial Laser Scanning (TLS) systems, shown by García et al. (2011a) and Seielstad et al. (2011) to be applicable for characterization of fuel properties. Therefore, the integration of TLS with ALS shows great potential for up scaling fuel type maps.

Accurate fire behaviour modelling depends on reliable maps of fuel types. Even under the same meteorological conditions and for the same ignition point, the fire perimeter does not grow the same if fuels change or are wrongly mapped as shown in the simulations in Fig. 22.1. Mutlu et al. (2008b) obtained very different fire perimeters and fire growth rates with FARSITE fire simulator using a fuel type map generated with Quickbird data alone versus a more accurate map that combined Quickbird and ALS data. Simulations like these, with adequate fuel types maps developed using ALS, can help to determine when and where fuel treatments are needed to reduce fire risk and its effects, planning prescribed burns and managing fire suppression activities, such as resources allocation and evacuation plans (Seielstad and Queen 2003).

A common pre-processing step to map fuel types using ALS discrete laser pulses is to stratify the laser point cloud into height intervals or bins (Mutlu et al. 2008a; García et al. 2011b). According to their X and Y values, each point is assigned to a raster grid cell and its Z (height) above the ground is calculated. Densities at each height interval are computed as the number of non-ground points relative to the total number of returns at each cell, in order to compensate for differences in scan density across flight lines. Additionally, the cell size needs to match the laser point density and its relation to the ALS scan pattern (Baltsavias 1999), so that enough points fall within each cell to build representative statistics. Mutlu et al. (2008a) chose 2.5 by 2.5 m cell size for an average 2.58 points/m^2 laser point density, which meant an average of approximately 16 points per cell. This number was not large enough for García et al. (2011b) who for a very similar laser point density average of 2.5 points/m^2 preferred a 6 by 6 m cell size instead, to obtain approximately 90 points per cell.

The strategy for defining the height intervals depends on the fuel types that are to be classified, the laser point density and the canopy height. Finer intervals are generally preferred to identify the surface fuels. Mutlu et al. (2008a) selected 0.5 m intervals up to 2.5 m to distinguish three NFFL fuel types classes defined according to their fuel height up to 0.3, 0.6 or 1.82 m. To characterize the forest canopy, they chose instead 2.5 m height intervals up to 10 m and finally 5 m intervals up to 30 m. Prometheus fuel types, adapted to the European Mediterranean basin, define surface fuel height limits up to 0.6, 2 or 4 m instead, so García et al. (2011b) chose the same 0.5 m intervals up to 4 m, and 1 m height intervals beyond.

To classify fuel types, variables extracted from ALS are often integrated with optical remote sensing data, since a fuel type is identified depending not only on its vertical fuel structure but also on its fuel horizontal spectral properties (García et al. 2011b). For example, ALS waveforms of two different fuel types, grasses and shrubs, can appear very similar if they have the same height with current ALS technology, but they can be distinguished by analysing their spectral response. On the other hand, two different fuel types, 0.6 and 2.0 m shrubs, can be identified with ALS waveform but not in the optical data (Fig. 22.2). Note that vegetation has a broader echo with returned energy from the top and within the canopy whereas the ground has a narrower one (Wagner 2010).

Multisource fuel type classifications with ALS and optical remote sensing rely on proper co-registration between sensors. The accuracy depends among other factors on the difference in the sensors' field of view and the laser point density in relation to the spatial resolution, together with the slope and vegetation height (Valbuena et al. 2011). In addition, these classifications combine variables whose residuals do not always satisfy the normal distribution assumption, hence non-parametric models are preferred to multivariate models to fully capture the co-variability structure between predictor and response (García et al. 2011b). To integrate ALS and Quickbird images, Mutlu et al. (2008a) applied minimum noise fraction to the input variables, an image processing technique based on principal component analysis, to separate information from noise, before running parametric Mahalanobis distance supervised classification that improved overall classification accuracy results by 3.9 %. Support Vector Machines (SVM), a non-parametric classifier, successfully combined ALS and airborne hyperspectral data to map land cover in a wildland-urban interface, useful to plan shrub clearing activities and determine the fire threat to houses (Koetz et al. 2008). SVM performed slightly better than maximum likelihood supervised classifier after merging ALS and airborne multispectral data, with a 5.7 % overall classification accuracy improvement (García et al. 2011b). In this case, authors

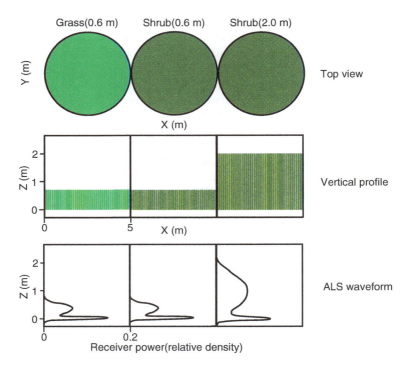

Fig. 22.2 *Top view, vertical profile* and *ALS waveform* of a 0.6 m grass, a 0.6 m shrub and a 2.0 m shrub canopies

selected the ALS and spectral variables that provided the greatest separability before running the SVM classification of six fuel groups. Over these groups, they applied decision rules based on ALS metrics only to classify Prometheus fuel types according to shrub height, shrub cover and vertical continuity between the understory and overstory.

These classification methods do not take advantage of contextual information. The relationship of an ALS cell and a pixel's spectral data with their neighbours is overlooked. Merging these simple units into larger objects, with the same fuel type, to then classify them, becomes especially relevant when ALS laser point density and optical data spatial resolution are much finer than the fuel type unit to map. Arroyo et al. (2006) applied this object-oriented classification technique to map fuel types, but processing only QuickBird optical satellite imagery.

When ALS vertical height profiles show gaps in the lower parts of the canopy resulting from a lack of laser points due to the interception by the upper canopy, confusion still occurs in classification of fuel types (García et al. 2011b). The intensity of the laser pulse returns and their transmission through the canopy depend on the configuration of the ALS flight: flying altitude, beam divergence, pulse repetition frequency, scan angle, pulse power concentration, lase pulse density and pulse geometry (Hopkinson 2007). Despite these ALS acquisition settings, Hopkinson (2007) suggests also that the penetration rate depends on the canopy reflectivity and its structure: canopy height, leaf density and orientation. For example, Chasmer et al. (2006) obtained a 13 % lower penetration rate for the taller conifer tree class with 100 kHz pulse repetition frequency. On the other hand, the presence of understory decreased the penetration rate in 7 % for similar or even lower tree heights. In all cases, increasing the laser point density would raise the chances of a more accurate characterization of the vertical profile (García et al. 2011b).

Another option would be to use full-waveform instead of discrete laser point instruments, something that would require other pre-processing steps. The systems used so far have a much larger footprint (several meters) so they do not provide as much detail of the canopy structure as the discrete systems described above. Another problem affecting large footprint systems is that for a steep terrain the complexity of the waveform is increased, making more difficult to resolve ground and canopy reflections within the waveform. The advent of new small footprint full waveform systems will overcome these limitations. Nevertheless, using full-waveform Geoscience Laser Altimeter System (GLAS) on board the Ice, Cloud, and Elevation Satellite (ICESat), Ashworth et al. (2010) discriminated two NFFL fuel types only different in the presence or absence of understory. Based on field recognition, they assigned each waveform to either fuel type and calculated their averages. For each fuel type, they picked only the range where waveform values were most similar within each fuel type, but least alike between fuel types. The authors selected two variables, a normalized laser intensity value in the upper canopy and another one for the lower canopy range, which were combined in a logistic regression model to classify these two fuel types with an overall prediction probability accuracy of 0.69.

22.3   Canopy Bulk Density (CBD)

Canopy bulk density (CBD), or crown bulk density when referring to an individual tree, is defined as the mass of available canopy fuel per unit canopy volume that would burn in a crown fire (Scott and Reinhardt 2001). CBD is an important characteristic for crown fire propagation and hence it is one of the input parameters of many fire behaviour and fire effects models. While the above definition of CBD is unanimously accepted, some differences can be found in the literature regarding the definition of available fuel. Thus, some researchers consider only the foliage biomass (Van Wagner 1977; Riaño et al. 2003; Keane et al. 2005), and others include foliage, lichen, moss and a portion of small branches that would be consumed in the flaming front (Scott and Reinhardt 2001; Keane et al. 2005). These differences in the definition of fuel load would lead to different CBD estimations.

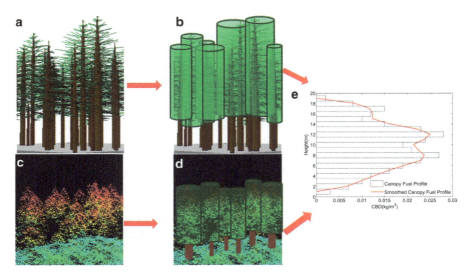

Fig. 22.3 Schematic representation of Sando and Wick's (1972) method for CBD estimation. (**a**) field sampling; (**b**) cylindrical shape assumption and fuel load distribution into height intervals, (**c**) ALS point cloud; (**d**) fuel load distribution based on ALS data; (**e**) fuel profile and CBD estimation

CBD is difficult to measure in the field. Moreover, unlike other canopy properties standardized field methods do not exist for CBD sampling (Keane et al. 2005). The methods proposed to quantify CBD are based on several assumptions, such as the equal distribution of the fuel load throughout the crown, or that the shape of the crown fits a given geometrical figure like a cone, a cylinder or an ellipsoid (see Figure 1 in Riaño et al. 2004). For uniform stands, CBD can be calculated as the fuel load divided by canopy depth (Keane et al. 1998); however, the assumption of uniformity is rarely fulfilled in most cases. To solve this limitation, Sando and Wick (1972) developed a methodology which first computes foliage and fine branch (<6 mm) biomass for every tree in the plot; the fuel load is then equally distributed into 1 ft (0.3048 m) intervals through the crown of the tree. Subsequently, the fuel load contribution of each tree for every height interval is summed providing a vertical fuel profile, representing the distribution of the fuel through the canopy. The resulting vertical profile of the fuel load is smoothed by applying a running mean filter. Finally, CBD is estimated as the maximum of the smoothed canopy fuel profile (Fig. 22.3).

Since CBD is essentially a 3-D variable that requires a description of the vertical distribution of fuel through the canopy, ALS data have great potential for estimating it thus improving the performance of fire behaviour models. Several studies using either discrete return or full waveform data have demonstrated this potential.

The most common approach using discrete systems is to derive ALS metrics related to canopy height, canopy closure and biomass which are subsequently used as independent variables in regression models (Andersen et al. 2005; Hall et al. 2005; Erdody and Moskal 2010; Skowronski et al. 2011). The pool of metrics derived varies largely between studies, ranging from 8 to 39 variables. As in many other forest ALS applications, percentiles are derived from the height distribution of the returns, which have been found to be suitable to estimate foliage and aboveground biomass (Lim and Treitz 2004), from which CBD is derived depending on its definition. The ratio of canopy returns to all returns is used as a measurement of canopy cover, which is related to the horizontal continuity of the fuels. Hall et al. (2005) derived a wide set of variables, some of them attempting to mimic variables derived from full waveform data, in an effort to evaluate the potential of metrics that could be derived from any type of ALS system. The models selected to estimate CBD included as explanatory variables metrics related to canopy height and canopy closure (Hall et al. 2005; Erdody and Moskal 2010; Skowronski et al. 2011). Variables related to canopy height and closure can be interpreted as a 3-D representation of canopy structure and thus, the distribution of the canopy fuels. The empirical approaches used in the studies described above yielded CBD estimates with R^2 values ranging between 0.77 and 0.94.

Skowronski et al. (2011) also used a different approach to estimate CBD. Although the final step of the method comprised a simple regression between field-derived and ALS-derived CBD, they derived the canopy height profile (CHP) as a representation of the distribution of fuel through the canopy and subsequently the ALS-CHP were regressed against the field-CHP. They performed this regression both considering the CHP as a whole and for each of the 1 m height intervals used to obtain the CHP. This approach has the main advantage of providing a 3-D representation of the CBD, instead of the 2-D given by the estimation of the maximum value of CBD derived from the canopy fuel profile obtained using Sando and Wick's method. As an alternative to simplify the CBD representation for each of their 20 height intervals, they proposed to fit Gaussian distributions to the CHP. These approaches yielded R^2 values of 0.82 for the CHP considering all layers; between 0.36 and 0.89 for the individual intervals; and 0.92, for the Gaussian function. It should be noted that both the CHP and the Gaussian functions provide qualitative estimates of CBD rather than actual quantitative values.

Riaño et al. (2003, 2004) derived CBD by dividing the foliage biomass by the crown volume, both previously derived from ALS data. Thus, foliage biomass was estimated using allometric equations using data derived from the ALS point cloud; whereas crown volume was derived as difference between canopy height and CBH (see Sect. 22.4 below for a detailed explanation on CBH), weighted by the canopy cover at different height intervals, that was derived from the CHP.

The convenience of using metrics that could potentially retain more information of the raw ALS data than the variables commonly derived from the height distribution was established by Zhao et al. (2011). These authors calculated a set of composite metrics that took into account the topological correspondence between the different type of returns (first, last and single for their study) associated to a given laser pulse. They combined these composite metrics with learning machine methods (SVM and Gaussian processes) that are more adequate for handling the higher dimensions of the newly derived metrics. Machine learning methods have shown their capability for capturing the implicit, potentially non-linear and complex relationships between dependent and independent variables. The use of machine learning techniques applied to the composite metrics slightly improved results obtained with linear regression techniques in the estimation of CBD with Root Mean Square Error (RMSE) values of 0.15 and 0.2 kg/m^3 respectively, and the correlation increased by about 10 %.

The benefit of integrating discrete return data with the spectral information of optical data was investigated by Erdody and Moskal (2010), who found only a slight improvement. This could be explained by the poor spectral information delivered by aerial photography, and by the fact that they provided little additional information for a 3-D variable. Nevertheless, the integration of multispectral and ALS data could be advantageous to apply species specific models after classification and ultimately to reconstruct the canopy as a 3-D multispectral object.

Full waveform systems have also shown great potential to estimate CBD. The higher vertical resolution of these sensors as compared to discrete systems could potentially provide more accurate estimates of CBD, as the waveforms can describe the distribution of canopy fuels, as long as the footprints of the new systems become smaller. Peterson (2005) estimated CBD using Laser Vegetation Imaging Sensor (LVIS), which was a medium-high altitude airborne ALS system developed by NASA. This system operated at 10 km altitude provided large footprint (25 m) full waveform records. From the energy recorded at 0.3 m height intervals several metrics were derived related to canopy height, canopy cover, canopy depth or the amplitude of the waveforms. Subsequently, a regression was carried out to estimate CBD as in other studies described above using discrete systems. This approach yielded poor results with $R^2 = 0.34$, which could be a consequence of the varied vegetation types of the study area and the structural differences between them. Development of species-specific models greatly improved the results with R^2 values ranging from 0.38 to 0.97, with an overall R^2 value between the collective observed and predicted CBD values of 0.71. In general, the selected variables were also related to canopy height and canopy cover.

García et al. (2012) developed a method to estimate CBD from the information provided by the waveforms themselves, based on the relationship between the energy recorded at each height interval and the amount of canopy material present at each interval. Thus the CHP was related to the vertical distribution of canopy fuels which subsequently served to estimate foliage biomass using the leaf area index derived from the CHP divided by the specific leaf area for the species present in their study area. Subsequently, the vertical fuel profile was smoothed using a 3 m running mean and the CBD was estimated as the maximum value of the smoothed fuel profile. This approach, yielded an R^2 of 0.78; and though originally developed to be used for ICESat/GLAS data, it could be applied to ALS full waveform data or even pseudo-waveforms derived from discrete return datasets.

22.4   Canopy Base Height (CBH)

Canopy base height (CBH), or crown base height when referring to an individual tree, is often defined from a forest fire point of view as the lowest height over the ground where there is enough canopy fuel to cause vertical fire spread (Scott and Reinhardt 2001) (Fig. 22.4). In the same context, other authors define CBH as a function of the amount of fuel required to spread fire into the canopy, normally by applying a minimum value of CBD as a threshold (see Sect. 22.3 for a detailed explanation on CBD), commonly set at 0.011 kg/m^3 (Sando and Wick 1972; Andersen et al. 2005; Reinhardt et al. 2006; Erdody and Moskal 2010). Other authors understand it as the vertical distance from the ground to the live crown base (Finney 1998), or distance to the crown base, whether live or dead (Ottmar et al. 2000). Accurate CBH estimation and other forest fuel variables can significantly improve the quality of the inputs in fire simulation models like FARSITE, Flamap (Andersen et al. 2005; Peterson et al. 2007; Popescu and Zhao 2008; Erdody and Moskal 2010), Wildfire Analyst or the Wildland-urban interface Fire Dynamics Simulator (WFDS) (Contreras et al. 2012).

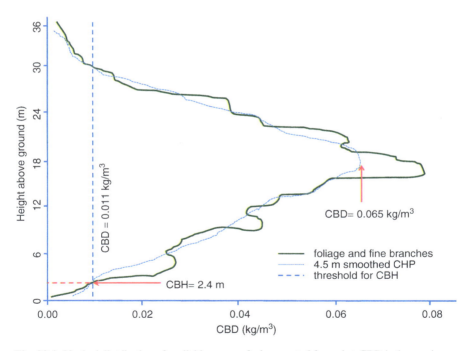

Fig. 22.4 Vertical distribution of available canopy fuel computed for a plot. CBD is the maximum of the smoothed fuel profile. CBH is the lowest point where smoothed profile reaches a CBD value of 0.011 kg/m^3 (Reinhardt et al. 2006)

Consideration of ladder fuels such as lichens, shrubs and small trees can provide more realistic representation of the danger of fire crown propagation. The methods for characterizing the ladder fuel risk can be divided into two approaches. One approach is qualitative and it uses field training experience to get a visual estimation of forest fire risk. In this context, Menning and Stephens (2007) developed a semi-qualitative/quantitative approach using a flowchart (ladder fuel hazards at site, LaFHAs) to direct technicians to classify forest ladder fuel risks (high, medium, low). The flow-chart gives consistency in assessing each site, systematizing the method and establishing up a range for possible results.

The other approach is quantitative and represents efforts to measure several forest properties (Sando and Wick 1972). The attempts to characterize ladder fuel risks involves obtaining these attributes for various spatial scales, where different fuel and vegetation can act as ladders from the ground to the canopy, becoming a complex scenario (Menning and Stephens 2007).

Despite multiple definitions and existing methods, the estimation of CBH is difficult to calculate because is not represented either by the lowest height in the crown nor the average CBH in the plot. To attain a more accurate CBH estimation, Sando and Wick (1972) considered CBH as the height where the minimum bulk density reached 0.037 kg/m^3. Moreover, Van Wagner (1993) simplified the scenarios by computing the CBH considering multiple forest layers such as only two layers.

TLS approaches have shown to be successful in arriving at CBH estimates. García et al. (2011a), used a voxel method to get a binary vegetation vertical profile, according to a canopy cover threshold, where the CBH was considered as the minimum height layer belonging to highest fuel layer. The results indicated some underestimation ranging from 0.6 to 6.9 m with an RMSE of 3.09 m. They also obtained the fuel strata gap which represents the distance between surface and canopy fuels. Despite being less commonly used than CBH in fire behaviour models, this variable has the potential to better represent the risk of transition from surface to canopy fires as it actually represents the distance between fire carriers (García et al. 2011a). Schilling et al. (2012) used a method that implied voxels and tree topology to get CBH estimates; they achieved differences of about 1.29 m showing underestimation between field data and TLS. Although results are promising, these experiments are time-consuming as they require extensive field campaigns to gather the representative spatial variability of the data.

Experiences with ALS have been more numerous and they can be classified between those using discrete pulse and full waveform systems. In the first instance, most of the studies have applied metrics from return height distributions to get CBH estimates. Riaño et al. (2004) used a cluster analysis to separate returns from two strata (overstory and understory). They then calculated metrics such as the first percentile of canopy hits and the minimum canopy laser height to estimate CBH at individual tree and plot level, getting 0.65 and 0.97 for R^2 respectively. Regression models based on these metrics usually give better performance at the plot level rather than at the individual tree level, mainly because the ALS point density is not large enough to characterize each individual tree crown and due to the occlusion by the upper part of the canopy. In addition, at the plot level there are structural features (from vegetation components to canopy openness) that contribute to the smoothing of the estimated metrics (Naesset and Økland 2002; Riaño et al. 2004; García et al. 2010). Naesset and Økland's (2002) tree model included metrics for first quartile with first and last returns, the model explained 53 % of variance and reaching 0.37 m for RMSE. Their plot level model explained 71 % with an RMSE of 0.25 m and included as predictor variables the 75th and 90th percentiles of the height of the last returns.

To overcome limited point density, Morsdorf et al. (2004) flew twice over the same area to estimate tree level CBH. The tree segmentation was based on a cluster analysis of the point cloud, which derived position and extension for each individual tree crown. To delineate individual tree crowns, Contreras et al. (2012) preferred instead to consider crown width as a percentage of tree height according to ranges of canopy cover. CBH was computed using a mobile window size (twice the crown width) centered at tree position determined throughout a stem identification algorithm (Popescu and Wynne 2004). The CBH value was then calculated as the mean height of all canopy height model (CHM) points inside the search window divided by the associated standard deviation of the heights. A different approach was used to segment the point cloud to isolate single tree crowns using a parabolic surface fit to the CHM (Holmgren and Persson 2004). Each tree crown was divided into slices of 0.5 m and those bins containing more than 1 % of the returns were marked as 1 whereas bins with less than 1 % of the returns were set to zero. CBH was considered as the distance from ground to highest bin set to zero. Using an algorithm involving tree position and vertical CHM profile to define crown boundary, Popescu and Zhao (2008) computed a pseudo waveform for each tree. CBH estimation used a polynomial fit over the waveform, finding height where intensity vertical profile showed its inflection point after the curve reached its maximum.

Discrete pulse ALS systems systematically overestimate CBH in both plot (Riaño et al. 2004; Hall et al. 2005) and individual tree levels (Holmgren and Persson 2004; Riaño et al. 2004; Popescu and Zhao 2008). The reason for this may lie in the inability to detect the first live branch due to occlusion by the upper canopy. This effect can be particularly noticeable when the canopy is very dense or the point density used is low (Holmgren and Persson 2004; Riaño et al. 2004). Overestimation could be reduced using higher point density or systems that record multiple echoes (Naesset and Økland 2002; Holmgren and Persson 2004). Other factors affecting CBH estimations are related to the vegetation characteristics (e.g. savannah vs. rainforest), such as canopy permeability or canopy cover which will limit the ability of the LiDAR to penetrate to lower parts of the canopy, thus producing an overestimation of CBH. Another important consideration is the difference between what is sampled in the field and what is actually measured by the ALS (Andersen et al. 2005).

Erdody and Moskal (2010) explored the potential of fusing multispectral data with ALS to obtain several canopy fuel metrics. The model providing the highest performance included ALS metrics (25th and 90th height percentiles) and information derived from multispectral images, such as the average value of the near-infrared band and homogeneity measure derived from the green band.

CBH models from fused datasets, albeit providing the best results, showed minor improvement respect to models obtained from the different datasets alone, as it can be expected since the aerial photography provided little additional information for CBH estimation.

Regarding waveform studies, Peterson et al. (2007) derived CBH data at plot level. They applied an exponential transformation to correct occlusion effects in the lower part of the canopy on the waveforms. Several metrics were computed as predictor variables for CBH estimation. CBH models explained 59 % variance with 0.57 m RMSE. Differences between the predicted CBH and field data could be caused again by the field methods to get the CBH. Only trees over 10 cm diameter breast height were measured, excluding a significant number of branch and stems that show in the lower part of density profile, leading to an erroneous higher CBH field estimation. Cuesta et al. (2010) derived CBH using a threshold method based on waveform return recognition, when a certain signal-to-noise ratio was exceeded. Threshold values were selected to avoid error probability (false alarm and no detection) in the vertical profile.

Similarly to discrete pulses, waveform tends to overestimate CBH as consequence of a full wave process, or depending on how the parameter extraction was defined from ALS data (Popescu and Zhao 2008). Apart from this, other sources of bias exist linked to the methods of obtaining field data, which often present biases affecting validation results (Peterson et al. 2007). Although full waveform processing allows the point density to be increased to improve the extraction of metrics, CBH estimation still shows overestimation associated with a limited pulse penetration on a denser canopy that generally is present in the young forests (Cuesta et al. 2010).

22.5 Surface Canopy Height (SCH)

SCH can be considered as the height of the fuel layer that is situated above ground fuels but beneath the canopy or aerial fuel (Scott and Reinhardt 2001). Moreover, Riaño et al. (2003) consider it as the height associated with the shrub strata that is generally below the forest canopy or growing without any layer above. In the context of fire behaviour, surface fuels (mostly grass, brush, slash and timber) contribute to intensity and fire spread (Finney 1998). Shrub fuel types are commonly classified according to height, which correspond to different fire behaviour. For example, European Prometheus fuel type classification system used by Riaño et al. (2002a), distinguished three different heights: surface shrubs (<0.6 m.), medium height-shrubs (0.6–2 m.) and tall-shrubs (2–4 m.)

SCH is traditionally measured from the field using transects, measuring tapes (Riaño et al. 2007) or survey plots (Mutlu et al. 2008a; Sankey and Bond 2011). However, this method only collects data at a reduced local scale (Sankey and Bond 2011), so this is where ALS plays an important role. When there is large height gap between canopies and the ground, with no understory vegetation, ground returns can be easily discriminated from those occurring inside the canopy; however, in areas where shrubs are predominant, height differences between ground and shrubs are very small, so laser pulse hits inside the canopy are frequently classified as "ground" (Axelsson 1999), increasing the uncertainty of surface fuel height (Riaño et al. 2007; Estornell et al. 2011a). For instance, Sankey and Bond (2011) showed an underestimation of about 25–40 cm when they compared heights from ALS to field data. These differences were produced by an insufficient pulse penetration into the canopy, by low point density used to represent canopy top, and by a ground overestimation in areas where the canopy is too dense to prevent a deeper pulse penetration, causing errors in the Digital Ground Model (DGM) interpolation process (Axelsson 2000). Similar results were shown by Glenn et al. (2011), when they used a discrete ALS to estimate shrub height in a semi-desert and steep terrain area. Based on individual isolation using ellipses, four methods were tested to develop shrub height models. The results showed that factors such as hill slope and vertical and horizontal ALS system accuracy ordinarily do not affect height estimation, which is conditioned instead by the error caused due to low point density, which prevent the shrub top height from being sampled. These results have led to the development of new methods that would increase the height accuracy for this type of vegetation. Most of these methods are based on algorithms that separate returns into canopy and ground clusters, and then apply an iterative process to reclassify irregular points as additional ground points, thus improving the DGM interpolation (Streutker and Glenn 2006).

The potential shown by using several data sources led Riaño et al. (2007) to combine discrete ALS with spectral data from colour infrared ortho-images to improve DGM point ground discrimination, using decision rule based on height threshold and Normalized Difference Vegetation Index (NDVI) values. This data combination improved correlations from 0.48 to 0.65 and RMSE from 0.18 to 0.15 m. In addition, Estornell et al. (2011a) also applied NDVI to correct CHM that presented unacceptably low values. With the improved DGM, the model showed 0.78 for correlation and RMSE was reduced to 0.13 m. Other variations based on fusion of ALS data and multispectral images have been tested. Mutlu et al. (2008a, b) used a height intervals approach on ALS data to derive vertical canopy height distribution, and they combined it with spectral bands from QuickBird imaging to derive a classification. Results improved fuel surface classification by 13 % respect to other datasets that did not apply data fusion.

Regarding the use of ALS data and hyperspectral images, Mundt et al. (2006) created shrub distribution maps in a semi-desert zone. Results showed that classifications made with fused ALS (roughness surface) and Mixture Tuned Matched Filtering (MTMF) algorithm, improved global accuracy by about 14 % when compared with classifications that applied only spectral data. There are few studies with full waveform in a shrub vegetation ecosystem. Riaño et al. (2003) proposed a waveform simulation from discrete pulse to improve shrub height estimations. This correction for lower canopy strata increased the understory cover estimation from 16 to 26 %, but their approach was not validated. Spaete et al. (2011) worked with ICE-Sat/GLAS data to estimate shrub height in a semi-desert zone with a methodology that could also be applicable to ALS full waveform data. Mean signed error (MSE) was preferred to compare ground height estimated from GLAS with those from ALS data. Maximum, minimum and mean vegetation height was obtained from the GLAS waveforms; linear regressions were used to explore relationships between vegetation height and raw GLAS waveform (skewness, amplitude, area, width) characteristics. Results showed an underestimation of 0.72 m for GLAS-derived vegetation height as compared to vegetation height derived from discrete ALS data.

Shrub height estimation has several shortcomings that need to be addressed in order to obtain better results. Most problems come from the DGM generation process and arise because ground and canopy hits are very difficult to separate for this type of vegetation. As a consequence, the DGM could mistakenly be generated including not only ground returns but also shrub returns (Streutker and Glenn 2006; Riaño et al. 2007; Estornell et al. 2011b). This result shows the need to have algorithms that minimize DGM errors, especially for lower vegetation and high slopes areas that are usually ignored by conventional ground filter algorithms (Estornell et al. 2011a; Glenn et al. 2011). Despite corrections made on data (DGM, CHM, slope, sample radius, point density), ALS data showed an underestimation for shrub height, caused mainly by insufficient point penetration in the canopy, low probability of detection of the top, errors in detection of the threshold needed to trigger a return, and ground overestimation in areas where the canopy is too dense, preventing return of a suitable number of samples from the lower layer (Sankey et al. 2010; Sankey and Bond 2011). Combination of ALS and multispectral data (Riaño et al. 2007; Varga and Asner 2008; Estornell et al. 2011a), hyperspectral (Mundt et al. 2006) and corrections for lower strata based in the waveform process (Riaño et al. 2007; Spaete et al. 2011) can improve SCH estimations.

22.6 Burn Severity

When a fire occurs, a detailed knowledge of the level of damage and its distribution within the burnt area is critical in order to quantify the impact, prioritize treatments and plan restoration and recovery activities (Patterson and Yool 1998; van Wagtendonk et al. 2004; De Santis et al. 2009). Lentile et al. (2006) suggested the term post-fire effects to quantify the damage and to distinguish between fire severity and burn severity as a function of time, from the immediate effects to the time that takes the vegetation to return to its pre-fire levels or function. Thus, the former term refers to the direct effects of the combustion process, e.g. vegetation consumption and mortality; the latter term refers to the impact of fire on soil and plants after the fire has been extinguished.

Several authors have proposed field methods to discriminate burn severity based on quantitative or qualitative criteria considering the physical, biological and chemical manifestation of combustion on vegetation (Miller and Yool 2002; Key and Benson 2006; Lentile et al. 2006). Thus, decrease in plant cover, depth of char, transformation of soil components to soluble mineral forms or tree mortality are evaluated (De Santis et al. 2009). The most widely used method to estimate burn severity is the Composite Burn Index (CBI), which defines burn severity from an ecological perspective. This method assigns scores from 0 (unburned) to 3 (completely burned), and although the CBI does not require quantitative measurements in the field, it provides a numerical estimation of burn severity for the site, which simplifies the statistical estimation derived from quantitative remotely sensed data (De Santis and Chuvieco 2007). De Santis and Chuvieco (2009) proposed a modification of the CBI, called GeoCBI which takes into consideration the fractional cover of each stratum to improve estimations of fire severity.

Burn severity has been commonly estimated using images obtained by passive sensors given the changes in reflectance resulting from vegetation removal, soil exposure and soil and vegetation moisture content. These changes are reflected as a decrease in near-infrared reflectance and an increase in short-wave infrared. Therefore, spectral indices such as the Normalized Burn Ratio (NBR) and their absolute (dNBR) and relative difference (RdNBR) have been related to burn severity as measured in the field.

Since fires will cause changes in forest structure at different degrees depending on fire intensity, ALS has the potential to be used for burn severity estimation. In addition, since CBI is structured in a hierarchal manner considering five strata (from substrate to trees higher than 20 m) and different degrees of severity levels that can occur at different strata, the relation between spectral indices and burn severity can be hindered, especially when only lower strata are affected. Some authors have found a non-linear relationship between dNBR and CBI for CBI >2 (Epting et al. 2005; Hall et al. 2008). Therefore, the use of ALS data either alone or combined with multispectral/hyperspectral data can help to improve burn severity estimations.

Post-fire effects can be characterized by changes in structural variables such as mean canopy height, fractional cover and vegetation fill or vertical cross-section occupied by vegetation as a result of biomass consumption (Wang and Glenn 2009; Wulder et al. 2009; Kwak et al. 2010). Thus, Wulder et al. (2009) evaluated whether these variables presented statistically significant differences between pre-fire and post-fire conditions in Canada as derived from a profiling ALS system. Whereas before the fire, segments inside and outside the fire perimeter did not show significant differences, after the fire the differences were significant (p-value <0.05). This results were confirmed both globally, for the whole ALS transect, and in a pairwise segment-based analysis. In addition, these variables showed a high correlation with post-fire conditions derived based on the NBR, dNBR and RdNBR obtained from Landsat-TM/ETM+ imagery. Goetz et al. (2010) have pointed out that although a decrease in canopy height can be observed after fire occurrence, it may not be a good overall indicator of fire disturbance since different levels of severity may be present within a burned area, including unburned patches.

When no multi-temporal ALS data is available ALS data can be integrated with multispectral data to provide a qualitative assessment of vegetation damage caused by fire. Kwak et al. (2010) evaluated the damage level of a fire in South Korea using airborne ALS and aerial photography. The level of damage was graded by combining the physical damage as a function of canopy cover and the biological damage as a function of NDVI. Each type of damage was divided into low and severe damage using a value obtained from the intersection of the probability density functions obtained from samples collected at burned and unburned areas as threshold. Finally the two damage types were fused into a single index with four values resulting from the combination of the values of each individual index.

Although less used than height related metrics, mean intensity and maximum intensity values associated with the returns have shown statistically significant differences between burned and unburned trees (Wing et al. 2010) and could thus potentially be used to assess post-fire conditions. On this basis, Kim et al. (2009) used intensity values in conjunction with field measurements to distinguish between live and dead standing tree biomass in a mixed coniferous forest in USA.

An important aspect to be considered when using structural information to estimate burn severity is the fact that CBI low-moderate severity classes can be obtained for surface fires that do not affect the canopy and so canopy height and canopy cover will not be the best indicators of post-fire conditions. In such a case structural variables should be related to understory vegetation.

22.7   Vegetation Regeneration After Fire

Vegetation regeneration after a fire refers to ecosystem recovery after this type of disturbance. Regeneration trends depend on fire severity, pre-existing vegetation, species traits adapted to fire and climatic and terrain conditions (Pausas and Vallejo 1999). All these factors need to be favourable for vegetation to return to its pre-fire state or otherwise extended recovery times or even permanent vegetation degradation can be expected (Frolking et al. 2009). Vegetation recovery can be measured in terms of how species composition, biomass, vegetation structure and height evolve after the fire, putting them in context of the pre-fire vegetation status (Frolking et al. 2009). Quantifying these parameters helps in (1) designing a resource allocation plan after the fire based on fire severity and ecological interest, (2) establishing a vegetation restoration plan to reduce soil erosion and promote vegetation establishment and (3) determining the needs for future pre-scribed burns (Keeley 2000).

To accomplish spatially comprehensive fire regeneration studies, supporting pre-fire field and remote sensing data is needed, which is always challenging except in the case of a prescribed fire. In addition, a number of repeated observations after the fire are required to track the regeneration process. The most common approach to follow the vegetation regeneration trajectory from remote sensing has been the NDVI time series analysis with Landsat-TM data (Díaz-Delgado et al. 1998). NDVI quantifies vegetation greenness as a proxy to biomass. With the Landsat TM/ETM+ archive freely available (landsat.usgs.gov, last accessed on February 16, 2014), it is possible to monitor pre and post-fire conditions to combine with ALS data. However, atmospheric and illumination inconsistencies between images, coupled with vegetation phenological shifts between image dates due to inter-annual and/or seasonal meteorological conditions differences, can mask the regeneration patterns (Henry and Hope 1998).

ALS can potentially track the recovery process since it measures, unlike optical remote sensing, vegetation height, biomass and structure directly. However, its application is limited due to the lack of data before a fire event and the cost of multiple ALS acquisitions over time to monitor the recovery. To partially overcome these difficulties, Angelo et al. (2010) compared a fire history map to predict time since fire using ALS vertical profiles in Florida (USA). Following a similar approach, Goetz et al. (2010) initially showed no clear trend in the relationship between vegetation height versus time since fire using ICESat/GLAS data in Alaska (USA). These approaches need to assume that pre-fire vegetation type, structure and fire severity was the same across fires (Goetz et al. 2010). Fusion with optical sensors is recommended to stratify the data by vegetation type. Goetz et al. (2010) divided their data according a MODIS Enhanced Vegetation Index (EVI) 0.5 threshold. Sites above this threshold were most likely deciduous forest with height regrowth of 0.21 m per year, but the less productive evergreens did not increase their tree height within the ~60 years study period. These authors also compared burned sites to unburned neighbour (control site) GLAS waveforms to assess fire severity, assuming as well that pre-fire vegetation type and structure was similar in both sites. This approach could also be applied to monitor regeneration mixing fires of multiple years and their corresponding control sites, a technique already applied to optical sensors (Díaz-Delgado et al. 1998; Riaño et al. 2002b). The limited use of ALS technology for monitoring vegetation regrowth could be a consequence of limited existence of space borne missions, unfavourable weather conditions and, the relatively high cost of new acquisitions (Gitas et al. 2012). As more and more ALS is available, more opportunities will arise to monitor vegetation regeneration before the launch of a full coverage space borne mission. To ensure an accurate regrowth measurement, special attention is needed to use similar ALS sensor with comparable flight configurations and to ensure the co-registration between multiple ALS acquisitions and with also any additional data sources from satellite-borne and airborne optical sensors.

Acknowledgements John Gajardo was supported by CONICYT Doctoral Fellowship, Government of Chile. Felix Morsdorf and Rubén Valbuena provided insightful review to improve this chapter. We would also like to thank Joaquín Ramírez and his team from Tecnosylva SL. for generating Fig. 22.1. Linguistic assistance from Richard Hewitt is as well acknowledged.

References

Albini FA (1976) Estimating wildfire behavior and effects. USDA, Forest Service, Intermountain Forest and Range Experiment Station, Ogden
Andersen HE, McGaughey RJ, Reutebuch SE (2005) Estimating forest canopy fuel parameters using LIDAR data. Remote Sens Environ 94:441–449

Anderson HE (1982) Aids to determining fuel models for estimating fire behavior. USDA, Forest Service, Ogden

Angelo JJ, Duncan BW, Weishampel JF (2010) Using lidar-derived vegetation profiles to predict time since fire in an oak scrub landscape in East-Central Florida. Remote Sens 2:514–525

Arroyo LA, Healey SP, Cohen WB, Cocero D, Manzanera JA (2006) Using object-oriented classification and high-resolution imagery to map fuel types in a Mediterranean region. J Geophys Res Biogeosci 111(G4):G04S04

Ashworth A, Evans DL, Cooke WH, Londo A, Collins C, Neuenschwander A (2010) Predicting southeastern forest canopy heights and fire fuel models using GLAS data. Photogramm Eng Remote Sens 76:915–922

Axelsson P (1999) Processing of laser scanner data—algorithms and applications. ISPRS J Photogramm Remote Sens 54:138–147

Axelsson P (2000) DEM generation from laser scanner data using adaptive TIN models. Int Arch Photogramm Remote Sens 33:111–118

Baltsavias EP (1999) Airborne laser scanning: basic relations and formulas. ISPRS J Photogramm Remote Sens 54:199–214

Burgan RE, Rothermel RC (1984) BEHAVE: fire behaviour prediction and fuel modeling system. USDA Forest Service, Ogden

Chasmer L, Hopkinson C, Smith B, Treitz P (2006) Examining the influence of changing laser pulse repetition frequencies on conifer forest canopy returns. Photogramm Eng Remote Sens 72:1359–1367

Chuvieco E, Riaño D, Van Wagtendonk JW, Morsdorf F (2003) Fuel loads and fuel types. In: Chuvieco E (ed) Wildland fire danger estimation and mapping. The role of remote sensing data. World Scientific Publishing Co. Ltd., Singapore, pp 120–142

Chuvieco E, Wagtendok J, Riaño D, Yebra M, Ustin SL (2009) Estimation of fuel conditions for fire danger assessment. In: Chuvieco E (ed) Earth observation of wildland fires in Mediterranean ecosystems. Springer, Berlin, pp 83–96

Contreras MA, Parsons RA, Chung W (2012) Modeling tree-level fuel connectivity to evaluate the effectiveness of thinning treatments for reducing crown fire potential. For Ecol Manage 264:134–149

Cuesta J, Chazette P, Allouis T, Flamant PH, Durrieu S, Sanak J, Genau P, Guyon D, Loustau D, Flamant C (2010) Observing the forest canopy with a new ultra-violet compact airborne Lidar. Sensors 10:7386–7403

De Santis A, Chuvieco E (2007) Burn severity estimation from remotely sensed data: performance of simulation versus empirical models. Remote Sens Environ 108:422–435

De Santis A, Chuvieco E (2009) GeoCBI: a modified version of the Composite Burn Index for the initial assessment of the short-term burn severity from remotely sensed data. Remote Sens Environ 113:554–562

De Santis A, Chuvieco E, Vaughan PJ (2009) Short-term assessment of burn severity using the inversion of PROSPECT and GeoSail models. Remote Sens Environ 113:126–136

Díaz-Delgado R, Salvador R, Pons X (1998) Monitoring of plant community regeneration after fire by remote sensing. In: Traboud L (ed) Fire management and landscape ecology. International Association of Wildland Fire, Fairfield, pp 315–324

Epting J, Verbyla D, Sorbel B (2005) Evaluation of remotely sensed indices for assessing fire severity in interior Alaska using Landsat TM and ETM+. Remote Sens Environ 96:328–339

Erdody TL, Moskal LM (2010) Fusion of LiDAR and imagery for estimating forest canopy fuels. Remote Sens Environ 114:725–737

Estornell J, Ruiz LA, Velazquez-Marti B (2011a) Study of shrub cover and height using LIDAR data in a Mediterranean area. For Sci 57:171–179

Estornell J, Ruiz LA, Velazquez-Marti B, Fernandez-Sarria A (2011b) Estimation of shrub biomass by airborne LiDAR data in small forest stands. For Ecol Manage 262:1697–1703

Finney MA (1998) FARSITE: fire area simulator – model development and evaluation. USDA Forest Service, Rocky Mountain Research Station, Ogden, RMRS-RP-4, p 47

Frolking S, Palace MW, Clark DB, Chambers JQ, Shugart HH, Hurtt GC (2009) Forest disturbance and recovery: a general review in the context of spaceborne remote sensing of impacts on aboveground biomass and canopy structure. J Geophys Res-Biogeosci 114:G00E02

García M, Riaño D, Chuvieco E, Danson FM (2010) Estimating biomass carbon stocks for a Mediterranean forest in central Spain using LiDAR height and intensity data. Remote Sens Environ 114:816–830

García M, Danson FM, Riano D, Chuvieco E, Ramirez FA, Bandugula V (2011a) Terrestrial laser scanning to estimate plot-level forest canopy fuel properties. Int J Appl Earth Obs Geoinfo 13:636–645

García M, Riaño D, Chuvieco E, Salas FJ, Danson FM (2011b) Multispectral and LiDAR data fusion for fuel type mapping using support vector machine and decision rules. Remote Sens Environ 115:1369–1379

García M, Popescu SC, Riaño D, Zhao K, Neuenschwander A, Agca M, Chuvieco E (2012) Characterization of canopy fuels using ICESat/GLAS data. Remote Sens Environ 123:81–89

Gitas I, Mitri G, Veraverbeke S, Polychronaki A (2012) Advances in remote sensing of post-fire vegetation recovery monitoring–a review. In: Fatoyinbo L (ed) Remote sensing of biomass – principles and applications. InTech, Rijeka, Croatia. http://www.intechopen.com/books/mostdownloaded/remote-sensing-of-biomass-principles-and-applications

Glenn NF, Spaete LP, Sankey TT, Derryberry DR, Hardegree SP, Mitchell JJ (2011) Errors in LiDAR-derived shrub height and crown area on sloped terrain. J Arid Environ 75:377–382

Goetz SJ, Sun M, Baccini A, Beck PSA (2010) Synergistic use of spaceborne lidar and optical imagery for assessing forest disturbance: an Alaska case study. J Geophys Res-Biogeosci 115:G00E07

Hall SA, Burke IC, Box DO, Kaufmann MR, Stoker JM (2005) Estimating stand structure using discrete-return lidar: an example from low density, fire prone ponderosa pine forests. For Ecol Manage 208:189–209

Hall RJ, Freeburn JT, de Groot WJ, Pritchard JM, Lynham TJ, Landry R (2008) Remote sensing of burn severity: experience from western Canada boreal fires. Int J Wildland Fire 17:476–489

Henry MC, Hope AS (1998) Monitoring post-burn recovery of chaparral vegetation in southern California using multitemporal satellite data. Int J Remote Sens 19:3097–3107

Holmgren J, Persson Å (2004) Identifying species of individual trees using airborne laser scanner. Remote Sens Environ 90:415–423

Hopkinson C (2007) The influence of flying altitude, beam divergence, and pulse repetition frequency on laser pulse return intensity and canopy frequency distribution. Can J Remote Sens 33:312–324

Keane RE, Garner JL, Schmidt KM, Long DG, Menakis JP, Finney MA (1998) Development of input data layers for the FARSITE fire growth model for the Selway-Bitterroot Wilderness complex, USA, vol GTR-3, General technical report RMRS. U. S. Department of Agriculture, Forest Service, Rocky Mountain Research Station, Ogden

Keane RE, Reinhardt ED, Scott J, Gray K, Reardon J (2005) Estimating forest canopy bulk density using six indirect methods. Can J For Res 35:724–739

Keane RE, Frescino T, Reeves MC, Long JL (2006) Mapping wildland fuel across large regions for the LANDFIRE Prototype Project. In: Rollins CK (ed) The LANDFIRE Prototype Project: nationally consistent and locally relevant geospatial data for wildland fire management, vol GTR_175, General technical report RMRS. USDA, Forest Service, Rocky Mountain Research Station, Frot Collins

Keeley JE (2000) Chaparral. In: Barbour MG, Billings WD (eds) North American terrestrial vegetation, 2nd edn. Cambridge University Press, Cambridge, UK, pp 204–253

Key CH, Benson NC (2006) Landscape assessment: ground measure of severity, the composite burn index; and remote sensing of severity, the normalized burn ratio. In: Lutes DC, Keane RE, Caratti JF, Key CH, Benson NC, Gangi LJ (eds) FIREMON: fire effects monitoring and inventory system. USDA Forest Service, Rocky Mountain Research Station, Ogden, General technical report. RMRS-GTR-164-CD: LA1-51

Kim Y, Yang Z, Cohen WB, Pflugmacher D, Lauver CL, Vankat JL (2009) Distinguishing between live and dead standing tree biomass on the North Rim of Grand Canyon National Park, USA using small-footprint lidar data. Remote Sens Environ 113:2499–2510

Koetz B, Morsdorf F, van der Linden S, Curt T, Allgower B (2008) Multi-source land cover classification for forest fire management based on imaging spectrometry and LiDAR data. For Ecol Manage 256:263–271

Koutsias N, Karteris M (2003) Classification analyses of vegetation for delineating forest fire fuel complexes in a Mediterranean test site using satellite remote sensing and GIS. Int J Remote Sens 24:3093–3104

Kwak D-A, Chung J, Lee W-K, Kafatos M, Lee SY, Cho H-K, Lee S-H (2010) Evaluation for damaged degree of vegetation by forest fire using lidar and a digital aerial photograph. Photogramm Eng Remote Sens 76:277–287

Lentile LB, Holden ZA, Smith AMS, Falkowski MJ, Hudak AT, Morgan P, Lewis SA, Gessler PE, Benson NC (2006) Remote sensing techniques to assess active fire characteristics and post-fire effects. Int J Wildland Fire 15:319–345

Lim KS, Treitz PM (2004) Estimation of above ground forest biomass from airborne discrete return laser scanner data using canopy-based quantile estimators. Scand J For Res 19:558–570

Menning KM, Stephens SL (2007) Fire climbing in the forest: a semiqualitative, semiquantitative approach to assessing ladder fuel hazards. West J Appl For 22:88–93

Merrill DF, Alexander ME (1987) Glossary of forest fire management terms. National Research Council of Canada. Committee for Forest Fire Management, Ottawa, p 44

Miller JD, Yool SR (2002) Mapping forest post-fire canopy consumption in several overstory types using multi-temporal Landsat TM and ETM data. Remote Sens Environ 82:481–496

Morsdorf F, Meier E, Kötz B, Itten KI, Dobbertin M, Allgöwer B (2004) LIDAR-based geometric reconstruction of boreal type forest stands at single tree level for forest and wildland fire management. Remote Sens Environ 92:353–362

Mundt JT, Streutker DR, Glenn NF (2006) Mapping sagebrush distribution using fusion of hyperspectral and lidar classifications. Photogramm Eng Remote Sens 72:47–54

Mutlu M, Popescu SC, Stripling C, Spencer T (2008a) Mapping surface fuel models using lidar and multispectral data fusion for fire behavior. Remote Sens Environ 112:274–285

Mutlu M, Popescu SC, Zhao K (2008b) Sensitivity analysis of fire behavior modeling with LIDAR-derived surface fuel maps. For Ecol Manage 256:289–294

Naesset E, Økland T (2002) Estimating tree height and tree crown properties using airborne scanning laser in a boreal nature reserve. Remote Sens Environ 79:105–115

Ottmar RD, Vihnanek RE, Wright CS (2000) Stereo photo series for quantifying natural fuels. Volume III: Lodgepole pine, quaking aspen, and gambel oak types in the Rocky Mountains. US Forest Service. National Wildfire Coordinating Group NIFC, Boise, p 85

Patterson MW, Yool SR (1998) Mapping fire-induced vegetation mortality using Landsat thematic mapper data: a comparison of linear transformation techniques. Remote Sens Environ 65:132–142

Pausas JG, Vallejo VR (1999) The role of fire in European Mediterranean ecosystem. In: Chuvieco E (ed) Remote sensing of large wildfires in the European Mediterranean basin. Springer, Berlin, pp 3–16

Peterson BE (2005) Canopy fuels inventory and mapping using large-footprint LiDAR. PhD dissertation, Faculty of the Graduate School of the University of Maryland, College Park

Peterson B, Dubayah R, Hyde P, Hofton M, Blair JB, Fites-Kaufman J (2007) Use of LIDAR for forest inventory and forest management application. In: Proceedings of the seventh annual forest inventory and analysis symposium, Portland, ME, USA, 3–4 October 2005

Popescu SC, Wynne RH (2004) Seeing the trees in the forest: using lidar and multispectral data fusion with local filtering and variable window size for estimating tree height. Photogramm Eng Remote Sens 70:589–604

Popescu SC, Zhao K (2008) A voxel-based lidar method for estimating crown base height for deciduous and pine trees. Remote Sens Environ 112:767–781

Prometheus SV (2000) Management techniques for optimization of suppression and minimization of wildfire effects. System validation. European Commission – contract number ENV4-CT98-0716

Pyne SJ, Andrews PL, Laven RD (1996) Introduction to wildland fire. Wiley, New York, USA

Reinhardt E, Scott J, Gray K, Keane R (2006) Estimating canopy fuel characteristics in five conifer stands in the western United States using tree and stand measurements. Can J For Res 36:2803–2814

Riaño D, Chuvieco E, Salas J, Palacios-Orueta A, Bastarrika A (2002a) Generation of fuel type maps from Landsat TM images and ancillary data in Mediterranean ecosystems. Can J For Res 32:1301–1315

Riaño D, Chuvieco E, Ustin SL, Zomer R, Dennison P, Roberts D, Salas J (2002b) Assessment of vegetation regeneration after fire through multitemporal analysis of AVIRIS images in the Santa Monica Mountains. Remote Sens Environ 79:60–71

Riaño D, Meier E, Allgower B, Chuvieco E, Ustin SL (2003) Modeling airborne laser scanning data for the spatial generation of critical forest parameters in fire behavior modeling. Remote Sens Environ 86:177–186

Riaño D, Chuvieco E, Condés S, González-Matesanz J, Ustin SL (2004) Generation of crown bulk density for Pinus sylvestris L. from lidar. Remote Sens Environ 92:345–352

Riaño D, Chuvieco E, Ustin SL, Salas J, Rodriguez-Perez JR, Ribeiro LM, Viegas DX, Moreno JM, Fernandez H (2007) Estimation of shrub height for fuel-type mapping combining airborne LiDAR and simultaneous color infrared ortho imaging. Int J Wildland Fire 16:341–348

Rothermel RC (1972) A mathematical model for predicting fire spread in wildland fuels. USDA, Forest Service, Ogden

Rothermel RC (1991) Predicting behavior and size of crown fires in the Northern Rocky Mountains. USDA, Forest Service, Ogden

Salas J, Chuvieco E (1995) Aplicación de imágenes Landsat-TM a la cartografía de modelos de combustibles. Revista de Teledetección 5:18–28

Sando RW, Wick CH (1972) A method of evaluating crown fuels in forest stands, vol 84, U.S. Department of Agriculture, Forest Service research paper NC. U.S. Department of Agriculture, Forest Service, North Central Forest Experiment Station, St. Paul

Sankey TT, Bond P (2011) LiDAR-based classification of sagebrush community types. Rangel Ecol Manage 64:92–98

Sankey TT, Glenn N, Ehinger S, Boehm A, Hardegree S (2010) Characterizing western juniper expansion via a fusion of Landsat 5 thematic mapper and lidar data. Rangel Ecol Manage 63:514–523

Schilling A, Schmidt A, Maas H-G (2012) Tree topology representation from TLS point clouds using depth-first search in voxel space. Photogramm Eng Remote Sens 78:383–392

Scott JH, Reinhardt ED (2001) Assessing crown fire potential by linking models of surface and crown fire behavior. U.S. Department of Agriculture, Forest Service, Rocky Mountain Research Station, Fort Collins

Seielstad CA, Queen LP (2003) Using airborne laser altimetry to determine fuel models for estimating fire behavior. J For 101:10–15

Seielstad C, Stonesifer C, Rowell E, Queen L (2011) Deriving fuel mass by size class in Douglas-fir (Pseudotsuga menziesii) using terrestrial laser scanning. Remote Sens 3:1691–1709

Skowronski NS, Clark KL, Duveneck M, Hom J (2011) Three-dimensional canopy fuel loading predicted using upward and downward sensing LiDAR systems. Remote Sens Environ 115:703–714

Spaete LP, Glenn NF, Shrestha R (2011) Estimating semiarid vegetation height from GLAS data. In: 34th international symposium on remote sensing of environment The GEOSS Era, 34 edn, Sydney

Streutker DR, Glenn NF (2006) LiDAR measurement of sagebrush steppe vegetation heights. Remote Sens Environ 102:135–145

Valbuena R, Mauro F, Arjonilla FJ, Manzanera JA (2011) Comparing airborne laser scanning-imagery fusion methods based on geometric accuracy in forested areas. Remote Sens Environ 115:1942–1954

Van Wagner CE (1977) Conditions for the start and spread of crown fire. Can J For Res 7:23–34

Van Wagner CE (1993) Prediction of crown fire behavior in 2 stands of jack pine. Can J For Res 23:442–449

van Wagtendonk JW, Root RR, Key CH (2004) Comparison of AVIRIS and Landsat ETM + detection capabilities for burn severity. Remote Sens Environ 92:397–408

Varga TA, Asner GP (2008) Hyperspectral and lidar remote sensing of fire fuels in Hawaii Volcanoes National Park. Ecol Appl 18:613–623

Wagner W (2010) Radiometric calibration of small-footprint full-waveform airborne laser scanner measurements: basic physical concepts. ISPRS J Photogramm Remote Sens 65:505–513

Wang C, Glenn NF (2009) Estimation of fire severity using pre- and post-fire LiDAR data in sagebrush steppe rangelands. Int J Wildland Fire 18:848–856

Wing MG, Eklund A, Sessions J (2010) Applying LiDAR technology for tree measurements in burned landscapes. Int J Wildland Fire 19:104–114

Wulder MA, White JC, Alvarez F, Han T, Rogan J, Hawkes B (2009) Characterizing boreal forest wildfire with multi-temporal Landsat and LIDAR data. Remote Sens Environ 113:1540–1555

Zhao K, Popescu S, Meng X, Pang Y, Agca M (2011) Characterizing forest canopy structure with lidar composite metrics and machine learning. Remote Sens Environ 115:1978–1996

Index

M. Maltamo et al. (eds.), *Forestry Applications of Airborne Laser Scanning: Concepts and Case Studies*, Managing Forest Ecosystems 27, DOI 10.1007/978-94-017-8663-8, © Springer Science+Business Media Dordrecht 2014

Lightning Source UK Ltd.
Milton Keynes UK
UKOW06n1816081114

241294UK00003B/37/P